doesn't define vernal equinox or explain theres ^(except in glossary)
12 hours of day & night
or Gibbous moon, waxing, waning
p. 15, 18, 16, 35, 54, 327, 28

Doesn't use umbra, penumbra,

no seasons, no precession

Doesn't explain that Venus has more CO_2 because
its too hot for oceans

no phases of Venus as evidence of
heliocentric

p56. "lower", not "higher"

Solar system not defined
why venus has greenhouse effect > earth's
p. 157
p. 109

Volcano p92
p. 233 — Balmer lines labled wrong
No diagram of Sun's structure

p. 187 — ~~H energy level spacing wrong~~
p. 303
poor discussion of pop I & II.

No Hubble time? Age of Univ?

fig 2-9: "a)" not labled
p. 47   typo
fig 13.2   axis unlabled

Appendix D & E etc use "apparent magnitude"
& abs.
which isn't explained.

"H-R diagram" not (cross referenced) in index

p. 351 expansion of Universe poorly explained

p. 348

ASTRONOMY

A Journey of Ideas

Michael Zeilik
The University of New Mexico

John Wiley & Sons, Inc.
New York   Chichester   Brisbane
Toronto   Singapore

Cover Photo: Lawrence Manning/Westlight

Acquisitions Editor	Cliff Mills
Developmental Editor	Barbara Heaney
Marketing Manager	Catherine Faduska
Production Supervisor	Katharine Rubin
Design	A Good Thing Inc.
Manufacturing Manager	Andrea Price
Copy Editing Supervisor	Marjorie Shustak
Photo Researcher	Hilary Newman
Illustration Supervisor	Anna Melhorn

This book was set in New Caledonia by York Graphic Services, Inc.
and printed and bound by Von Hoffmann Press.
The cover was printed by Lehigh.

Library of Congress Cataloging in Publishing Data:

Zeilik, Michael.
    Conceptual astronomy : a journey of ideas / Michael Zeilik.
      p.   cm.
    Includes index.
    ISBN 0-471-50996-5 (pbk.)
    1. Astronomy. I. Title.
QB45.Z429 1993
520–dc20   92-36354
                             CIP

Printed in the United States of America

10 9 8 7 6 5 4 3 2 1

CONCEPTUAL ASTRONOMY

CONCEPTUAL

3. Heat flows
from interior
to surface
of planets

Concepts in the Evolution of the Solar System

To my son,
Zachary Alexander·Zeilik,
who was born while this book was crafted.

Preface

Science is not about control. It is about cultivating a perpetual condition of wonder in the face of something that forever grows one step richer and subtler than our latest theory about it.

Richard Powers, *The Gold Bug Variations*

Students are fascinated by astronomy but frustrated in their study of it. In the United States, colleges and universities typically offer an introductory astronomy course for non-science majors—a course I will call "Astronomy 101." This course lasts but a semester and covers "all" of the universe, or at least selected pieces of it. The students taking this course are usually new to the university (and may well be older in age), and choose Astronomy 101 over other disciplines (such as physics!) that may also satisfy some form of a science requirement for graduation. At large universities, Astronomy 101 tends to be offered as a large lecture course with a few hundred students. Most of these students are novices, with little practical experience with science or much careful thought about it.

Novice students in all areas suffer two common problems. One, they lack concrete experience. Science seems especially hard to them because they have not done it themselves or observed scientists carrying it out. Hence, the students lack even the exposure to the intellectual framework and the practical aspects of "doing science." They tend to see science as completed rather than in process, frozen facts rather than a vital adventure. Two, they see lots of trees and rarely a forest. I am an astronomer because of the beauty of the unity of the universe. Students see this and that planet, this and that star, this and that galaxy—but have a hard time seeing how it all hangs together. Astronomy makes matters worse by being at such a distance; once we leave the solar system, it is hard to present astronomical objects concretely. This problem reinforces their preconception that science is a large bunch of unrelated "facts," which are memorized for exams and then quickly forgotten. The *Project 2061 Panel Report on Physical and Information Sciences and Engineering* of the American Association for the Advancement of Science addresses many of these issues in a concise way. Project 2061 has as its overarching goal the development of science literacy; the panel report specified the knowledge, skills, and attitudes students in science and engineering should acquire. It has shaped the structure of this book.

Less Is More: A Conceptual Approach

What to do? We should meet the needs of most of the students in Astronomy 101 while conveying to them the true nature of the scientific enterprise—and the beauty of the universe. I have evolved my teaching in a direction that attempts to grapple with the problems that baffle novice students. First, I have opted for "less is more" by covering less astronomical material. In particular, you will find very little in this book about the history of astronomy. Also, I have given scant attention to instrumentation; most of that is in Appendix H on Telescopes.

Second, I have tried to emphasize the conceptual story—how astronomers understand the universe by building models of its parts and pieces and also of the whole cosmos. The glue for the astronomical story is some essential physics, without which we have no connections. I believe you cannot teach introductory astronomy with any understanding unless you provide selected physical concepts in the proper context.

Third, I have made explicit the astronomical and physical concepts that provide the foundation to our understanding. The slant here is on the astrophysics rather than just the astronomy. My experience is that the many facts fog the concepts for novice students, so why not bring the concepts right out in front?

The story line revolves around 25 conceptual themes that are specified at the end of the Preface. For most of these, at least two sections are listed in which these themes come across most strongly. If any theme resounds the strongest, it is that about energy—its varied forms, fast and slow transformations, and careful conservation. The flow of heat from hotter to cooler regions marks an essential insight in understanding the functioning of the universe.

Specifying and Unifying Concepts

Conceptual Astronomy unfolds the concrete form of my philosophy. I have presented the concepts at many levels to help students see their interconnections. The book is split into three parts: Part 1, Concepts of the Solar System; Part 2, Concepts of Stellar Evolution; and Part 3, Concepts of Cosmology. Note that we start with the familiar (the earth and sky) and move to the more abstract (and generally more distant!) pieces of the universe. Each part deals with interrelated material. Each also begins with a statement of *Unifying Concepts* that ties the information together; their selection has been guided by the *Project 2061 Panel Report*.

Within the parts, each chapter starts with a *Central Concept* statement that aims to unify the material. Within the chapter, *Chapter Concepts* are identified by color bars. Now, some concepts are more powerful and more important than others; these appear in more than one chapter. So concepts appear explicitly at many levels in this book, just as they do in astronomy.

Integrating Concepts

How to help the student pull these concepts together into the big picture? Each chapter ends with a section called *The Larger View*. These are not intended to act as summaries; instead, I have tried to highlight overarching concepts and point ahead to their connections with other key notions. The theme of model building is repeated here in the context of different topics.

Also at the end of each chapter, after *The Larger View*, you will find carefully selected *key terms* (**boldface** in the text) that are integrated with *key ideas* (relating to the chapter concepts). This integration often calls out a technical term more than once, sometimes from previous chapters—and that's the idea, for it shows the interrelations of concepts. (Technical terms are introduced only if they are used again in this book; they are not introduced for their own sake and then never used again. The expanded *Glossary* actually contains more astronomical terms than are presented in the text.)

The end-of-chapter material also contains *Review Questions*, which can be answered in a word or two (at most a short sentence) and *Conceptual Exercises*, which require more thought and a longer answer (a few sentences). Finally, many (but not all) of the chapters have a *Conceptual Activity* to provide a hands-on activity relating to one particular concept in the chapter.

Astronomical concepts are highly enjoyable when combined with speculation. Adapting material first prepared by Sheridan A. Simon, a colleague at Guilford College, I have added a section called "The View from . . . " to each part. These are based on astronomy and physics, but with a twist. I hope that you find them fun. You might want to read each twice—before and after each part.

Math and Data

Because this is a conceptual approach rather than a quantitative one, I have deemphasized numbers and equations. Within the text, measured quantities are typically given to no more than two significant figures. In tables, the number of significant figures are greater, generally to the accuracy of the measurement. The "At a Glance" tables provide brief summaries of information at appropriate places in the text. Almost every equation in this book, which are few and carefully selected, is set off as a *Math Concept*. Equations are intended to be a shorthand expression of physical ideas rather than a tool to manipulate numbers or symbols.

 ## The Illustration Program

The art program was carefully reviewed by Dennis Schatz (Pacific Science Museum) and myself to provide a human perspective and scale whenever possible. In astronomy, that is sometimes not possible. We also evaluated the illustrations from the view of a novice student with an aim to simplify and clarify illustrations as much as possible. Color has been used in only those cases in which we thought it would aid the student in understanding the material; for example, in color-coding the interior structure of earthlike planets. The endpaper illustrations provide visual summaries of two key conceptual themes: the origin and evolution of the solar system (front) and the orgin and evolution of a star like the sun (back).

Many of the color photographs are computer-generated in false colors. Because students tend to think that the colors are "real"—those that would be seen by eye—I have tried to be very explicit in the figure legends to identify the false-color images; Appendix G offers additional explanations about this kind of image processing.

 ## Differences from Astronomy: The Evolving Universe

I believe that this clarity, emphasis, and interweaving of essential concepts will reveal the core of astronomical thinking to students. It also marks the main difference between this book and *Astronomy: The Evolving Universe* (6th edition, John Wiley & Sons, 1991), which is a full-featured book for a more in-depth introductory course. The story told in *Conceptual* is one of building up a model of the cosmos (and its parts) from the standpoint of contemporary as-

tronomy. So I have limited the historic development and the people involved with it; for that story, you should turn to *Evolving*. You may also notice that some traditional astronomical topics, such as apparent and absolute magnitudes, are not covered in this shorter book. (Overall, *Conceptual* has about 40 percent fewer words than *Evolving*, and 19 rather than 22 chapters.) I have judged that such topics are not essential to the conceptual development. My students have not missed them!

 ## Supplements

An innovative package of supplemental materials is available to assist in teaching an introductory course with this book. It includes:

- *Instructor's Manual* by J. Wayne Wooten of Pensacola Junior College. This contains detailed chapter outlines and overviews, additional discussion topics, class demonstrations, answers to Review Questions and Conceptual Exercises in the text, and additional resources.

- *Test Bank* also by Professor Wooten. This resource has over 1000 statistically tested multiple-choice questions.

- *Computerized Test Bank*. This computerized classroom management system (for both MS-DOS and Macintosh computers) offers the *Conceptual Astronomy* test bank to develop tests and answer keys.

- *Overhead Transparencies*. These four-color illustrations from the text can be projected in class.

- *Dance of the Planets* (planetarium version). Adopters of *Conceptual Astronomy* will receive a coupon for 15% off the purchase price of this MS-DOS based program. This fascinating software provides a physical simulation of the dynamics of the solar system.

- *Astronomy Video*. Developed in cooperation with the Astronomical Society of the Pacific, this videotape provides short, single-concept segments, mostly as simulations and animations. I selected the segments and wrote the commentary for them.

Acknowledgments

Many people helped with the development of this book. Kimberly Zeilik assisted with the illustration manuscript. Claudia Smith-Porter did the bulk of the photo research. Dan Weeks, University of New Mexico, provided informal comments.

Astronomers and teachers with advice and input on various aspects of the book included:

Jeffrey J. Braun
University of Evansville

David H. Bruning
Stellar Research & Education

Martin Burkhead
Indiana University

Will Chamberlain
David Lipscomb University

George W. Crawford
Southern Methodist University

Heinrich Eichhorn
University of Florida

Benjamin C. Friedrich
Jersey City State College

Solomon Gartenhaus
Purdue University

Charles Hagar
San Francisco State University

Dean Hirschi
University of Central Arkansas

Thomas Hockey
University of Northern Iowa

Darrel Hoff
Center for Astrophysics

James R. Houck
Cornell University

Robert Kennicutt
University of Arizona

Anthony Lomazzo
University of Bridgeport

Dinah Moché
Queensborough Community College

Nancy Montague
Sierra College

George Patsakos
University of Idaho

Donald F. Ryan
SUNY Plattsburgh

Dennis Schatz
Pacific Science Center

Norm Siems
Juniata College

Norman Sperling
Oakland, CA 94618

Walter G. Wesley
Moorhead State University

J. Wayne Wooten
Pensacola Jr. College

Arthur Young
San Diego State University

Robert Zimmerman
University of Oregon

 ## Feedback

Any errors in the text are my responsibility. Please write to me if you find any. It is a little-known fact that minor corrections and changes *can* be made in future printings of this edition. I take a special effort to update new printings as they occur. (You can check the bottom of the copyright page to find out which printing you have in hand.) *I remind reviewers to mention this fact when they comment upon any errors; also, please send them to me so I can fix them!* Your feedback can improve this book! Any comments are appreciated; send them to me:

Michael Zeilik
Department of Physics and Astronomy
The University of New Mexico
800 Yale Blvd. NE
Albuquerque, New Mexico, 87131, USA.
Internet e-mail: Zeilik@chicoma.lanl.gov

Note:
Abbreviations are often used for the names of major observatories and agencies (particularly in the figure captions). These are: NASA—National Aeronautics and Space Administration; NOAO—National Optical Astronomy Observatories; KPNO—Kitt Peak National Observatory; CTIO—Cerro Tololo Interamerican Observatory; NSO—National Solar Observatory; NRAO—National Radio Astronomy Observatory; ESA—European Space Agency; ESO—European Southern Observatory; JPL—Jet Propulsion Laboratory of the California Institute of Technology; AATB—Anglo-Australian Telescope Board; NCSA—National Center for Supercomputer Applications.

Fundamental Conceptual Themes

The story line of *Conceptual Astronomy* revolves around the following conceptual themes, which provide the foundation of our understanding of astronomy.

- Scientific models (Section 2.1)
- Heliocentric model of the solar system (Sections 2.3–2.5)
- Kepler's laws of planetary motion (Section 2.5)
- Motion: speed, velocity, and acceleration (Section 3.1)
- Newton's laws of motion (Section 3.2)
- Newton's law of gravitation (Sections 3.4–3.5)
- Energy: its forms and its transport (Sections 4.1 and 10.4)
- Magnetic fields and their interactions with ionized gases (Section 4.5)
- The conservation of energy (Sections 4.6 and 9.2)
- Doppler shift (Sections 5.1 and 9.5)
- Angular momentum and its conservation (Sections 6.5 and 8.4)
- The origin of the solar system (Sections 8.4–8.5)
- Fundamental nature of matter (Sections 9.1 and 18.5)
- Spectra and spectroscopy (Sections 9.1–9.3)
- Atoms and light (Sections 9.3–9.4)
- Ordinary (Section 10.2) and extraordinary gases (Sections 13.5 and 14.1)
- Fusion reactions, nucleosynthesis, and the conversion of matter to energy (Sections 10.5, 13.4, and 14.5)
- The origin of the sun and stars (Sections 12.3–12.4)
- Einstein's theory of general relativity (Sections 10.7, 14.7, and 18.1)
- Inverse–square law for light (Section 11.1)
- The Hertzsprung–Russell diagram (Sections 11.5 and 13.1)
- Hubble's law and the expansion of the universe (Section 16.4)
- Dark matter in the universe (Sections 15.2 and 16.7)
- The origin of the universe in the Big Bang model (Section 18.2)
- The origin of life (Section 19.2)

Contents

Part 2 Concepts of the Stellar Evolution 175

Part 3 Concepts of Cosmology 319

13

T

1

Concepts of the Solar System

.

Unifying Concepts

The earth is one of the planets orbiting the sun. All solar system bodies follow universal laws of motion and gravitation, and so their motions are predictable. The planets and moons of the solar system have diverse physical characteristics, produced by evolutionary processes acting over the billions of years that have elapsed since the formation of the solar system. Our recognition of these processes unifies our view of the solar system.

1
Motions in the Solar System

■

Central Concept

The motions of astronomical objects you can see by eye follow distinctive patterns and cycles in the sky over both short and long periods of time.

■

If you watch the sky often for a year, you can observe just by eye the behavior of the stars, planets, sun, and moon. The sun rises and sets; the seasons flow. The moon's illumination changes nightly; different constellations appear as the seasons change. Planets move majestically and sometimes oddly among the stars. Careful study allows you to detect the pattern and timing of these movements and events.

This chapter deals with such observations, naked-eye observations. From these you can sense the regular cycles of motions in the heavens. Long-term observations over months and years can establish the repeating periods of celestial cycles with amazing precision. Most of the brightest naked-eye objects lie in the solar system. Of these, the planets follow the strangest motions of them all.

1.1 The Visible Stars

Before we tackle the motions of the planets, sun, and moon, let's examine the basic backdrop of the sky: the stars.

Constellations and Angular Measurement

If you take the time to study the stars, you'll find that they fall into patterns, designs imposed by your mind (Fig. 1.1). These patterns, such as Orion, are called **constellations**. The official constellations used today (88 in all) are established by international agreement of astronomers. (Appendix I contains sea-sonal star charts of constellations as seen from the midlatitudes of the Northern Hemisphere.) A special set of the constellations, twelve in all, makes up the **zodiac** (Table 1.1). See also p. 12

If you observe nightly, you'll see that the shapes of the constellations don't change. In fact, if you watched them for your whole life, you wouldn't notice any change. The stars appear to hold fixed positions relative to each other, which we can measure.

How far apart do stars appear in the sky? A sighting and measuring device, such as the extent of your fist held at arm's length, will allow you to measure the angle between one star and another; this angle is the **angular separation** or **angular distance** between two stars. Angular measurement is based on counting by 60: a circle is divided into 360 degrees (°), each degree into 60 minutes of arc (*arcmin or '*),

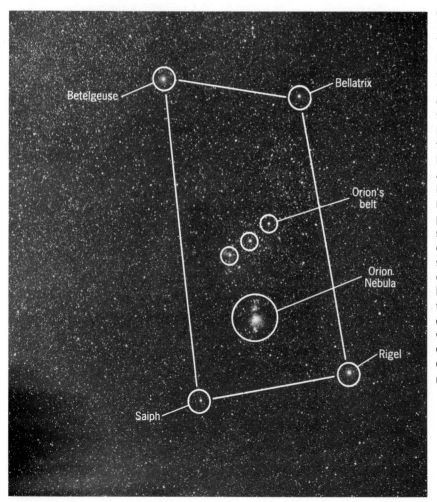

FIGURE 1.1

Time exposure of the stars in the constellation Orion and others nearby. The three bright stars making a diagonal line pointing to the top right form Orion's belt. Below the center star is the fuzzy patch of the Orion Nebula. The very bright star at bottom right is bluish Rigel; to the upper center is reddish Betelgeuse. This photo shows many faint stars that are invisible to the eye. An outline is superimposed of the pattern of Orion, with names of the brightest stars and main features. See the constellation charts for winter in Appendix I to find Orion. (Courtesy Dennis di Cicco, *Sky & Telescope* magazine)

Table 1.1

The Sun's Location Within The Zodiacal Constellations

Constellation	Traditional Pattern	Approximate Date
Aries	Ram	April 30
Taurus	Bull	May 30
Gemini	Twins	July 5
Cancer	Crab	July 30
Leo	Lion	September 1
Virgo	Virgin	October 11
Libra	Balance	November 9
Scorpius	Scorpion	December 3
Sagittarius	Archer	January 7
Capricornus	Goat	February 8
Aquarius	Water bearer	February 25
Pisces	Fishes	March 27

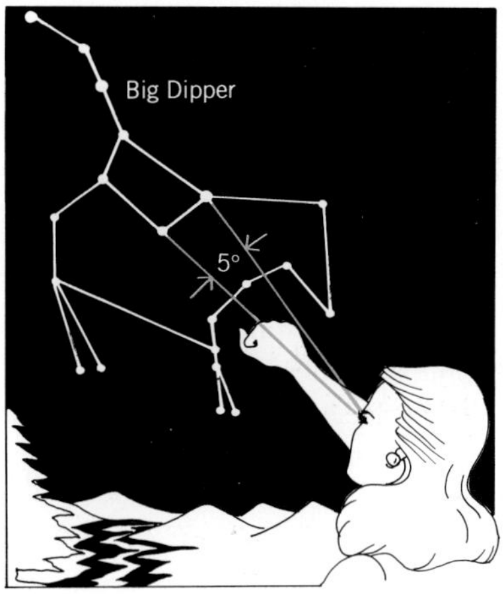

FIGURE 1.3

Measuring angles on the sky with a hand at arm's length. The angle between the two end stars in the bowl of the Big Dipper is about 5°, or half a fist at arm's length. The Big Dipper is a part of the constellation Ursa Major, the Great Bear.

and each minute into 60 seconds of arc (*arcsec* or ″). An *arc* is any part of the circumference of a circle. One such circle is the **horizon**, where the earth and sky appear to meet. (Your actual horizon is probably cluttered with buildings and trees; out at sea, you would see a more ideal one.) Note that the angular distance around the horizon is 360°; from the horizon to the point overhead, the angle is 90°.

At arm's length, your fist covers about 10° of sky; when your fingers are spread, each fingertip covers roughly 1° (Fig. 1.2). So you can use your hand to measure the *angular size*—the diameters in angular units—of the sun and moon (both about ½, or half a fingertip) and the separation of any stars, such as the pointer stars in the Big Dipper (about 5°, or half a fist; Fig. 1.3).

The stars appear to be laid out with fixed angular distances relative to each other in stable patterns on the sky.

FIGURE 1.2

Angular measurements made with a hand extended at arm's length. The angles are roughly typical for an average adult for extended fingers, a tight fist, and the forefinger. You will need to calibrate your own hand to find its specific dimensions. You can do this by measuring how many of your fists make up the angular distance from the horizon to overhead, which spans 90°.

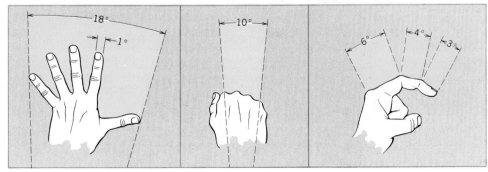

Motions of the Stars

Stay out one night and watch the stars from dusk to dawn. They move relative to your horizon—rising in the east, slowly traveling in arcs against the sky, and setting in the west. In an hour, the stars move about 15° westward with respect to the horizon. If you live in the Northern Hemisphere and look north, you'll find that some stars never drop below your horizon. Instead, they trace complete circles above it (Fig. 1.4). As these **circumpolar stars** swing around, they trace concentric circles like the rings of a bull's-eye. The center of these rings marks the **celestial pole**, the point about which the stars seem to pivot. A modestly bright star called **Polaris** lies close to the north celestial pole. Polaris is now the **north pole star** (Fig. 1.4). (No bright star now falls close to the south celestial pole, so for now, there is no south pole star.)

The constellations that are visible to you change with the seasons. For example: in winter in the Northern Hemisphere, at 8 P.M., you can view Orion in the south. (Refer to the winter star map in Appen-

dix I.) Look south on following nights at the same time. Orion moves slowly to the west, toward the sun. (See the spring star map in Appendix I.) In summer, you can't see Orion at all at night—because it's next to the sun, it is up during the day. (See the summer star map in Appendix I.) In winter, a year later, Orion again lies south at night. A constellation takes one year to return to its initial place in the sky relative to the sun. So, with respect to the sun, a constellation drifts about 1° westward per day.

Note that this gradual *yearly* change relative to the sun is much slower than *daily* motion relative to the horizon (from east to west). How fast an object covers a certain angular distance with time is called its **angular speed**, in angular units divided by time units, such as degrees per day. The two angular speeds here are very different: *1° per day* with respect to the sun, and *15° per hour* with respect to the horizon.

The stars move, daily, from east to west with respect to the horizon, and also yearly, at a much slower angular speed, with respect to the sun.

FIGURE 1.4

Circumpolar motions of the stars above Star Hill Inn, New Mexico. The motion of the stars around the north celestial pole in this long time exposure traces out arcs, which are progressively larger the farther the star lies from the pole. The brightest star near the center of the stellar circles is Polaris; note that it has traced a small arc, which indicates that it is not exactly at the north celestial pole. The trails among the buildings at the bottom were made by flashlights carried by astronomers. (Courtesy Dennis di Cicco, *Sky and Telescope* magazine)

1.2 The Motions of the Planets

Naked-eye observing soon singles out fairly bright objects that move, and so don't belong to the constellations. Five of these wander in regular ways through the stars of the zodiac: the **planets** Mercury, Venus, Mars, Jupiter, and Saturn. (Uranus, Neptune, and Pluto cannot be seen without a telescope.) Viewed by eye, the planets look pretty much like stars, though they twinkle less than stars do. At times, some planets are brighter than the brightest stars; all planets slowly vary in brightness. Their motions with respect to the stars, however, really separate the planets from all other objects in the sky.

Retrograde Motion

Suppose you observe Mars every night for a few months near the time when Mars appears brightest

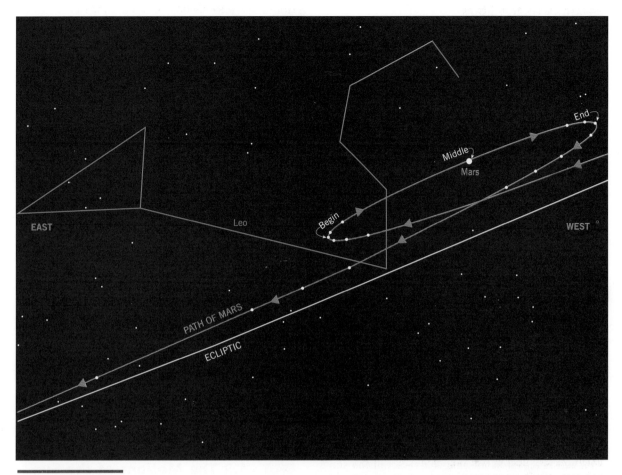

FIGURE 1.5

The retrograde motion of Mars, relative to the stars, during its 1995 opposition. The line across the center marks the ecliptic; shown is the outline of the zodiacal constellation Leo. "Begin," "Middle," and "End" mark the points on Mars' path in the sky when the retrograde motion will begin (January 1995), reach its midpoint (February 1995), and end (March 1995). Note that during its retrograde motion Mars moves from *east to west*; otherwise it moves from west to east. The path of Mars is marked with 7-day intervals. (Diagram generated by *Voyager* software for the Macintosh, Carina Software)

in the sky. (This occurs about every two years.) At first Mars moves slowly eastward (with respect to the stars) through the zodiac (Fig. 1.5). In this respect, it moves through the stars like the sun and the moon. But later, Mars falters in its eastward motion with respect to the stars and stops for a short time. Next, for about three months, Mars moves *westward*— opposite its normal motion. Toward the end of its westward motion, Mars appears to slow down and stop. Then Mars resumes its normal eastward course. The planet's backward motion to the *west* is called **retrograde motion**. In the middle of its retrograde motion, Mars shines at its brightest.

> Planets loop through the stars of the zodiac with periodic retrograde (*westward*) motion, in contrast to their general eastward motion.

In addition to these motions, the planets move daily from east to west with respect to the horizon. So the planets display three motions: (1) daily, rising in the east and setting in the west; (2) eastward, in general, through the zodiac; and (3) occasionally westward in retrograde loops. Note that a planet's motion with respect to the stars is much slower than its daily motion with respect to the horizon—and that a planet has more than one motion at the same time!

Elongations, Conjunctions, and Oppositions

The alignment of the sun with a planet in the sky at the time of retrograde motion divides the visible planets into two groups. Mercury and Venus make up one; they never stray very far (in angular distance) from the sun. Because they stick close to the sun, they are visible only as morning and evening "stars." That means that they hover over the western horizon after sunset and pop above the eastern horizon before sunrise. (But keep in mind that they are *not* stars: they shine by reflected light.)

You can use your fist to measure the maximum angular separation of Mercury or Venus from the sun. For Mercury, the average maximum separation is 23° (about 2½ fists); for Venus, about 46° (about 4½ fists). When either planet is at its greatest angular separation from the sun, it reaches **maximum elongation** (Fig. 1.6). (Mercury's maximum elongation varies a lot; it can be as large as 28°. The figure used here is its *average* value.)

Mercury and Venus begin their retrograde motions after they have swung farthest east of the sun as evening stars. They then move westward, pass the sun, and reappear as morning stars west of the sun. When Mercury or Venus lies close to the sun in the sky, the planet and the sun are aligned in **conjunc-**

tion. (Two celestial objects are in conjunction whenever they come close together in the sky.)

The second group of planets consists of Mars, Jupiter, and Saturn. In contrast to Mercury and Venus, these planets move freely around the sky with respect to the sun. And they retrograde when they stand in **opposition** to the sun: opposite the sun in the sky. At opposition, the sun and the planet are separated by 180° (just as the moon is at full phase). Then the planet rises as the sun sets. If you point one arm to Mars at opposition and the other at the sun, your arms would be 180° apart (they'd make a straight line). When at opposition, a planet crosses the middle of its retrograde loop, shines its brightest, and is highest in the sky near midnight.

Mercury and Venus are *never* in opposition; they retrograde after passing their greatest eastern elongation; Mars, Jupiter, and Saturn retrograde *only* at opposition.

The distinctive motions of the planets, especially the puzzle of retrograde motion, marks the central theme of Chapter 2.

1.3 The Motions of the Sun

The sun also moves in the sky. Its daily motion—rising above the eastern horizon, tracing an arched path in the sky, and setting below the western horizon—sets the most basic cycle of time: day and night. Midway between sunrise and sunset, the sun ascends to its highest point relative to the horizon, which defines **noon**. The interval from one noon to the next sets the length of the 24-hour **solar day**. The angular height of the sun at noon varies with the seasons and sets the length of the year. **Altitude** is the name given to the angular height of the sun (or any celestial object) above the horizon.

Motions Relative to the Horizon

A short stick placed vertically in a flat place on the ground (Conceptual Activity 1) acts as an instrument to study the sun's daily and seasonal motion relative to the horizon. The tip of the shadow marks the end

FIGURE   1.6

Measuring the greatest elongation of Venus at sunset. At this time the angle between Venus and the sun is about 46°. The same observation for Mercury gives a maximum angle of about 23°.

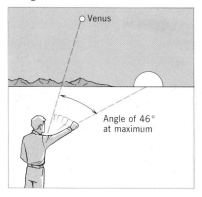

of a line that connects the shadow's tip, the top of the stick, and the sun. Or, you can look at a shadow cast by a flagpole, telephone pole, or yourself! The shadow points opposite from where the sun is in the sky and the length of the shadow tells the altitude of the sun. When the sun hangs lower in the sky, the shadow is longer. At noon the shadow has its shortest length for that day. Also at noon in the midnorthern latitudes, the shadow points due north—so the sun lies due south. If you face away from the sun, right across your shadow, you face north and your back is to the south.

Observe your shadow at noon throughout a year while measuring the altitude of the sun with your fist (Fig. 1.7). You'll find that the altitude of the sun at noon varies with the season. During the summer, the shadow falls the shortest at noon (Fig. 1.7a) on the **summer solstice** (around June 21), the day with the longest duration of daylight hours. At the summer solstice, the noon sun hits its greatest altitude for the year. In winter at noon the shadow stretches longest (Fig. 1.7c) on the **winter solstice** (around December 21), the day with the fewest daylight hours. The noon sun has dropped to its lowest point altitude for the year. On the first day of spring and fall, the stick casts a shadow with a length between its summer minimum and winter maximum (Fig. 1.7b)—the **equinoxes** (around March 21 for spring and September 21 for fall). The cycle of the shadow defines the second basic unit of time: the year of seasons. (*Note*: The dates given here are for the Northern Hemisphere; the seasons are reversed in the Southern Hemisphere.)

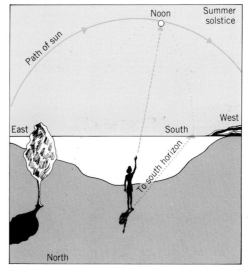

(a)

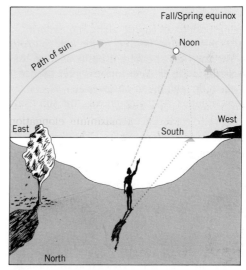

(b)

FIGURE 1.7

Noon shadows and the position of the sun in the sky at noon: east is to the left, west to the right. The higher the sun relative to the southern horizon, the shorter the shadows. The shortest noon shadow of the year occurs on the day of the summer solstice (a), the longest on the winter solstice (c) for a midnorthern latitude. At the spring and fall equinoxes (b), the noon shadow has a length between those for winter and summer. By using your fist, you can measure the angular height of the sun above the southern horizon at noon. The total angular change from summer to winter solstice is 47°.

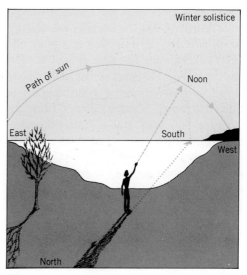

(c)

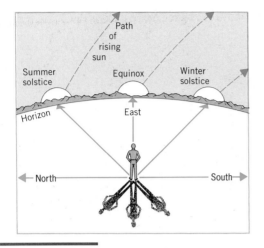

The changing position of sunrise along the eastern horizon with the seasons: north is to the left, south to the right. The sun rises at different positions from the summer solstice (north of east) to the winter solstice (south of east), as seen by an observer at midnorthern latitudes. At the equinoxes, the sun rises due east.

At noon in midlatitudes, the sun is highest in the sky at the summer solstice, lowest at the winter solstice, and in between at the equinoxes.

The seasonal change in the sun's noon position relates to a cyclical change in the sunrise and sunset points. Consider a series of observations at sunrise (Conceptual Activity 1). At the summer solstice, for midnorthern latitudes, the sun rises at a point the farthest north of east for the year (Fig. 1.8). For about four days, the sun seems to rise at the same place on the horizon—its rising point appears to stand still. (The word "solstice" comes from the Latin words for "sun" and "stand.") After the summer solstice, the sunrise position moves southward. At the winter solstice, the rising point reaches as far south of east as it will get for the year. The sun's rising point again stands still for a few days. Then it moves northward, slowly at first and then more quickly. Between the solstice points lies the point due east for sunrise at the equinoxes, when the sun's rising point has the greatest angular speed along the horizon.

The rising (and setting) points of the sun vary with the seasons: due east at the equinoxes, farthest north at the summer solstice, and farthest south at the winter solstice.

Motion Relative to the Stars

The sun also moves with respect to the stars—a motion hard to observe, for you cannot see the stars during the day. (You can get a rough idea near sunrise and sunset by observing which constellations lie near the sun.) Relative to the stars, the sun appears to move to the east. In one year the sun returns to the same position relative to the stars, so it makes a circuit of the whole sky—360° in a year, or about 1° a day. Relative to the stars, the sun moves at an angular speed of roughly 1° per day to the east.

Imagine that you recorded the sun's position among the stars for a year. If you drew an imaginary line through these points, you'd trace out a complete circle around the sky: It is called the **ecliptic** (Fig. 1.9). The traditional 12 constellations through which the sun moves define the *zodiac*. (Note that the sun travels through a part of a thirteenth constellation called Ophiuchus, which is not part of the traditional zodiac.) This eastward motion of the sun with respect to the stars in a year is much slower than the westward motion of the sun relative to the horizon in a day!

The sun's position along the ecliptic is labeled with reference to the constellations of the zodiac. (Table 1.1 gives approximate dates for the sun to be located about in the center of the constellation.) For instance, to say that the sun is "in Taurus" specifies a place along the ecliptic. The zodiac probably arose from a desire to mark the sun's position with respect to the stars. The planets also move through the zodiac, on or close to the ecliptic. So to say that a planet is "in Taurus" also gives its approximate position along the ecliptic.

The sun's location in the zodiac also roughly indicates the time of year. Twice yearly, in spring and fall, the equinoxes occur. The summer and winter solstices mark two other key times. Currently, the sun lies in Pisces in the spring, Gemini in the summer, Virgo in the fall, and Sagittarius in the winter (Fig. 1.9).

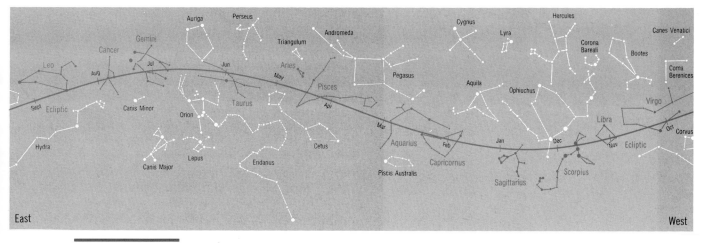

East   West

FIGURE 1.9

The ecliptic and zodiacal stars. If there were no atmosphere, you could see the sun's changing position with respect to background stars. The sun's path is called the *ecliptic*. This diagram indicates the sun's position by date along the ecliptic and shows the major constellations. The zodiacal constellations are indicated. Note that the sun moves *west to east* among the stars (see Table 1.1).

'in blue

The sun moves annually, from west to east through the zodiac, along the ecliptic.

 ## 1.4 The Motions of the Moon

If you watch the moon carefully, you can spot two of its celestial motions. First, like the sun and the stars, the moon rises in the east and sets in the west. Second, the moon also journeys eastward against the backdrop of the zodiacal stars. Here's how to observe this eastward motion (Fig. 1.10). Wait until the moon appears close to a bright planet or star and fix this position in your mind. On the same night, observe the moon and planet again a few hours later.

FIGURE 1.10

The motion of the moon relative to Venus as both *set* over Tulsa, Oklahoma, in 1988; the sequence runs from upper left to lower right for later times. The view is toward the west. Note how the moon moves eastward (upward and to the left) with respect to Venus; so it lies close to the ecliptic as Venus is near the ecliptic. (Courtesy William Sterne, Jr., Sterne Photography)

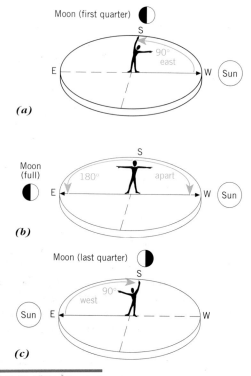

(a)

(b)

(c)

FIGURE 1.11

The orientation of sun and moon in the sky for different phases of the moon. At first quarter (*a*) the moon is 90° east of the sun (due south as the sun sets). At full (*b*) it is 180° away (in opposition, so the full moon rises as the sun sets). At last quarter (*c*) the moon is 90° west of the sun (due south as the sun rises). The view here is facing south from a midnorthern latitude. An observer pointing at the sun and the moon in each case would measure the angles shown.

The moon will have moved to the east with respect to the planet (both will have moved westward in the sky with respect to the horizon). If you measure the moon's change in position over time, you'll find an angular speed of about $\frac{1}{2}°$ per hour. At this rate, the moon circuits the zodiac in a bit more than 27 days. (Although the moon's path does not fall right on the ecliptic, it lies close to it; so the moon stays within the zodiac.)

Watch the moon for a few nights; you'll note that the amount of its surface that is illuminated—its **phase**—follows a regular sequence. When the moon rises at sunset, its face is completely illumi-nated—a *full moon*. About $14\frac{1}{2}$ days later, the moon is *new* and is not visible in the sky. A few days later, the moon reappears at sunset partially illuminated—a *crescent moon*. About a week later, the moon appears at *first quarter*. And about a week after that, we return to a full moon. A complete cycle of phases—say, from one full moon to the next—takes about $29\frac{1}{2}$ days. It defines a third basic unit of time: the month of phases. (Note that the month defined by a cycle of phases is longer than the time the moon takes to move once through the stars of the zodiac.)

The different phases of the moon relate to specific alignments of the sun and the moon in the sky be-cause the moon reflects sunlight. At new moon, the angular separation of the sun and moon is small—less than a few degrees. At first quarter, the moon lies 90° east of the sun (Fig. 1.11). At full, the moon is nearly 180° from the sun; at last quarter, it is 90° west of the sun. First and last quarters refer to the position of the moon in the sky—one-quarter of a full circle away from the sun—*not* to the amount of illumination; the moon at quarter-phase looks half full.

Daily, the moon rises in the east and sets in the west with respect to the horizon and moves eastward with respect to the stars of the zo-diac, while changing the amount seen illumi-nated.

Hasn't yet said that the moon orbits the earth!

1.5 Eclipses of the Sun and Moon

An eclipse of the sun (a **solar eclipse**) occurs when the moon passes in front of the sun at new moon (Fig. 1.12). Although the moon is actually a smaller body than the sun, it's closer to the earth by just the right amount needed to make the **angular diame-ters** of the sun and moon appear to be almost the same—about $\frac{1}{2}°$. So the moon can just cover the sun's disk when it passes directly between the sun and the earth, as it may do at new moon (Fig. 1.13).

Why don't eclipses happen each month? Mainly

moon orbits in one month

FIGURE 1.12

The total eclipse of the sun in July 1991. As viewed from the earth, the moon covers the sun's visible disk, so the sun's outer atmosphere (its "corona") is visible. (Courtesy NSO–Sacramento Peak/NOAO)

because the moon's path in the sky relative to the stars does *not* coincide exactly with the ecliptic; it is tilted at an angle of some 5°. The ecliptic and the moon's path cross at two points. Only at or near these points will the positions of the sun and moon be so close to overlapping that an eclipse occurs. For instance, consider a solar eclipse. If the moon is more than $\frac{1}{2}°$ above or below the sun, it will pass by without blocking out the sun's disk. No eclipse will occur. When the sun and moon are exactly lined up, the new moon completely covers the sun, and we have a *total solar eclipse*. at some place on earth

During a total solar eclipse, the moon's shadow can have a width of only about a few hundred kilometers (abbreviated *km*; see Appendix A on units). Only people in this narrow band on the earth will see a total eclipse as the shadow sweeps by. Those just outside the central band see a partial solar eclipse— that is, the moon does not completely cover the sun. There will be total solar eclipses on November 3, 1994 (visible from South America) and October 24, 1995 (visible from Iran, India, and Southeast Asia).

Note that solar eclipses prove that the moon must be closer than the sun because the moon blots out the sun but the reverse never happens. So we can reason as well that the sun must be larger in actual size than the moon.

An eclipse of the moon (a **lunar eclipse**) occurs when the moon passes directly through the shadow cast by the earth at full moon (Fig. 1.14). Then the

FIGURE 1.13

Alignment of the moon, sun, and earth for a total solar eclipse. The moon must be new and on or very close to the ecliptic. The length of the moon's dark inner shadow is often long enough to hit the earth. People in the path where the central shadow moves along the earth will see a total eclipse.

Moon's inner shadow — Moon's outer shadow — Path of totality

Sun's rays

Moon

Earth

Total eclipse of sun

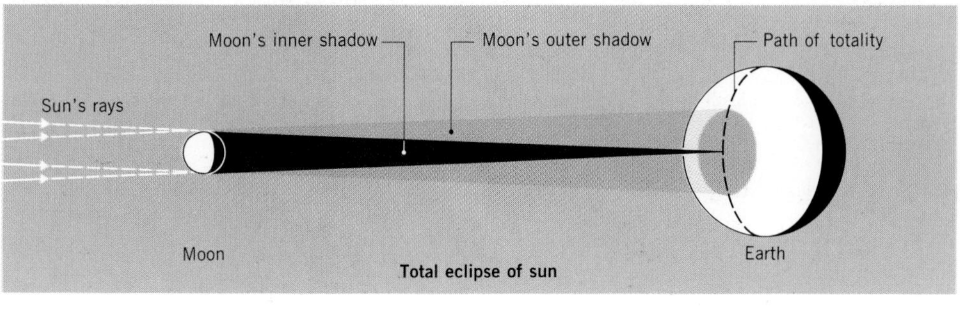

how long one lasts

FIGURE 1.14

A total lunar eclipse. The full moon stands in the center of the earth's shadow in this time exposure taken in New Mexico in 1989. The reddish color comes from dust in the earth's atmosphere, which scatters blue light but allows red light to pass. (Courtesy J. Riffle, Astro Works Corporation; photographed with an Astromak™ telescope)

sun's illumination is cut off from the moon. A total eclipse of the moon takes place only when the moon is *full*—when the earth lies directly between the sun and the moon (Fig. 1.15). Also, the moon must be close to the ecliptic; otherwise it will miss the earth's shadow. There will be total lunar eclipses in 1993 on June 4 and November 29—and the second will be visible in parts of the United States.

Lunar eclipses have a large potential audience—all people on the night side of the earth—because the entire moon is in the earth's shadow. During your lifetime from one location on the earth, you could expect to see roughly 50 lunar eclipses, about half of them total. In contrast, solar eclipses are staged only along narrow bands on the earth's surface. So a total eclipse of the sun occurs rarely for any one location on the earth. Eclipses take place less frequently than the other celestial events described here, but they can be predicted with great accuracy.

Solar eclipses take place only at *new* moon; lunar eclipses at *full* moon; they are predictable from the motions of the sun and moon.

Now you can see how the ecliptic got its name: Only when the moon lies on or close to the *ecliptic* can *eclipses* occur.

FIGURE 1.15

Alignment of the sun, moon, and earth for a total lunar eclipse. The moon must be full and on or close to the ecliptic. The dark, inner shadow of the earth produces the total eclipse as the moon passes through it. The total lunar eclipse is visible to everyone on the night side of the earth.

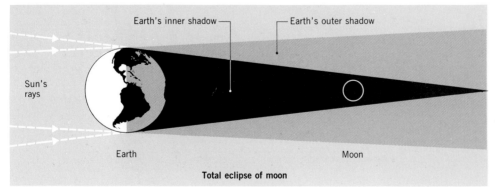

The Larger View

Simple and persistent naked-eye observations reveal the basic cycles in the sky. The sun, moon, and planets each have their special motions with respect to the stars within the band of the zodiac. And all celestial objects move daily with respect to the horizon. The stars, though arrayed in constellations, do not appear to move with respect to each other. They anchor the backdrop for the motions in the solar system.

Such observations were made with care by ancient peoples throughout the world and by traditional societies (such as the Pueblo Indians of the U.S.

Southwest) today. The heavenly cycles regulated human activities, such as the proper times to plant, to harvest, and to celebrate with dances and songs. The rhythms are not calendric abstractions; they are tied to observations of the sky.

You can do your own observing if you can get away from city lights, so that the sky seems like a starry room. Try scanning the heavens on a regular basis over at least a few months. You will start to see patterns that will make you wonder and want to look deeper than what you can see.

Key Ideas and Terms

angular diameter
angular distance
angular separation
angular speed

1. Naked-eye observations of the angular positions of visible objects reveal that the celestial bodies participate in cyclic motions. Some are short term (daily, weekly) and others long term (monthly, yearly, and longer) in duration.

celestial pole
circumpolar stars
constellations
horizon
north pole star
Polaris

2. Relative to the horizon, the stars rise in the east and set in the west every day. The stars make up fixed patterns, called constellations, that do not change over a human lifetime. Different constellations are visible in the night sky at different seasons. Circumpolar stars do not rise or set.

altitude
ecliptic
equinox
horizon
noon
solar day
solstice
zodiac

3. Relative to the horizon, the sun rises roughly in the east and sets roughly in the west daily. The exact position of the sunrise changes with the seasons. The sun attains its highest point in the sky at noon. The altitude of the noon sun varies with season—it is highest at the summer solstice, lowest at the winter solstice, and midway at the equinoxes. Relative to the stars, the sun moves eastward along a path called the ecliptic that cuts through the constellations of the zodiac; the sun completes one eastward circuit of the zodiac in a year.

month
phases (moon)

4. Relative to the horizon, the moon rises in the east and sets in the west. Relative to the stars, the moon moves eastward (completing one circuit of the sky in about a month). As the moon moves around the sky relative to the sun, it goes through a cycle of phases, which depends on the angle between the sun and the moon.

conjunction
horizon
maximum
  elongation
opposition
planets
retrograde motion

5. Relative to the horizon, the planets rise in the east and set in the west daily. Generally, the planets move on or close to the ecliptic eastward with respect to the stars; occasionally, the planets loop westward relative to the stars in what is called retrograde motion. Of the five planets visible without a telescope, Mars, Jupiter, and Saturn retrograde only when in opposition to the sun, Mercury and Venus when in conjunction with the sun and moving from evening "star" to morning "star"; also Mercury and Venus never move far from the sun.

angular diameter
eclipse (lunar/solar)

6. A lunar eclipse occurs when the full moon passes through the earth's shadow. A solar eclipse occurs when the new moon passes between the sun and the earth. Eclipses happen because the angular diameters of the sun and moon are about the same. Solar eclipses show that the moon is closer to the earth than the sun is.

wrong !

Review Questions

1. What celestial objects can you see without a telescope?
2. What is a short-term motion that all celestial objects have in common?
3. Which group of constellations never rises or sets?
4. What is the name of the star that is now the north pole star?
5. In what direction is the sun's daily motion? Relative to what?
6. In what direction is the sun's annual motion relative to the stars?
7. For midnorthern latitudes, on what day of the year is the sun lowest in the sky at noon?
8. In what direction does the moon move relative to the stars?
9. When the moon lies in opposition to the sun, what is its phase?
10. Relative to the stars, in what direction does a planet move during its retrograde motion?
11. Relative to the stars, does the moon move slower or faster than the sun?
12. What is the phase of the moon during a total solar eclipse?

Conceptual Exercises

1. Tell how you can find roughly the position of the ecliptic in the sky. Tell how you can find the constellations of the zodiac.
2. Draw a schematic diagram of the retrograde motion of a planet relative to the stars. Be sure to indicate the directions east and west clearly.
3. What celestial bodies *never* show retrograde motion?
4. Into what two groups can the planets be divided on the basis of their retrograde motion?

5. When Mars is at opposition, at what time will it rise? Set?

6. For what *two* reasons can you argue that the moon was closer to the earth than the sun?

7. (a) You go outside at 9 P.M. and face south. The moon is up and off to your right, near the horizon. Is it rising or setting? What is its phase?

   (b) The next night you go out again at 9 P.M. Where is the moon? Is it higher, lower, or not up at all? Did it move east or west with respect to the stars? Has its phase changed? If so, how?

8. Describe the changing position of the rising sun on the eastern horizon throughout a year, with special emphasis on the solstices and equinoxes.

9. What phase must the moon be in for a solar eclipse? A lunar eclipse? What else must happen for an eclipse to occur?

10. At what time of year would the points on the horizon at which the sun rises or sets change their location from one day to the next the most rapidly? The least?

11. Consider observing the sun at sunset from a midnorthern latitude. How would the setting point change from winter to summer solstice?

12. Explain the following apparent anomaly: the month of phases is longer than the time the moon takes to make one circuit through the stars.

Conceptual Activity 1   Observing the Sun's Motions

Here are two specific observations of the sun, which demonstrate its seasonal motion.

First, you can make your own horizon calendar by tracking the sun's different seasonal positions at sunset along the horizon. Find a place that has a clear view of your horizon but also has a few prominent landmarks. Make a sketch of the general horizon features. Standing in the same spot each day, make a mark on the drawing of the sun's setting position; also write down the date and time. Do this at least weekly for at least three months centered around the solstice or the equinox. (As an option, you can take a series of sunset pictures just before the sun dips below the horizon.) Be persistent! You need careful long-term observations to come to sure conclusions.

From your drawing, answer the following questions. In what direction did the point at which the sun sets move (north, south, or both) along the horizon? Was the change uniform in the same time interval? If not, how did it vary? Can you explain the observed motion? Can you guess what was happening to the sunrise point during the period in which you made your sunset observations?

Now observe the sun at noon. For this, you'll need to measure a shadow. Get a straight stick 10 to 15 cm long and devise a way to hold it upright. Find a level piece of ground out of the way of people and buildings. (Use a carpenter's level to check that the ground is not tilted.) The stick must be placed on a piece of cardboard so that it is vertical (Fig. F.1). The simplest way to check is to take

Using the length of the sun's shadow at noon to find the sun's angular height above the horizon.

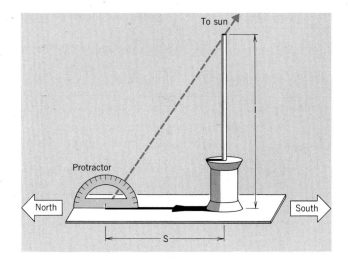

a string with a small weight and hang it from the top of the stick. The weight will make the string hang vertically; align the stick with the string. Secure the cardboard so it won't move during the daily observing and can always be replaced in the same position.

At least one hour before noon, mark the position of the tip of the shadow shown on the cardboard; write the time next to each mark. Repeat every 10–15 minutes, until at least an hour has passed after noon. Now draw a smooth curve through the points. Noon occurred when the shadow was the shortest; when was noon? (It might not be 12:00 noon by your watch.) In what direction was the shadow pointing at noon? You can check the altitude of the noon sun by measuring the angle with a protractor positioned at the end of the shadow (Fig. F.1).

About a month later, do the same series of observations on the same piece of cardboard. Did the sun's height increase or decrease? Are you closer to winter or to summer?

2
Models of the Solar System

■

Central Concept
Scientific models of the solar system can explain and predict the motions of celestial bodies, especially those of the planets.

■

Chapter 1 described naked-eye astronomical observations. As you read, you probably felt the urge to place these observations into a grand design, a model for the operation of the heavens. This natural desire is what gave birth to *scientific models*, conceptual plans designed to explain the workings of nature. Scientific models lie at the heart of the workings of modern science.

This chapter examines basic scientific models of the solar system. One crucial development in the Western scientific tradition is that models had to relate to actual observations. You'll see how two different models both can make sense of the motions seen in the sky, and you can contrast them in various aspects as scientific models. The focus is on how to build a model that deals well with the motions of the planets in the solar system.

2.1 Scientific Models

A **scientific model** evokes a mental picture that tries to explain by analogy what we see in nature—an attempt to visualize the invisible. You can clearly identify the patterns in the heavens. How are they explained? A model tries to provide that explanation.

In astronomy, models are usually based on geometric ideas, physical concepts, aesthetic notions, and basic assumptions. First, observations provide patterns that excite our curiosity (Fig. 2.1). Then the mind tries to interpret this input by using geometry, physics, and aesthetics. The choices here are filtered by whatever assumptions apply.

The elements of a model interact to give it form. *Geometry* outlines a visual framework for the model or the shapes of the objects involved. *Physics* deals with the motions and interactions of various parts of the model. And by *aesthetics*—innate judgments of what seems beautiful—we select the simplest, most pleasing models from the many imagined. The principle that the simplest model wins—all else judged equal—knocks out the competition.

FIGURE 2.1

Schematic flowchart of the process of modern scientific model making in astronomy. Initial observations (and assumptions) go into the model, which has geometry, physics, and aesthetics at its base; the model makes predictions that can be compared to new (or old) observations. A discrepancy requires a revision of one or more aspects of the model, until the match to observations is acceptable.

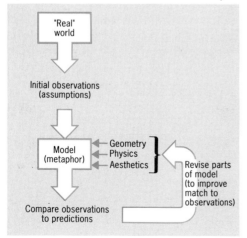

Astronomical scientific models have elements of geometry, physics, and aesthetics.

A scientific model has two key goals: to *explain* observations and to *predict* future observations accurately. The building of the model provides its power to explain. If constructed well, it should also provide specific predictions. A model's predictions must relate directly to observations and do so accurately enough to be convincing and acceptable. How well do the features of the model tally with observations? If this comparison turns out well (within the errors of observations), the model is confirmed as workable. If not, various aspects of the model are modified to get a better fit. If this revised version still does not work, the model must be abandoned completely and a new one built up. Models must be verified by accurate observations. This public act of verification distinguishes science from other human ways of knowing.

Scientific Models At a Glance	Explain	current observations
	Predict	future observations
	Verifiable	by a variety of observations
	Changeable	to match observations better

Every observation has an error attached to it. For instance, at best the human eye can see an angle as small as 1 arcminute. So all unaided observations of the sky include an error of at least this amount, which is about one-thirtieth the angular size of the moon. Instruments such as telescopes can improve on this natural limitation; but even the best equipment today generates errors in every observation.

Scientific models are mental pictures that explain key observations in nature and predict new ones with sufficient accuracy to be acceptable. A model that cannot be verified by observations must be changed or replaced.

A model's power of prediction prompts the drive to verify it. This endless search ensures that *all scien-*

tific models are ultimately tentative. Scientific models may need to change if new or more accurate observations contradict predictions. Or, if two models explain the same observations (with their errors), we have to examine other aspects of the models, besides prediction, to choose between them—such as which is simpler.

2.2 A Model of Motions in the Solar System

Let's build a basic model of the solar system that explains the motions observed in the sky. We'll approach a model of the planetary motions as needing the simplest, most effective solution we can imagine. To start, divide the observable motions into two kinds: major ones that our model *must* explain and minor ones that at first seem less essential. All these motions have *periods*—they repeat in regular intervals.

Major Motions in the Sky

Here's a summary of the major naked-eye motions from Chapter 1.

1. The entire sky moves daily from east to west with respect to the horizon.
2. The moon moves eastward with respect to the zodiacal stars in about a month.
3. The sun moves eastward with respect to the zodiacal stars in a year.
4. The naked-eye planets (Mercury, Venus, Mars, Jupiter, and Saturn) generally move eastward through the zodiac; however, during their times of retrograde motion (which last for weeks or months), they move westward.

Note that the first is a short-term motion; the others have longer periods. How to explain them? Assume that the earth lies at the center of these motions; after all, that's the way it appears. Let's start with the eastward motion of the moon relative to the stars. We can explain this observation by having the moon move around the earth to the east (counterclockwise as seen from the north) once a month. The simplest path for the moon would be a circle centered on the

earth (Fig. 2.2a), a pleasing geometric idea. The moon would move at a uniform rate around the circle, so that its angular speed, as seen from the earth, would not vary.

FIGURE 2.2

Geometric models of the motions of the moon and sun. (*a*) For the monthly eastward motion of the moon around the zodiac, place the moon on a circle with the earth at center and have it move around the circle once a month—its period. The moon's motion is counterclockwise as viewed from the north. (*b*) For the sun's annual motion, set the sun on a circle (larger than the moon's) and have it move counterclockwise around the earth. The sun completes its period in a year.

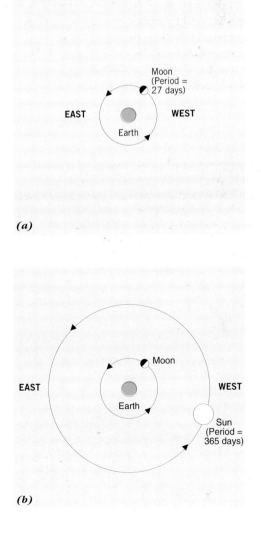

(a)

(b)

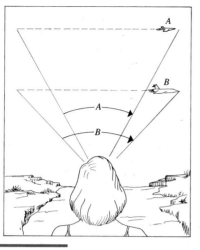

FIGURE 2.3

Judging relative distances from a comparison of angular speeds. In the same time, plane *A* covers a smaller observed angle than plane *B*. So plane *B* must be closer, if both planes fly at the same speed.

Now, the long-term motion for the sun. We know from solar eclipses that the moon is closer to the earth than to the sun (Section 1.5). We also observe that the sun takes a longer time to move once around

the zodiac. So by analogy to the moon's motion, we can place the sun on a larger circle, also centered on the earth, moving to the east (counterclockwise) with a period of one year (Fig. 2.2*b*).

We can now extend this geometric model building to the planets. Let's first ignore their retrograde motions and focus on their eastward motions through the zodiac, on or near the ecliptic. The amount of time each planet takes to circle the zodiac provides a clue to the relative distances of the planets from the earth. Assume that the planets move at the same actual speed (they actually don't!) on circles centered on the earth. The planet that appears to move at a faster angular speed is closer to earth than another that moves at a slower angular speed. The slowest-moving planet is most distant from the earth; the swiftest, the nearest.

Here's an analogy (Fig. 2.3). Suppose you are watching the lights of two airplanes at night and wish to estimate their relative distances. Assume that both planes fly at the same speed. The one that appears to move faster must be the closer of the two. Apply the same argument to the moon, sun, and planets: the fastest (the moon) is closest to the earth; the slowest (Saturn) is the farthest. A reasonable general order is: moon, Mercury, Venus, sun, Mars, Jupiter, and Saturn. (Remember we are viewing these motions as

FIGURE 2.4

Simplified geocentric model of the naked-eye sky. The stars appear to form a background sphere called the *celestial sphere*, with the earth at its center. Planets, sun, and moon move on circles from west to east (counterclockwise). The celestial sphere turns east to west (clockwise) once a day and carries all the other celestial objects with it.

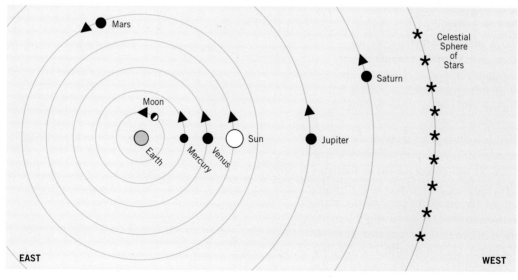

uniform around circles that are centered on the earth.)

Well, we've skipped the daily motion of the sky with respect to the horizon—the stars and everything else. We can take care of this aspect of our model by simply having all the heavens rotate *clockwise* around the earth once in a day. In this model, the earth does *not* rotate. We then have invented a simple model (Fig. 2.4) that is **geocentric**— centered on the earth. Note that we have arrived at the concept naturally from simple observations.

Our natural inclination, based on observations from the earth, is to view the solar system as geocentric.

What about retrograde motions in this geocentric model? That will take a bit more effort and additional geometry. Let's look at one planet, Mars, sticking with the notion that all motions are uniform along circles.

Imagine that Mars moves around on a small circle, the center of which moves around the earth on a larger circle (Fig. 2.5a). Allow Mars to swing along the smaller circle in the same direction (counterclockwise viewed from the north) that it moves along the larger circle. As Mars swings inside the larger circle, it appears to move in the *opposite* direction (westward and clockwise) to its normal motion. Set the time it takes for Mars to complete one revolution on the small circle equal to the interval between retrograde motions for Mars (780 days); for the period of the center of the smaller circle around the larger circle, set the average time it takes Mars to circuit the zodiac once (687 days). Then we have a simple explanation using two circles for both the normal and retrograde motions of Mars (Fig. 2.5b). While Mars is in the middle of its retrograde motion, it is closest to the earth and shines the brightest, as is observed.

Note that we can't really say how far Mars is from the earth. We have reasoned that it is closer than Jupiter but farther than Venus. We can adjust its speed around the two circles to fit the observations

FIGURE 2.5

Geometric model for retrograde motion. (a) Mars is attached to a small circle (solid-line circle) whose center (at A) rides on a larger circle (dashed circle). The earth lies in the center of the larger circle. The radius of the small circle turns in the same direction as the radius of the larger circle. So when Mars moves on the inside of the larger circle (from M_1 to M_2), it moves *opposite* its normal motion with respect to the stars; this is its retrograde motion. The motion around the small circle creates a rotary motion on the larger one. (b) The motion of Mars, as seen from the earth, produced by the combined motions. Note that Mars is closest to the earth in the middle of its retrograde loop, and so it appears at its brightest then. The period of the large circle is the average time it takes Mars to circuit the ecliptic; the period of the small circle is the time between retrograde motions.

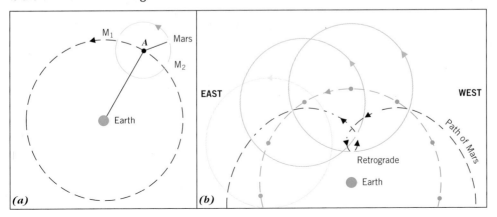

for any reasonable range of distances. If we move it closer, we slow down its speed along the circle; if we push it farther away, we speed them up. We can then apply the same geometric ideas to lay out the circles for Jupiter and Saturn.

Note that while building this model, we have made two basic assumptions: that the solar system is *geocentric* (and the earth lacks motions), and that planetary motions are *uniform* along circles. And we've ignored any *physical* concepts so far. Our model is geometrical rather than physical.

Uniform motion around a circle is the simplest motion for celestial objects to follow.

Minor Motions in the Sky

Several observed cycles or variations of cycles are not explained by our model so far.

1. The general eastward motions of the moon, sun, and planets are *not* uniform in their angular speeds, even excluding retrograde motion.

2. Mars, Jupiter, and Saturn retrograde at opposition to the sun. In contrast, Mercury and Venus retrograde at conjunctions with the sun, and neither is ever very far from the sun.

3. The retrograde motions of the planets are *not* uniform in their durations or their paths relative to the stars; and there are other variations in the eastward motion of the planets. (See Conceptual Activity 2.)

Adding explanations of these observations requires a modification of our initial model. For the retrograde motions, we will treat the motions of Mars, Jupiter, and Saturn differently from for those of Mercury and Venus, since they appear to be different. Take Mars. To retrograde at opposition means that Mars must appear opposite the sun in the sky and on the inside of its motion on its small circle. So we adjust for this fact by lining up the radius of the small circle for Mars (and Jupiter and Saturn) with the radius of the circle for the sun so that the two are parallel as they move (Fig. 2.6). We've added a little more complexity to the model.

The first minor motion needs the introduction of a new geometric technique. So far, we have kept the earth at the center of the main circles on which the sun, moon, and planets move. Instead, set the centers *away* from the earth (Fig. 2.7). When the sun or a planet moves around on the side closer to the earth; it will appear to move faster. On the side away from the earth, it will seem to go more slowly rela-

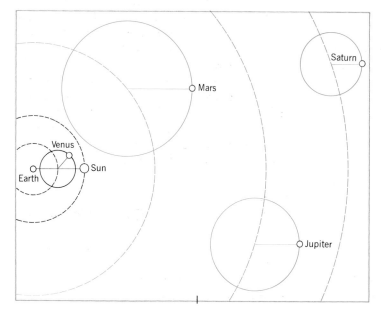

F I G U R E   2 . 6

Retrograde motions and the layout of a geocentric model. To account for the observed sizes of the retrograde loops, the small circles carrying the planets outside the sun decrease in size, so that Mars has the largest and Saturn the smallest. The radii of these circles must align with the earth–sun radius for the retrograde motions to occur only at opposition. The center of its small circle for Venus must lie on the earth–sun line.

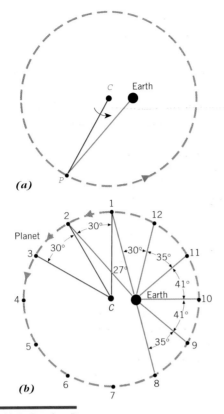

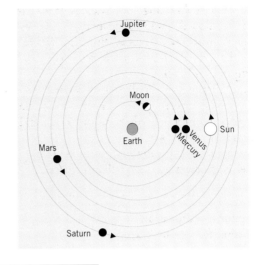

FIGURE 2.8

Overview of a simplified geocentric model, emphasizing the offsets of each planet's large circle from the earth as a center. The small circles are not shown.

FIGURE 2.7

Geometric device for nonuniform motion of the planets and the sun in a geocentric model. (*a*) Planet *P* revolves with uniform circular motion about the center (*C*) of its path. The earth is displaced from the circle's center, so the planet's motion as seen from the earth is not uniform. (*b*) The effect of this layout for modeling a planet's motion. As seen from the circle's center (*C*), the planet moves through 30° angles in equal times. But as seen from the earth, the planet covers different angles in the same times, and so appears to move at different speeds. Using the sun instead of a planet gives the same results for the sun's motions.

tive to the stars. Actually, the sun or planet is moving at a constant speed around the center of the circle; we have just shifted the observing point—the earth—away from the center. We allow a different amount of offset from the center for each celestial body. Now the model looks a lot more complex (Fig. 2.8). Do you like it?

The last observation is pretty tough to model in a simple way. Perhaps we can judge it as a very minor

point and ignore it. If we are not satisfied with that approach, we are forced to make the model more complex. So far we have kept the motions of the planets *uniform and circular about the center of a circle*. Let's break with this notion. Consider an imaginary point away from the center of the circle and opposite the earth's position (Fig. 2.9). Now require the planet, as seen from this imaginary point (point *E* in Fig. 2.9), to move at a uniform rate. Then from the earth, the motion will appear nonuniform and will follow the observed motion. But, from the center of the circle, the motion is also nonuniform! We have gotten a better match between our model and observations by violating one of our basic assumptions; namely, that all celestial motions are uniform around circles. Note, however, that we have said very little about the physics behind this model.

Describing Basic Observations with a Geocentric Model

To help grasp the main features of this model, let's try to connect it with real observations of the sky. Suppose you go out at night and face south. You observe Mars rising in the east just as the sun sets in

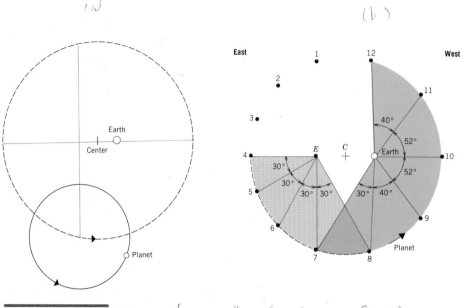

(a)   (b)

FIGURE 2.9 *non-uniform motion (except as seen from E)*

Geometric device to explain variations in retrograde motion in a geocentric model. (a) Here the earth is at a point offset from the center of the circle. Consider an imaginary point at *E* on the opposite side of the center from the earth. (b) A planet moves on the large circle so that as viewed from *E*, it would appear to cover equal angles (30°) in equal times, but as seen from *C* or the earth, it moves through different angles in equal times. The motion illustrated here ignores the swing around the small circle that generates retrograde motion.

the west. Jupiter is 45° up in the eastern sky. You also observe Venus, in the western sky, 46° east of the sun (greatest eastern elongation). The moon is at first quarter phase and due south. The scene is shown in Figure 2.10.

To explain this scene with our model, let's draw a picture. Put the earth at the center and draw five circles around it for the paths of the moon, Venus, the sun, Mars, and Jupiter. Imagine that you are looking at the earth from above the north pole. Put the sun off to the right, because (if you face south) that's where it is setting. Next mark the center of Venus' small circle (which must always be on the earth–sun line). Now put Venus on its small circle at the point at which a line from the earth is tangent to it, because this orientation gives the greatest angle of Venus away from the sun (greatest elongation). Since Mars is just rising at sunset, it must be opposite to the sun (at opposition). Because the radius of its small circle must be parallel to the earth–sun line, the center of its small circle must be on the same direct line as that from the earth to Mars.

FIGURE 2.10

Evening scene of the sky from midnorthern latitudes on the earth. The view is to the south. The sun has just set; Mars is just rising. The moon, at first quarter (90° from the sun), lies between Venus and Jupiter. Venus is about half the angular distance of the moon from the sun; Jupiter about halfway between the moon and Mars.

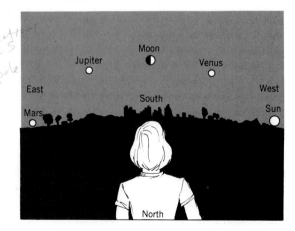

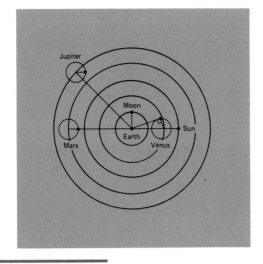

Geocentric model for the scene of Figure 2.10. Note that the layout depends mainly on the observed angles: Venus is about 45° from the sun; the moon 90°; Jupiter about 145°; and Mars is at 180° (in opposition to the sun).

The moon is observed to the south halfway around the sky between the sun and Mars, and so must be on its circle 90° around from the earth–sun line. What about Jupiter? Draw a line 45° clockwise from the earth–Mars line. Now place the center of Jupiter's small circle on its larger one such that the radius to Jupiter is parallel to the earth–sun line. You should now have something like Figure 2.11. Try to imagine the view by putting yourself back on the earth.

In a geocentric model, simple geometric ideas can explain most of the observations of the planets, sun, and moon in the sky.

2.3 Building an Alternative Model

You might be feeling a bit uneasy at this stage. I've developed a pretty complete geocentric model. You probably have the notion that the solar system is really centered on the sun—**heliocentric**. Let's see

if we can build a heliocentric model; you'll see that it is *not* at all obvious from naked-eye observations. Start with two assumptions: (1) all planetary motions are uniform around circles, and (2) the solar system is centered on the sun, so that the planets move around it. Now let's try to explain our seven observations.

Major Motions

First, consider the daily motion of the stars, sun, moon, and planets from east to west. Simple enough: just let the earth rotate daily from west to east (counterclockwise). Then *everything* in the sky appears to move in the opposite direction (clockwise) with respect to the horizon. We have a clear advance compared to a geocentric model—one motion of the earth replaces all the daily motions of the planets, sun, moon, and stars. But, we need to come up with some explanation (a physical one!) of how objects stay on a rotating earth (or a firm observational proof of the earth's rotation). Since we skipped physics before, let's be fair and bypass it again for now.

The next, the moon's monthly motion with respect to the stars is also easy: have the moon orbit the earth, counterclockwise, with a period of one month.

The third motion, the sun's yearly passage along the ecliptic, will be a bit harder because we have to shift our point of view if we assume that the earth moves around the sun. How does the sun appear from such a moving platform? Here's an analogy. Imagine yourself walking slowly counterclockwise around a lamppost. Look in the direction of the lamppost and note the background behind it. You'll see the background slowly change; the lamppost appears to move counterclockwise (the same direction as you're going) with respect to background objects.

Now imagine that the lamppost is the sun, you are the earth revolving around the sun, and the background is the stars of the zodiac (Fig. 2.12). From the earth you see the sun in a constellation, say Leo. As the earth revolves counterclockwise, the stars behind the sun change. After one month the sun appears in Virgo, one constellation to the east. The sun seems to have moved, relative to the stars, counterclockwise. Actually, the earth has moved; it's the *line of sight* from the earth to the sun that has changed. So if the earth's period around the sun is

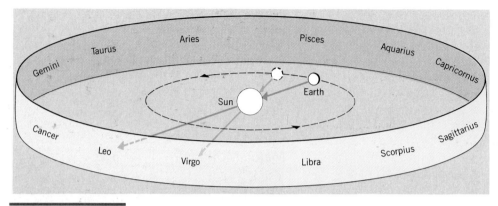

FIGURE 2.12

The sun's apparent eastward motion through the zodiac along the ecliptic in a heliocentric model. As the earth travels around the sun (counterclockwise as seen from the north; or *west to east*), the line of sight to the sun and toward the background stars moves in the *same* direction (west to east). For example, start with the sun in Leo. A month later the earth will have moved far enough eastward that Virgo lies behind the sun. The sun seems to move *eastward* on the ecliptic through the zodiacal stars at the rate of about one constellation a month or about one degree per day.

one year, it will appear that the sun moves along the ecliptic with a period of one year or at about a degree per day.

Basically, we have shifted the positions of the earth and sun that we established for a geocentric model. For the heliocentric model, the order of the solar system is: sun, Mercury, Venus, Earth (with its moon), Mars, Jupiter, and Saturn.

In a heliocentric model, simple geometric ideas can explain most of the observations of the planets, sun, and moon.

The fourth major observation relates to this change in the line of sight. If we assume that all the planets move around the sun in the same sense (counterclockwise), then, from the earth, we'd expect to see the planets move eastward with respect to the stars. And they do! But what about the retrograde motions? These arise from the planets chasing one another around, the faster (inner) planets regularly passing the slower (outer) ones. An analogy: when you pass a car on a highway, the slower car appears to move backward with respect to background objects. Similarly, as the earth speeds past a planet, that other planet seems to move westward (back-

ward) against the backdrop of stars. When the earth passes any of the outer planets or when the earth is passed by the inner ones, retrograde motion occurs. The *passing* is the key to understanding retrograde motion in this model.

Note that retrograde motion is based on the assumption that the farther a planet is from the sun, the more slowly it moves around the sun. Given an inferred order (Mercury, Venus, Earth, Mars, Jupiter, and Saturn), we also can deduce the *distances* from the sun, relative to the earth–sun distance

Table 2.1

Distances and Periods in a Heliocentric Model

Planet	Distance from Sun (AU)	Period Around Sun (years)
Mercury	0.39	0.24
Venus	0.72	0.62
Earth	1.00	1.00
Mars	1.5	1.9
Jupiter	5.2	12
Saturn	9.5	29

(which we call one **astronomical unit**, or AU), and the *periods* of the planets around the sun as seen from the sun (Table 2.1).

Now let's see in detail how retrograde motion works in a heliocentric model. Take Jupiter as an example (Fig. 2.13). Its retrograde motion takes place at opposition. In a heliocentric model, opposition of a planet occurs when the earth is exactly between that planet and the sun. As the earth approaches, the line of sight from the earth to Jupiter moves eastward. But as the earth passes Jupiter, the line of sight swings westward relative to the stars. As the earth moves on, the line of sight eventually moves eastward again. The observer undergoes the illusion of retrograde motion as the earth passes Jupiter.

In a heliocentric model, retrograde motion is an illusion, caused by the passing of *any* two planets.

The same chase-and-pass scenario results in the retrograde motions of Venus and Mercury and implies that these planets lie closer to the sun than the earth does. Assume that this is true. These planets then pass the earth because they move faster around the sun. Consider Venus. When Venus moves around the back of the sun (from greatest western elongation), it appears to move eastward as seen

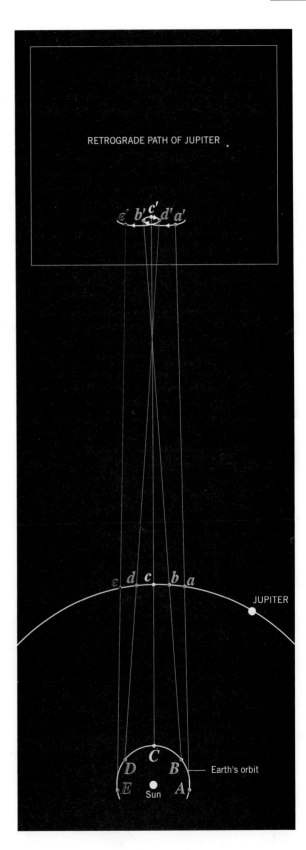

RETROGRADE PATH OF JUPITER

JUPITER

Earth's orbit

Sun

FIGURE 2.13

Retrograde motion of Jupiter in a heliocentric model. As the earth comes around the same side of the sun as Jupiter, it is moving faster along its path (from *A* to *C* to *E*) and so passes Jupiter (at point *C*). During this passing interval, Jupiter appears to move backward (to the west) with respect to the stars (*b* to *c* to *d*). Note that Jupiter appears in the middle of its retrograde motion (at point *c*) just as the earth passes it. Jupiter is in opposition to the sun as viewed from the earth. Also, it is closest to the earth, and so appears its brightest in the sky. Each position of Jupiter and the earth are at 50-day intervals. The same basic diagram applies to Mars and Saturn, which undergo retrograde motion when the earth passes them. The retrograde motion results from the passing situation.

from the earth. But as Venus catches up to the earth (from the east side of the sun) and passes it, Venus appears to move westward with respect to the stars (toward the west side of the sun). So for the two inner planets, retrograde motion arises from relative motion, but with the earth moving slower.

Minor Motions

Let's tackle the first of the minor observed motions: the eastward motions are *not* uniform. For the moon, set the earth a little away from the center of the moon's orbit. For the planets (including the earth), set the centers of their circles away from the sun. These offsets are small but needed. The result looks very much like the geometry we built for the

FIGURE 2.14

Simplified views of a heliocentric model drawn to scale for distances. (a) The planetary paths for Mercury, Venus, Earth, and Mars. Note how much the sun is offset from the center of Mercury's path. (b) The paths of Earth, Mars, Jupiter, and Saturn. Of these, Mars has the greatest offset.

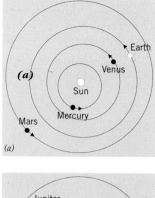

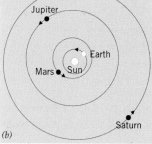

geocentric model; we have switched the positions of the earth and sun, however.

The layout of the heliocentric model naturally explains the second minor motions. Mercury and Venus have orbits interior to the earth's orbit; so their angular distance from the sun never becomes very large, and they move retrograde at conjunctions. Mars, Jupiter, and Saturn are in opposition to the sun during their retrograde motions; so they are opposite the sun in the sky, on the same side as the earth. The earth is passing them, and they naturally exhibit retrograde motion. Also, they are closest to the earth and are fully illuminated (like a full moon at opposition), and so appear at their brightest in the sky.

The final minor motions are subtle ones and more difficult to explain in a heliocentric model than a geocentric one. We could, by analogy to the geocentric model, eliminate the assumption that the motions of the planets are uniform around the centers of circles. Then we could imagine a point offset from both the sun and the centers of the planets' circles from which the motion is seen as uniform. This is just what we did in the geocentric model, but we've shifted the sun and the earth. Not a great idea, but let's adopt it as a temporary explanation. We have added complexity to the basic heliocentric design (Fig. 2.14); compare it to the geocentric one.

Describing Basic Observations with a Heliocentric Model

Consider again the following observational situation: Mars rising at sunset, Venus at greatest eastern elongation (an evening "star"), the moon at first quarter phase, and Jupiter 45° above the eastern horizon (Fig. 2.10). How to explain this view with a heliocentric model?

Put the sun at the center of a piece of paper. Draw four circles around it (if you want a good scale model, draw them with radii in the ratio of 0.7, 1.0, 1.5, and 5.2). Place the earth on its path (the second circle). From the earth, draw a line tangent to the first circle, and place Venus at the tangent point. Mars, since it is at opposition, must be placed on the third circle exactly on an extension of the earth–sun line. Draw a line, making a 45° angle with the sun–earth–Mars line, outward from the earth to the fourth circle, and place Jupiter at the intersection.

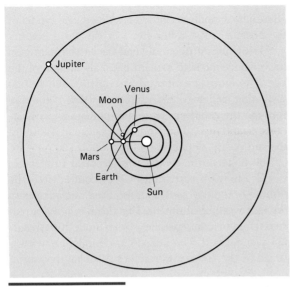

FIGURE 2.15

Simplified heliocentric model for the night scene in Figure 2.10. Note that all the angles from the sun, as seen from the earth, are the same as those in Figure 2.11.

Finally draw a small circle around the earth and place the moon on it 90° counterclockwise from the earth–sun line.

You should now have something that looks like Figure 2.15. Compare this to the drawing of the same observations in the geocentric model (Fig. 2.11.) What do you see? *They look very much the same except for the centers of the motions.* Still, the heliocentric model has two simplifications: the rotation of the earth replaces the rotation of the sky, and the passing of planets (to explain retrograde motion) replaces the very small circles of the geocentric model. In these two aspects, it is simpler and so more pleasing aesthetically. Is that enough? No! How, then, to decide on the proper model? By also considering *physics*—the aspect of scientific models that has been glossed over so far. For these solar system models, that physics is the physics of motion.

From the geometric aspects alone, geocentric and heliocentric models explain the naked-eye motions in the solar system equally well.

2.4 Models of the Solar System: A Historical Perspective

As you might expect, the first serious models of the solar system were geocentric—the most plausible explanation. The geocentric model in the Western tradition comes mostly from Greek culture. The Greeks viewed the nature as a **cosmos**, an orderly and harmonious system. Plato (427–347 B.C.) saw the perfection of the universe in the form of a sphere. In keeping with this symmetrical shape, he assumed that all the heavenly bodies moved at a uniform rate around circles. The Greek philosopher Aristotle (384–322 B.C.), the most famous of Plato's pupils, devised a complex geometric model based on the idea of uniform, circular motion. But the model did not describe the celestial motions very well.

A distinguished Greek astronomer who lived and worked at Rhodes from 160 to 127 B.C. was Hipparchus. He added the geometric devices of eccentrics, epicycles, and deferents to explain aspects of planetary motions that had previously been shrugged aside. The large circle for each planet is the *deferent* (dashed circle in Fig. 2.5a); when the earth is offset from the deferent's center, it becomes an *eccentric* (dashed circle in Fig. 2.7a). The small circle that each planet actually moves on is the *epicycle* (solid circle in Fig. 2.5a). Each device accounted for observed features of these motions: the epicycle and deferent together explained the retrograde motions (Fig. 2.5b); the eccentric, the nonuniform motion of the planets and sun through the zodiac (Fig. 2.7b). I described these devices in Section 2.2 without naming them as such; I will not use them beyond this chapter.

Two and a half centuries after Hipparchus, Claudius Ptolemy (Fig. 2.16) worked at the library in Alexandria, Egypt. He molded the existing astronomical traditions into a comprehensive model that would endure for centuries. Ptolemy's most influential astronomical work was the *Almagest*, the first professional astronomy textbook. Ptolemy was the first to design a complete system that accurately predicted planetary motions, with errors of usually not more than 5° and often less.

Ptolemy did invent one important geometric device, which he called the *equant*. This is the imaginary point, *not* the center of a circle, about which the motion of a planet is uniform and circular (point *E* in

FIGURE 2.16

Claudius Ptolemy: ". . . we shall only report what was rigorously proved by the ancients. . . ." (Sixteenth-century French woodcut, the Granger Collection)

Fig. 2.9*a*). As in our model (Fig. 2.9*b*), this procedure upsets the assumption of uniform, circular motion. Why did Ptolemy violate this aesthetic ideal? Probably because he demanded that his model fit observations reasonably well. With the equant, he could match the model's predictions better to the observations, especially the annoying variations displayed by the planets in their many retrograde cycles. This match apparently struck him as more important than the traditional precept of uniform, circular motion. As seen from the equant point, the motion *is* uniform, though the motion along the circle is *not*.

Ptolemy's Model ------- **At a Glance**	Geocentric No forces for heavenly motions Finite in size, stars near compared to earth–sun distance Variations in eastward motions: eccentrics Retrograde motions: deferents and epicycles

But the geocentric model was not the only one imagined in ancient times. Long before Ptolemy, the Greek scientific tradition centered on the library at Alexandria. Here worked the astronomer Aristar-

chus, who actually proposed a heliocentric model of the cosmos. Aristarchus lived in the third century B.C. His heliocentric model had the earth rotating on its axis once a day to explain the daily motion of the sky. The earth also moved around the sun in one year; this explained the annual motion of the sun through the zodiac, as we did in our model. Aristarchus' model was all but forgotten because it was not worked out in detail.

The heliocentric model was attacked in classical Greek times on two fronts: (1) it contradicted the *physics* of the day, and (2) it required a *stellar parallax* that was not observed. For these two main reasons—physics and parallax—the model of Aristarchus languished. The heliocentric model always was vexed by these two notions; let's see what they were.

Physics and Models

Aristotle had viewed the cosmos as divided into two distinct realms: a region of change near the earth and an eternal region in the heavens. The realm of change contained bodies made of four basic elements: earth, air, fire, and water. Each element had its own natural motion toward its natural place of rest in the cosmos: the earth to the center, the fire to the greatest heights below the moon, the air below fire, and water between the earth and the air. In contrast, the heavens were made of an immutable, transparent substance. The outermost shell of this material contained all the stars.

Each realm had different versions of **natural motion**: motion not maintained by forces. In the heavens, the celestial spheres rotated naturally. So *no* forces were needed to move the planets around the earth. In the terrestrial realm, earth, air, fire, and water each had its natural motion. For example, the natural motion of earthy material was toward the center of the cosmos. But here **forced motions** also could occur. For instance, to keep a cart moving, a person must keep pushing it. Once this force is removed, the cart rolls a bit but will soon stop. Such motions, Aristotle reasoned, required a **force**, a push or a pull, to keep them going.

These physical ideas shaped Greek models of the cosmos. The earth must be stationary and in the center of the universe. How so? First, the natural motion of earthy material to seek the center of the cosmos explained the location of the earth there.

Second, if the earth moved, bodies thrown upward would not drop back to their point of departure. Yet, heavy objects thrown upward do return to their starting place. So the earth did not move; the geocentric nature of the cosmos had a natural physical explanation. Third, the natural motion of the heavenly materials was rotation; this explained all the motions of heavenly bodies around the earth.

In early Greek geocentric models, the heavenly spheres were seen as actual physical spheres; the natural motion of these spheres—rotation—was believed to drive all the heavenly motions; no force was required.

The geocentric model was also successful because it accounted for the lack of a stellar parallax. Aristotle noted that if the earth moved around the sun, the stars must display an annual shift in their positions, called *stellar parallax*. No one observed this change, and Aristotle concluded that the earth did not move around the sun. Let's explore this idea in a bit more detail.

Parallax and Models

Stellar parallax is an apparent shift in the positions of a star (or stars) because of the change in position of the observer. In a heliocentric model, because stellar parallax arises from the earth's revolution around the sun, it is called **heliocentric parallax**. The details of heliocentric parallax differ, depending on whether the stars in space are confined to a thin shell (as in the Greek picture) or spread throughout space (as in modern concepts).

Parallax occurs when the observations take place from two separate locations. Consider, for example, a merry-go-round. As it turns, a rider on it views nearby objects to shift in their positions relative to more distant ones. That shift in angle is *parallax*.

Imagine that the stars are stuck in a thin shell, such as in the model of Aristarchus (Fig. 2.17). Pick out two stars close together on the celestial sphere (A and B in Fig. 2.17). Observe them when they are due south at midnight (position 1 in Fig. 2.17); they will appear some angular distance α apart. Just after sunset three months later, observe the stars again

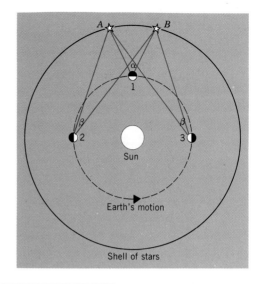

FIGURE 2.17

Stellar parallax in a finite, heliocentric model: A and B are two stars fixed on the stellar sphere. The earth moves around the sun in a year, causing a change in the observed angle between the stars. The largest angle, α, occurs at position 1; an intermediate angle, β, at positions 2 and 3.

(position 2 in Fig. 2.17); they will appear closer together (angle β is less than angle α), partly because you're now seeing the stars at an angle rather than face-on. Observe them again six months later (position 3 in Fig. 2.17); their angular separation is again β.

Consider viewing these stars over a six-month cycle: from positions 3, 1, and 2. You'd see the stars close together (3, angle β), farther apart (1, angle α), and then closer together again (2, angle β). This cyclical shift in angular position (from β to α to β) is heliocentric stellar parallax for a model with the stars all at the same distance from the sun. Note that the size of the earth's path compared to the size of the shell of stars determines the size of the shift: the smaller the ratio, the smaller the shift. Of course, if the earth were in the center of the stellar shell and did not move, no shift would occur.

We now know that heliocentric parallax is too small to detect without a telescope, because the stars are very far away and at different distances (Section 11.2). Ancient Greek astronomers, who were *not* able to observe this parallax, viewed the heliocentric model as inconsistent with observations and in part rejected it on this basis.

Stellar parallax is a natural result of a heliocentric model, but it amounts to a very small angle, far too small to observe without a telescope.

 ## 2.5 A Heliocentric Model Reborn

Astronomy languished with the decline of Greek civilization in the first few centuries after Ptolemy. But the essential Ptolemaic model worked well enough; practicing astronomers lacked good reasons to discard it. The Polish astronomer Nicolaus Copernicus (1473–1543) initiated an intellectual revolution by developing a new heliocentric model of the cosmos when the old, geocentric one seemed adequate—and so reinventing the model of Aristarchus. The revolution in astronomy after Copernicus (Fig. 2.18) marked the first major shift in our concept of the earth's place in the cosmos: from geocentric to heliocentric. This revolution also injected the concept of a *physical* force working in the heavens.

FIGURE 2.18

Nicolaus Copernicus: "In the center rests the sun." (Courtesy O. Gingerich)

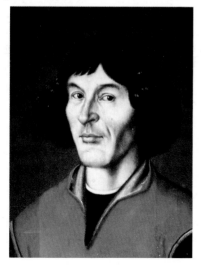

The Heliocentric Copernican Model

To build his model, Copernicus stayed with the assumption that all celestial motions must be uniform, circular motions. He viewed Ptolemy's equant as a violation of this ideal and so attacked the Ptolemaic model. Recall that the equant required the center of a planet's epicycle to move *nonuniformly* along a circle and as seen from the earth. The new model would reinstate uniform circular motion, which Ptolemy, in practice, had given up.

Copernicus took 20 years to develop his model; yet his predictions generally came out no better (and sometimes worse!) than those based on the Ptolemaic model. Copernicus also sidestepped the fact that his model violated Aristotelian physics. He did not offer new physical ideas to support his model; Copernicus had created a new *geometric* model, not a *physical* one.

The heliocentric Copernican model had better aesthetics than the geocentric Ptolemaic one, but it was not much better in predicting planetary positions.

Note that the Copernican model did *not* include a force between the sun and the planets—an idea that we are inclined to project upon it. Planetary motions arose from the natural motions of the celestial spheres—namely, uniform rotation. As in the Ptolemaic model, such natural motion needed no forces to persist. Yet, Copernicus had no good physical ideas to explain a spinning earth.

The Copernican Model	Heliocentric
	No forces for heavenly motions
	Finite in size, stars farther than earth–sun distance
At a Glance	Variations in eastward motions: earth's path eccentric
	Retrograde motions: illusion from the passing of planets

Kepler and the Physics of Planetary Motion

After Copernicus, Johannes Kepler (1571–1630) of Germany forged the new physical ideas about plane-

FIGURE 2.19

Johannes Kepler as a young man: "Astronomy has two ends, to save the appearances and to contemplate the true form of the edifice of the world." (Courtesy O. Gingerich)

The line through the foci to both sides of the ellipse (R_a to R_p) is called the *major axis*. Half this length is the **semimajor axis**, usually designated by a. The major axis has length $2a$. The distance from the center of the ellipse to a focus is often designated c, so the distance between the two foci is $2c$.

An essential property of an ellipse is its **eccentricity**, or "flatness" (how much its shape differs from a circle). Imagine that you took the two tacks at the foci and moved them closer together. The ellipse would become more circular. In fact, when the two tacks coincide, the ellipse becomes a circle. Its eccentricity then is zero. Moving the two tacks farther apart increases the eccentricity until the tacks reach the limits of the string. The eccentricity is then 1.

The paths of the planets are ellipses with different semimajor axes and eccentricities.

Kepler's Laws of Planetary Motion

Kepler's fame rests mostly on his three laws. And rightly so, for they broke with the tradition of uniform motion on circular orbits. Let's look briefly at each law in modern terms.

LAW 1 · **Law of Ellipses** The orbit of each planet is an ellipse, with the sun at one focus (Fig. 2.20). The other focus is located in space and lacks physical import. Note that the distance from the sun to the planet varies as it moves along its elliptical orbit. At its closest to the sun, the planet is at its **perihelion** point. When at its greatest distance, it lies at **aphelion**. The amount of time it takes for a planet to complete one orbit is its **orbital period**.

LAW 2 · **Law of Equal Areas** A line drawn from a planet to the sun sweeps out equal areas in equal times as the planet orbits the sun. Physically, this law notes that the orbital speeds—the rate at which a planet moves—are nonuniform but vary in a regular way: the farther a planet is from the sun, the more slowly it moves in its orbit (Fig. 2.21).

LAW 3 · **Harmonic Law** This law, which points out that the planets with larger orbits move more slowly around the sun, suggests that the

tary motion that helped to jell the modern ideas about them. Kepler (Fig. 2.19) shaped the Copernican model into a truly *physical* one in which the sun determines the planetary orbits by a force—an interaction—between the sun and the planets. Kepler also simplified the model and perfected its usefulness for the prediction of planetary positions.

Kepler pondered the idea that planetary motions were driven by a force from the sun. He imagined the sun as a fountain of magnetic force—an idea we now know to be incorrect. But Kepler did hit upon the correct notion that the orbits of the planets were *elliptical* rather than *circular*.

Ellipses have basic properties that are fairly easy to grasp, and to understand Kepler's laws, you need to have in mind their fundamental geometrical properties. You can draw an ellipse by taking a loop of string and holding it down to a board with two tacks. Keeping the string taut with the tip of a pencil held against it, draw a curve around the tacks. That curve is an **ellipse**. The tacks mark the two **foci** (F_1 and F_2) of the ellipse. For all points on the ellipse, the sum of the distances from the two foci (F_1 to P and F_2 to P in Fig. 2.20) is the same. (The singular of "foci" is **focus**.)

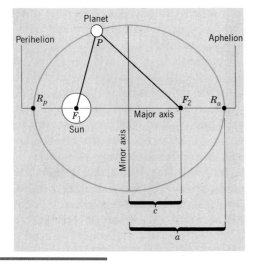

FIGURE 2.20

An ellipse and Kepler's first law. The sum of the distances of any point on the ellipse from the two foci (F_1 and F_2) is a constant. The shape of the planetary orbits is elliptical (greatly exaggerated here), with the sun located at one focus of the ellipse. The closest point along the orbit to the sun is called the *perihelion*; the farthest, the *aphelion*.

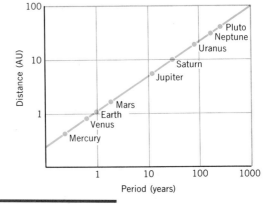

FIGURE 2.22

Kepler's third law. A planet's orbital period (once around the sun) is related to its average distance from the sun. This diagram shows for each (now known) planet a plot of its orbital period (in years) *squared* to its distance (in AU) *cubed*. The line indicates the values expected from Kepler's third law. Note how well the points for the planets fall along the same straight line—a strong indication that the law is correct.

FIGURE 2.21

Kepler's second law. Consider two equal time intervals: one when the planet is closest to the sun (*AB*), and one when it is farther away (*CD*). At *AB*, the planet–sun distance is shortest, and the planet moves fastest in its orbit. At *CD*, the planet moves more slowly because it is farther from the sun. In both cases, the areas (*AB* to the sun and *CD* to the sun) swept out by the line drawn from the planet to the sun are equal.

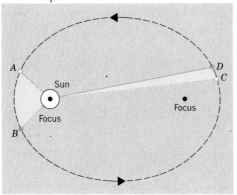

⟨MATH CONCEPT⟩

Law 3 can be written algebraically as follows:

$$\frac{P^2}{a^3} = k$$

where P is the orbital period, a the average distance from the sun, and k a constant. If P is in years and a in astronomical units (AU), then k equals 1, and we write:

$$\frac{P^2}{a^3} = 1$$

Law 3 can be used to find—for any body orbiting the sun (even spacecraft!)—the average distance from the period or the period from the average distance. For example, if $a = 4$ AU, then $P = 8$ years. This result depends only on a; no eccentricity of the orbit has any influence on the period.

sun–planet force decreases with distance. Specifically, the *square* of the orbital period of a planet is directly proportional to the *cube* of its average distance from the sun (Fig. 2.22).

What do these laws mean? Law 1 states that the shape of the orbits is elliptical, so a planet's distance from the sun varies. (Note that the sun does *not* lie in the center of the ellipse but at one focus.) Law 2 notes that as a planet's distance varies, so does its orbital speed: the closer a planet is to the sun, the faster it goes. You get the sense that the sun pulls a planet toward it; then the planet whips around the sun, and the sun slows it down as it moves away— hinting that a force from the sun acts on it. Law 3 states that the farther a planet's orbit is from the sun, the slower its average orbital speed will be. Laws 2 and 3 imply that a force between the sun and the

planets weakens with increasing distance. Hidden in law 3 is an exact description of how the sun–planet force decreases with more distance, but Kepler was unable to figure it out.

Kepler's laws provide a concise and simple description of the motions of the planets.

Using his laws, Kepler was able to predict planetary positions about 10 times better than either Ptolemy or Copernicus. Kepler finally broke the ancient spell of perfect circles and uniform motion that had mesmerized astronomers for centuries. The antidote was ellipses and a physical force. The correct nature of the force, though, would be discovered later by Sir Isaac Newton (Chapter 3).

The Larger View

We create models to understand by analogy patterns in nature. Scientific models include physical, aesthetic, and geometric ideas that are designed to explain and predict observations. How well they match observations lies at the heart of the process of verifying them. Scientific models lie at the heart of modern science.

Early geocentric models of the solar system relied mainly on geometric and aesthetic ideas; the physics played a minor role. Largely, these early models had a geocentric view of the solar system. In Europe during the sixteenth century, a heliocentric model emerged—but it required new physical ideas about motions in the heavens. The debate between the geocentric and heliocentric models was a passionate one. At stake was nothing less than an entire world view and our place in the cosmos. Kepler brought physics to

the forefront in this debate—a physical force between the sun and the planets. But Kepler did not specify the correct force. That was provided by Isaac Newton: gravitation (Chapter 3).

Battles over scientific models flare up today. We have much to understand about the universe, and there is much that we may unconsciously assume. How the earth evolved, whether or not other planets lie beyond the solar system, how the universe began—these and many other topics we astronomers investigate and debate. This book will feature some of these controversies. In all cases, we'll use basic concepts to build scientific models from data and physical laws. But if observations persistently contradict a model, we must be prepared to modify or even discard it. That is the vital process of science.

Key Ideas and Terms

scientific model   1. Scientific models are the core of the scientific process. Such models are designed to explain and predict what is observed; astronomical models contain geometric, physical, and often aesthetic elements. Scientific models are never complete or final—they evolve continually.

retrograde motion

2. The retrograde motions of the planets vary in duration and shape from planet to planet and for the same planet. Explaining these phenomena was a most vexing puzzle for early models, which resorted to a variety of geometric devices, mostly using circles.

cosmos
geocentric model

3. The ancient Greeks made the first models based on aesthetic notions and later scientific ones based on a desire to match observations. The aesthetics included harmony, symmetry, simplicity, and the use of geometry to describe the observed motions.

force
forced motion
natural motion

4. The ancient Greeks developed a complete geocentric model based on physical ideas; different ideas of motion applied to the terrestrial and celestial realms; no forces were needed to turn the celestial spheres.

heliocentric model
heliocentric parallax

5. Aristarchus proposed the concept of a heliocentric model, which never gained headway because it violated Aristotle's physics and predicted a stellar parallax, which was not observed.

heliocentric model

6. Copernicus proposed his heliocentric model mainly for aesthetic reasons; he especially disliked the equant in the Ptolemaic model, which violated the traditional precept that heavenly motions be composed of uniform motions about circles.

astronomical unit
heliocentric model
retrograde motion

7. A heliocentric model views the daily motion of the sky as arising from the earth's rotation, the sun's eastward motion along the ecliptic as a reflection of the earth's revolution around the sun, and retrograde motion as an illusion when one planet passes another. The order of the planets is set by their periods of revolution about the sun; the planets' distances are set by observations and simple geometry.

heliocentric model
stellar parallax

8. Copernicus' model did not predict planetary positions much better than the Ptolemaic model, nor was it much simpler. It violated the accepted physics of the day (that of Aristotle) and required a stellar parallax that was not observed.

aphelion
ellipse
eccentricity
focus
orbital period
perihelion
semimajor axis

9. Kepler found that planetary orbits were elliptical, having the sun at one focus (first law), that the planets move around the foci nonuniformly but predictably (second law), and that the farther a planet is from the sun, the longer it takes to orbit (third law). Kepler believed that the sun exerts a force that keeps the planets in their orbits, but he did not find the right force.

aphelion
ellipses

10. Kepler's model predicted planetary positions much more accurately than either the Ptolemaic or Copernican model because Kepler broke with

Kepler's laws
perihelion

uniform, circular motion for celestial objects. Planetary motions and the layout of the solar system arose from a force between the sun and the planets.

Review Questions

1. What is one main feature of a scientific model?
2. In what one way does the retrograde motion of a planet change?
3. Give one example of how the concepts of symmetry and simplicity influenced a cosmological model.
4. Provide a common example of a forced motion.
5. How is stellar parallax a natural consequence of a heliocentric model?
6. What aspect of the equant offended Copernicus?
7. What is the earth's location in a heliocentric model?
8. How did the Copernican model violate the physics of Aristotle? Give one example.
9. Kepler discovered that planetary orbits had what geometric shape?
10. In Kepler's model, do the planets move at a uniform speed along their orbits?

Conceptual Exercises

1. How did a geocentric model explain (a) the apparent lack of motion by the earth, (b) the daily motion of the stars, (c) the annual motion of the sun along the ecliptic?
2. How did the ancient Greek philosophers argue against a heliocentric model on physical and aesthetic grounds?
3. Name the observation that was explained in a geocentric model with each of the following geometric devices: (a) epicycle, (b) eccentric, (c) equant.
4. State one strength of a geocentric model. Then give one strength of a heliocentric model.
5. How does a geocentric model differ from a heliocentric one in its prediction of an annual stellar parallax?
6. How does a heliocentric model account for retrograde motion? Explain with a simple diagram as well as in your own words.
7. If you were on Mars, when would you see Jupiter retrograde? The earth?
8. According to Kepler's second law, a planet travels slowest at what point in its orbit? Fastest?
9. Imagine a planet positioned between earth and Mars. Use Kepler's laws to describe its general motions.

Conceptual Activity 2 Retrograde Motion of Mars

Table F.1 gives the positions of Mars, relative to the stars, over part of a year. The positions are listed for 10-day intervals around the ecliptic; starting with 0° at the vernal equinox, the positions are given in degrees along the ecliptic (the longitude), increasing in the eastward direction up to 360°, which marks the vernal equinox again.

On graph paper along the horizontal x-axis, plot each longitude from starting to ending dates, clearly labeling each position by date, in 10-day intervals. Note that some dates will overlap when Mars makes its retrograde motion. You might want to show these by a different mark or in a different color. When you have completed the graph, answer the following questions as best you can, using the graph and the table:

1. At what date did Mars' retrograde motion begin? End?

2. When did Mars reach the middle of its retrograde motion? On that date, where in the sky would you expect to see Mars?

3. What is the total angular size in degrees, along the longitude axis, of the retrograde loop of Mars?

Table F.1

Motions of Mars

Date	Longitude
1990 Sep 5	63°
1990 Sep 15	67°
1990 Sep 25	71°
1990 Oct 5	74°
1990 Oct 15	75°
1990 Oct 25	74°
1990 Nov 4	72°
1990 Nov 14	68°
1990 Nov 24	66°
1990 Dec 4	62°
1990 Dec 14	59°
1990 Dec 24	57°
1991 Jan 3	57°
1991 Jan 13	57°
1991 Jan 23	60°
1991 Feb 2	63°
1991 Feb 12	67°

3
Gravitation and the Solar System

■

Central Concept
Newton's laws of motion and gravitation explain, predict, and unify the motions of the bodies in the solar system.

■

A complete and satisfying model of the solar system must include physics as well as geometry and aesthetics. The key question is: What makes the planets go 'round? According to Isaac Newton, it's gravitation.

In the seventeenth century, Newton erased the old separation between the earth and the heavens by linking gravity to the orbital motion of the planets. The links he forged were his laws of motion and of gravitation, from which he built a new model of the solar system. With Newton's ideas, the universe took on a new appearance. In Newton's grand vision the universe grew to an infinite expanse, driven by a single force: *gravitation*.

3.1 The Physics of Motion

The solar system is an arena of constant motion. Planets revolve around the sun; and moons around the planets. Comets plunge toward the sun, and then fly to the outer reaches of the solar system. To understand these myriad motions, you need to be familiar with the area of physics called **mechanics**, which describes the details of motion produced by applied forces. This study requires an important conceptual tool: a description of the motion of falling bodies, that is, the motion of masses under the influence of gravity. Let's try to describe motion more precisely so that we can understand it more clearly.

Acceleration, Velocity, and Speed

When you step on a car's accelerator, the vehicle does just that: it accelerates. If you start out at rest, the car goes faster and faster, as you can tell from the speedometer. Or if you are cruising on the highway, you pass a slower car by stepping on the accelerator. In both instances your velocity changes (Fig. 3.1a); that's the meaning of **acceleration**: *the rate of change in velocity*. It is measured in distance per unit of time per unit of time, such as meters per second per second (m/s/s)

But what is velocity? It's not simply *speed*, which tells you how fast you're going. Speed is measured in distance per unit of time—miles per hour (mph) or kilometers per hour (km/h), for instance. Suppose that you travel from Albuquerque to Santa Fe, New Mexico; you go from one place to another at some *speed* in some *direction*. **Velocity** involves direction and speed: it is the *rate of change of position in a certain direction*. For example, if you drive from Albuquerque to Santa Fe, a distance of about 100 km, in one hour, your speed is about 100 km per hour. When you drive back at the same speed, your velocity is different because you're heading in a *different direction*.

Imagine now that you're driving around a circular track at a constant speed. Is your speed changing? No! Are you accelerating? Yes, because your velocity is constantly changing as you turn around the circle (Fig. 3.1b). Though your speed does not change, your *direction* does. So your velocity is changing, and you are accelerating.

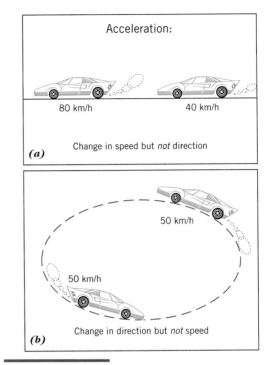

FIGURE 3.1

Velocity and acceleration. (*a*) You must accelerate the car you are driving to a higher speed. The direction of motion is the same, but the speed has increased. (*b*) If you drive a car around a circular track at a constant speed, you are accelerating because your direction is changing even though your speed remains the same.

Acceleration occurs if speed changes while direction remains the same, if direction changes while speed remains the same, and if *both* speed and direction change.

Note that speed and velocity have different physical meanings. **Speed** is the average rate of travel of something. Velocity is speed with something added: direction. Acceleration is the rate at which *velocity* (not speed!) changes.

Natural and Forced Motion

Motions fall into two categories: forced and natural. An example of forced motion is the throwing of a

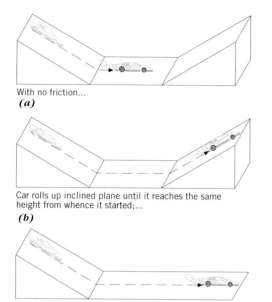

With no friction...
(a)

Car rolls up inclined plane until it reaches the same height from whence it started;...
(b)

or if it rolls onto a flat surface, it keeps going
(c)

FIGURE 3.2

Experiment about inertia (natural motion). The trick is to ignore friction in this experiment of a toy car rolling down an inclined plane. (*a*) If the toy car rolls down the plane onto a flat surface, it will keep moving on the flat section at some speed. (*b*) If it rolls up an incline, it will slow down and stop at the same height at which it started. (*c*) If it rolls onto a long flat surface, it will keep going at a constant speed.

rock. Now after you have thrown an object, its motion will continue for a while even after the force has been removed. This tendency for the motion to continue, even when the force has stopped, is called **inertia**. (Another aspect of inertia is that a body at rest will remain at rest unless a force acts on it.) The concept of inertia relates directly to that of natural motion. It also relates to *mass*, the amount of matter in an object.

The downward motion of objects results from an attractive force: *gravity*. In addition, the horizontal motion of objects flying through the air is due to their inertia. This inertial motion is a natural one and would continue if no forces, such as air friction, acted.

Consider an example (Fig. 3.2). If a perfect toy car (no friction in its wheels!) were placed on a hard, flat surface that sloped, the car would roll down the slope forever if the surface were infinitely long. The greater the slope, the faster the car would roll. Also, such a car traveling up a slope would eventually stop. If, however, the surface had no slope and the same perfect car were placed on the level with no horizon-

tal part of the velocity, the car would remain at rest. A car on a level surface, if pushed, would continue to move straight ahead at a constant velocity forever, since it is not slowed by an ascent or speeded up by a descent.

Inertia is the natural tendency for a body in motion to remain in motion, or one at rest to remain at rest.

With this concept of inertia, let's examine falling bodies, whose motion is from a force: gravity. Experiments show that such motion takes place at a *constant acceleration*, with the object's velocity changing at a *constant rate* as it falls. (The acceleration of gravity at the earth's surface has a value of 9.8 m/s/s [32 ft/s/s] and is usually denoted by *g*. This means that for every second of fall, an object gains 9.8 m/s of speed.) Now, *all* falling masses (without air friction) have the *same* acceleration at the earth's surface. When dropped, they reach the same velocity

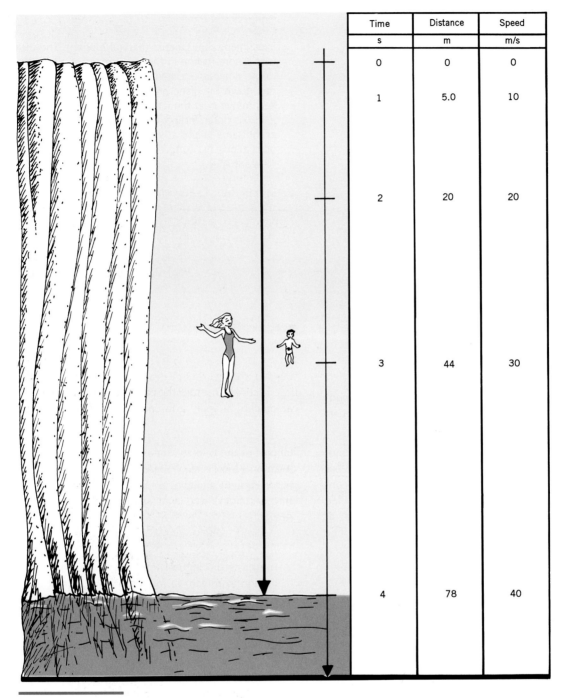

Time	Distance	Speed
s	m	m/s
0	0	0
1	5.0	10
2	20	20
3	44	30
4	78	40

FIGURE 3.3

Two different masses falling at constant acceleration. Imagine two divers, one more massive than the other, stepping off a cliff at the same time. Even though they have different masses, both would fall the same distance in the same time (here, about 78 m to reach the water in 4 s) and hit the water at the same speed (about 40 m/s). Note that the values of the speeds and distances are rounded off to two figures.

after the same time; they also fall the same distance in the same time (Fig. 3.3).

For constant acceleration, such as occurs with falling bodies, the speed continually increases as time passes, while the direction (down!) remains the same. So a dropped object falls with greater and greater speed as it gets closer to the ground. When you throw a ball near the earth's surface, its path describes an arc (Fig. 3.4) that results from a combination of its falling motion (which occurs with a constant acceleration) and its horizontal motion (which occurs at pretty much constant speed because air friction has little effect on the ball during its time in the air). The curved path results from a force in one direction only: that pulling the ball to the ground, toward the center of the earth.

A thrown object's path results from a combination of two types of motion: a natural, inertial one in the horizontal direction, and a forced one vertically.

Many of these notions of mechanics were developed in more or less modern form by the Italian scientist Galileo Galilei (1565–1642), who was a contemporary of Kepler. Galileo believed in the Copernican model and promoted it. He realized, though, that the model lacked a physical basis. He tried to develop one by examining the motions of bodies on and near the earth—especially falling bodies. Galileo hoped

that these could be applied to the motions of celestial bodies. He failed to achieve this goal, which later was reached by Newton.

3.2 Motion and Forces

Sir Isaac Newton (1642–1727), the great British scientist, emerged as the genius destined to explain motions in the celestial realms (Fig. 3.5). The publication in 1687 of Newton's *Principia Mathematica*, containing his concept of gravitation, resulted in a truly physical view of the universe. Newton's genius in discovering gravitation involved solving the old puzzle of the motions of the planets and rebuilding the Copernican model.

Accelerated Motions

Let's first reflect on our experiences with forces. Imagine a toy car that moves along a flat surface with little friction. Consider the car at rest. Then push the car quickly to apply a force for a brief time. What

FIGURE 3.5

Sir Isaac Newton: "For the whole burden of philosophy seems to consist of this—from the phenomena of motions to investigate the forces of nature." (Courtesy the Bettmann Archive)

FIGURE 3.4

Path of an object thrown into the air. Gravity pulls it straight down, and its inertia keeps it moving horizontally until it hits the ground. The result is a curved path (called a *parabola*).

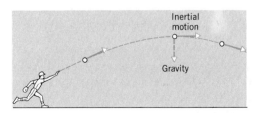

happens? First, the car accelerates from zero velocity to some velocity. Second, the direction of the acceleration (and the resulting velocity) is the *same* as the direction of the initial force. So this example shows that forces produce accelerations (rate of change in velocity, remember!)

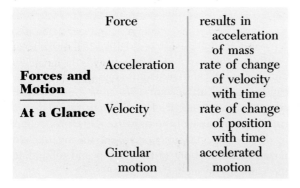

	Force	results in acceleration of mass
Forces and Motion	Acceleration	rate of change of velocity with time
At a Glance	Velocity	rate of change of position with time
	Circular motion	accelerated motion

What happens when you push the car, again at rest, with a greater force than you used the first time? Right—the car will have a greater acceleration and will reach a greater speed than before. In fact, if you applied twice the force, the car would accelerate twice as much. (The acceleration is still in the same direction.) If you used three times the force, the car would accelerate three times as much, and so on. Hence, the acceleration of any object is *directly* related to the force exerted on it.

Now return to our initial force, but this time add weights to the car, doubling the car's original mass. What occurs? The acceleration is still in the same direction as the applied force, but it is *less* because more mass has to be moved. If you applied the same amount of force you used the first time, the cart will have *half* the acceleration for *twice* the mass. With three times the mass and the same force, the resulting acceleration would be one-third as much, and so on. **Mass** is a measure of an object's resistance to acceleration. (The standard unit of mass is the **kilogram**; see Appendix A.) The acceleration of any object is *inversely* related to its mass—the greater the mass, the less the acceleration (for the same amount of force).

The acceleration of an object depends directly on the strength of the force and inversely on the mass of the object.

Note here that "strength of force" means the *net* force on the object. Why net force? Because more than one force may act on an object, and such forces may balance if they act in opposite directions. Consider a tug-of-war. If two matched teams pull on the rope with equal strengths of force, but in opposite directions, the net force is zero because the opposite, equal forces balance. This example emphasizes that forces have *direction* and can work together or against each other.

Newton's Laws of Motion

In the *Principia*, Newton describes his three **laws of motion**. These famous laws are stated in modern terms as follows.

LAW 1 · **The Inertial Law** A body at rest or in motion at a constant speed along a straight line remains in that state of rest or motion unless acted upon by a net outside force (Fig. 3.6).

LAW 2 · **The Force Law** The rate of change in a body's velocity from an applied net force is in the same direction as the force and proportional to it, but is inversely proportional to the body's mass (Fig. 3.7). $F = ma$

LAW 3 · **The Reaction Law** For every applied force, a force of equal strength but opposite direction arises (Fig. 3.8).

Newton's first law states that uniform motion is the *natural* state of moving mass anywhere in the uni-

---

FIGURE 3.6

Newton's first law. An object pushed by an astronaut moves away at a *constant* speed along a straight line (there is no air friction to slow it down). It travels the same distance during each minute because it moves at a constant speed.

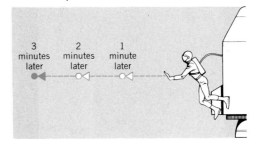

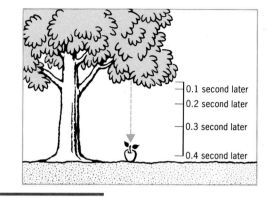

FIGURE 3.7

Newton's second law. Here, gravity acts on an apple and accelerates its descent to the ground. Because the apple falls at a constant acceleration, it moves faster—it covers a greater distance in the last tenth of a second than it did in the first.

verse. The first law gives you a way to judge whether a net force is acting on an object: you are told to look for a change in an object's speed or in the direction of its motion, or in both speed and direction.

The second law extends the recognition from a force to the recognition of its results; the direction of the change in motion is in the same direction as the applied force. Also, the amount of acceleration—the rate of change in the object's velocity—depends directly on the strength of the force. So exerting a force (which you can think of simply as a push or a pull) accelerates an object; that is, it slows it down,

speeds it up, or changes the direction of its motion.

For example, suppose you are floating in space next to a bunch of small objects. You push one. It accelerates and moves away, and it travels in the direction in which you pushed it. You can measure its acceleration, its change in velocity. Now push another mass with the same force. You measure its acceleration and find that it attains half the acceleration of the first. Since you applied the same force to both objects, the second must have twice the mass of the first. Newton's second law provides a means to measure mass along with the consequences of applying forces to masses.

--- **MATH CONCEPT** ---

In algebraic form, Newton's second law is

$$F = ma$$

where F is the net applied force, m the object's mass, and a the acceleration resulting from the force. Note that forces, like velocities, have directions, and so do accelerations. An object will accelerate in the *same* direction as an applied force. That is,

$$a = \frac{F}{m}$$

and substituting for F and m allows you to see what happens to a.

FIGURE 3.8

Newton's third law. Imagine you are in space next to a spacecraft having a mass larger than yours (a). Push the spacecraft momentarily; it pushes back on you with a force equal to the force you apply to it, but in the opposite direction. You and the spacecraft move apart (b). If you push off from a spacecraft that is more massive than you, you end up moving with a greater velocity than the spacecraft because your lower mass is being accelerated by the same force that has been applied to the massive spacecraft.

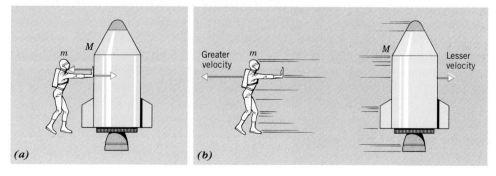

The third law recognizes that forces are interactions and must act in simultaneous pairs. If you were at rest in space and pushed against a massive spacecraft, the ship would react to your applied force with an equal but oppositely directed force, pushing you away, as described by Newton's third law. Now the forces applied to you and the spacecraft are the same, but the resulting accelerations are different. According to Newton's second law, the acceleration is greater for you than for the spacecraft, because your mass is less. So you would move away from the spacecraft quickly, while it would hardly budge. Also, you and the spacecraft would be moving in *opposite* directions.

With his third law, Newton saw gravitation as an *interaction* between the sun and the planets, and all the planets with each other.

3.3 Motion and Gravitation

From these ideas about force and motion, Newton attacked the problem of the motions of the planets by devising the law of gravitation. To do so, Newton combined his laws of motion with Kepler's planetary laws (Section 2.5) to arrive at a law of universal gravitation. Kepler believed the force was magnetic; Newton proved it was gravitation.

Newton verified that the type of force that causes the elliptical orbits of Kepler's first law is a **central force**, one directed to the center of the motion. He proved that the gravitational force between the sun and a planet causes a planet to move in an ellipse with the sun in one focus. Also, he showed that planets (and other bodies) moving under the influence of any central force obey Kepler's second law. Finally, Newton proved from the geometric properties of ellipses that the force may be described by a *specific type* of force law (an *inverse square* law) and then derived Kepler's third law.

Centripetal Acceleration

How do the moon and an apple enter into this scheme? Newton knew that gravity causes the ap-

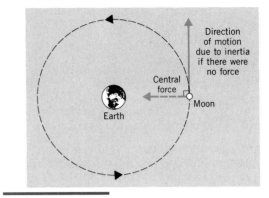

FIGURE 3.9

Central force and the moon's orbit. Since the moon does not move along a straight line, we know from Newton's first law that a force must be acting on it. This force pulls the moon toward the earth in a direction toward the center of the moon's orbit. The orbit is idealized here as a circle: it is actually elliptical.

ple's fall. Might the earth's gravity, pulling on the moon, keep the moon in its orbit? Now, the direction of the moon's orbital motion changes constantly: the moon stays on a curved path rather than moving along a straight line away from the earth (Fig. 3.9). Newton's first law tells us that a force is acting on the moon, which according to the second law results in an acceleration toward the earth's center. Such a centrally directed force is called a **centripetal force**. (*Centripetal* means "directed toward the center.") The resulting acceleration is called **centripetal acceleration**.

You experience centripetal force and acceleration when you turn a corner in a car. Obviously, the faster you go around the corner, the greater the acceleration. But the size of the turn also affects the acceleration. For instance, if you make a long, gentle curve—even at 90 km/h—you can change the car's direction with very little force on the steering wheel. But if you take a sharp turn at 90 km/h, you must exert a much greater force.

Centripetal force depends on both the speed of an object around a circle and the radius of the circle.

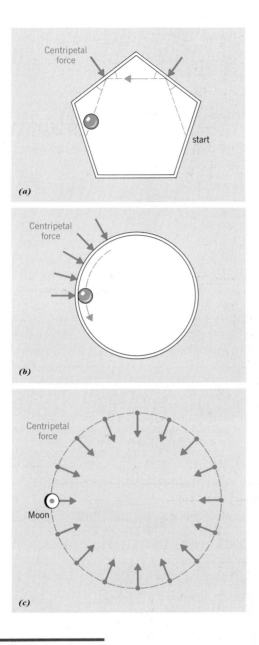

FIGURE 3.10

Centripetal force and circular motion. (a) A ball moving on a table with five sides strikes the center of each side. The force at each impact is centrally directed, that is, centripetal. (b) On a circular table, a centripetal force is exerted on the ball at every point of contact in its motion around the circumference. (c) Like the ball, the moon orbiting the earth has a centripetal force acting on it at each and every point along the orbit; if this were not true, the moon would not stay in its orbit.

Let's apply the concept of centripetal force to the motion of the moon. As an analogy, consider first a billiard ball colliding into the middle of one after the other of the side cushions of an imaginary square billiard table. At each collision, the direction of the ball's velocity changes; therefore, it must be accelerated by an applied force. The direction of the force that is deflecting the ball is toward the center of the table for every collision. Imagine now a pentagonal table (Fig. 3.10a): the force of collision at each side still points to the center, but the angle of strike and rebound is smaller than for a square. Now imagine a circular table, one with an infinite number of sides (Fig. 3.10b). The ball touches the cushion continuously (the angle between the cushion and the path is zero), and at every point, the force on the ball points to the center of the circle, and so its acceleration is also in that direction. Finally, imagine the moon's orbit as circular. At every point in the orbit, a centripetal force acts on the moon (Fig. 3.10c).

Newton's Law of Gravitation

Every body attracts every other with a gravitational force. This is a simple statement of Newton's **law of gravitation**. What does this law mean? First, it says that *all* masses in the universe attract all other masses. Second, if you consider just two masses, the amount of the gravitational force depends directly on the amount of material *each* mass has. So if you doubled the mass of one and kept the distance between the two the same, the force would also double.

Third, masses at greater separations exert *less* gravitational force than those closer together, and this drop-off of force with distance happens in a special way—as the inverse square of the distance (Fig. 3.11). That is, the correct description is that of an **inverse-square law**. Consider, for example, two masses 1 meter apart. A certain amount of gravitational force attracts one to the other along a line between their centers. Now move the masses so that they are 2 meters apart. The force is less strong. How much less? By $\frac{1}{2}$ squared, or $1/(2 \times 2)$, that is, by one-fourth as much as when the masses were 1 meter apart. At 3 units apart, the force is one-ninth as much; at 4 units, one-sixteenth as much. (See Conceptual Activity 3.)

FIGURE 3.11

Gravity and the inverse-square law. (a) Imagine three masses—all the same—placed at three different distances from the earth. The one at A is one unit of distance away; the one at B, half a unit; the one at C at 2 units of distance. If the gravitational force between the mass at A and the earth has a strength of one unit of force, then that at B is 4 units and that at C is a quarter-unit. The arrows show the direction and relative strength of the gravitational force at the three positions. (b) Graph illustrating how the strength of the gravitational force changes with the distance between masses as shown in (a). At a separation of 1 unit of distance, the force has 1 unit of strength. When the masses are twice as far apart, a distance of 2, the force has only one-fourth the initial strength.

(a)

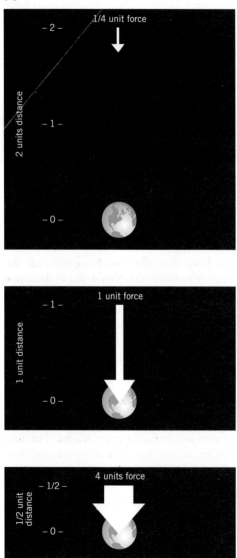

(b)

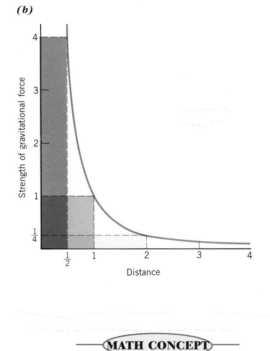

MATH CONCEPT

Newton's law of gravitation is

$$F = \frac{G\,m_1 m_2}{R^2}$$

where F is the strength of the gravitational attraction between two spherical bodies m_1 and m_2, whose centers are separated by a distance R. The symbol G is a constant; its value depends on the units used and relates the size of the force to the sizes of the other quantities. The value for G in SI units (the International System of units, see Appendix A) is 6.67×10^{-11}. In this system, a unit of force is called a *newton (N)*—the amount of force needed to accelerate a mass of one kilogram (kg) at a rate of one meter per second per second (m/s/s).

Newton's law of gravitation is an inverse-square law; gravitational attraction *decreases* in a specific way as the distance *increases*.

Newton used motions of the moon and an apple to test the validity of his law of gravitation. He knew that the earth's gravity at its surface (1 earth radius from the center) caused the apple to fall. The earth–moon distance is about 60 earth radii, so if an inverse-square law correctly describes gravitational forces, the acceleration of the moon toward the earth must be $1/60^2$ or 1/3600 as much as the acceleration of the apple.

Newton then compared his *predicted* centripetal acceleration with the centripetal acceleration derived from *observations* of the moon's orbit to find that the accelerations were pretty nearly the same. Newton concluded that the cause of the moon's centripetal acceleration is the same as that of the apple's: the earth's gravity. This force, extended out to the distance of the moon, keeps the moon in its orbit. In other words, the moon's motion results from two independent actions: its natural, inertial motion (along a straight line) and its centripetal acceleration from the earth's gravity. An apple orbiting the earth at the same distance as the moon would have the *same* centripetal acceleration as the moon does (though the *forces* would differ)!

Consider free-fall from Newton's standpoint. Imagine dropping a cannonball and a tennis ball from a tall building. The earth exerts a much greater gravitational force on the cannonball than on the tennis ball because of the cannonball's much greater mass. But, when the two objects are dropped, they fall side by side and land simultaneously (if you neglect air friction). So gravity's effect has been the same on both objects; more precisely, the *acceleration* of each is the same, because they speed up at the same rate. Although the forces are different, the accelerations turn out to be the same. Why? Because the extra force on the cannonball is exactly offset by its greater inertia to changes in motion.

Finally, note that Newton's third law requires that the gravitational force of the earth on the moon exactly equal that of the moon on the earth. They are *interacting* bodies.

3.4 Cosmic Consequences of Universal Laws

The gravitational force of the earth causes the centripetal acceleration of the moon. This was Newton's central discovery: his three laws resulted in a new understanding of the planetary orbits. The most obvious: the sun's gravity locks the planets in their elliptical orbits. Having found the physical interaction between the sun and the planets first sought by Kepler, Newton rebuilt the Copernican model into a truly physical one.

The Newtonian Model **At a Glance**	Heliocentric
	Gravitational forces for heavenly motions
	Infinite in size, nearby stars farther than earth–sun distance
	Variations in eastward motions: Earth's orbit around sun elliptical
	Retrograde motions: illusion from the passing of planets

Newton's laws also allow us to figure out the masses of the planets (if the planet has at least one satellite or influences the motion of any nearby mass) and of the sun—quantities never known before! Newton answered in detail the ancient question of how the planets moved. And he answered it precisely: his predictions of planetary positions were far better than previous ones—and, if calculated correctly, they were exact.

One important result of Newton's work is that Kepler's third law takes on a new form that makes it a bit more complex but a lot more useful. The new formulation includes the masses of the two orbiting bodies and Newton's constant of gravitation, G. This new version will serve as a tool for finding the masses of celestial bodies—planets, stars, and even galaxies!

Newton's laws of motion and law of gravitation allow us to find out the masses of celestial objects.

MATH CONCEPT

By using his laws of motion and gravitation, Newton reworked Kepler's third law so that it had the form

$$P^2 = \frac{4\pi^2}{G(M_{\text{sun}} + m_{\text{planet}})}a^3$$

where m_{planet} is the mass of the planet, and M_{sun} the mass of the sun. Compare this to Kepler's third law in the form

$$P^2 = ka^3$$

with $k = 4\pi^2/G(M_{\text{sun}} + m_{\text{planet}})$. Note that the constant relates to the masses of the two bodies involved and includes Newton's constant of gravitation, G.

We now have a powerful mathematical tool that can be used, for instance, to find the masses of celestial bodies. An example: use the earth's orbit to find the sun's mass. The earth–sun distance, a, is 1.50×10^{11} meters. The earth's period, P, is 365.26 days, or 3.16×10^7 seconds. Because the mass of the sun is much larger than that of the earth, we can approximate $m_{\text{earth}} + M_{\text{sun}}$ by M_{sun}. Then

$$M_{\text{sun}} = \frac{4\pi^2}{G}\frac{a^3}{P^2}$$

$$M_{\text{sun}} = \frac{4\pi^2}{6.673 \times 10^{11}}\frac{(1.50 \times 10^{11})^3}{(3.16 \times 10^7)^2}$$

$$= 1.99 \times 10^{30} \text{ kg}$$

not defined

The Earth's Rotation and Revolution

One objection to the heliocentric model was that objects not tied down should fly off the earth because of its rotation. Newton noted that gravity holds things down. Another objection was that dropped objects should land behind their starting position, because the turning earth should leave them behind. Newtonian physics explained that objects on the earth have inertia. When falling, they do not lose their forward inertial motion but continue to move with the ground beneath them. So they land at their starting points.

Here's how. When you throw an object straight up, it moves in the direction of the earth's rotation while it is in the air with the same speed it had while in your hand. So the object aloft keeps up with the turning earth, because no force acts on it to change its eastward velocity. When it descends, it returns to your hand. An analogy: consider riding in a car going north at 50 km/h with a ball in your hand. If you toss the ball straight up, it is still going north at 50 km/h while in the air and so lands right back in your hand.

Newton's laws of motion give a plausible explanation for the earth's rotation and eliminated a major physical objection to a heliocentric model.

Newton's laws show that the sun has roughly 3.3×10^5 the mass of the earth. Now, Newton's third law of motion requires equal gravitational forces between the sun and the earth—the strength of the force of the sun on the earth equals that of the earth on the sun. However, the second law demands that the earth's acceleration be much greater than the sun's (in fact, 3.3×10^5 times greater!). This happens only if the motion of the sun is small, so that it attains a small acceleration in one year. Thus the earth orbits the sun rather than the other way around.

In other words, the sun and earth interact gravitationally, and because of their mutual attraction, both orbit around a common point called the *center of gravity* or the *center of mass*. As seen from this point, the earth's centripetal acceleration is 3.3×10^5 times greater than the sun's; so the earth has greater change of velocity than the sun.

The **center of mass** is the balance point of two objects connected together. Consider two unequal masses at the end of a rod (Fig. 3.12). The balance point is closer to the more massive of the two. If you threw the rod in the air, the two masses would spin around the center of mass. Gravity works by binding two masses together, as the rod does here, and two masses linked by gravity also have a balance point, a center of mass. For the earth and sun, the center of mass is very close to the sun's center because the sun is much more massive than the earth. For Jupiter

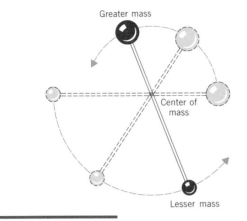

FIGURE 3.12

Motions of two masses at the end of a rod as the rod is thrown in the air. Both revolve in the same period of time around their common *center of mass*. Note that the larger mass is closer to the center of mass than the smaller one.

and the sun, the center of mass lies outside the sun's body.

Gravity and Orbits

To add to its solar system achievements, Newton's physics correctly described the elliptical orbits of comets. Newton and his colleague, Edmond Halley (1656–1742), decisively proved that comets orbit the sun following the law of gravitation and laws of motion. In fact, Halley correctly predicted the return of the comet that bears his name, but he did not live to see the event. (More on comets in Section 8.2.)

Newton's ideas also ultimately led to the discovery of Neptune in 1846, long after the initial publication of the *Principia*. Neptune was the first planet to be found by its gravitational effects on other bodies (in this case, Uranus). Newton's laws predicted the existence of Neptune *before* the planet itself was observed.

The discovery of Neptune rested on observed irregularities in the orbit of Uranus (discovered by accident with a telescope in 1781). Astronomers had noted small but significant discrepancies between the observed positions of Uranus and those predicted from Newton's laws. Such irregularities occur because all planets attract one another. For instance, because Jupiter's tug influences the orbit of Uranus,

this orbit differs from the orbit we would expect if Uranus and the sun were the only mutually attracting bodies. The undiscovered Neptune revealed itself when Uranus' motion deviated from the elliptical path predicted from the effects of the known planets. In the twentieth century, alleged discrepancies in Neptune's orbit stimulated a search for other planets, which resulted in the accidental discovery of Pluto (Section 7.7).

An equally dramatic discovery, which extended the validity of Newton's laws beyond the bounds of the solar system, was the observation of binary star systems in the late eighteenth century. (A *binary star system* consists of two stars, held together by their mutual gravity, orbiting each other.) Calculations of the orbital periods of binary star systems and the separation between the stars confirmed that their orbits were elliptical and their motions followed Kepler's laws. These laws can be derived from Newton's laws of motion and gravitation. And the motions permit us to measure the masses of the stars.

> The motions of binary star systems, which obey Kepler's laws, confirm that Newton's laws apply beyond the solar system.

For almost all the astronomical situations you will encounter, Newton's concepts will give clear and correct explanations. But you should be aware that in the twentieth century, Albert Einstein (1879–1955) created a new theory of gravity, called the *general theory of relativity*. This theory, which has been accepted as the replacement of Newton's law of gravitation, makes predictions that differ from those of Newton in special conditions (usually where gravity is very strong.). Observations have confirmed these predictions, thus serving to verify Einstein's notions. You won't need to know these in any detail until Chapter 14, but be aware that Newton did not have the last word about gravity.

Orbits and Escape Speed

From Newton's laws of motion and gravitation arises the concept of **escape speed**, the minimum speed an object needs to escape the gravitational bonds of another.

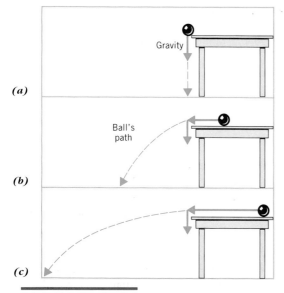

FIGURE 3.13

Motion of a ball near the surface of the earth. (a) Dropped from the edge of a table, the ball falls straight to the ground. (b) When pushed along the table before it reaches the edge, the ball moves horizontally (along a parabola) before it hits the ground. (c) When pushed harder, it moves farther before it strikes the ground. In each case, the ball takes the same amount of time to hit the ground.

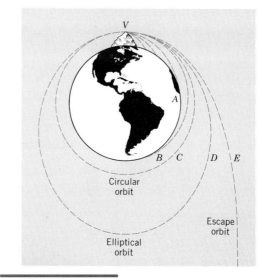

FIGURE 3.14

Launching earth satellites, according to Newton. From the top of a high mountain (V), cannonballs fired at a low speed hit the earth after traveling some short distance (A and B). A certain minimum speed places them in a circular orbit (C); a somewhat greater speed produces an elliptical orbit (D). Above the escape speed, the result is an escape orbit (E).

Consider a ball at the edge of a table (Fig. 3.13a). Let it drop and it falls straight to the ground. Now push the ball (Fig. 3.13b). As it passes the end of the table, it falls down, but the push makes it travel some distance horizontally before it hits the ground. Push it harder and it goes farther (Fig. 3.13c).

Now imagine, as Newton did, a giant cannon placed on the top of a very high mountain and aimed parallel to the ground (Fig. 3.14). Fire a cannonball. It travels some distance before reaching the ground. If you use more powder in the cannon, the ball travels farther along the earth before it hits the ground. If you put a large enough charge in the cannon, the ball goes completely around the earth in a *circular* orbit, returning to the cannon. The inertial motion of the ball just compensates for the falling from gravity. (This explanation also applies to the motion of the moon around the earth.)

The circular orbit defines a crucial path around the earth, for it requires a specific speed—no more, no less—parallel to the surface for that orbital path. A somewhat higher speed results in an elliptical

orbit. The position of the top of the mountain marks the farthest point from the earth in such an orbit— the *apogee* point.

Now dump in a larger charge. Increasing the starting speed makes the orbit more elliptical. The point at the top of the mountain will be the point in the orbit closest to the earth—the *perigee*. Eventually the orbit becomes so elongated that its semimajor axis is infinitely long, and it would take the ball an infinite time to return. So it never returns. This speed is called the *escape speed*. Any speed larger than the escape speed produces the same effect: the ball leaves the earth, never to return.

An object's mass and its radius determine the escape speed needed from its surface; greater mass results in a greater escape speed, greater radius in a smaller one.

What determines the escape speed from the surface of an object? Consider the earth. At its surface, the escape speed is 11 km/s. Rockets need at least this speed to escape the earth. Suppose that you kept the earth at its present radius but increased its mass. The escape speed would increase. Now suppose that you kept the earth's mass the same but decreased its radius. The escape speed from the surface would again increase.

The Larger View

Isaac Newton used new ideas of force and gravity to build a new and needed foundation for the Copernican model. The finished work resulted in the final overthrow of the geocentric model. Now the earth was seen as really another planet, orbiting the sun once a year and rotating around its axis once a day. We found our place in the universe—on a planet, third from our sun.

In terms of a scientific model, Newton's solar system was complete. Its predictive success was astounding. Time after time, the numbers, not just the general ideas, came out right. Planetary positions could be predicted precisely, and also the motions of planetary satellites. And Newton's ideas were *universal*. No longer were separate laws needed for falling bodies, the moon, or even binary stars. One simple force binds them all: gravity!

You will see that Newton's concepts apply far beyond the local space—to distant stars, galaxies, and even aspects of the universe. We will use Newton's laws time and again to understand motions in the cosmos—motions controlled by matter, both visible and invisible.

Key Ideas and Terms

gravitation
mechanics
Newton's laws of motion

1. With his laws of motion and law of gravitation, Newton abolished the physical distinction of motions in the terrestrial and celestial realms. This view resulted in a physically unified cosmos.

acceleration
force
inertia
mass
speed
velocity

2. Newton's laws of motion described natural motion and forced motion. He viewed natural motion as that of a body at rest or moving at a constant speed along a straight line. Forces result in accelerations (changes in velocities) and act in mutual pairs that balance; any net force results in accelerations.

central force
centripetal acceleration
centripetal force
law of gravitation

3. From these ideas about motion, Newton concluded that planetary orbits resulted from a centrally directed or centripetal force. That force is the gravitational force between the sun and the planets. He used the moon's orbit to check that the force of gravity changes with the inverse square of the distance between the two masses.

center of mass
inverse-square law

4. Newton's law of gravitation describes the gravitational force as depending directly on the product of the masses of the two objects and inversely as the

square of the distance between their centers. Newton's third law requires that such bodies in orbit revolve around a common center of mass.

ellipses
scientific model

5. Newton's physics justified the Copernican model, as revised by Kepler, with the correct force, gravity. The earth's rotation and revolution were naturally explained. Newton's work also resulted in very accurate predictions of planetary motions along elliptical orbits; the result was simple, pleasing, and accurate.

center of mass
gravitation

6. The discovery of binary stars whose motions follow Kepler's laws verified the validity of Newton's law of gravitation beyond the solar system. Such stars revolve around a common center of mass.

escape speed

7. Newton's laws of motion and gravitational result in the important astronomical idea of escape speed, the minimum speed an object needs to get completely beyond the gravitational attraction of another mass.

Review Questions

1. In what sense are Newton's laws *universal*?
2. What is Newton's concept of the natural motion of a mass?
3. What happens to the velocity of a car as it goes around a corner at a constant speed?
4. In what direction is the centripetal force that keeps the moon in orbit?
5. If two masses are moved apart, does their gravitational force increase or decrease?
6. What one objection to the Copernican model's rotating earth does Newton's physics eliminate?
7. What holds a binary star system together?
8. If two masses are an infinite distance apart, what is the gravitational force between them, according to Newton's law of gravitation?

Conceptual Exercises

1. You have two spheres of the same size and shape. One is lead, the other wood. You drop them together. What happens? Explain.
2. Imagine that you're out in space and push away from you an object having the same mass as you. What happens? Explain.
3. Use Newtonian physics to argue that the earth rotates on its axis and revolves around the sun, answering the main physical objections to the Copernican model.
4. Give a simple example, different from those in the text, illustrating each of Newton's laws of motion.
5. Imagine that you hold a ball in your hand, with your arm out to your side. You walk along quickly toward a target on the floor. You release the ball just

as it reaches a point above the target. Use Newton's concept of inertia to predict where the ball will land: behind, on, or in front of the target.

6. Consider an object shot upward from the earth with *less* than escape speed. Describe its motion.

7. Consider a special satellite sent into space to orbit the earth just as the moon does. How would the centripetal acceleration of such a satellite compare to the centripetal acceleration of the moon?

8. When you see Mars at opposition to the sun, the earth and Mars lie closest together on the same line from the sun. Consider the gravitational attraction of Mars on the earth at that time. How does it compare to the gravitational attraction of the earth on Mars at the same time?

9. What would happen to the earth's escape speed if you kept its mass the same but imagined that its radius increased?

Conceptual Activity 3   The Inverse-Square Law

The relationship between the strength of the gravitational force between two masses and the distance between them is an essential aspect of Newton's law of gravitation. (You will see later that light has a property that also follows an inverse-square law.) However, this kind of relation is not common to your experience. Let's try to get a feel for it by a graph.

Get a piece of graph paper that has fairly fine divisions, about 10 to the centimeter. Set the paper vertically. Label the horizontal axis "Distance"; start at zero and go out to 5 units, using the entire length of the axis. The size of your distance units doesn't really matter. Label the vertical axis "Strength of force." Start at the top with 1.0 and put zero at the bottom. Then going down to zero, place tic marks on the axis at $\frac{1}{2}$ (0.5), $\frac{1}{4}$ (0.25), $\frac{1}{9}$ (0.11), $\frac{1}{16}$ (0.063) and $\frac{1}{25}$ (0.040). Be sure that you have fine enough divisions on the y-axis to do this correctly!

Now plot a dot for each entry in this table:

Distance	Amount of Force
1.0	1.0
2.0	1/4 (0.25)
3.0	1/9 (0.11)
4.0	1/16 (0.063)
5.0	1/25 (0.04)

As best you can, draw a smooth curve freehand between the dots. Can you identify its shape? (It will not be a straight line.) What would it look like if you brought the masses closer together rather than moving them farther apart? (*Hint*: What would be the strength of the force at a distance of $\frac{1}{2}$?)

4

The Earth: A Model Planet

■

Central Concept

The dynamic earth is a highly evolved planet, built over many, many years by geological processes driven by the slow outflow of internal heat.

■

The earth is more than a small mass orbiting the sun. As our home, the earth now seems an optimal planet (Fig. 4.1). Its current atmosphere, climate, landmasses, and vegetation suit us well. Yet, in our planet's past, its environment was much different; our earth has changed tremendously since its birth. It's a dynamic world now, as it was in the past.

This chapter looks at the physical makeup and evolution of the earth. Our present understanding of the earth shows that it has dramatically altered in its physical structure since it formed. We will use our home planet as the basis of comparison for understanding the makeup and evolution of the other earthlike planets in the inner part of the solar system.

FIGURE 4.1

The earth from space. Note North America near the center amidst the heavy cloud cover, which is typical for the earth. The spiral cloud patterns arise from the global circulation of the atmosphere. (Photo from Apollo 16, courtesy NASA)

4.1 Planetary Evolution and Energy

The earth is the planet we know the best. And it's changing, sometimes violently. Hurricanes churn the sea and whip the land. Volcanoes blast materials high into the atmosphere. Earthquakes rend the ground and kill thousands of people.

Amid this violence, continental masses spread like slow rafts. Water and wind weather the land. The climate undergoes long spells of hot and cold. Heat flow from the earth to space drives these processes. For other planets, too, heat flow to space determines the processes of planetary evolution. To understand the earth's evolution, you need to know a bit about energy and heat.

Energy and its transformations mark a central idea of modern science. The universe and all its parts are built of energy and matter. It's the energy that motivates the matter—and us, when we say we are energized. **Energy** is ability to do work.

Energy exists in many different forms. Three common ones are (1) *kinetic energy*, that of motion; (2) *potential energy*, that stored up by position under an applied force; and (3) *radiative energy*, that carried by light. All forms of energy are measured in the same unit: the **joule**. If you lift an apple from the ground to a position above your head, you've done about one joule's worth of work. (Table A.3 in Appendix A compares a variety of energies.) You have also changed the apple's energy; if you release it, it will fall to the ground. The apple has acquired potential energy (from your work on it), which would be transformed to kinetic energy as it fell.

Energy is the ability to do work; it comes in many specific forms.

Kinetic energy is that associated with motion. To stop a moving object requires an obvious effort, and the more massive the object or the faster its motion, the greater the effort (work!) needed to stop it. For instance, it is easier to grab and stop a bicycle traveling at 5 mph than one moving at 10 mph. But you'd not want to try that with a car, even at 5 mph!

The kinetic energy of a moving object depends on its mass and its speed in a special and precise way: directly as the mass and as the *square* of the speed. So a bicycle moving at 10 mph has four times the kinetic energy of one that travels at 5 mph; and it has nine times the energy if it goes three times as fast, at 15 mph compared to 5 mph.

MATH CONCEPT

In algebraic form, the expression for kinetic energy is

$$KE = \tfrac{1}{2} m v^2$$

where *KE* is the kinetic energy, in joules (J), m the mass (kg) of the moving object, and v its speed (m/s), which is squared.

You must distinguish between the kinetic energy of a large object, such as a baseball, and the *average* kinetic energy of the particles that make it up. A baseball's kinetic energy allows it to go somewhere but doesn't affect the microscopic motions of its atoms, which move small distances at random within the ball. Their motion makes up the **thermal energy** of the ball—a measure of the total kinetic energy of a large collection of particles, the internal thermal energy a body contains. That's the difference between a *fast* baseball and a *hot* baseball.

When you heat a gas, there is an increase in the average value of the random speeds of the molecules that make it up. What have you done? You have used a hot flame, say, to raise the thermal energy of the gas. **Heat** is the thermal energy that is transferred from one body to another. Heat always flows from a hotter body to a colder one—and thermal energy is moved around.

Temperature measures the average kinetic energy of particles in, say, a gas. That, in turn, depends on the average speed of the particles. The temperature scale that shows this average speed most directly is called the **Kelvin** scale; at 0 kelvin (0 K); all microscopic motions have reached a minimum. (See Appendix A for temperature units.) Astronomers generally use the Kelvin scale.

Heat is the flow of thermal energy from one hotter body to another, cooler one; temperature measures the average kinetic energy from the random motions of the particles in a body.

As you will see, each major part of the earth—interior, crust, atmosphere, and oceans—has a temperature different from the others. So heat flows between these parts (from hotter to cooler) and activates their changes. What are the sources of the thermal energy? Some heat was deposited in the earth's interior at the planet's formation; some is generated by radioactive decay. All the internal energy gradually flows to the surface. There, some energy is added by sunlight to the air, ground, and oceans. Finally, this energy flows to space, which is much colder than the earth.

4.2 The Earth's Mass, Density, and Interior

The two most important physical properties of a planet are its mass and density. These, along with the internal temperature, determine the internal structure. How to find the mass and density for the earth—or any planet?

Mass and Density

Recall (Section 3.1) that bodies at the earth's surface undergo the same acceleration from gravity, g, of about 9.8 m/s/s. Newton's law of gravitation and his second law (Section 3.3) relate this acceleration to the earth's mass, radius, and the gravitational constant, G. So if we know G (which can be measured in the lab), the earth's radius, and g (which can be measured at the earth's surface), we can figure out the earth's mass. Or, we can place a satellite in orbit around the earth. From its orbital period and its distance from the center of the earth, we can use Newton's version of Kepler's third law (Section 3.4) to calculate the earth's mass. It comes out to about 6×10^{24} kg—or about 3×10^{21} passenger cars.

Newton's law of gravitation and second law of motion or Kepler's third law allow us to measure a planet's mass.

The Earth At a Glance		
	Radius	6378 km
	Mass	5.97×10^{24} kg
	Bulk density	5520 kg/m^3
	Escape speed	11.2 km/s
	Atmosphere	nitrogen, oxygen

Knowing both the earth's mass and its volume (from its radius), we can find out the second key physical property: its *bulk density*. **Density** is a measure of how well matter is packed into a given space—mass per volume. If you divide the earth's mass by its volume, you obtain a bulk density of roughly 5500 kg/m^3, or 5.5 times the density of water. This bulk density indicates that the earth has both rocky and metallic materials. (Most rocks have a density between 2000 and 3700 kg/m^3; the density of iron is 7800 kg/m^3.) A simple way to compare densities is relative to water (which is 1000 kg/m^3); iron, for instance, is 7.8 times that of water. Iron is more dense than water; wood, less so. This fact can be tested by placing both materials in water: wood floats, and iron sinks.

Rocks near the earth's surface average 2400 kg/m³, about half the average density of the earth. This difference implies that the core of the earth is denser than the average, perhaps 12,000 kg/m³. This high density implies that the core contains very dense metals, such as iron. The weight of the overlying layers compresses the core, causing the material there to have a higher density than it would have if it were uncompressed.

A solid or liquid object's bulk density gives us information about that object's overall composition.

The Structure of the Interior

Based on current models, the interior of the earth falls into three distinct layers: the core, the mantle, and the crust (Fig. 4.2). As we move inward toward the center, each region is denser and hotter. The **core**—mostly of iron—makes up the central zone and extends more than halfway to the surface; its radius is about 3500 km. Because the pressures are so great, the inner core is solid even though the outer core is molten. The temperature here may be as high as 5000 K.

Above the core extends the lower-density **mantle**, some 2900 km thick. The mantle material is rock made up of iron and magnesium combined with silicon and oxygen. The temperature here varies, from about 2500 K at the base of the mantle to about 1500 K at its top. The mantle is solid rock, but it can flow slowly, in part because of its temperature.

Encasing the mantle is the layer of lowest density—the **crust**, the solid surface layer, which varies in depth from 8 to 70 km. Most ocean crustal material consists of rocks that have solidified from molten lava (they are called **igneous rock**). These rocks are *basalt*, a dark-colored rock made of oxygen, silicon, aluminum, magnesium, and iron. The ocean basins and the subcontinental sections of the crust are comprised of basalt. The continental masses are mostly light-colored *granite*, made of oxygen, silicon, aluminum, sodium, and potassium. Because granite has a lower density than basalt, the continental plates float slowly on the basalt. And, the entire crust floats on the mantle. Rocks are generally made of *silicates*,

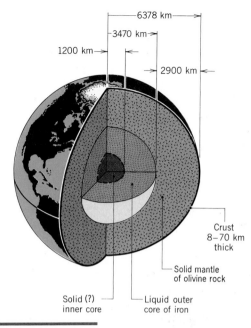

FIGURE 4.2

Scale model of the earth's interior, showing core, mantle, and crust. Each region has a different composition—the mantle and crust are basically rocky, the core metallic. The core and mantle have roughly constant thickness. The crust differs in thickness by almost a factor of 10. It is thinnest under the ocean basins (about 8 km) and thickest under the continents (up to 70 km). Note how large the radius of the core is compared to the overall radius of the earth.

minerals containing compounds of silicon and oxygen.

The earth's interior is **differentiated**; it has layers with the least dense materials at the surface and the most dense at the center. Basically, the interior has two zones: one metal-rich (the core), the other silicate-rich (the mantle and crust). This separation naturally occurs in a mixture of materials of different densities that have become fully or partially melted.

Lower-density materials naturally float on higher-density ones; this happened with materials of different densities in the earth's differentiated interior.

What heated the interior so that it could separate? One old source was internal energy deposited at formation (Section 4.6). Another past and current source: the heat generated by radioactive decay. As radioactive materials decay, they produce fast-moving particles that collide with the matter in the rocks and so transfer kinetic energy to it. Thus the temperature of the rocks increases. In the past, the earth had more radioactive material than now, and the heating produced by radioactive decay melted the earth's interior enough to cause the material to flow and separate into regions of different density. The heat outflow of such internal energy to the surface has driven the crustal evolution of the earth.

Today, the total outward heat flow averages a mere 0.06 joule per square meter (J/m^2) at the surface. If you could capture it all, it would take about two weeks to raise a cup of water to boiling temperature, 100 °C! Yet, because the earth has a large surface area, the amount flowing out per second is large. The earth must still store a vast amount of thermal energy to keep evolving as it does.

The interiors of planets are differentiated, which implies current or past heating of these regions.

4.3 The Evolution of the Crust and the Earth's Age

Clues to the earth's crustal evolution show most clearly in the ocean basins. Zones of active volcanoes and frequent earthquakes lie along chains of young mountain ranges and islands. This activity shows up in the ocean basins, especially in the Atlantic, which contains a **midoceanic ridge**, an almost continuous submarine mountain chain that twists through the ocean basins. Changes in the midoceanic ridge reveal the crustal evolution in the ocean basins.

Continental Drift

Our facts about the evolution of the earth's crust are unified in the model of **continental drift**, including the idea that the present continents were once a uni-

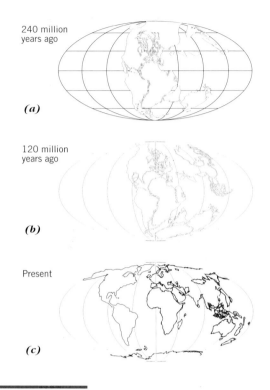

240 million
years ago

(a)

120 million
years ago

(b)

Present

(c)

FIGURE 4.3

Continental drift. These computer-generated models show the trend of continental drift from 240 million years ago (*a*), through 120 million years ago (*b*) to the present (*c*). The position of the continents is based on magnetic evidence. The present form of the continents will change just as dramatically during the next few hundred million years. (Based on maps prepared by A. M. Ziegler and C. S. Scotese of the University of Chicago's Paleographic Atlas project)

fied landmass that fragmented and drifted apart (Fig. 4.3). Supporting this notion are remarkable geological connections, such as similar rock formations and fossils, found on lands on opposite sides of the Atlantic. The face of the earth looked quite different only 100 million years ago, when the present continental masses were more crowded together. (Before the breakup 200 million years ago, the continents moved around in other, mostly unknown, shapes.)

Evidence from the ocean floors near ridges supports the continental drift model. As the continents move apart, the seafloor between them spreads.

When lava solidifies to form igneous rock, the iron minerals in the rock align with the earth's current magnetic field. The directions of the earth's magnetic field have changed in the past, and these reversals are preserved in the rock. On both sides of a ridge, the reversal patterns appear identical—each side is a mirror reflection, where upward-flowing lava pushes older material aside in both directions. When the lava solidifies, the rock records the current magnetic field alignment. So the seafloor rocks act as a magnetic tape (a very slow-moving one!) that preserves the past changes in the earth's magnetic field and gives the rate at which the seafloor spreads. The speed is about 3 cm per year (the amount your fingernails grow in a year) at its fastest across the mid-Atlantic ridge.

Plate Tectonics

The ocean plains are renewed from material that oozes from the earth's interior, an ongoing process.

This marks one example of **volcanism**, the process by which molten material (produced by internal heating) rises through a planet's crust to the surface. The continental plates float like large rafts on the basaltic basin material. Where one continental plate crashes into another, the impact raises mountains; this process renews mountains while erosion wears them down. Often, one plate forces another to fold under and descend into the mantle in *seafloor trenches* (Fig. 4.4). This process subtracts material from the crust. Earthquakes and volcanoes emerge along the lines of the plates' collisions.

This model of the earth's crustal activity and evolution is called **plate tectonics**. Six large (and many smaller plates), each about 100 km thick, cover the earth's surface. The geological action takes place where plates collide. Here earthquakes happen along fault lines (such as the San Andreas fault), and volcanic mountains (such as Mount St. Helens) thrust above the ground. In continental masses where plates diverge, **rift valleys** form—a distinctive feature of plate tectonics.

F I G U R E   4 . 4

Model for the interactions of oceanic and continental plates. The oceanic plates gain material from the outflow of magma at oceanic ridges. As these plates expand, they collide with continental plates and in some places flow under them. Here mountain building and earthquakes occur. Note that most of the heat is generated in the lower mantle, which also releases plumes of material that flows upward into the upper mantle and crust.

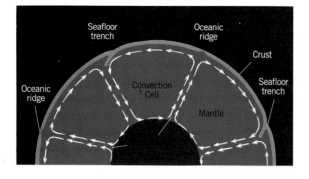

FIGURE 4.5

Model of convection in the mantle. Hot material rises from the lower mantle, flows horizontally, and then descends to set up large *convection cells* in the mantle. The upflow adds material to the crust at the oceanic ridges and pulls it down at the seafloor trenches. The horizontal flow moves the crustal plates.

What moves these plates? Heat flow! The process of **convection** transfers energy in fluids. Consider a hot blob of fluid material surrounded by cooler material. The hotter material's density is lower, so it rises. Denser, cooler material descends to replace it, and a convective pattern is established (Fig. 4.5). So a heat flow is set up between the lower, hotter regions and the cooler, upper ones.

In any fluid, convection carries heat from hotter to cooler regions by mass motions of blobs of material.

One model pictures the mantle as divided into large convective whirls of flowing rock (Fig. 4.5). The mantle's plasticity allows a slow flow upward, horizontally, and downward. At the upper region of horizontal flow, friction between the plate and the mantle drags the plate along with the mantle's flow. The up-welling magma supplies new materials to the plates. Note that the earth's ocean basins, because they have been recently formed (no more than 200 million years old), are the youngest parts of the earth's crust, while the centers of continents are usu-

ally the oldest. Other earthlike planets may show evidence of their crustal evolution by similar processes.

The earth's crust has evolved since its formation and is changing right now by plate tectonics, driven by internal heat flow.

The Earth's Age

From **radioactive dating,** which uses the natural decay rate of radioactive materials, geologists estimate the earth's age at 4.6 billion years. The crucial point is the *rate* of decay. You cannot estimate when any one atom will decay because the process is random. But given many atoms, you can find an overall rate of disintegration. An analogous random process is the popping of popcorn. You can't predict which kernel will pop next, but you can estimate when the entire batch will be finished.

Half a sample of uranium-238 decays to lead in 4.5 billion years, half again in the next 4.5 billion years, and so on. The length of time required for half the material to disintegrate is called the **half-life** of the element (Fig. 4.6). So you can calculate the amount of original uranium left at any time, even though the decay time for any one uranium atom cannot be specified. (See Conceptual Activity 4.)

By reversing this idea, you can estimate the age of a rock. Take a rock sample containing the radioactive

FIGURE 4.6

The radioactive decay of uranium into lead, which illustrates the concept of half-life. Note that the time axis is in units of billions of years. If we start with 1000 g of uranium-238, only half remains after 4.5 billion years pass, and a quarter after another 4.5 billion years.

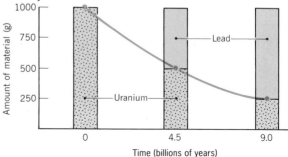

isotope ^{238}U (238 is the number of protons plus neutrons in the nucleus; see Appendix F) and lead. Knowing the half-life of ^{238}U (4.5 billion years), we can calculate the age of the sample by noting how long it would take to form the observed amount of lead by radioactive decay. (To use this technique requires you to be able to estimate the amounts of various isotopes of lead originally in the rock at formation.) Whatever radioactive elements are used, the derived age is the time elapsed since the rocks last solidified. The dating clock resets if the rock melts.

The natural decay of radioactive materials provides a clock to estimate the ages of rocks.

With all the action at the earth's crust, it is hard to find rocks that have survived since the earliest times. The most ancient known igneous rocks, found in western Australia, have been dated at 4.2 billion years. Geologists estimate that about a half-billion years had to elapse for the crust to melt and then cool to form the first rocks. So the earth's age is that of the oldest rocks plus the time to form them.

This estimate falls close to that for meteorites (4.55 billion years) and lunar rocks (4.6 billion years), determined by similar radioactive dating techniques. The near-coincidence of these ages strongly suggests that the solar system formed about 4.6 billion years ago (Chapter 8). The exact times are still being debated, but most scientists agree that the earth is 4.5 to 5 billion years old, with 4.6 billion as the best estimate. Other methods support this value.

The age of earth and the age of the solar system have been determined from radioactive dating techniques.

4.4 The Earth's Atmosphere and Oceans

Our atmosphere provides oxygen for breathing, shields out the cancer-causing ultraviolet radiation of the sun, furnishes a blanket to keep the surface warm, and spreads the heat around the earth. The earth's atmosphere provides information useful for the study of other planetary atmospheres in the solar system.

No other planet in our solar system has oceans of water! The earth's favorable environment for life results from its atmosphere and oceans, which store and move solar energy around the globe. But the atmosphere and the oceans have changed over time, influenced by and influencing the earth's biological evolution.

Composition and Structure of the Air

By volume, our atmosphere contains two major gases: about 78 percent molecular nitrogen and 21 percent molecular oxygen. Basically, the **atmospheric composition** is nitrogen and oxygen. The other gases are really traces—such as carbon dioxide (0.03 percent) and water vapor (variable amounts, sometimes as much as 2 percent near the surface)—but these small amounts turn out to be important.

The weight of the upper atmospheric layers makes the lower region denser than the upper. After all, air has mass (about one kilogram in one cubic meter at the earth's surface), and gravity pulls it down. The gas pressure at sea level on the earth's surface, where the entire atmosphere is piled above it, is called *one atmosphere* (1 atm) of pressure. (The **pressure** of a gas is the force it exerts on an area of surface.) At sea level, the pressure is about 10^5 newtons per square meter (or about 15 pounds per square inch). A car tire has an inflated pressure of about twice this amount. The atmospheric pressure and density decrease with height, rapidly at first, then more slowly.

High in the atmosphere, molecules and atoms can escape into space. Particles of low mass are more likely to leave because, at a given temperature, they will, on the average, have the highest speeds. If a particle heads outward with at least the earth's escape speed (Section 3.4) and if it doesn't bump into another particle, it can leave the earth's gravitational grip.

In the outer regions of a planet's atmosphere, atoms and molecules that move at escape speed (or faster) disappear into space.

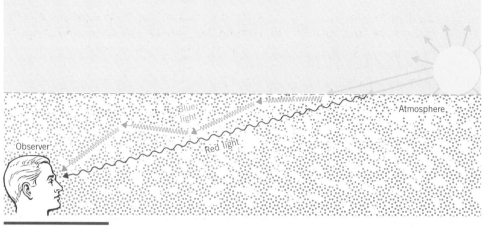

FIGURE 4.7

Scattering of sunlight and the blue sky. Air molecules let red light from the sun pass through relatively unhindered, but blue light is scattered in all directions in the atmosphere. Looking toward the sun, you see red light directly. Because blue light scatters all around the air, in every direction you look you see the scattered blue light and so a blue sky.

FIGURE 4.8

Reddening of sunlight. At sunrise, the scattering of blue light occurs to the largest extent and the sun appears distinctly reddish. The reddening is enhanced when the atmosphere is dusty or smoggy. In this sequence of exposures, the sun rises over Tulsa, Oklahoma. (Courtesy William Sterne, Jr., Sterne Photography)

The atmosphere absorbs and scatters some light that penetrates it. This reduction of light is called **atmospheric extinction**. The closer an object appears to the horizon, the greater the thickness of the atmosphere through which the object's light passes, and so the dimmer it becomes. Because of atmospheric extinction, the rising sun or moon appears dimmer than when overhead. It also appears redder. How?

All light has wavelike properties. For visible light, the wavelengths run from shorter (violet) to longer (red), and air molecules (and dust) scatter shorter-wavelength blue light from the sun more than red. The sky looks blue because when air depletes a beam of light of its shorter (bluer) wavelengths, these are scattered throughout the sky. In any direction you look, you see blue light, and so the entire sky appears blue (Fig. 4.7). The red light of longer wavelengths reaches you directly along the line of sight. The sinking sun appears reddish because its light passes through much atmosphere before reaching you. Along this path, most of the blue light scatters out, leaving mostly red light to strike your eye (Fig. 4.8). This process is called **atmospheric reddening**.

Air scatters blue light more than red; that's why the sky is blue.

Planets shine by reflected sunlight, as does the earth when viewed from space. Different materials reflect light in different amounts. Look back at Figure 4.1. You'll note that the oceans appear the dark, landmasses not as dark, and the clouds the brightest. Clouds (and snow and ice) are good reflectors of sunlight; water absorbs much more than it reflects.

A body's overall reflecting ability is called its **albedo**, the ratio of the light reflected to the incoming light. If an object reflected all the light that struck its surface, its albedo would be 1. If it absorbed it all, the albedo would be 0. The clouds in the earth's atmosphere greatly reflect visible light, and about 35 percent of the incident light reflects back into space; so the earth's average albedo equals 0.35, or 35 percent. The atmosphere and surface absorb the other 65 percent.

Light carries radiative energy. When it is absorbed, it must be converted to another form. For a solid body, absorbed light adds to the internal thermal energy, so the object heats up.

A solid body reflects some of the light that hits it and absorbs the rest, which heats the object.

Sunlight strikes the ground, heating both it and the air in contact with the surface. But, if direct sunlight were the only source of heat, the temperature at the ground would be well below freezing. The average temperature is actually much higher, about 288 K (+15 °C or about 60°F).

What causes this excess heating? The earth's surface emits infrared radiation—light of low energy and longer wavelength than red. It is invisible to our eyes, but our skin senses it as heat. If infrared radiation simply flowed out to space, the earth would be too cold for life. But this radiation doesn't escape completely; some is absorbed by the earth's atmosphere (mostly by water vapor and carbon dioxide). The atmosphere heats up by absorbing infrared radiation. A little goes off into space; the rest radiates back to the ground and heats it more. The atmosphere acts like a blanket, insulating the ground from space and so helping to warm the earth.

This process is often called the **greenhouse effect** by analogy to one process that keeps a greenhouse warm. Glass is transparent to visible light but opaque to infrared. So sunlight enters the greenhouse and warms the interior, which emits infrared.

This heat can't radiate through the glass, so it helps to warm up the interior. The same process helps to heat a car's interior on a sunny day.

Any planetary atmosphere that is more or less transparent to sunlight but opaque to infrared radiation will keep a planet's surface warmer than it would be if a planet lacked an atmosphere.

About 30 km up in our atmosphere, the oxygen strongly absorbs ultraviolet light from the sun. (Ultraviolet light has a shorter wavelength than violet light that we see.) The absorbed energy promotes the binding of three oxygen atoms to make the ozone molecule, and an *ozone layer* forms. Some ultraviolet light does not make it to the earth's surface because of this absorption, which generates protective ozone. Most life on the earth has evolved under the shelter from ultraviolet light provided by the ozone layer. Even mild exposure to ultraviolet light results in a painful sunburn for many people. Small doses over a long time can promote skin cancer.

Human activity seems to have resulted in a sharp drop in the ozone content of the atmosphere. As measured from Antarctica from the 1950s to the 1980s, the amount of total ozone decreased by half, resulting in the so-called ozone hole above that region. Chlorofluorocarbons (abbreviated CFCs) released in aerosol propellants, refrigerants, and other industrial applications are the probable culprits in this depletion. In a process that takes a year or two from first release of the gases, each chlorine atom in each CFC destroys some 100,000 molecules of ozone. Every year some one million tons of CFCs are emitted as a result of human actions; these materials remain in the atmosphere for about 100 years. Because of the longevity of the CFCs, large amounts of gases that will damage the ozone layer are already in the air.

Atmospheric Circulation

Sunlight heats the ground and drives the winds; and global wind and cloud patterns result from the earth's rotation. These processes work to drive the atmospheric circulation on any planet. The general pattern is then affected by land and water masses.

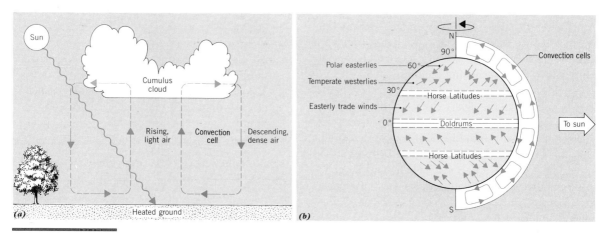

F I G U R E   4 . 9

Atmospheric circulation and solar heating. (*a*) Convection cells cause the development of cumulus clouds in the earth's atmosphere, as heated (hence less dense) air rises and displaces cooler air, which has a higher density and so sinks. Overall, heat moves from hotter to cooler regions. (*b*) Solar heating and the earth's rotation result in a global air circulation pattern. The large convection cells have their air flow twisted by the earth's rotation to create an overall flow of winds near the earth's surface.

Differences in heating of the ground result in *convection*, which carries energy from hotter to colder regions (Fig. 4.9*a*). Such convective currents work on both global and local levels. Generally, the air over the poles is cold and dense; that over the equator is hot and less dense, because the tropics receive more solar energy than the poles. The temperature differences result in a north–south circulation, and convective cells are set up in these regions (Fig. 4.9*b*) The earth's rotation then produces spiral motions of these flows: counterclockwise in the Northern Hemisphere, clockwise in the Southern Hemisphere. The earth's rapid rotation, combined with convective flows, fixes the atmospheric circulation. The same applies to other planets.

Atmospheric circulation carries energy around a planet by convection.

Evolution of the Oceans and Atmosphere

The earth lacked oceans when it formed 4.6 billion years ago. The primeval surface may have been fairly

hot (Section 4.6), about a few thousand kelvins—certainly too hot for water to be liquid! When the surface had cooled to 373 K (100 °C), water could condense. Perhaps one or two continents existed on the surface, and the rest was the initial ocean basin, which contained very little water then.

The rest of the oceanic water probably came from the earth's interior. When hot, molten rock breaks through the crust, it carries a variety of gases, mostly water vapor (80 percent) and carbon dioxide (12 percent). This action, called **outgassing**, is part of the process of volcanism. The steam arises from water that was trapped in the solid earth when the planet formed. Upon cooling, the steam condenses and falls to the surface. The present rate of water gassing out from the interior over billions of years accounts for the water in the oceans today. In the past, when the earth was hotter, the outgassing may have been much faster.

The outgassing from the earth's interior that created the oceans also is the source of atmospheric nitrogen and carbon dioxide. But not oxygen! Biological activity produced (and maintains) the high abundance of atmospheric oxygen. Geological evidence suggests that the transformation to an oxygen-rich atmosphere began roughly 2 billion years ago, when plant activity bloomed. The oxygen probably increased gradually. About 1 billion years ago the atmosphere may have contained only 10 percent of the present amount of oxygen. About 600 million years ago, however, the oxygen content suddenly rose to present levels, with a dramatic increase of life.

The earth's surface temperature relates to the evolution of the atmosphere. How hot it gets at the earth's surface depends on how much energy it receives from the sun and how effectively the greenhouse effect operates. Less solar energy results in lower temperatures. A larger greenhouse effect (more carbon dioxide and water vapor in the atmosphere) fosters higher temperatures.

People have disturbed the natural carbon dioxide balance by extracting fossil fuels from the earth, burning them for energy, and so adding to the carbon dioxide in the atmosphere. Our activities have a net result of increasing the percentage of carbon dioxide in the earth's atmosphere. Observations in Hawaii indicate an increase of 5 percent over the past 30 years. Although there is still much controversy over the potential impact of this increase, it may result in a temperature increase of about 2 °C by the year 2020 if the general conditions of water vapor and cloudiness do not change.

An earthlike planet with a variable percentage of carbon dioxide (and water vapor) in its atmosphere will have a variable greenhouse effect.

4.5 The Earth's Magnetic Field

The earth and some other planets have a global magnetic field. You can visualize the earth's magnetic properties by imagining a giant bar magnet located

---

FIGURE 4.10

Model of the earth's magnetic field. Note that the magnetic axis (*MN* to *MS*) is *not* aligned with the geographic spin axis (*GN* to *GS*), but is tilted about 12°. The magnetic field lines connect the two magnetic poles with no breaks; the field originates in the earth's core. The convention that is used in this book is that the magnetic field lines point toward the north magnetic pole and out of the south magnetic pole.

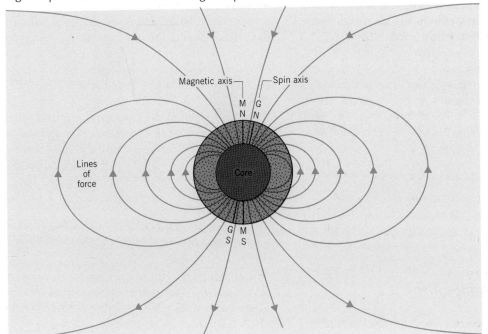

in the core. The magnetic field protrudes from the *south magnetic pole* in the Southern Hemisphere and returns to the *north magnetic pole* in the Northern Hemisphere. The magnetic axis, which connects the magnetic poles, tilts about 12° from the spin axis (Fig. 4.10). The part of this magnetic field that is parallel to the earth's surface orients a compass needle so that it points to the north and south magnetic poles. Such a two-pole field is called a **dipole field** (Fig. 4.11).

Magnetic fields arise from electrical charges in motion. In a bar magnet (Fig. 4.11*a*), the circulation of electrons in iron atoms inside the magnet sets up its magnetic field. In an electromagnet, an electric current (a flow of electrons) passing through a loop of wire creates a magnetic field. Dipole fields can be pictured as arising from a giant bar magnet buried in planet (Fig. 4.11*b*), although they really come from circulating electric currents.

To visualize a magnetic field, take a very small compass and move it around a magnet. The changing direction of the compass needle shows the direction of the magnetic lines of force, called **magnetic field lines**. We usually draw the lines of force so that the spacing of the lines indicates the relative strength of the magnetic field: the closer the spacing, the stronger the field. The farther you are from the magnet, the weaker the field, as indicated by the spreading out of the magnetic field lines. Magnetic field strengths are measured in the SI unit the *tesla* (abbreviated T). The strength of the earth's magnetic field at its surface is only about 0.00004 T. In contrast, the field near a small bar magnet is about 0.01 T.

Moving charged particles produce magnetic fields, which commonly show a dipole structure.

Origin of the Earth's Magnetic Field

The source of the earth's magnetic field is its metallic core, which is both liquid (in part) and a good electrical conductor. Perhaps it acts like a giant electrical generator—a dynamo—and an electromagnet. The liquid core generates electricity and creates a magnetic field. (An electrical dynamo in a generating plant uses the reverse process—a spinning magnet moves electrons to make an electrical current.)

The dynamo model implies that any planet with a strong magnetic field has a substantial fluid, conducting core and rotates rapidly.

Basically, organized fluid motions in a conductor will generate a magnetic field. The hot, liquid part of the earth's core contains convective flows, where hotter material rises and cooler material falls. Because it is liquid, the inner regions are carried

---

FIGURE 4.11

Magnetic field configurations. The spacing of the field lines indicates the strength of the field: the closer the lines, the stronger the field. (*a*) The magnetic field of a bar magnet is a dipole. (*b*) The earth's magnetic field has a dipole-like form with north and south poles.

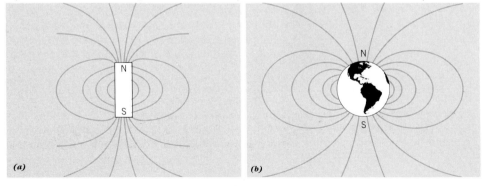

around faster than the outer one. This kind of motion causes the convective regions to curl around rather than flow up and down. These curling motions generate magnetic fields. The earth's rotation helps to stir flows in the core that whirl around to generate electrical currents—the basic **dynamo model** for the earth's magnetic field.

The Earth's Magnetosphere

Earth-orbiting satellites have detected two regions enclosing the earth that contain many protons and electrons. These two belts of protons and electrons trapped by the magnetic field are called the *Van Allen radiation belts*, after their discoverer, James A. Van Allen. The sun supplies the charged particles that are trapped in the Van Allen belts.

The Van Allen belts mark one aspect of the interaction of the earth's magnetic field with the *solar wind*, which consists of charged particles that stream from the sun. The earth's magnetic field directs the flow of charged particles for many hundreds of earth radii out into space, a region called the earth's **mag-**

netosphere (Fig. 4.12). The magnetosphere acts like a buffer between the earth and the solar wind, which flows over and around it. In turn, the solar wind compresses the earth's field on the day side and stretches it on the night side in a long magnetic tail.

The formation of a magnetosphere for the earth (and other planets) results from the interaction of charged particles (from the sun) and planetary magnetic fields.

Magnetic Fields and Charged Particles

Charged particles cannot easily cross magnetic field lines. When charged particles move parallel to the field lines, they feel no force. But if some part of their motion is perpendicular to the field lines, they will experience a force at right angles to both the field lines and their motion. This force results in the particles spiraling along the field lines. The direction

---

FIGURE 4.12

Model of the earth's magnetosphere, created by the interaction of the earth's magnetic field with the flow of charged particles in the solar wind. A long magnetic tail forms downstream, with a thin *plasma sheet* in its core and a *plasma mantle* around the earth, both trapped by the magnetic fields. Note that the magnetic fields have one polarity (arrows) above the plasma sheet and the opposite below it. The magnetic field lines cross polarity at the *magnetic neutral point*. The *Van Allen belts* are dense regions of trapped charged particles close to the earth. Note how the impact of the solar wind compresses the Van Allen belts on the sunward side.

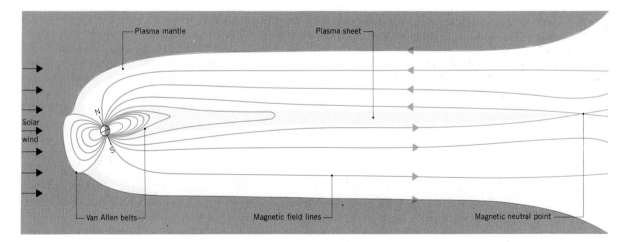

of the spiral twist depends on whether the particle is positively or negatively charged.

This linking of magnetic fields and charged particles explains what happens to a magnetic field that is immersed in an ionized gas, called a **plasma**. (An atom in a gas becomes an *ion* when it loses one or more electrons; it then acquires an overall charge: negative.) Since all, or almost all, of the particles in a plasma are ionized, it contains equal numbers of positively-charged particles (ions) and electrons. As a plasma moves through a magnetic field, the charged particles can be trapped in spiral paths along the magnetic field lines. The particles can move along the field lines but cannot cross them. So the plasma becomes bound to the magnetic field.

At the same time, the plasma captures the magnetic field and holds it. Once an organized flow of charged particles has been set up, it generates its own magnetic field (as does an electromagnet). This field from the moving particles maintains the magnetic field that first caused the current flow. The original field is reinforced. Then if a plasma moves in bulk, it carries the magnetic field lines with it.

Magnetic fields control the flow of plasmas; in turn, plasmas carry along magnetic fields.

Spacecraft flying downstream of the earth have found that the magnetosphere interacts with the solar wind. The plasma seeps into the magnetosphere along its boundary and builds up in its tail. Eventually this plasma breaks off and flows away with the wind. The magnetic field lines in the magnetosphere reconnect in new configurations.

A drippy faucet provides an analogy to the interaction of the solar wind and the magnetosphere, with the surface tension of the water acting like force of magnetic field lines on a plasma. As water drips out of a faucet (Fig. 4.13a), a drop builds up, elongates, and breaks off. The water remaining attached to the faucet springs back to its original shape. Similarly, as the solar wind injects plasma into the magnetosphere (Fig. 4.13b), it builds up in mass and energy until a critical amount is reached. Then the field lines in the central part of the tail pinch off and join together again—a process called **magnetic reconnection**.

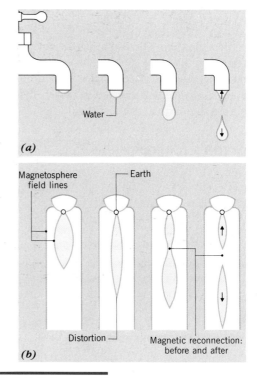

(a)

(b)

Magnetosphere field lines — *Earth*

Water

Distortion — *Magnetic reconnection: before and after*

FIGURE 4.13

The process of magnetic reconnection. (a) As an analogy to the magnetic process, consider water dripping from a faucet. It breaks into drops when the weight of the water overcomes the surface tension. (b) As a magnetosphere fills with plasma, the tension of the magnetic field lines breaks at the neutral point, releasing a blob of plasma downstream into the solar wind flow.

Magnetic reconnections drive the sudden release of magnetic energy and the acceleration of charged particles.

In this model, field lines inside the breakage point snap back toward the earth, injecting high-energy particles into the atmosphere to create an *aurora*. These particles collide with molecules in the air and excite them so that they emit light (Fig. 4.14). Auroras at the north and south magnetic pole regions appear as shimmering crowns that make visible large disturbances in the entire magnetosphere.

(a)　　　　　　　　　　　　*(b)*

FIGURE 4.14

The earth's aurora from the ground and space. (*a*) The aurora as seen from the ground in Alaska. (Courtesy Ichi Akasofu, Geophysical Institute, University of Alaska) (*b*) View of the auroral emission from the space shuttle, May 1991. (Courtesy NASA)

 ## 4.6 An Overview of the Earth's Evolution

The earth is the most evolved of the planets; an active world now (powered by internal energy), it shows no sign of quitting. Geological evidence implies that the earth has been active since birth—4.6 billion years ago. Let's scan its overall evolution. But before we do so, let's look again at energy, so we can understand how the earth formed hot.

The Conservation and Conversion of Energy

The concept of energy has fundamental importance because, although energy comes in many forms, the total amount of energy that an object has can be described and assigned a number. If you have a bunch of objects, the sum of their individual energies is the total energy of the group. The fact that the total energy of a group of objects, isolated from the rest of the world, has a constant value is known as the law of **conservation of energy**.

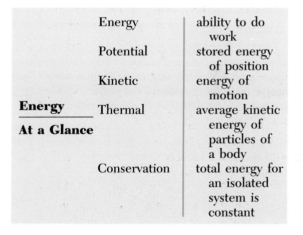

Energy At a Glance	Energy	ability to do work
	Potential	stored energy of position
	Kinetic	energy of motion
	Thermal	average kinetic energy of particles of a body
	Conservation	total energy for an isolated system is constant

Potential energy is the stored energy of position. One important example of potential energy relates to gravitational attraction. On the earth, all masses feel the earth's pull, although some masses, such as an apple held in your hand, do not immediately re-

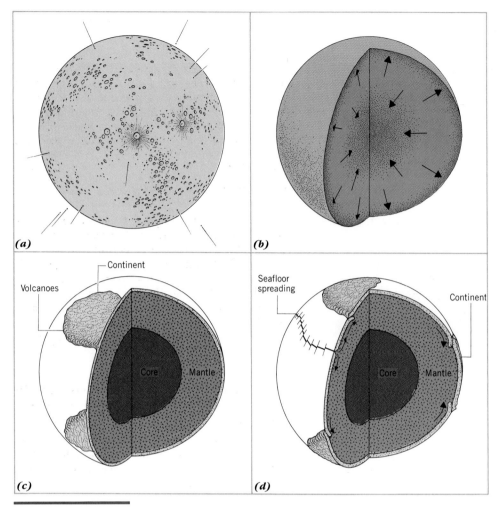

FIGURE 4.15

One model for the main stages in the overall evolution of the earth. (a) The earth forms by the accretion of small, solid bodies; late-arriving ones crater the surface. (b) The interior, heated by radioactive decay and impacts, differentiates. (c) Water condenses, the first continents form, and intense volcanic activity occurs. (d) The crust thickens, and crustal plates slowly roam across the surface.

spond to the force. When you drop the apple, however, it gains kinetic energy, and the higher the point from which you drop it, the greater the kinetic energy it has attained by the end of its fall. The position above the earth's surface relates to the total amount of potential energy the object has. Energy is transformed from one form to another (potential to kinetic), yet the total energy of the apple remains constant, for as it loses potential energy, it gains kinetic energy.

Now imagine a collection of solid debris in space, with a range of sizes. Consider what happens when gravity brings the pieces together. Their potential energy is transformed to kinetic energy. They collide

The conservation of energy means that energy cannot be created or destroyed; it may be transformed, but the total does not change in an isolated (closed) system.

and stick together—a process called **accretion**. The resulting mass gains internal energy, and so its temperature increases. If the earth (or any planet) formed by accretion, it was originally hot. The energy flows out to space by infrared radiation, but some is retained even now. Note that in the later stages of accretion, once a surface has formed, late-arriving debris punches out impact craters. The infalling rubble strikes at high speeds—at least the escape speed—and so carries considerable kinetic energy.

Stages of Evolution

The earth's evolution resulted from diverse processes. Many specifics of our current model are still tentative. But the sequence can be divided into three general stages.

One The earth accreted (Fig. 4.15a) from smaller bodies in the cloud of gas and dust that eventually formed our solar system. The buildup probably took place quickly, in only a few million years. Debris bombarded the surface, heating it. Later, the crust solidified and debris formed impact craters. This left a pockmarked planet of almost uniform composition (since each body was made of about the same combination of materials). The atmosphere at the time of accretion was rich in hydrogen and helium gases.

Two Some tens of millions of years later, impact and radioactive heating melted the interior, which differentiated (Fig. 4.15b). Denser materials sank to the core, lighter ones rose to the crust. The original atmosphere of hydrogen and helium was lost, escaping into space. Perhaps an intense solar wind helped to blow off the primeval atmosphere. This atmosphere was replaced by an outgassed one containing much water, methane, ammonia, sulfur dioxide, and carbon dioxide. How? By volcanic activity caused by interior heating. The earth's surface cooled enough for rain to fall and the oceans to form (Fig. 4.15c).

Three About a billion years later, the first continents appeared. Plate tectonics began; mountains grew, only to fall to the weathering of wind and rain. Slowly the atmosphere evolved. Roughly 2.2 billion years ago, crustal cooling thickened the crust enough to allow plate activity as we see it today. At the end of this stage, 600 million years ago (Fig. 4.15d), the earth's structure looked much as it does today, constantly but slowly reshaped by plate tectonics.

That's one model of how our planet was built. The earth's structure and evolution, which we know in more detail and depth than for any other planet, serve as models for the investigation of similar bodies within the solar system, those earthlike worlds called the **terrestrial planets**: Earth, Moon, Mercury, Venus, and Mars.

The Larger View

The earth is a laboratory for us to discover how a planet is made and evolves. Though the model still has gaps, we are fairly satisfied with it. The earth's atmosphere, crust, and interior have changed greatly since the earth's formation some 4.6 billion years ago. The interior's evolution is driven by internal energy: some from radioactive decay, some left over from the earth's formation by accretion. The internal energy caused the interior to melt and form layers of different materials. Part of the metal core is still molten and generates the earth's global magnetic field.

As the heat flows outward (and finally out to space), it drives the crustal movement and evolution. Sunlight heats the surface, warming our planet (aided by the greenhouse effect) and driving atmospheric convection. The terrestrial planets, each to a different degree, follow the evolutionary model of the earth. Mercury, Venus, the moon, and Mars have been modified by similar processes in their interiors, on their surfaces, and in their atmospheres. Energy and its flow as heat impel this evolution of planets. When the heat is all gone, a planet dies.

Energy plays a primary role not only in planetary evolution but throughout the universe. It can take different forms besides heat, but it cannot be created or destroyed. The conservation of energy provides a bank account that cannot be overdrawn even though the account can exchange many types of currencies.

Key Ideas and Terms

core
crust
density
differentiated
mantle
terrestrial planets

1. The interior of the earth is divided into three zones of different densities—core, mantle, crust. The deeper they lie below the surface, the denser these zones are. The core, the densest region, is partly molten and probably consists mostly of iron. Both the crust and the mantle are made of rocky materials. Other terrestrial planets have a similar construction.

half-life
igneous rock
radioactive dating

2. From radioactive dating of the oldest surface rocks, we estimate that the earth's age is 4.6 billion years.

atmospheric composition
atmospheric pressure
escape speed
temperature

3. The earth's atmosphere consists mainly of nitrogen molecules and oxygen molecules. The pressure at the surface (1 atmosphere) comes from the weight of all the air above it. In its outermost region, the atmosphere slowly leaks into space, where particles have at least escape speed.

atmospheric extinction
atmospheric reddening
convection
heat

4. The atmosphere affects the light that enters it from space. Ultraviolet is absorbed, forming the ozone layer. Blue light is scattered (making the sky blue). Most of the visible light makes it to the surface, heating the ground, which in turn heats the atmosphere to create convection within it.

albedo
greenhouse effect

5. Water vapor and carbon dioxide trap the infrared emitted by the ground and so keep the earth warmer than it would be if such radiation were to escape into space; this heating results in the greenhouse effect.

continental drift
convection
midoceanic ridge
plate tectonics
rift valleys
thermal energy
volcanism

6. The earth's crust evolves by seafloor spreading and the motions of crustal plates, driven by convective flows in the mantle that promote volcanism. The ocean basins are the youngest parts of the earth's crust; the continents are older.

escape speed
outgassing
volcanism

7. The water in the oceans and the gases in the atmosphere probably came from the release of materials from the earth's crust by outgassing; these two forms of matter interact with each other and have changed because of the development of life on our planet. The evolution of the atmosphere is influenced by outgassing from the crust, escape of gases into space, biological activity, and radiation from the sun.

dipole field
dynamo model
magnetic field lines

8. The earth's magnetic field is probably generated by motions in its fluid metallic core. The field interacts with the plasma of the interplanetary solar wind to create a magnetosphere. Magnetic reconnection in the

magnetic reconnection
magnetosphere
plasma

magnetosphere generates auroras high in the atmosphere.

conservation of energy
energy (joule)
heat
kinetic energy
potential energy
temperature (kelvin)

9. A key rule in the operation of the universe is that energy is conserved, although it can be converted into many forms. The potential energy of gravity transforms into the kinetic energy of a falling body. Temperature is a measure of the average kinetic energy of the particles within a body. Heat is the flow of thermal energy from a hotter body to a cooler one.

accretion
heat
terrestrial planets
thermal energy

10. The interior and crust of the earth have been greatly modified since the formation of the planet 4.6 billion years ago. The flow of thermal energy as heat drives the evolutionary processes of the earth. Some of this internal energy was provided by gravitational accretion of material at the earth's formation, some from radioactive decay. Our understanding of the earth as a dynamic planet serves as the model for building an understanding of other planets in the solar system, especially ones like the earth—the terrestrial planets Mercury, Venus, Mars, and the moon.

Review Questions

1. How does the density of the earth's crust compare to that of its mantle?
2. How does the composition of the earth's core differ from that of the crust?
3. After one half-life has elapsed, how much of the original amount of a radioactive isotope remains?
4. What is the main constituent of the earth's atmosphere?
5. What is the minimum speed a molecule of the atmosphere needs to move out into space and not return?
6. Does the moon at the horizon appear redder or bluer than when it is viewed overhead?
7. What gases in the earth's atmosphere contribute most to the greenhouse effect?
8. What part of the earth's crust is the youngest?
9. How can volcanoes affect the evolution of the earth's oceans and atmosphere?
10. What one factor most affects the evolution of the earth's atmosphere now?
11. Has the configuration of continents on the earth's surface always appeared the same?
12. How do we know that the earth has a magnetic field?

Conceptual Exercises

1. Explain how you could determine the earth's mass by jumping off a building. (Explain it; don't do it!)

2. Discuss uncertainties in the statement, "The earth's age is 4.6 billion years."

3. Make a simple argument to demonstrate that the earth's core is denser than its crust.

4. Describe two effects that the earth's atmosphere has on sunlight passing through it.

5. Suppose the amount of water vapor in the atmosphere suddenly increased by a large percentage. What might happen to the temperature at the earth's surface? What if cloud cover increased sharply as a result of the added moisture?

6. What was the likely composition of the earth's first atmosphere? What happened to it? What was the likely composition of the earth's second atmosphere? Where did it come from? What happened to it?

7. Present two ways in which the oceans affect the earth's atmosphere.

8. What is the probable source of the oceans' water?

9. In what way is the earth's magnetic field able to trap the particles of the solar wind?

10. Imagine the earth's rotation is very gradually slowing down, so that billions of years from now, the earth will rotate much more slowly than now. What will happen to the earth's magnetic field then, if the dynamo model is correct?

Conceptual Activity 4   Half-Life

The half-life of a radioactive isotope of an element is a key property, but probably not one that you have had the opportunity to measure. At the University of New Mexico, we have some fast-decaying isotopes with short half-lives. We can measure the relative amount of an isotope by tracking its activity with a Geiger counter and measuring the counts per minute (cpm) with it as time passes.

Here's a table of typical results.

Time (minutes)	Radioactivity (cpm)
Start (0)	1380
5	1150
10	1000
15	900
20	720
25	600
30	520

Draw a graph of these results. Put "Time (min)" along the horizontal axis, starting from zero; place tic marks at intervals of 5 minutes. Label the vertical axis "Radioactivity (cpm)"; set it to zero at the bottom and 1400 at the top. Place tic marks at intervals of 100 (with small divisions of 10).

Plot a dot for each entry in the table and draw a smooth, freehand curve between them. From the curve, estimate the half-life of this isotope.

Note: Every measurement has an error. Thus the curve may not look exactly like the ideal!

5

Venus and Mars: Earthlike Worlds

■

Central Concept
The evolutions of Venus and Mars, planets with hot interiors and substantial atmospheres, have been driven by processes similar to those that have cast the earth.

■

Venus and Mars resemble the earth more than any other planets—they are truly *terrestrial planets* (Table 5.1). Their compositions and internal structures are similar. But their differences are also vital. Venus is blistering hot at its surface, which has undergone recent modification by volcanism. Here lava flows, not water, have shaped the surface. On Mars, volcanoes are scattered and probably extinct. Mars now appears as a cold desert; the water that once sculpted the surface ceased to flow tens of millions of years ago.

How did Venus and Mars end up so different from the earth? What forces shaped their evolution? We cannot answer these questions for certain, but we do have many hints using the earth as a basis for comparison to build up models for other terrestrial planets—a process called **comparative planetology**.

TABLE 5.1

Comparison of the Terrestrial Planets

Planet	Diameter (earth = 1)	Mass (earth = 1)	Density (water = 1)	Surface Pressure (earth = 1)	Main Gases in Atmosphere
Mercury	0.38	0.056	5.4	10^{-15}	Helium, sodium
Venus	0.95	0.82	5.2	95	Carbon dioxide
Moon	0.27	0.012	3.3	Essentially none	Essentially none
Earth	1.0	1.0	5.5	1.0	Nitrogen, oxygen
Mars	0.53	0.11	3.9	0.006	Carbon dioxide

5.1 Venus: Orbital and Physical Properties

Venus has an unbroken swirl of clouds, which reflect 77 percent of the incoming sunlight (Fig. 5.1). This cloud cover blocks any view of surface features through optical telescopes. We have pierced Venus' cloudy veil with radar beams and remote landers to uncover the surface environment. Venus has gained the reputation of being the earth's twin. In terms of size, average density, and mass, the match is appropriate. In most other ways, however, Venus has an environment tremendously different from the earth's—indeed, a genuine hell!

Revolution and Rotation

The second planet from the sun at an average distance of 0.72 AU, Venus completes one orbit in 225 days. The earth and Venus can come as close as 40 million kilometers. Venus is the closest planet to the earth (after the moon).

Before 1961, astronomers lacked any real idea of Venus' rotation rate because they could not see surface features. Using the *Doppler shift* (see below), radar astronomers in the 1960s found that Venus rotates once every 243 days, *retrograde*. In **retrograde rotation,** the planet spins from east to west, rather than west to east like the earth.

This long retrograde rotation (combined with the planet's orbital motion) results in a solar day on Venus of 117 terrestrial days. Because of the clouds, you could never see the sun's disk directly from the surface. It would be light during the day, as on an overcast day on earth, but the sun's intensity on Venus is only about $\frac{1}{100}$ that at the earth's surface.

Size, Mass, and Density

Venus has a physical diameter of about 12,100 km—smaller than the earth by only 5 percent. We know the mass of Venus with accuracy because Newton's form of Kepler's third law (Section 3.4) has been applied to data from space craft orbiting Venus: the planet's mass is 0.82 times the earth's.

Venus	Radius	6052 km
	Mass	4.87×10^{24} kg
At a Glance	Bulk density	5240 kg/m^3
	Escape speed	10.4 km/s
	Atmosphere	carbon dioxide, nitrogen

From the mass and size, we find a bulk density of roughly 5200 kg/m^3, almost the same as that of the

FIGURE 5.1

Venus, photographed in blue light at crescent phase. Note the unbroken cloud cover. (Courtesy Palomar Observatory and California Institute of Technology)

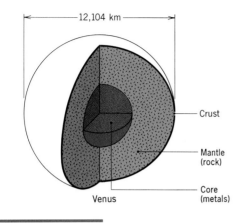

← 12,104 km →

Crust

Mantle (rock)

Core (metals)

Venus

FIGURE 5.2

Scale model for the interior of Venus; note how closely it resembles that of the earth with a large, rocky mantle and metallic core. The crust appears to be mostly volcanic rock.

earth. We guess that the interior of Venus closely resembles the earth's interior (Fig. 5.2): a rocky crust (which the Venus landers confirmed), a large rocky mantle, and a metallic core. Because Venus has a bit lower bulk density than the earth, we infer that it has a somewhat smaller core. The core probably is liquid because the interior of Venus is hot.

> Planets with similar mass and bulk density are likely to have similar compositions and interior structures.

Magnetic field

If the metallic core of Venus is liquid in part, there should be a planetary magnetic field. Because Venus rotates 243 times more slowly than the earth, we expect its internal dynamo to be weaker and so the magnetic field less intense than the earth's (Section

> A dynamo model predicts that a slowly rotating planet will have a weaker magnetic field than a faster rotating one with similar mass and density.

4.5). No probe to date has detected *any* internal magnetic field on Venus—if one exists, it is at least 10,000 times weaker than ours! That's *much* weaker than expected from a simple dynamo model, so the model needs modification for Venus.

A very weak magnetic field means that the interaction of Venus with the solar wind differs from that of the earth. Without the buffer of a magnetic field, the solar wind runs right into the upper atmosphere of Venus. The solar wind carries off some of the upper atmosphere, as the wind flows around the planet and creates downstream a long magnetic tail that resembles the magnetic tails of comets (Section 8.2). In contrast to the earth, Venus does *not* have a magnetosphere (Section 4.5).

The Doppler Shift and Planetary Rotation

see p 192

Radar observations of the Doppler shift from Venus allows us to measure the planet's rotation rate. Let's see how.

Radio and radar are types of light; one fundamental aspect of all kinds of light is that light has wave

FIGURE 5.3

Wavelength and relative motion for water waves. Moving into the waves makes their wavelength appear shorter and their frequency greater; moving with them, their wavelength seems longer and their frequency lower.

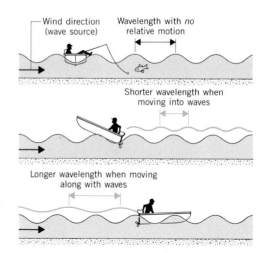

Wind direction (wave source)

Wavelength with *no* relative motion

Shorter wavelength when moving into waves

Longer wavelength when moving along with waves

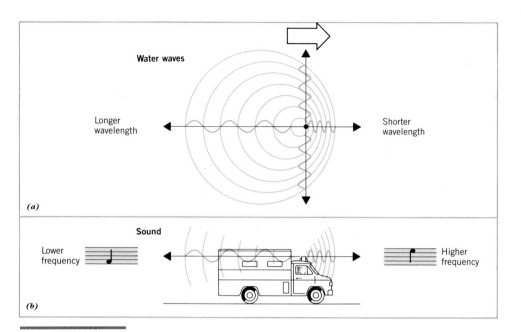

FIGURE 5.4

The Doppler shift for water and sound waves. (*a*) In water waves, a moving source compresses the waves in the direction of motion and spreads them out in the opposite direction. An observer at the right finds a higher frequency and shorter wavelength; one at the left, a longer frequency and lower wavelength. (*b*) Sound waves from the horn of a moving truck undergo a similar effect, increasing and decreasing in frequency (pitch) as the truck approaches and then passes a stationary observer.

properties. The Doppler shift occurs with waves of all kinds—light, sound, and even water waves. Imagine you are out fishing in a small motorboat (Fig. 5.3). You have been sitting in one spot with little luck, so you decide to move on. First you go into the wind (the wave source). You notice that you bob up and down *more* frequently than when you were at rest; the wavelength—the distance between the crests of the waves—appears to have gotten *shorter*. If you drive the boat away from the wind, you discover that your bobbing is less frequent; the waves seem to you to have a *longer* wavelength. The explanation: when you moved away from the wind in the direction of the waves, they had to catch up with you; when you went into the wind, you were meeting the oncoming waves.

That's the essence of the **Doppler shift**: when you are moving *toward* a wave source, the waves appear more frequent and shorter; in contrast, when you move *away* from a wave source, the waves appear less frequent and also longer.

This concept for water waves (Fig. 5.4*a*) also applies to a Doppler shift with sound waves (Fig. 5.4*b*). As a truck approaches you with its horn blasting, its pitch is higher than it would be if the truck were at rest. Just as the truck passes you, its pitch (frequency) is unshifted. Then as the truck moves away, its horn seems to put out a lower pitch. The same effect is observed with radar and waves, which we can detect with a radio receiver. Such waves are transmitted at specific frequencies, such as those in the AM and FM radio ranges. We can measure the Doppler shift in such waves by the *change in their frequency*. We do so by tuning the receiver to a slightly different frequency: the signal shifts toward higher frequencies (more waves passing per second) for an object approaching you, and toward lower frequencies (fewer waves passing per second) for one that is receding.

For light, it's only the *relative velocity along the line of sight*—called the *radial velocity*—that causes the Doppler shift (for speeds much slower than that

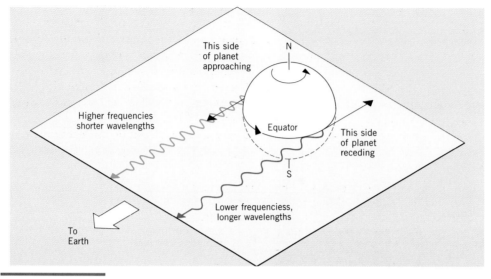

FIGURE 5.5

Doppler shift from a rotating planet. Viewed along the equator, the left side of the planet is approaching the observer, and the right side receding. Radar waves striking the left side are shifted to shorter wavelengths and higher frequencies; those hitting at the right to longer wavelengths and lower frequencies.

of light). Since velocities are relative, it makes no difference whether you're moving or the source is moving (or both). The Doppler shift is an apparent shift in the received wavelength or frequency when the source and receiver have a relative radial velocity.

The Doppler shift provides a cosmic speedometer that allows us to measure the speeds of objects moving toward or away from us.

Now apply this Doppler concept to radar observations of a planet. We send radar signals of an exactly known frequency from the earth. They strike a rotating planet all over its surface, and some waves bounce off and return to the earth for reception. Imagine viewing the planet along its equator. Then some reflected waves come from the edge of the planet that is moving toward us as the planet rotates (Fig. 5.5); these are shifted to higher frequencies and shorter wavelengths. The other edge moves away from us; waves reflected here have lower frequencies and longer wavelengths.

When the reflected waves are received at the earth, the radar telescope has to be tuned to slightly different frequencies to pick up the waves reflected from both sides of the planet. The *difference* in frequency between the original signals and the received ones tell us how fast the planet is rotating. Note that radar waves (which penetrate the clouds) are used for Venus because we cannot see the surface with visible light.

The Doppler shift provides a way to measure the rotation rate of planets.

The Doppler shift is named after the Austrian physicist Christian J. Doppler (1803–1853), who first noticed the effect in sound waves. You will come across astronomical applications of the Doppler shift throughout this book; Section 9.5 explains in detail how it works with visible light.

5.2 The Atmosphere of Venus

The atmosphere of Venus differs remarkably from ours—a key clue to understanding the planet's evolution. Substantial atmospheres directly affect changes in the surface of a terrestrial planet. An atmosphere's composition reflects some aspects of a planet's history. Venus' atmosphere contains about 96 percent carbon dioxide, 3 percent nitrogen, some argon, and traces of water vapor (varying from 0.1 to 0.4 percent) and sulfur dioxide. Note the high percentage of carbon dioxide compared to the earth.

The surface pressure is 95 atm and the sunlit surface temperature about 740 K (about 470°C). The pressure on the surface of Venus is about the same as the pressure at a depth of 1 km in the earth's oceans; the temperature about that of an electric oven dur-

?
see →

FIGURE 5.6

Motions from winds in the upper cloud layers of Venus. The clouds circle the planet once in about 4 days. In this sequence of that circulation imaged by the Pioneer Venus orbiter, north is at the top, east to the right. (*a*) May 2, 1980. Note the vague Y-shaped marking in the clouds; it lies on its side and opens to the west side (left) of Venus. The stem of the Y lies as a broad band above the equator. (*b*) Nine hours later, the vertex of the Y is near the west side of the planet. (*c*) Four hours later, the vertex is still visible. (*d*) On the following day, the arms of the Y have changed shape. Note the stable caps at the polar regions. (Courtesy NASA)

(a)

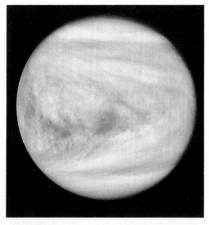

(b)

(c)

(d)

ing its self-cleaning cycle. This high temperature results from the trapping of surface heat (Section 4.4), because the so-called **greenhouse gases**—mainly carbon dioxide and water vapor—absorb infrared radiation well. Because only about 3 percent of the sunlight hitting the upper atmosphere of Venus makes it to the surface, an extreme greenhouse effect keeps Venus very hot.

Planets with dense atmospheres and a high percentage of greenhouse gases trap heat, making their surface temperatures higher than we would expect from direct solar heating alone.

The yellowish-white clouds of Venus reach about 65 km above the surface. (The highest clouds on the earth go up to 16 km.) These cloud tops flow with the upper atmosphere, in patterns similar to jet streams of the earth (Fig. 5.6). Ringing the equator, the clouds circle the planet in only 4 days. The winds blow from the day to the night side and from equatorial to polar regions. The air flow carries thermal energy, which helps to keep the temperatures fairly constant—they vary about 10 K or less from day to night.

The clouds float in two broad layers at heights of 50 and 60 km (Fig. 5.7). Below 50 km, the clouds gradually thin out; below 30 km, and down to the surface, the atmosphere is clear of any particles. The upper level clouds are concentrated solutions of sulfuric acid. (Sulfuric acid, found in most car batteries, is extremely corrosive.) The clouds contain a solution of 90 percent sulfuric acid mixed with water.

Although the atmosphere and the clouds of Venus do contain some water vapor, it doesn't amount to very much compared to all the water on the earth. All the water in the atmosphere of Venus (none exists on the surface now because it's so hot) would amount to a layer only 30 cm thick, about 10,000 times less than contained in the earth's oceans and atmosphere.

5.3 The Active Surface of Venus

The surface of Venus has been investigated by landing probes to examine it close up, and by analysis of

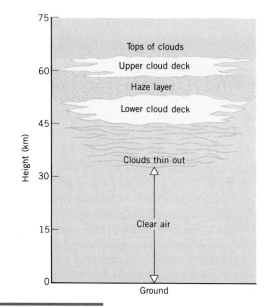

FIGURE 5.7

Model of the structure of the clouds in the atmosphere of Venus. The clouds, consisting of sulfuric acid and water, make up two decks (extending from 45 to 65 km) with a layer of haze between them. The air is clear from the surface up to the lower cloud deck. (Based on NASA data)

the reflections obtained by bouncing radar (which penetrates the clouds) off the surface. Radar mapping has taken place both from the earth and Venus orbiters. The goal has been to find evidence for processes like volcanism and plate tectonics that modify the surface. (*Note:* Features on Venus bear the names of famous women or female mythological characters.)

The Soviet Union landed four spacecraft that sent back close-up photos. The pictures showed slabby rocks about tens of centimeters in size (Fig. 5.8). A few rocks have small holes that were once filled with gas; this implies a volcanic origin. The rocks rest on loose, coarse-grained dirt. Most of the rocks are volcanic basalt, like those lining the earth's ocean basins, or in continental rift zones (such as in Africa), or in the earth's ocean floor near midoceanic ridges. These provided evidence of local volcanism.

Highlands and Lowlands

What about the general lay of the land? We have a good idea from radar mapping, which reveals a varied terrain: mountains, high plateaus, canyons, volcanoes, ridges, and impact craters. Overall, Venus looks fairly flat (Fig. 5.9). Elevation differences are small, only 2 to 3 km, except for a few highland regions. The continents there reach up to only some 10 km, compared with a difference of 25 km on Mars and 20 km on the earth.

The southern and northern halves of the mapped face of Venus differ remarkably. The northern region is mountainous, with uncratered *highland plateaus*; these resemble continents on earth. The southern part has relatively flat *rolling terrain*, which appears to be vast lava plains.

The great northern plateau is called *Ishtar Terra*, which measures some 1000 km by 1500 km (Fig. 5.10). (That's larger than the biggest highland plateau on the earth, the Himalayan Plateau.) This great plateau may have been built up from thin lava flows over an uplifted section of older crust. Mountain ranges border Ishtar on the east, northwest, and north. These three mountain ranges may have folded and risen from moving plates in Venus' crust, making them similar to mountains built from plate tectonics on the earth. The eastern range, called *Maxwell Montes*, contains the highest elevations on Venus: up to 11 km. A volcanic cone lies near Maxwell's center.

Most of the surface consists of flat, volcanic plains that are marked by tens of thousands of volcanic domes and shields. The southern half of Venus' rolling lava plains is punctuated by craters, both large

(a)

(b)

Earth

Venus

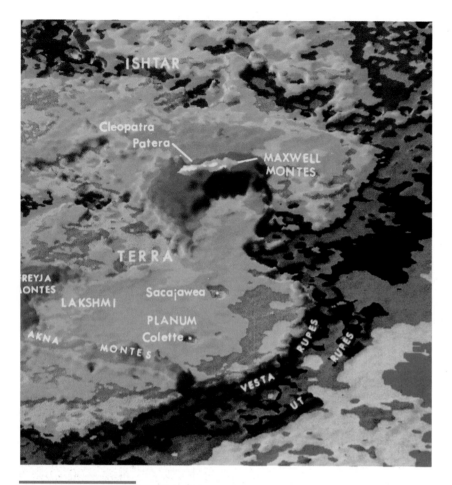

FIGURE 5.10

False-color surface contour map made by radar (see Appendix G) of the Ishtar Terra region on Venus. Reds indicate the highest elevations; blacks and blues the lowest. Maxwell Montes is a large shield volcano. The Lakshmi plateau rises about 5 km above the mean level of Venus; it is bordered by mountain ranges to the north and northwest. On Ishtar's southern flank, the Ut and Vesta cliffs descend to the vast lowlands. The feature called Colette may be a collapsed volcanic crater. (Courtesy NASA)

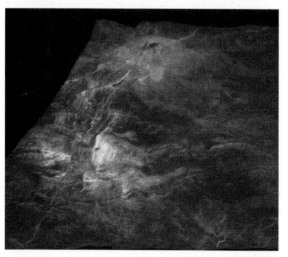

FIGURE 5.11

Beta Regio (longitude 280°, latitude 30°) is made up of two large volcanoes flanking a rift in the crust. This computer-generated view shows the region's two large volcanoes, Theia Mons at the top and Rhea Mons at the bottom. The artificial viewing angle is from the north above the horizon; vertical relief has been exaggerated to bring out the different levels. (Image courtesy A. S. McEwen, U.S. Geological Survey; observations at Arecibo Observatory by G. H. Pettengill, D. B. Campbell, and H. Marusky)

(up to 100 km in diameter) and smaller ones (around 10 km in size). These probably are impact craters created by solid bodies from space crashing into the surface. (Craters less than 10 km do not form on Venus because the dense atmosphere completely vaporizes small incoming debris before it reaches the ground.) In general, the craters are shallow as a result of erosion by the thick atmosphere.

Volcanoes and Tectonic Activity

Beta Regio, which contains at least two separate volcanoes, is an enormous volcanic complex that formed along a great north–south rift zone (Fig. 5.11). The volcanoes here bear gentle slopes; they are called **shield volcanoes**. (Instead of consisting of a sharply uplifted cone, shield volcanoes are relatively flat, like an armor shield.) Often shield volcanoes have a collapsed central crater at the summit; these two appear to have them. The two volcanoes in Beta Regio (Fig. 5.11) are called *Theia Mons* (the southern one) and *Rhea Mons* (the northern one). Theia Mons has a diameter of 820 km, a height of 5 km, and a summit crater 60 km by 90 km. In contrast, the island of Hawaii (which is a shield volcano island) is 200 km across and 9 km high.

Whether any volcanoes are now active is a subject of debate. Volcanoes discharge gases from inside a planet. On earth (Section 4.4), this outgassing material contains water vapor, carbon dioxide, nitrogen, and sulfur gases—all found in the atmosphere of Venus. Yet, the Magellan orbiter (see below) has not found firm evidence of current activity.

Venus' surface has highland plateaus, mountains, rift valleys, large volcanoes, and other evidence of volcanism; much of the surface is covered by lava plains.

For its variety, the surface of Venus is remarkably flat compared to the earth: only some 10 percent of the mapped surface extends above 10 km. In contrast, about 30 percent of the earth's surface reaches above 10 km (from the bottom of the ocean basins). Now, the large difference on the earth between ocean basins and continental masses is continually kept up by plate tectonics. Tectonic work on Venus seems to be concentrated in zones spread around the planet, especially near highland regions.

Magellan at Venus

In August 1990, the Magellan spacecraft settled in an orbit around Venus and one month later began its

---

FIGURE 5.12

Magellan images of impact craters on Venus. (*a*) Three large impact craters visible in the Lavinia region; they range in diameter from 30 to 50 km. Note the bright material, ejected from the impact, surrounding each crater and the central peak in each. In the lower right-hand corner are several domes of volcanic origin. (*b*) The double-ring Mead crater has a diameter of 275 km. Its original interior has been filled in by lava. (Both courtesy Jet Propulsion Laboratory and NASA)

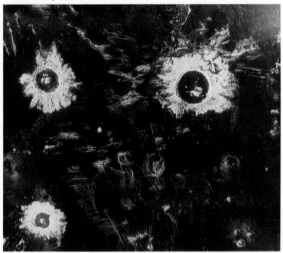

(a)

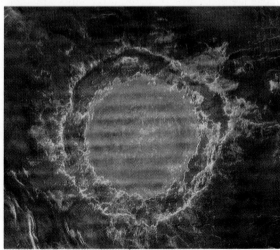

(b)

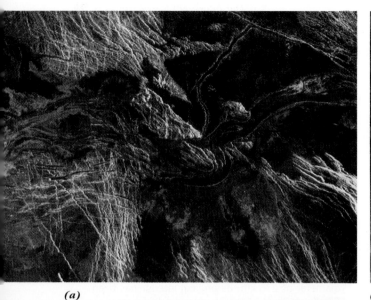

(a)

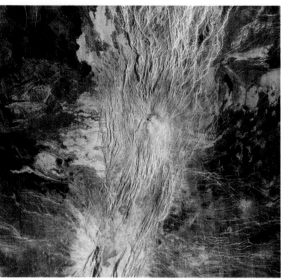

(b)

(c)

FIGURE 5.13

Magellan images of volcanism and tectonic movements. (*a*) Running across the central part of this image is part of a lava channel some 1200 km long and 20 km wide. Very fluid lavas carved this channel in the Lada Terra region at 51° S latitude. The image covers about 150 km across. (*b*) Devana Chasm at the equator is a rift zone with multiple faulting and deep fractures. It is part of a wider zone of surface deformation. The lava flows in the center are about 70 km by 280 km and flow down one flank of the rift valley. (*c*) A view of Maat Mons, an 8-km-high volcano, with foreground lava flows. (All courtesy Jet Propulsion Laboratory and NASA)

mission of high-resolution radar mapping—100 meters at best, or about 10 times better than previously! This mission lasted almost a year, and it will take many more years to analyze the data. We have images of the surface of Venus with a clarity never before achieved, detailing a surface molded by extensive, recent volcanism and tectonic activity.

Amid the volcanic landforms lie a few (perhaps a thousand) but stunningly fresh impact craters (Fig. 5.12*a*) They generally show central peaks, terraced walls, shocked surfaces, and flooded floors. But they lack extensive ray systems of exploded debris, because of the dense atmosphere; for the same reason,

no crater smaller than 3 km in diameter has been seen. The largest known crater, called Mead (after Margaret Mead), spans 275 km in an unusual double-ring structure.(Fig. 5.12*b*). Its interior lacks a central impact peak; instead, it is filled to the brim with lava.

Overall, Venus appears a fresher, more violent world than earlier images had allowed us to infer—a planet with a largely young surface, no more than a few hundred million years. The degree of volcanism seems more widespread than on the earth (Fig. 5.13*a*). The entire surface may have been reshaped over the past several hundred million years by extremely runny lava flows. Rift valleys, which form at the separations of crustal plates, show extensive surface fracturing (Fig. 5.13*b*). Also visible are folded and faulted regions that resemble mountain-building regions on the earth. Such features indicate tectonic activity. The present high degree of activity on Venus (Fig. 15.13*c*), should continue for eons to come.

Most of the surface of Venus is relatively young; it has been modified by extensive volcanism and tectonic activity during the past few hundred million years.

The Evolution of Venus

The concept of comparative planetology includes comparison of the evolution of planets as well as their current physical properties. The evolution of terrestrial planets is primarily linked to size and mass, such that the *larger, more massive planets undergo more activity for a longer time.* We expect the earth and Venus to be very similar, yet in some important respects, they are not. How did the differences develop, when Venus has about the same size, density, mass, interior composition, and structure?

Although Venus has a high degree of overall volcanism and local outbreaks, it does not show the extensive global pattern as it does on the earth. The crust of Venus has experienced some localized crustal plate movement, especially near the equator. But the plate movement has *not* been a planetwide process, as it has been on the earth. Still, the tectonic features imply that Venus now undergoes a vigorous convective upwelling of mantle material.

We can speculate about the history of Venus using the earth as a guide. We infer that Venus formed about 4.6 billion years ago with the other terrestrial planets (Section 8.5). Venus' interior became differentiated as a result of internal radioactive and impact heating. During the first 500 million years, a crust formed and solidified. About 3 to 4 billion years ago, large masses bombarded the surface (as they did the earth) and fractured the crust. Volcanoes erupted. Then smaller bodies from space cratered the surface; this heavy bombardment ended about 3 billion years ago. That ancient surface is gone, having been followed by plate movement, which helped to push up some highland regions. Huge volcanoes vented

through cracks in the surface; their cones formed the shield volcanoes visible today. Lava flows have crossed hundreds of kilometers. Volcanism and tectonics have reworked the surface almost completely over the past 500 million years.

Venus may well resemble the earth at an early age, from 4.5 to 2.5 billion years ago. The Venus of today appears like a view of the earth in its distant past.

Venus seems to have evolved in a sequence similar to that of the earth, but more slowly, with recent volcanism and tectonic activity.

5.4 Mars: General Characteristics

Viking and other spacecraft have disclosed Mars as a lifeless planet with ancient craters, giant canyons, and huge volcanoes (Fig. 5.14).

Mars has a diameter of some 6800 km, about half the earth's size. Mars has two moons, so we can use Kepler's third law (Section 3.4) to find its mass from their orbits: it is only 6.4×10^{23} kg, about 11 percent of the earth's mass. Mars' bulk density is roughly 3900 kg/m^3, only a bit higher than the moon's (3300 kg/m^3) and much less than the earth's (5500 kg/m^3). This comparatively low density implies that Mars' interior (Fig. 5.15) differs from the earth's. In particular, its metallic core is smaller and probably has a lower density than the materials in the earth's core. The Martian mantle probably has the same density as the earth's and a similar composition.

The interior structure of Mars resembles that of the earth but with a smaller metallic core.

Mars **At a Glance**	Radius	3397 km
	Mass	6.42×10^{23} kg
	Bulk density	3940 kg/m^3
	Escape speed	5.0 km/s
	Atmosphere	Carbon dioxide, nitrogen

FIGURE 5.14

True color mosaic of Viking images of the northern hemisphere of Mars. Valles Marineris (about 5000 km long) lies across the lower middle and center, and the Tharsis ridge with its huge shield volcanoes (three brown circular regions) is at the far left. Flood waters bursting at the eastern end of Valles Marineris (right) poured northward, eroding deep channels (now dry) along the way. Faint white clouds are visible at the upper left. (Image created by A. S. McEwen, Arizona State University and the U.S. Geological Survey, from NASA photographs)

FIGURE 5.15

One model of the Martian interior; note the small size of the metallic core relative to the rocky mantle. (Based on a model by J. S. Lewis)

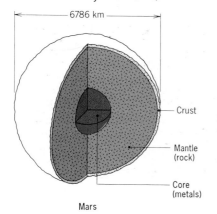

Mars has an extremely weak planetwide magnetic field, only $\frac{1}{500}$ the strength of the earth's. That presents a puzzle if the dynamo model (Section 4.5) correctly describes planetary magnetic fields. Mars rotates as fast as the earth. Though the core is smaller, it should contain metals. We have no direct evidence that the core is liquid, but the evidence for past volcanic activity implies a hot mantle and so probably a hot, liquid core. We would expect Mars to have a moderately strong field, but it does not. Again, the dynamo model is called into question by the weak field of an earthlike planet.

Martian Atmosphere and Surface Temperature

Mars possesses a thin atmosphere with an average surface pressure of roughly $\frac{1}{100}$ the earth's surface pressure. (You'd have to travel 40 km up in the earth's atmosphere before the pressure falls that low.) This thin atmosphere consists of 95 percent carbon dioxide, 2 to 3 percent molecular nitrogen, about 1 to 2 percent argon, and 0.1 to 0.4 percent molecular oxygen.

The atmosphere of Mars is very similar in relative composition to that of Venus (but nowhere near as dense).

The Viking orbiters measured the water vapor in the atmosphere of Mars and found the greatest amounts in the high northern latitudes in the summer. Peak concentrations were about 10,000 times less than the atmospheric water vapor of the earth. Mars is a very dry planet now. However, it is common to have water ice on Mars, either on the surface or in the clouds. Some evidence suggests the presence of water in a permanent frost layer beneath the surface in most regions.

Although the atmosphere contains mostly carbon dioxide, its low density limits its greenhouse effect. Most of the infrared radiation can escape, so temperatures are low and temperature variations are large. At the Martian equator, when Mars is closest to the sun, the difference between noon and midnight is almost 100 K. Rarely does the ground exceed the freezing temperature of water; a typical maximum temperature in the middle latitudes is 270 K (−3 °C).

A thin, low-density atmosphere, even if rich in greenhouse gases, will produce only a small greenhouse effect.

 ## 5.5 The Martian Surface

Spacecraft visits have found that Mars was once an active world geologically but now is a wind-swept, cold desert with little large-scale change.

Light reddish regions make up almost 70 percent of the Martian surface to give Mars its striking appearance. These areas contain substantial rusted iron combined with water—perhaps as much as 1 percent of the surface is water bound up with minerals. The Martian surface is covered with rusty sand (Fig. 5.16).

The reddish sand—finer than that on the earth's beaches—is blown up by fierce winds, faster than 100 km/h, which create planetwide dust storms. These disturbances occur most violently when Mars is closest to the sun. These global storms sandblast the land.

Canals and Polar Caps

In 1877 Giovanni Schiaparelli (1835–1910) recorded what he thought to be real Martian surface features in great detail. He charted many dark, almost straight features, which he called *canali*, Italian for "channels." The word was mistranslated into English as "canals," however, which implied to some people that they were artificial structures.

These so-called canals ignited the curiosity of the American astronomer Percival Lowell (1855–1916). To pursue his interest in Mars, Lowell (Fig. 5.17) in 1894 founded an observatory near Flagstaff, Arizona, to take advantage of the excellent observing conditions there. Shortly afterward, he published Martian maps showing a mosaic of more than 500 canals. In a series of popular books, Lowell argued that the canals were artificial waterways, constructed by Martians to carry water from the polar caps to

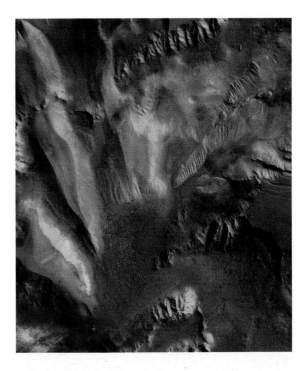

FIGURE 5.16

Candor Chasm, Mars. This image has been processed from Viking orbiter photos and color enhanced to bring out the ground features and variations in the surface colors. The variety of colors is a bit richer than that seen by the eye. Candor Chasm is one part of the great Valles Marineris. The surface has been shaped by wind erosion and collapse of the sides of the valley walls. The plateau regions are several kilometers high. (Image by A. S. McEwen, Arizona State University and the U.S. Geological Survey, from NASA photographs)

FIGURE 5.17

Percival Lowell: ". . . the solidarity of the Martian land system points to an efficient government. . . ." (Courtesy Yerkes Observatory)

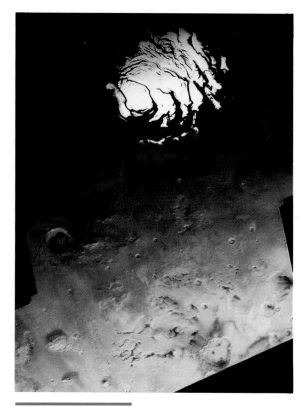

FIGURE 5.18

The residual south polar cap on Mars in the summer; note the layered terrain. The cap is about 400 km across and covered with a coating of dry ice. (Temperatures are colder here than in the north polar region.) Water ice exists in the underlying terrain. (Courtesy U.S. Geologic Survey and NASA; image produced by T. Becker)

irrigate arid regions for farming. These speculations have turned out wrong.

We know now that the polar caps *do* consist mostly of water ice, especially the residual cap left in the summer, which ranges in thickness from year to year from a meter to a kilometer (Fig. 5.18). The outer reaches of the caps, prominent in winter, consist of carbon dioxide ice, which condenses at a lower temperature than water ice. (At Martian surface pressures, water ice condenses at 190 K, carbon dioxide ice at 150 K.)

Water does exist on Mars—but it does not flow freely on the surface now because the temperature is too cold and the pressure too low. If all the water in the polar caps could cover the surface as liquid, it would form a layer only about 10 meters deep.

General Surface Terrain

The two Martian hemispheres have different characteristics: the southern hemisphere is relatively flat, older, and heavily cratered; the northern hemisphere is younger, with extensive lava flows, collapsed depressions, and huge volcanoes. Near the equator lies a huge canyon, called *Valles Marineris* (Fig. 5.14). This chasm is 5000 km long (about the east–west length of the United States) and some 500 km wide in places.

Seen close up, the Martian surface is bleak and dry. Large rock boulders are strewn about, amid gravel, sand, and silt. The boulders are basaltic; so is most of the surface. Some of the boulders contain small holes (Fig. 5.19) from which gas apparently has escaped; the holes make the rock look spongy. On earth, such basalts originate in frothy, gas-filled lava; the Martian rocks probably had a similar origin.

Most of the Martian surface is covered with rusty, basaltic materials of volcanic origin.

Channels and Outflows

Many sinuous *outflow channels* have been cut in the surface of Mars by running water (Fig. 5.20). The largest ones have lengths up to 1500 km and widths as great as 100 km. (These channels are *not* the canals seen by Lowell and others; they are too small to be visible from the earth.) These channels resemble the **arroyos** commonly found in the southwestern part of the United States, in which water flows only occasionally; most of the time such channels are dry.

For several reasons, we think these channels were cut by water: the flow direction is downhill; the flow patterns meander; tributary structures show where several flows merged to form a larger one; and sandbars are cut by smaller flow channels, as in dry river beds on the earth. Most such channels originate within the cratered highlands, just north of Valles Marineris. Some sections of the ancient terrain in the southern hemisphere show networks of small valleys, which resemble terrestrial runoff channels.

The Martian channels suggest extensive running water for at least a short time. Since Mars does not have liquid water on its surface now, proper condi-

FIGURE 5.19

Viking 2 lander view of the surface of Mars. Many of the rocks have small holes, which indicates a volcanic origin—basaltic rocks. The rocks are 10 to 25 cm in size. The sandy soil, blown by the wind, piles up between the rocks. (Courtesy NASA)

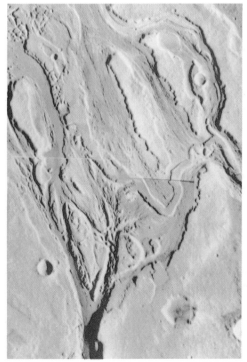

FIGURE 5.20

Arroyos on Mars. Water flows in the past sapped the terrain to create channels in the eastern region of Mangala Valles. The region here is only 20 km across; features as small as 20 meters are visible. These dry river beds, which may have been a major river junction, are part of an extensive network of branching channels in the northern hemisphere. (Courtesy NASA)

tions—that is, a warmer climate and a denser, more humid atmosphere than now—must have occurred in the past. Brief water and mud flows during this wet period could have cut the arroyos and valleys we see today.

Many channels on Mars were cut in the past by water, which implies a wetter, denser atmosphere.

Volcanoes

The most awesome Martian features are the shield volcanoes clustered on and near the Tharsis ridge. The largest is *Olympus Mons*, some 600 km across at its base (Fig. 5.21). The cone reaches 27 km above the surrounding plain, and its base would span the islands of Hawaii, which are made of several volcanoes.

Olympus Mons crowns a string of volcanoes situated on the Tharsis ridge (look back at Fig. 5.14).

FIGURE 5.21

Olympus Mons. This mosaic made from Viking photographs shows how the volcano rises above the surrounding terrain; the base spans some 600 km, the ridge around the base is 6 km high; and the summit lies 27 km above the plain. The central depression at the summit is about 3 km deep and 25 km across. Old lava flows drape over the sides of the volcano. (Courtesy A. Allison and A. S. McEwen, U.S. Geological Survey)

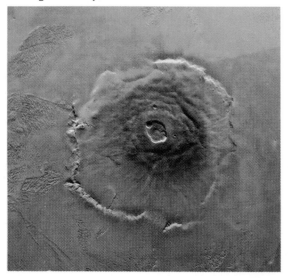

Occasionally, thin ice clouds decorate the tops of the volcanoes there. These clouds might result from erratic spurts of outgassing. On earth, volcanic activity spews forth gases (including water vapor) from the earth's mantle. Such outgassing by the giant Martian volcanoes may have contributed significantly to the Martian atmosphere in the past.

The Tharsis ridge stamps Mars' northern hemisphere, which formed about 1.5 billion years ago. The ridge rises about 10 km above the average surface height for the planet and contains many volcanic structures. Very few impact craters are visible. In contrast, the southern hemisphere is basically a desert pockmarked by old, eroded craters. The geological inference is that about 3 billion years ago in the northern hemisphere, a huge mass of lava oozed from under the surface, creating the volcanic plains and the volcanoes over a long period. This flow wiped out the older deserts and craters in this region. Other flows have taken place in this region since the first.

The volcanic, northern highland regions of Mars mark the youngest terrain.

Craters and the Southern Hemisphere

Every terrestrial planet and satellite displays *impact craters*, formed by solid debris from space bursting onto a solid surface. Once produced, craters are obliterated by wind and water erosion (if the air is dense enough), impacts of later craters, and volcanism. Now, you will see that the early solar system had an episode of **intense bombardment** some 4 billion years ago for which we see good records on airless, inactive worlds such as the moon (see Section 6.3, which also gives more details about the cratering process). Evolution of ancient surfaces modifies the cratering record and gives us an indirect means to date the surface. A general rule of thumb is that a cratered surface is an ancient one, relatively unmodified; a volcanic surface is a relatively young one, recently modified. The very active worlds of the earth and Venus have all but wiped out their ancient cratering records; the moon and Mercury have preserved theirs well. Mars serves as the intermediate case.

FIGURE 5.22

Argyre Planitia in the Martian southern hemisphere. Argyre is a large impact basin (left of center). Many relatively uneroded craters are on the surface. Atmospheric haze is visible at top. (Courtesy NASA)

The southern hemisphere of Mars has a cratered terrain. The landscape contains impact craters that range in size from huge, lava-filled basins down to some only a few meters across (Fig. 5.22). Many are filled with windblown dust. Wind scours the craters and piles the dust in dunes within the craters' bowls.

> The cratered southern region of Mars marks the oldest terrain; it shows evidence of some erosion.

A History of Mars

Putting this information together, the past we infer for Mars relies on the earth as a model. Because its mass is less than that of the earth, Mars has less internal thermal energy. And since Mars is smaller, it does not retain its heat as well as the earth does. So Mars had a shorter episode of surface evolution driven by the heat outflow.

As for the earth, we propose that the formation of Mars resulted by accretion, and impact craters originally covered the surface. Shortly afterward, the planet differentiated (as did the earth and Venus) to form a crust, a mantle, and a core. Regions of thicker crust rose to higher elevations. In the second phase,

thin regions of the crust fractured, and the Tharsis ridge uplifted, cracking the surface around it. During this time, a primitive atmosphere, more dense than now, and warmer, held large amounts of water vapor from the volcanic outgassing. Rainfall may have eroded the surface in furrows and then percolated to a depth of a few kilometers. Decreasing temperatures formed ice at shallow depths. When heated (perhaps by volcanic activity), this ice melted, leading to the formation of flow features. Planetwide water erosion carved the surface.

Next, extensive volcanic activity occurred, especially in the northern hemisphere, some 1.5 billion years ago. The Tharsis region continued to uplift, generating more faults. Valles Marineris may have formed at this time. Finally, volcanism—most of it concentrated on the Tharsis ridge—broke the surface and spewed out great flows of lava. Since that last time of great eruptions, wind erosion mainly has sculpted the Martian surface. A few small impact craters probably have formed from time to time.

In contrast to the earth, we have limited evidence of global plate tectonics (Section 4.3) on Mars. Volcanoes on earth tend to form in chains, where one crustal plate encounters another. On Mars, volcanoes pretty much cluster in one highland region. On earth, plates also show up as continental masses with basins in between; Mars does not have such crustal configurations. However, Valles Marineris may well be a rift valley, indicating one region of crustal movement. The implication is that because of its smaller mass and size, Mars cooled quickly and developed a thick, inactive crust (and mantle). Its mantle lacks the large convective cells that drive the plate tectonics on the earth.

> Mars, because of its smaller size and mass, never progressed as far in its evolution as did the earth and Venus.

 ## 5.6 The Moons of Mars

Two natural satellites encircle the planet Mars; they are named *Phobos* and *Deimos* ("Fear" and "Panic"), after the companions of the god Mars.

(a)

(b)

FIGURE 5.23

(a) Phobos photographed by the Viking 1 orbiter. This side of Phobos is the one that faces Mars. The largest crater is Stickney, 10 km in size. (b) Phobos viewed from a distance of 120 km. The smallest craters are about 10 meters in diameter. (Both courtesy NASA)

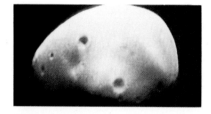

(a)

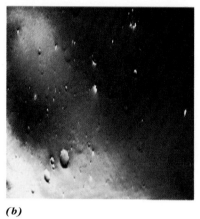

(b)

FIGURE 5.24

(a) Overall view of Deimos. The largest crater is 1.3 km in size. The illuminated part of the moon is about 12 by 8 km. (b) Close-up view of Deimos, showing a region 1.2 km by 1.5 km and features as small as 3 meters. The boulders are about the size of a house. (Both courtesy NASA)

Both lie close to Mars and orbit the planet rapidly. Deimos, the outer moon, circles Mars in 30.3 hours; Phobos, the inner one, takes a mere 7.3 hours. Phobos actually orbits Mars *faster* than that body spins. So while Deimos rises in the east and sets in the west, as our moon does, Phobos rises in the west and sets in the east! Like the earth's moon, each Martian moon keeps the same face to the planet.

Spacecraft observations have found that Deimos and Phobos have the same general oblong shape. Phobos, the larger, has axes about 27, 21, and 19 km long; Deimos's axes are only 15, 12, and 11 km. Photographs also show that Phobos (Fig. 5.23) and Deimos (Fig. 5.24) have cratered surfaces. The sizes and numbers of the craters indicate that the surfaces of these satellites are at least 2 billion years old.

The surfaces have low albedos. They reflect only 2 percent of the light that strikes them—much less than our moon, which reflects about 7 percent (less than coal). Because of a high carbon content, Phobos and Deimos rank among the blackest objects in the solar system and are likely to be captured asteroids (Section 8.1).

The moons of Mars are dark, rocky, cratered bodies, which resemble asteroids.

The Larger View

Our comparative planetology has begun with the terrestrial planets, those that most closely resemble the earth. Our model of the earth's structure and evolution serves as a baseline for comparison. Like the earth, the other terrestrial planets—Venus, Mars, Mercury, and the moon—are rocky and metallic bodies. Their evolution involves change to their interiors, surfaces, and atmospheres.

Investigating other planets confirms and enlarges our view of the evolutionary processes of the earth. We see surfaces that are modified by several processes: impacts of interplanetary debris, outflow of internal heat, volcanism, erosion by wind and water, and crustal movements (if the mantle is sufficiently hot). The atmospheres change from interaction with sunlight, escape into space, outgassing (from a hot mantle), and life (if it exists elsewhere).

These changes are driven mostly by the planet's internal heat—the chief agent in any planet's evolution. The more massive the planet, the more internal heat it generates (from early impacts and radioactive decay), the longer it retains this heat, and the further it evolves. The earth proceeds to evolve with vigor as the most evolved of the terrestrial planets. The others had less initial thermal energy and lost it more rapidly. The time span of their activity is shorter than the earth's and less vigorous.

Key Ideas and Terms

core
crust
density
differentiated
mantle
terrestrial planets

1. Venus and Mars, the terrestrial planets most like the earth, have bulk densities close to that of earth and so have somewhat similar interior structures. (Venus is more like the earth than Mars.)

atmospheric
  composition
atmospheric
  pressure

2. Both Venus and Mars have atmospheres with essentially the same composition: almost all carbon dioxide. The Martian atmosphere is very thin compared to that of Venus and so has a much lower surface pressure.

albedo
greenhouse effect
greenhouse gases
surface temperature

3. The average surface temperature on Venus is very high (700 K) compared to that on the earth (288 K); in contrast, Mars' surface is quite cold (225 K). Venus is so hot because of a supergreenhouse effect produced by its thick, carbon dioxide atmosphere.

tectonic activity
igneous rock
impact craters
plate tectonics
rift valleys
shield volcanoes
volcanism

4. The surface of Venus is divided into lowlands (majority) and highlands (minority). It displays mountains, volcanoes, high plateaus, rolling plains, large valleys, lava flows, and impact craters. Most of the surface is very flat, with shallow lowlands made of lava flows; these regions are filled with volcanic landforms, such as volcanoes. The surface shows recent tectonic activity.

comparative
  planetology
heat
mass
thermal energy

5. Venus, because it has almost the same mass and internal heat as the earth, most likely went through a similar evolutionary sequence. Mars, with less mass and a smaller size, has not evolved nearly as far.

arroyos
igneous rock
impact craters
intense
  bombardment
rift valleys
shield volcanoes

6. The surface of Mars presents giant volcanoes (now extinct), arroyos cut by water (in the past), a huge rift valley system, and impact craters. The northern hemisphere is higher, mostly volcanic, and younger than the southern. The southern hemisphere shows many eroded craters, some from an era of intense bombardment about 4 billion years ago.

arroyos
atmospheric
  pressure
outgassing
shield volcanoes
volcanism

7. Mars is now a desert world, with much less water than exists on the earth. Most of the water is probably in the form of ice in the cores of the polar caps and under the surface. Eroded channels and craters indicate that in the past, Mars had an atmosphere extensive enough to permit liquid water to exist on the surface (now the surface pressure is too low and the temperature too cold).

dipole field
dynamo model
magnetic field lines
magnetosphere

8. Both Venus and Mars have much weaker magnetic fields than expected from the dynamo model developed for the earth. As a result they lack the magnetospheres like the earth, which has such a region as a result of the interaction of its magnetic field with the solar wind.

comparative
  planetology
energy
heat
thermal energy
terrestrial planets
volcanism

9. Terrestrial planets go through similar evolutionary sequences, with mass mostly determining how far a planet will evolve (distance from the sun plays a secondary role). The less mass, the less internal heat and the faster a planet loses it: these are the factors that limit a planet's evolution.

Doppler shift
retrograde rotation

10. The Doppler shift occurs for light and radio waves when a receiver and a source are moving relative to each other. For planets, the Doppler shift can be used to determine rotation rates, especially for planets whose surfaces we cannot see visually.

Review Questions

1. How does the bulk density of Venus compare to that of the earth?
2. In what *one* major respect is the atmosphere of Mars similar to that of Venus?
3. How does the surface temperature of Venus compare to that of the earth?
4. What kinds of crater are found on the surface of Venus?

5. In what one respect is Venus most like the earth in basic physical properties?
6. Where does the water on Mars now reside?
7. Which hemisphere of Mars shows the most impact craters?
8. What feature on the surface of Mars most directly shows that water flowed there in the past?
9. Comparing earth, Venus, and Mars, which planet is the most evolved? The least evolved?
10. If you observe a higher pitch than normal as a Doppler shift, is the source of sound moving toward you or away from you?

Conceptual Exercises

1. How is the Martian surface similar to that of Venus? Different?
2. In what one respect is Venus most like the earth? Most different from the earth?
3. How is Mars most like the earth? Most different from the earth?
4. Suppose you are kidnapped by an evil alien creature who threatens to drop you on Venus or Mars with limited life-support supplies. Which planet would you prefer, and what are the reasons for your choice?
5. Under what conditions might the Martian arroyos have formed?
6. Using your knowledge of terrestrial volcanoes, outline how the Martian volcanoes may have affected the evolution of the Martian atmosphere.
7. Compare the contents of the Martian polar caps in summer and in winter.
8. What one surface feature can be used to argue that Venus has a very young surface?
9. Would you expect very many or very strong earthquakes on Mars now? Why or why not?
10. Which planet would show the *greatest* Doppler shift because of rotation from a reflected radar signal, earth, Venus, or Mars?

Conceptual Activity 5

The Doppler Shift and Planetary Rotation

Let's apply the Doppler shift to find out the rotation period of Venus from radar observations. The signal from the earth is sent out at a specific frequency (call it ν_0) and aimed at Venus. Assume that the beams strike Venus right on the equator; then the receding side of the planet will produce a redshift to lower frequencies, and the approaching side a blueshift to higher frequencies (Section 9.5). The signal received at the earth is spread out to both higher and lower frequencies; such observations are shown in Figure F.2.

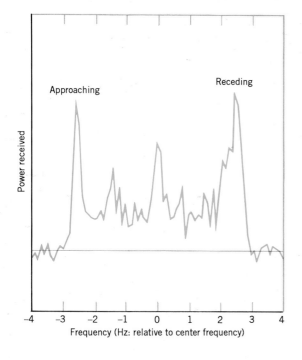

FIGURE F.2
Doppler-shifted radar signals reflected from Venus. The curve shows the power of the signal received as a function of frequency relative to the center (unshifted) frequency. The unit of frequency is the Hertz (Hz), one cycle per second. Note the two peaks on each side of the center; these are the Doppler-shifted signals from the approaching (negative) and receding (positive) edges of the planet. (Adapted from a diagram by G. H. Pettengill and I. I. Shapiro)

Examine the figure; note that there are two peaks in the received signal from the two sides of the planet. These appear at negative and positive frequencies (shown on the horizontal axis) relative to the original frequency. The spread defines the measured $\Delta\nu$, and the Doppler shift relation is

$$\frac{\Delta\nu}{\nu_0} = \frac{V}{c}$$

where V is the relative velocity of one side of the planet with respect to the other, and c is the speed of light.

To find the rotation period, P, you will need to measure the total spread in frequency, $\Delta\nu$. To do so, first measure (in millimeters) the distance between the two peaks. Then use the scale on the horizontal axis to find out how many millimeters correspond to one hertz (1 Hz). Divide the number of millimeters you measured between the peaks by this number to determine the difference in hertz between the peaks. (Your answer should lie between 5 and 6 Hz.)

Take your result for $\Delta\nu$ and plug it into the following formula for the rotation period, P, in days for Venus:

$$P = \frac{1262}{\Delta\nu}$$

What is your result? (It should come out near 243 days.)

6

The Moon and Mercury: Fossil Worlds

■

Central Concept
Because of the rapid loss of their internal heat to space, the limited evolution of the small, airless worlds of the moon and Mercury contrasts sharply with ongoing evolution of the earth.

■

Through a small telescope, the moon strikes you as a stark world with a blasted surface. Mercury shows a similar face. Like the moon, Mercury is small and airless, pockmarked with many old craters, scoured by intense sunlight. Both the moon and Mercury are dead worlds. Their interiors are now cooler than that of the earth; no internal heat drives surface activity. Their heyday of evolution as terrestrial planets has passed.

This chapter provides an insight into the evolution of these tiny worlds by comparing them to each other and to the earth. The emphasis falls on our moon because of the lunar missions, which give sound clues that enable us to reconstruct the moon's history and to infer that of Mercury in comparison.

6.1 The Moon's Orbit, Rotation, Size, and Mass

The moon revolves around the earth in an elliptical orbit at an average distance from the earth's center of some 30 earth diameters (Fig. 6.1), or about 384,400 km. Because the moon's orbit is fairly eccentric, the point of closest distance of the moon from the earth, called *perigee*, is as small as 356,400 km. The point of greatest distance, called *apogee*, reaches as far as 406,700 km.

The moon now always shows the same face toward the earth. So it must rotate on its axis (Fig. 6.2) with a period of 27.3 days—the same as its period of revolution around the earth with respect to the stars. This matching of the moon's rotation and revolution rate is called **synchronous rotation**. To see how the moon must rotate once per revolution to keep the same face to the earth, place a globe on a table. Circle the table, imagining your head to be the moon. As you circle, keep your eyes on the globe. Your head (and body) need to turn around once for one revolution around the table. (Synchronous rotation is the pattern for almost all moons in the solar system.)

All solar system bodies rotate with respect to the stars; that rotation looks different according to whether it is observed from another planet or from the sun.

FIGURE 6.1

The moon's orbit relative to the earth. In this view from the north, we are looking down on the moon's orbital plane. The *minimum* perigee and *maximum* apogee distances are shown for this fairly eccentric orbit. (The actual values of apogee and perigee vary from month to month.) The earth and moon are not drawn to the same scale as the distances, but they do have the correct relative sizes.

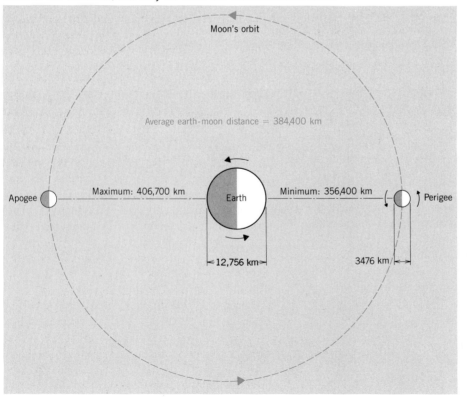

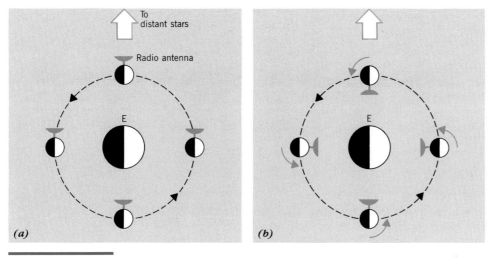

The moon's rotation. The moon keeps the same face toward the earth because its rotation period is the same as its period of revolution with respect to the stars. Imagine a radio antenna on the moon. (*a*) If the moon did not rotate with respect to the stars, the antenna would not always point at the earth. (*b*) But because it *does* rotate at a rate equal to its revolution, an antenna always points to the earth. The moon exhibits synchronous rotation.

The moon keeps the same face to the earth, but not to the sun. It rotates once with respect to the sun in 29.5 days, so the lunar day is 29.5 earth-days long. This is the time interval between noon on the moon and the next noon, which equals the interval between identical phases of the moon, such as from one full moon to the next.

The moon is a small planet. Its diameter is roughly 3500 km, about one-fourth the earth's diameter. If the earth were the size of your head, the moon would be about the size of a tennis ball. On the same scale, the diameter of its orbit would be about 12 meters. The moon's mass is only $\frac{1}{81}$ times the mass of the earth, or 7.4×10^{22} kg. From the mass and radius, the bulk density comes out as 3300 kg/m^3, about the same density as the rocks in the earth's mantle. This low density implies that the moon's interior contains only a small percentage of metals.

The Moon		
	Radius	1738 km
	Mass	7.35×10^{22} kg
The Moon	Bulk density	3340 kg/m^3
	Escape speed	2.4 km/s
At a Glance	Atmosphere	helium, argon (extremely thin)

Tides and the Moon

You may be familiar with the periodic rise and fall of ocean tides, which are caused by the moon. Typically two high tides and two low tides occur each day. Newton's law of gravitation provides a general explanation of tides.

Imagine that the earth's surface is level and covered completely with a layer of water. Consider for a moment the moon's gravitational attraction at three points lined up with the moon (Fig. 6.3*a*). Recall that the force of gravity decreases as the inverse square of the distance between masses. So the strength of the moon's gravitational force must be greater at *A* than at *B* and greater at *B* than at *C*. The stronger the force acting on the same mass, the greater its acceleration. So a mass at *A* experiences more acceleration than a mass at *B*, and a mass at *B* undergoes more acceleration than a mass at *C*.

The *difference* between these accelerations is crucial. Because of the difference, the water at *A* bulges ahead of the earth (point *B*), and the water at *C* lags behind the earth and forms a bulge on the side of the earth opposite the moon. So we have two high tides: one on the side of the earth toward the moon, and one on the opposite side. At any point in the earth's

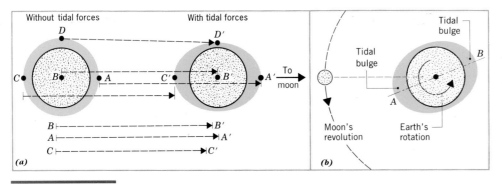

FIGURE 6.3

Tides and the moon. (*a*) Imagine a smooth earth covered with water. Consider the water at *A*, *C*, and *D* pulled by the moon's gravity. How far the water is pulled in depends on its distance from the moon; water at *A* is pulled farther toward the moon than water at *C*. Water at *D* is pulled the same distance as the earth's center (*B*), because the acceleration is the same in each case. So as the earth moves toward the moon (from *B* to *B'*), water on one side moves a bit more (from *A* to *A'*) and on the other side a bit less (from *C* to *C'*). Water flows from *D* to supply the bulges; the depth at *D'* is less than at *A'* or *C'*. (*b*) The tidal bulges on the earth and the moon's motion. The bulges spin ahead of the moon because the earth rotates faster than the moon revolves and friction carries the bulges with the earth's surface.

oceans, these two high tides take place a little more than a half-day apart. **Tidal forces** are the differences between gravitational forces at two different locations along the same radial line. Both the solid earth and the water on the earth's surface are affected by the moon's tidal forces.

Tides result in *tidal friction* on the earth. As water moves about in the ocean basins, both high tide peaks are in line with the moon (one on the side of the earth under the moon, and one on the side opposite). However, the tidal bulges are not exactly on a line to the moon (Fig. 6.3*b*); the closer bulge (*A* in Fig. 6.3*b*) lies ahead of the moon, the farther bulge (*B* in Fig. 6.3*b*) lags behind. This displacement occurs because of the earth's rotation and the friction of the water flowing in the ocean basins (especially in shallow ones). Because the earth rotates faster than the moon revolves around it, the friction pulls the line between the tidal bulges ahead of the line between the center of the earth and moon. In return, that friction slows the rotation rate of the earth.

The *difference* in gravitational forces accounts for the tides; this tidal force tends to pull bodies apart.

The orbit of the moon has changed because gravity ties the earth and the moon together. Tidal interactions between earth and moon slow the earth's rotation (as inferred from the fossil record from oceans). This decrease results in an increase of the earth–moon distance (Section 6.5). In the future the moon will move so far from the earth that the length of the month (longer than now) will equal the length of the day (also longer: 55 present days long.) About a billion years ago, the moon may have been much closer to the earth—perhaps only 18,000 km away. About 400 million years ago, the moon may have orbited at half its present distance.

6.2 The Moon's Surface Environment

Because the moon has a smaller mass and radius than the earth, its surface gravity is less—one-sixth that of the earth—so objects weigh one-sixth as much. As a result of the low gravity and therefore low escape speed, the moon holds essentially no atmosphere—just traces of helium and argon. The moon's escape speed is only 2.4 km/s. At typical lunar temperatures, even high-mass gas particles at

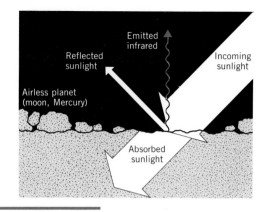

FIGURE 6.4

The energy budget for an airless planet. The ground reflects and absorbs incoming sunlight. The absorbed light heats the ground, which emits mostly infrared radiation to space. The balance of loss and gain fixes the surface temperature. The relative amounts of absorbed and reflected sunlight are shown by the widths of the respective arrows.

the surface attain escape speed and are lost to space. Most gases have escaped from the moon since its formation.

Bodies with a small mass and low surface gravity quickly lose any atmosphere to space.

We can estimate an airless planet's *surface temperature* from its energy budget. Suppose that only sunlight heats the planet's surface (Fig. 6.4). The surface absorbs some incoming sunlight and reflects some back into space. The moon's albedo is 7 percent; the surface absorbs the 93 percent of the incoming light. The absorbed sunlight heats the surface, which then it reradiates mostly in the infrared.

The balance between the incoming sunlight and outgoing infrared determines the surface temperature of the sunlit side. At night, solar heating stops, and the infrared—because no atmosphere traps it—radiates away into space. The temperature plummets to 125 K at night; in the day, by noon, it hits 375 K (about the boiling point of water at the earth's surface). The large noon-to-midnight temperature difference occurs because the moon lacks an atmosphere and rotates slowly. We expect the same large

temperature variation from any airless, slowly rotating planet.

The balance between the incoming sunlight and outgoing infrared determines the surface temperature of the sunlit side of a planet; at night, the loss by infrared determines the temperature.

6.3 The Moon's Surface: Pre-Apollo

Until Apollo astronauts walked on the lunar surface and sampled it, the study of the moon was limited to viewings from the earth, from orbiting spacecraft, or from lunar landers. Viewed through a small telescope, the moon reveals a fascinating terrain (Fig. 6.5). Lunar craters give the moon a pockmarked face. Bright rays flower from some craters. The moon impresses you as a rough, old world, where violence has carved a splotchy surface.

Maria and Basins

The darker areas on the moon are called **maria** (Latin for "seas"; the singular form is **mare**.) Orbiter photos show that most of the maria lie in the northern half of the moon's hemisphere on the side toward the earth; few are on the far side. Their dark irregular stretches compose the face of the "man in the moon."

Most maria look circular and have smooth surfaces compared with the brighter, cratered regions (Fig. 6.6*a*). The maria have lower elevations—by about 3 km—than the rest of the surface. So the maria are named the lunar **lowlands**; the other areas, the **highlands**. Most of the moon's total surface are highlands.

The vast, flat expanses of the maria are solidified lava flows. The lunar maria fill up large, shallow **basins** on the surface. One very striking basin is Mare Orientale (Fig. 6.6*b, c*), located on the far side. The outer rim of its mountains has a diameter of about 970 km and rises to a height of 7 km. Surrounding this basin for about 1000 km outward lies a blanket

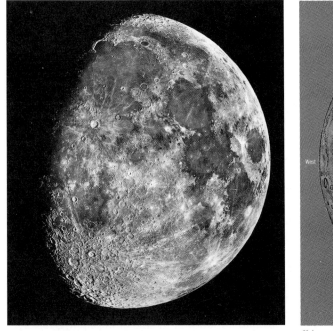

(a)

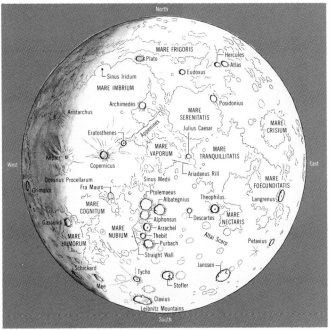

(b)

FIGURE 6.5

Main features of the moon. (*a*) The moon as seen from the earth through a telescope just before full. The darker regions are lowlands; the lighter areas, highlands. Note that most of the craters appear in the highlands. Rays emanate from some of the larger craters. (Courtesy Lick Observatory) (*b*) Map of the major features on the moon's near side.

FIGURE 6.6

Lunar basins. (*a*) Mare Imbrium, in the moon's northern hemisphere on the near side (Fig. 6.5*b*). Note the few craters on the mare's surface and the flooded, semicircular region at the top. The basin's overall diameter is about 1200 km. (Courtesy Yerkes Observatory.) (*b*) Close-up view of the Orientale Basin on the edge of the moon's far side. The Cordillera Mountains ring the basin like a bull's-eye, an indication of the impact process of the basin's formation. The diameter across the outer ring is about 600 km. The horizontal lines are an artifact of the image processing. (Courtesy NASA) (*c*) Wide-angle view of the Orientale Basin (near center) and part of the moon's near side (to right) and far side (to left). At the lower left is the Aitken Basin, about twice the size of Orientale but much older and more degraded. (Courtesy NASA)

(*a*)

(*b*)

(*c*)

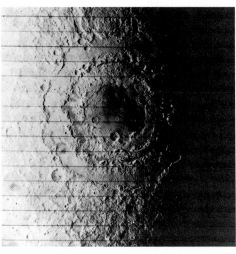

of lighter material covering the older lunar surface—evidence of formation by impact. Beneath the maria lie unusual concentrations of mass called **mascons,** whose density is *greater* than the average density of the moon. Some widespread process deposited large amounts of dense material under the maria to fill in the lowland basins.

Craters

Sa p 92

Craters—round depressions—litter the moon almost everywhere (Fig. 6.7). These features range widely in size, from countless ones smaller than a pinhead to a few with diameters larger than 200 km. The heights of the rims of lunar craters are small compared with the diameters, and the floors are depressed compared with the surrounding landscape.

Most (perhaps all!) lunar craters were formed by objects from space that slammed into the surface (Fig. 6.8). An impact produces a crater, often with undulating slopes, usually covered by debris, with a rippled terrain around it. If the debris and crater wall were put back into an impact crater, they would fill it up. Material blasted out may fall in streaks, leaving a raylike pattern. Large chunks of thrown-

FIGURE 6.7

The crater Aristarchus, about 60 km wide, with its ray system. The ripples in the surface just outside the crater were formed from the shock of impact, as was the central peak. (Courtesy NASA)

FIGURE 6.8

Formation of a crater by impact of a solid body. (*a*) A meteoroid strikes the surface at a speed of some tens of kilometers per second. (*b*) The initial explosion ejects surface material at a high speed. (*c*) Shock waves through the rock compress and fracture it. The rebound of the rock throws material out of the crater. (*d*) The rim of the crater is folded back by the rebound. Ejected material can form rays and smaller craters surrounding the main impact.

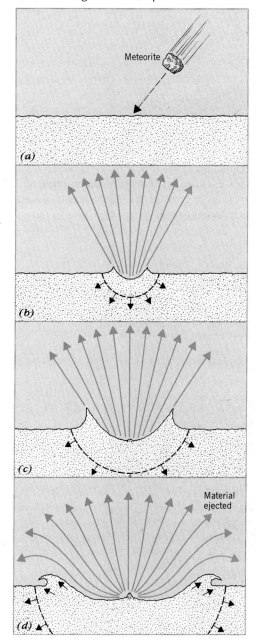

out material can create small secondary craters around the main impact crater.

Cratering

Every terrestrial planet and nearly every satellite in the solar system is scarred by *impact craters*. Let's look at a few major points about the process of cratering, which is influenced by the atmosphere and surface evolution.

> Impact cratering by solid objects from space is a key process in the modification and evolution of the surface of a solid planet.

On most solid surfaces, smaller craters greatly outnumber the larger ones. The size of a crater depends on the incoming projectile's mass and speed—its kinetic energy. The largest projectiles were rocks greater than 100 km in diameter. These objects strike planetary surfaces with speeds of 10 km/s or more. They bang into the surface with enormous amounts of kinetic energy: some 10^{23} J for the mass that blasted the crater Copernicus on the moon. Large craters involve explosions a *million* times greater than the eruption of Mount St. Helens in May 1980, which blew up with an energy of 10^{17} J, an explosive yield equivalent to 35 megatons of TNT! (*Note:* Appendix A begins with an explanation of the "powers of 10" convention in which exponents are used to indicate multiplication or division by 10.)

Upon impact, a projectile's kinetic energy is converted into an explosion at a depth only a few times the diameter of the infalling mass (Fig. 6.8). A shock wave from the impact spreads through the rock, deforming it and throwing it outward. The ejected material forms a ray motif around the crater. Because the explosion occurs below the ground, even projectiles hitting at oblique angles leave round craters and an ejected ray pattern.

Even on an airless world, craters can be wiped out by (1) later impacts, (2) materials thrown up from younger, nearby craters, and (3) lava flows. Large craters are harder to obliterate by these processes than the small ones, so generally the largest craters on a surface are the oldest. Overall, the oldest part of the surface will have the highest density of craters.

Our moon's craters tell us that a little more than 4 billion years ago the cratering rate was 1000 times greater than now and that this intense influx tapered off, reaching close to the present rate about 3 billion years ago. The moon preserves the record of the *intense bombardment* that occurred throughout the solar system during its first billion years, so it is truly a fossil world!

> We can rank the evolution of the planets and moons according to the relative amounts of cratering on their surfaces.

6.4 Apollo Mission Results

The Apollo program obtained about 400 kg of lunar material from six missions to different sites on the moon's near side. These samples allow us to infer many physical details of the moon and its history.

The Lunar Surface

The Apollo samples reveal the physical and chemical nature of the lunar surface. The very top layer (1 to 20 m deep) is a porous, somewhat adhesive layer of debris. It consists of fine particles (called the **lunar soil**) and larger rock fragments. The soil samples contain a large amount of mostly round pieces of glass (Fig. 6.9), which make the surface slippery.

FIGURE 6.9

Close-up of the lunar soil. Note the glass spheres that form when the surface melts upon impact, and droplets of silicates cool quickly. The course rock fragments are only 1 to 2 mm in diameter. The scale at the lower right is marked in millimeters. (Courtesy NASA)

FIGURE 6.10

The main lunar rock types. (*a*) Typical mare basalt. The holes were air bubbles in the lava. The pit in the center resulted from an impact by a small meteorite. (*b*) Rock containing anorthosite (white area). This sample from Apollo 15 was dubbed the "Genesis Rock." (*c*) Lunar breccia. The dark areas are melted rock. Large pieces of other rock are visible embedded in it. (All courtesy NASA)

(a)

(b)

(c)

The moon rocks mostly fall into three categories: dark, fine-grained igneous rocks (Fig. 6.10*a*) similar to terrestrial basalts (magnesium-iron silicates) called **mare basalts**; light-colored igneous rocks (Fig. 6.10*b*) with large grains called **anorthosites** (aluminum–calcium silicates)—by far the most common rock on the surface; and **breccias**, rock and mineral fragments fused together by heat from impacts (Fig 6.10*c*).

What do these characteristics imply? Igneous lunar rocks formed from the solidification of lava. The rate at which lava cools determines the grain sizes: fast cooling results in small grains, slow cooling in large ones. So the small-grained rocks (found in the lowlands) cooled faster than the large-grained ones (found in the highlands). We infer that the anorthosites formed from partial melting and slow cooling of low-density silicates at a relatively low temperature inside the moon. In contrast, the mare basalts were hotter, cooled more rapidly, and formed from more dense materials.

In a few key ways, moon rocks differ radically from the earth's igneous rocks. They contain more titanium, uranium, iron, and magnesium. Compared with earth rocks or meteorites (Section 8.3), they are depleted of elements that would condense at temperatures of less than 1300 K. These elements, called **volatiles,** include sodium, potassium, copper, argon, and chlorine. The moon rocks also contain *no* water. The volatiles were lost before or during the moon's formation.

In one critical way, on a nuclear level, the moon resembles the earth. Consider the amount of oxygen (in relative abundances of the isotopes oxygen-16, -17, and -18); all these isotopes have the same number of protons—8—but they differ in the number of neutrons in their nucleus. It has been found that the amount of oxygen in the lunar samples is the *same* as that for the earth. The similarity for the earth and moon shows that these bodies formed in the same general region of the solar system.

Ages of Lunar Samples

Moon rocks are dated by the same radioactive decay techniques used to date earth rocks (Section 4.3), which resulted in an age of the earth of 4.6 billion years.

A few individual rocks and some fragments from the lunar soil have ages as great as 4.6 billion years. The anorthosites from the highlands generally are the oldest: 4 billion years. The mare basalts are the youngest: some only 3.2 billion years old and a few as old as 3.8 billion years. These ages imply that the moon formed about 4.6 billion years ago—at about the same time as the earth. After formation, the highlands solidified, about 4 billion years ago. By 3 billion years ago, the lava flows that made the maria had taken place.

The Moon's Interior

The Apollo missions found the moon to be an inactive world. Few moonquakes occur, and those that do release only about 10,000 J (10^4 J), barely a tremble by our standards. If you stood directly over the strongest moonquake so far recorded, you would not even feel the ground shake. (The great San Francisco earthquake of 1906 released about 10^{17} J!) This low activity indicates that the moon is largely cold and solid in its crust and upper mantle (Fig. 6.11), so no volcanism can occur now. The mantle consists of silicates a little more dense than those at the surface.

FIGURE 6.11

FIGURE 6.11

Model of the interior structure of the moon inferred from Apollo seismic and heat flow measurements. Note the relatively large core of iron-rich silicates, the rocky mantle and relatively thick crust, and the asymmetry along the line to the earth. Scale has been exaggerated for clarity.

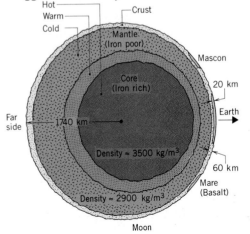

Moon

Encasing the mantle is the crust. On the surface lie the rocks and soil sampled in the Apollo missions.

The bulk density of the moon is a mere 3300 kg/m^3, which implies that the moon overall contains a low percentage metals and probably does *not* have a large metallic core (Fig. 6.11). We don't know for certain whether the moon has a well-defined core. If it does, this feature probably makes up the inner 500 km and may be hot and molten in whole or in part. Evidence for a hot core comes from the Apollo measurements of the heat flow through the lunar surface at a rate of about one-third that of the earth. We know that the upper mantle is cool, so the core may be the source of this thermal energy.

Small, low-mass bodies lose internal heat quickly and so become geologically inactive in a short time.

Lunar History

From the Apollo results we can build a scenario of the moon's history since its formation. The inferred sequence of events (Fig. 6.12 and Table 6.1) relies heavily on the dating of the lunar rocks. Keep in mind that because of its smaller mass, the moon's initial internal heat was less than that of the earth. And because of its smaller size, the moon lost the heat more rapidly, providing a shorter episode of evolution.

About 4.6 billion years ago, the moon formed by the accretion of chunks of material. Such pieces continued to plunge into the moon after most of its mass had gathered. During the first 200 million years after formation, these projectiles from space bombarded the surface and heated it so that it melted. Less dense materials floated to the surface of the molten shell; volatile materials were lost to space. The crust began to solidify from this melted shell about 4.4 billion years ago. From 4.4 to 4.1 billion years ago, the crust slowly cooled as the intense bombardment from space tapered off. The debris from this later bombardment made the many craters now found in the highland areas.

Below the surface, the moon's material remained molten, so volcanism could take place. About 4 bil-

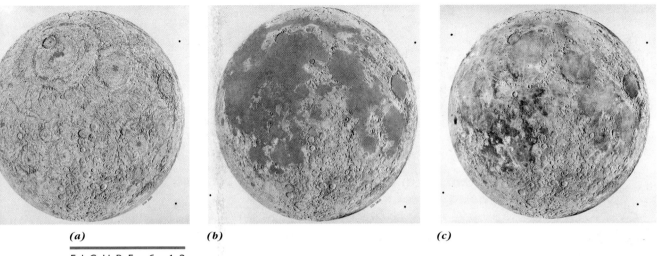

(a)   *(b)*   *(c)*

F I G U R E   6 . 1 2

Model for the moon's evolution. (*a*) 4.0 to 3.9 billion years ago: the surface has solidified and is cratered by the infall of solid bodies; most of the surface is saturated with craters. (*b*) 3.8 to 3.0 billion years ago: fractures in the surface allow magma from the interior to flow out and fill the lowland basins with material darker than the highlands. (*c*) Now: many of the large rayed craters blossomed on the surface after the formation of the maria. Parts of the maria have grown lighter in color from the material blown out by crater formation. (Paintings by D. Davis and D. Wilhelms, U.S. Geological Survey)

lion years ago, a few huge chunks smashed the crust to produce basins that later became maria—mostly on the side that faces us. For example, the Mare Orientale basin formed some 4 billion years ago when an object about 25 km across smashed into the moon. Only later did the basins fill in. As the crust lost its original heat, short-lived radioactive elements (which decay rapidly) reheated sections of it. From 3.9 to 3.0 billion years ago, lava from the radioactive

reheating punctured the thin crust beneath the basins, flowing in to make the maria.

For the past 3 billion years, the crust has been inactive because of the rapid internal heat loss. But small particles from space have incessantly plowed into the surface since it solidified. These sand-sized grains scoured the surface, smoothed it down, and pulverized it. Continued bombardment by larger bodies churned the fragmented surface. The moon's

Table 6.1

A Model for the Evolution of the Moon

Event	Time (billions of years ago)	Processes
Formation	4.6	Accretion of small chunks of material.
Melted shell	4.6–4.4	Melting of outer layer by heat from infall of material and/or radioactive decay; volatile elements lost.
Cratered highlands	4.4–4.1	Cooling and solidification of crust while debris still falls in to crater it.
Large basins	4.1–3.9	Reduced infall but formation of basins by impact of a few large pieces; outflow of basalts from lava below solid crust.
Maria flooding	3.9–3.0	Flooding of basins by lava produced by radioactive decay.
Quiet crust	3.0–now	Bombardment by small particles to pulverize and erode surface.

surface today resembles a heavily bombarded battle-field—constantly fragmented and stirred up.

The moon's time span of activity lasted only about 1.5 billion years; much of the surface preserves a fossil record of its evolution.

6.5 Angular Momentum and the Earth–Moon System

Gravity couples the earth and moon together. The evolution of this twin body system is dominated by its *angular momentum*. You can think about momentum in general like this: once you get a mass moving, you have to exert a force to stop it. An object moving along a straight line exhibits the *linear momentum* for that mass.

Consider a spinning object, such as the earth rotating on its axis. What keeps it spinning? Its inertia about its spin axis. The faster it spins and the more mass is spinning, the harder it is to stop. This spinning momentum is one kind of **angular momentum**. Another kind involves the earth orbiting the sun.

You can think of angular momentum as the tendency for bodies, because of inertia, to keep spinning (rotating) or orbiting (revolving). Angular momentum of a body is determined by its mass, velocity (around the center of motion), and radius (the distance of the mass from the center of motion). For a *single* particle moving in a circle, the angular momentum equals the product of the mass, the circular velocity, and the radius. For a body such as the earth, which is made up of many particles moving at different velocities at different distances from the axis of rotation, we must add up the angular momentum of all particles.

Now here's the key point: angular momentum is *conserved*. If no twisting forces, called *torques*, act on an object, its angular momentum remains the same. (A torque is a force applied with a lever, such as a torque wrench used to tighten spark plugs in a car.)

You can test this idea by performing a simple experiment (Fig. 6.13). Tie a ball to the end of a string and whirl it around your head at a constant speed.

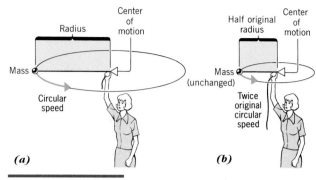

F I G U R E   6 . 1 3

Illustration of the concept of the conservation of angular momentum. (*a*) A person swings a mass in a circle at the end of a string. (*b*) When the length of the string is decreased, the circular speed of the mass increases.

Now, with your free hand, grab the end of the string. Very slowly pull the end through your hand and shorten the string until it is at half the distance you started with. The ball will move with double its circular speed. Note that as the distance from the center of spin decreases, the rate of spin increases. But no torques have been applied, so the angular momentum is the same even with the string at the different lengths. This is an example of the **conservation of angular momentum**.

In an isolated system, with no torques applied, the total angular momentum is conserved.

The total angular momentum of the earth–moon system has two parts: the spin angular momentum (of the earth and the moon rotating about their axes) and the orbital angular momentum (of the moon revolving about the earth). The conservation of angular momentum says that the sum of these must remain constant. Because the moon rotates slowly and its mass is small, the moon's spin adds very little to the total and can largely be ignored. So we need to consider only the spin of the earth and the revolution of the moon.

Because of tidal friction, the earth's rotation is slowing down, so its spin angular momentum decreases. For the system's total angular momentum to remain constant, the moon's orbital angular momentum must increase. This can happen only if the moon moves away from the earth. Measurements have shown that such movement is now occurring about 3 cm per year.

The conservation of angular momentum requires the distance of the moon from the earth to increase as the earth's spin decreases; in this way, the total angular momentum of the earth–moon system remains constant.

6.6 The Origin of the Moon

Older models for the moon's origin fall into three broad categories: fission model, binary accretion model, and capture model. To these has been added a hybrid idea called the giant impact model, which is the strongest contender to date. Let's examine the key elements to build a model for the birth of the moon.

The **fission model** envisions that the earth was spinning more rapidly in the past than now. If the earth were molten, its equatorial speed may have become so great that friction and gravitational attraction would no longer have been able to hold equatorial material to the earth. As a result, a chunk of mantle detached, which spiraled out from the earth, cooled, and formed the moon.

The fission model runs into major difficulties. One, the angular momentum of the earth–moon system must be conserved. If the moon were joined to the earth now, the combined mass would spin about four times faster (once every 6 hours) than the earth does now. But that rate is *not* fast enough to separate the lunar mass from the earth. Two, the lunar basalts differ critically in chemical composition and in water content from the terrestrial basalts that line the ocean basins. So the fission model, though once intriguing, is not a promising idea now.

The **binary accretion model** views the moon as created out of the same cloud of material as the earth. Dust particles grew from a gradual condensation of gas. These eventually accreted into the young moon and earth. The moon formed so close to the earth that earth's gravity held the moon in a close orbit around the earth's equator. The leftover bits and pieces from the formation of the two planets fell into the moon to heat and crater its primitive surface.

The composition of lunar minerals discredits this idea. The moon and the earth are somewhat different in composition. And the moon lacks water. In addition, the moon lacks a mostly iron core. If the two bodies had the same embryonic environment and materials, how did the earth accumulate iron and the moon not? On the other hand, the oxygen isotopic composition is the same for the earth and the moon; this fact supports some binary accretion models.

The **capture model** (Fig. 6.14) pictures the moon formed in some part of the solar system far from the earth. By chance, the moon traveled close enough to the earth to be captured gravitationally. The earth caught the moon in a highly eccentric orbit in which the moon orbited the earth *opposite* the direction of the earth's rotation. The earth's tidal bulge acted to slow the moon down at its closest approach until its orbit decreased in size. The tidal bulge also acted to change the inclination of the moon's orbit, tipping it over toward the direction in which the earth rotates. Eventually the orbit flipped over, and now the tidal bulge works to speed up the moon on its closest approach and so make the orbit larger.

The basic problem with any capture model is the capture! Physical laws severely limit the conditions under which the earth could capture a body as massive as the moon. Basically, a third body needs to be involved; the situation gets rather messy and so the model is not very compelling.

Finally, we come to the **giant impact model**, a hybrid that relies heavily on the binary accretion model and computer simulations of the possible circumstances. This model imagines a glancing impact of a Mars-sized body into the young earth at a speed comparable to the earth's escape speed (Fig. 6.15). The tremendous impact vaporizes the earth's mantle and spews some of this material (and a large fraction the impacting body's material) into orbit around the earth. The metal-rich core of the impacting body remains intact and joins up with the earth; what was its rocky mantle remains in orbit. The orbiting material condenses, creating a ring of debris, and the moon accretes from this material. The essential events are over in about one earth-day.

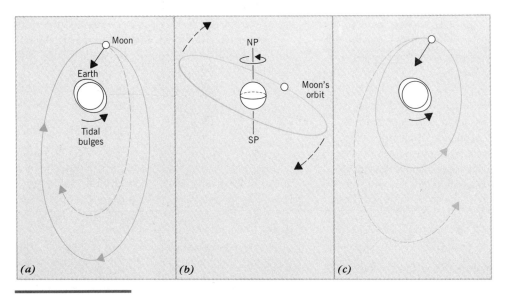

F I G U R E 6 . 1 4

Capture model of the moon's origin. (*a*) The moon, formed in a part of the solar system away from the earth, passes close to the earth and is captured in a clockwise orbit. The earth's tidal bulges pull on the moon. (*b*) Tidal forces flip the orbit over the earth's poles, with the result that the orbital direction is reversed to counterclockwise (as it is now). (*c*) Continual tidal interactions cause the moon to spiral away from the earth into its current orbit.

F I G U R E 6 . 1 5

Supercomputer simulation of the impact of two bodies to represent the formation of the moon by giant impact. The incoming projectile has a mass about that of Mars and a speed of 8 km/s. The target body represents the young earth. Both bodies have metal cores (red and pink) and rocky mantles (brown and green). Following the collision (*a*), a plume of projectile and target material spreads out (*b* and *c*). Eventually the plume condenses into dust particles that accrete to form the moon at an initial distance of about 10 earth radii. Gravitational attraction brings together the clumps of material. (Courtesy M. E. Kipp, Sandia National Laboratories, and H. J. Melosh, University of Arizona)

(a) *(b)* *(c)*

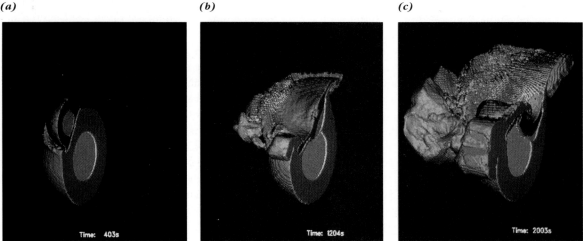

Note that the chemical differences and similarities of the earth and moon are cleverly taken care of: the similarities arise from the ejected terrestrial material, the differences from those of the impacting body. The vaporization releases volatiles from the material that forms the moon, thus explaining their absence from lunar samples. The material ejected from the earth's mantle ensures that the oxygen isotopic abundances match. The angular momentum is accounted for properly if the body strikes the earth on a glancing blow at a speed of about 10 km/s.

The giant impact model addresses major issues in a plausible way and is the best bet for a basic explanation of the moon's origin, given current evidence.

6.7 Mercury: Orbital and Physical Characteristics

The Mariner 10 mission in 1974 and 1975 (the only spacecraft mission to Mercury so far) revealed that Mercury's surface (Fig. 6.16) resembles that of the

FIGURE 6.16

The cratered surface of Mercury photographed by Mariner 10. Note the heavy cratering, which resembles that of the light-colored regions on the moon. The horizontal lines are artifacts of the imaging process. (Courtesy NASA)

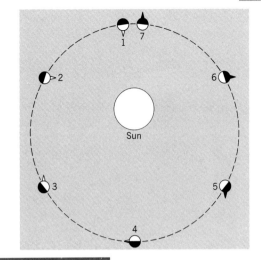

FIGURE 6.17

Mercury's rotation and solar day. Radar measurements in 1965 showed the rotation period is 59 days with respect to the stars. Relative to the sun, a feature on Mercury at noon (1) ends up at midnight (7) after one revolution. After the next revolution, it is at noon again (1). So the solar day on Mercury lasts twice the orbital period, or 176 days. The ratio of the rotational period to the orbital period is 2 to 3.

moon in many ways and opened a direct comparison of these inactive, fossil bodies.

Mercury races around the sun once every 88 days in a very eccentric orbit. We measure the rotational speed using the Doppler shift of radar signals (Section 5.1): Mercury rotates once in about 59 days. This rotation period is two-thirds its orbital period. That ratio results from tidal forces between the sun and Mercury (much as the tidal forces of the earth and moon have resulted in the moon keeping one side toward the earth). The solar day on Mercury— the time from noon to noon—is about 176 terrestrial days, just twice the length of Mercury's year (Fig. 6.17). (See Conceptual Activity 6.)

Mercury turns out to be a tiny world, only some 4900 km in diameter. When Mariner 10 sped past Mercury, its acceleration from Mercury's gravity was accurately measured. From that we have a good value for Mercury's mass: 0.055 that of the earth, or about 3.3×10^{23} kg. Mercury's bulk density is high: at roughly 5400 kg/m^3, it is virtually the same as the earth's. Comparing Mercury's interior with that of

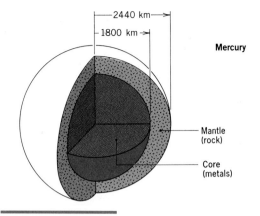

FIGURE 6.18

Model for Mercury's interior. Note the large metallic core (relative to the size of the planet) and the small, rocky mantle.

the earth, we expect a large metallic core (Fig. 6.18) and a rocky mantle.

Mercury At a Glance	Radius	2439 km
	Mass	3.30×10^{23} kg
	Bulk density	5430 kg/m³
	Escape speed	4.3 km/s
	Atmosphere	sodium, potassium (very thin!)

Mercury's Surface Environment

At Mercury's equator at perihelion at noon, the surface temperature reaches 700 K (about 430°C)! It drops more to 425 K at sunset and reaches about 100 K at midnight. This range of temperatures is the widest known in the solar system. On earth, at sea level, the day/night variation in surface temperature rarely exceeds 20 kelvins.

The long, hot solar day and low escape speed (4.2 km/s) make it unlikely that Mercury has much of an atmosphere. Mariner 10 detected an atmosphere (of sorts) of helium and hydrogen (probably picked up from the solar wind). The surface atmospheric pressure was *very* small, a little less than 10^{-15} that of the earth. Recent observations show that sodium

and potassium vapors exist in the atmosphere on the day side. These elements probably are released from the rocks when they absorb ultraviolet light from the sun. Overall, the sodium vapor makes up the major part of Mercury's very thin atmosphere.

Little atmosphere means no insulation from space; that's why the range of noon-to-midnight temperatures on Mercury (and on any airless planet) is so severe.

Craters, Basins, and Scarps

Mercury's surface mainly resembles our moon's surface. There are differences, though: no mountains; many shallow, scalloped cliffs, called **scarps** (Fig. 6.19), reaching lengths of hundreds of kilometers and rising as high as to 1 km; fewer basins and large lava flows; and more relatively uncratered plains amid the heavily cratered regions.

Mercury's highlands are riddled with impact craters like the moon's bleak highlands (Fig. 6.20). Some craters are more than 200 km wide, comparable to the biggest lunar craters. What about large basins? Mercury's largest, the Caloris Basin (Fig. 6.21), has a diameter of some 1300 km and probably was formed by the impact of a large mass. The basin is bounded by rings of mountains about 2 km high.

FIGURE 6.19

This scarp on Mercury's surface (arrow) is more than 300 km long; it extends diagonally from upper right to lower left. (Courtesy NASA)

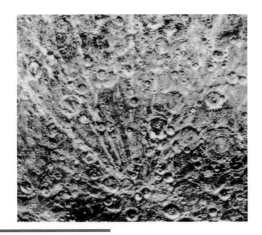

FIGURE 6.20

Impact craters on Mercury in the south polar region. The craters with the brightest rays are the youngest. The largest craters in this photo are about 200 km across. (Courtesy NASA)

FIGURE 6.21

The Caloris Basin (arrow). Only a part of this feature shows in the left half of this photo; the rest is in shadow. The basin is about 1300 km across and rimmed by mountains 2 km high. (Courtesy NASA)

In size and structure, the Caloris Basin resembles the moon's Mare Orientale. The Caloris Basin has a crinkled floor, fractured from rapid cooling of lava that filled the basin. About 20 other basins are known on the mapped surface.

Mercury's scarps vary in length from 20 km to more than 500 km and have heights from a few hundred meters to one kilometer. Individual scarps often travel over different types of terrain. The scarps imply that Mercury's radius has shrunk some 1 to 3 km. How? Probably from cooling of its core, its crust, or both, much as the skin of a baked apple wrinkles as it cools.

Mercury's Magnetic Field

Mercury's magnetic field is a dipole, almost aligned with its spin axis, but only about 0.01 times the strength of the earth's magnetic field. Small as this sounds, it's sufficient to carve out a magnetosphere in the solar wind (Fig. 6.22). Here the magnetic field deflects the solar wind around the planet.

Mercury, like the earth, probably has a metallic core. But it's presumed to be relatively cold and solid now because a small planet loses heat quickly. Recall (Section 4.5) that the earth's magnetic field is thought to arise from swirling motions in its hot, liquid metallic core. Because Mercury rotates much more slowly than the earth and it is thought to have a cool core, no one really expected it to have a planet-wide magnetic field. What's the explanation? No one knows for sure. Perhaps part of the core is fluid. Or, perhaps the field is left over from an earlier time—a fossil field like a permanent magnet frozen in the core.

6.8 The Evolution of the Moon and Mercury Compared

What forces first shaped the surfaces of Mercury and the moon? Impact cratering! Both the moon and Mercury lack dense atmospheres; weathering does not erode the surfaces. Both are tiny worlds with

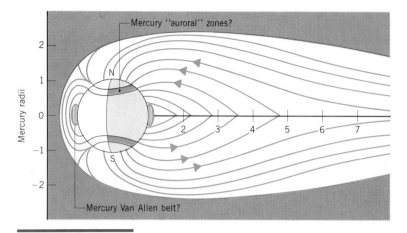

FIGURE 6.22

Mercury's magnetosphere, inferred from Mariner 10 observations, may trap solar wind particles to form Van Allen type radiation belts. Reconnection events may cause storms that form auroral zones on the surface of the planet because of the lack of an atmosphere to stop the particles. (Adapted from a diagram by D. Baker, J. Borovsky, G. Gisler, J. Burns, and M. Zeilik)

cool interiors compared to the earth's interior. So neither has much (if any) volcanic activity now, and neither has undergone the continual surface evolution that the earth experiences from the shifting of crustal plates (Section 4.3) and from weathering.

The lack of an atmosphere and the short period of crustal evolution relate to the small masses of Mercury and the moon. The surface gravities of these bodies are so low that most gases reach escape speeds, and an atmosphere is lost. The small masses also imply that internal heating from radioactive decay would be less than that for the earth, and the flow of heat outward would be so fast that both bodies would cool off quickly. Any era of volcanism could last but a short time.

We don't have samples of Mercury's surface as we do for the moon. But using lunar analogies, we can speculate about the evolution of Mercury. Once formed, Mercury probably went through the following general stages: (1) heating of the surface (by impacts or radioactive decay) and formation of a

solid crust, (2) heavy cratering, (3) formation of impact basins, (4) filling in of basins by volcanism, and (5) low-intensity cratering.

What about Mercury's large metallic core? Mercury has a *much* higher percentage of metals than models for the origin of the solar system (Chapter 8) predict. As a differentiated terrestrial planet, Mercury has too small a mantle relative to its core. A giant impact early in Mercury's history could have stripped off a rocky mantle, leaving mostly a metallic core. Hence, violent impacts in the early days of the solar system may have influenced the formation of the moon and of Mercury.

The moon and Mercury are cool fossil worlds, their dramatic evolution ended; they are the least evolved of the terrestrial planets.

The Larger View

Both the moon and Mercury are worn-out worlds without protective atmospheres. Surface temperatures range widely. Bodies from space plunge into their surfaces—the only source of erosion. Their interiors are cool, so planetwide evolution has ceased. They lack evidence of tectonic activity now or in the past. Lava flows took place early on, when these planets were first heated by accretion and radioactive decay.

These planets at first appear quite different from the earth, yet their internal structures are similar. And once again it is the flow of energy as heat—or lack of it—that mainly shapes the evolutionary histories. Mercury and the moon originated with less heat and have lost it faster than the earth.

Now they are dead worlds—their stories finished 1 to 2 billion years after formation, quite early by astronomical standards. By applying our model of terrestrial evolution, we have been able to infer the histories of these two terrestrial planets.

Along with the concepts of energy and heat flow, we have applied the ideas of density, gravity, tidal forces, and angular momentum to understand the nature of the moon and Mercury. These concepts are the essential ingredients to build models from observations. With these concepts and others, we can make sense of the solar system and approach the question of its origin (Chapter 8). Before we do so, we need to explore an alien aspect of the solar system: the Jovian planets (Chapter 7).

Key Ideas and Terms

bulk density
comparative planetology
terrestrial planets

1. The moon and Mercury are the two smallest terrestrial planets. Their bulk densities indicate that they are made of mostly rocks (moon) and metals (Mercury).

basins
impact craters
lunar soil
mare (pl. maria)
mascons

2. The most prominent surface features of the moon are craters and the basins of maria, formed by the impacts of solid bodies from space some 4 billion years ago. The lunar soil was produced by these impacts.

albedo
escape speed
greenhouse effect

3. The moon has essentially no atmosphere and no greenhouse effect, so the surface temperature ranges widely, from 375 K at noon to 125 K at midnight. Any atmosphere has been lost because even low-mass molecules have escape speed.

angular momentum
conservation of angular momentum
synchronous rotation
tidal forces

4. The tidal interactions of the moon and the earth have, over time, forced the moon into synchronous rotation, so that the same side of the moon always faces toward the earth. The tidal forces raise the ocean tides and cause the moon to move away from the earth at the rate of a few centimeters each year.

breccias
crust
highlands
lowlands
lunar soil
mare basalts
radioactive dating

5. The main regions of the moon's crust are the highlands (covered with low-density, light-colored rocks) and the lowlands (covered with darker, denser rocks). The Apollo samples, dated by radioactive decay techniques, show that the highlands are older than the lowlands; the maria are lava flows that filled in huge basins. The moon formed about 4.6 billion years ago.

anorthosites
igneous rock
mare basalts
volatiles

6. Moon rocks differ from earth rocks as follows: they contain more titanium, uranium, iron, and magnesium; they have fewer volatiles; they contain no water; and they are similar in their densities and isotopic abundances of oxygen.

core
crust
density
differentiated
heat
mantle
thermal energy

7. The moon's physical structure is differentiated into core, mantle, and crust. The mantle is inactive. The core is not metallic like the earth's; it consists dense rocks with few metals, and it may be warm. The current heat flow at the moon's surface is much less than that at the earth's surface.

binary accretion
    model
capture model
fission model
giant impact model

8. Four models contend for the explanation of the origin of the moon: capture, fission, binary accretion, and giant impact. The current giant impact model has the fewest problems and appears to be the likely general explanation.

core
crust
density
differentiated
mantle

9. We infer that Mercury's internal structure is differentiated into core, mantle, and crust. Mercury has a bulk density almost the same as that of the earth, so it probably has a large metallic core; a giant impact may have knocked off most of the mantle. The differentiation of Mercury's interior indicates that in the past it was hot enough to be liquid.

albedo
escape speed
greenhouse effect

10. Mercury's extremely thin atmosphere of helium, sodium, and potassium does not offer enough insulation to provide a greenhouse effect; thus temperature extremes range from 700 K at noon to 100 K at midnight.

basins
comparative
    planetology
craters
heat
scarps
thermal energy

11. The most prominent features on Mercury's surface are craters (formed by impacts), basins (formed by the impacts of very large objects), and scarps. The scarps probably were formed by the shrinkage of the planet as it cooled. The surface of Mercury generally resembles that of the moon, so the processes that shaped it and the timing of the sequence of important events probably follow those of the moon.

magnetic field
magnetosphere
plasma
solar wind

12. Mercury has a much stronger magnetic field than would be expected from a basic dynamo model. This field produces a magnetosphere that interacts with the solar wind. The moon has no global magnetic field now.

Review Questions

1. How does the bulk density of Mercury compare to that of the earth?
2. How does the bulk density of the moon compare to that of the earth?
3. What is one telltale sign that a crater has been formed by impact rather than vulcanism?
4. A planet's surface shows many craters. The planet has no atmosphere. Is the surface relatively young or relatively old?
5. How strong is the greenhouse effect for a planet with a thin atmosphere?
6. What general effect do tidal forces exert on a planetary body?
7. In general, are the lunar highlands *younger* or *older* than the lunar lowlands?
8. From radioactive dating of lunar samples, approximately what age do we infer for the maria?
9. Compared to earth rocks, lunar rocks contain far less of what important group of materials?
10. Relative to the earth's core, is the moon's core hotter or cooler?
11. What model for the origin of the moon appears the best to date?
12. Do we expect Mercury's core to be large or small relative to the overall size of the planet?
13. Compared to the earth's atmosphere, is that of Mercury's thinner or denser?
14. How does Mercury's rotation rate compare to its period of revolution around the sun?
15. Do the scarps on the surface of Mercury imply that the planet was hotter or cooler in the past?
16. Compared to the earth's magnetic field, is that of Mercury stronger or weaker?
17. In terms of the overall sequence of planetary evolution, do you expect Mercury to be more like the moon or more like the earth?

Conceptual Exercises

1. Suppose you were on the moon's surface. You look slowly around, at both the ground and the sky. How does what you see now differ from what you would see on the earth?
2. Suppose you were on the surface of Mercury. How would the scene differ from that on the earth?
3. Argue from a comparison of bulk density that the moon cannot have a metallic core like the earth's. Argue that Mercury should have a metallic core.
4. How were most of the craters on the moon formed? On Mercury? Back up your statement with specific evidence.
5. What specific evidence do we have that the moon's lowland regions (maria) formed *after* the highlands?

6. You are writing a grant proposal to NASA to do research on the origin of the moon. Describe the model you plan to support in the best light possible.

7. Compare the characteristics of the Orientale Basin on the moon to the Caloris Basin on Mercury.

8. In *one* sentence, describe how the surfaces of the moon and Mercury *differ*.

9. Neither the moon nor Mercury has a substantial atmosphere. Why not?

10. In *one* sentence, describe the difference between the interiors of the moon and Mercury.

11. In *one* sentence, compare the evolution of the surface of the moon and the surface of Mercury.

12. Mariner 10 photographed only about half of the surface of Mercury. Imagine that in the future a new flyby mission photographs the rest, finding a region almost *devoid* of craters. What statements can you make about the age and history of this region?

Conceptual Activity 6 The Solar Day on Mercury

Mercury's rotation and revolution result in its solar day (noon to noon) being equal to its year (one orbit around the sun). Let's see how this works out. The key fact is the Mercury's period of rotation with respect to the stars is two-thirds its period of revolution around the sun ($\frac{2}{3} \times 88$ d = 59 d).

On a sheet of paper, draw a circle with a radius of 5 cm; this circle represents Mercury's orbit (which is really an ellipse). Starting at the top of the circle, place a mark every 120° to divide the circle into three equal parts. Now place a dime at each point, and trace the edge of the coin, to represent Mercury at the three positions. Use a quarter at the center of the circle to trace out another circle representing the sun. If we are viewing Mercury from above its north pole, then the planet orbits *counterclockwise* and also rotates *counterclockwise*. Label the positions on the orbit 0 (at top) and then 1 and 2 going counterclockwise. See Figure 6.17 (which is not to scale) for the basic setup.

Imagine you are standing at Mercury's equator at noon with the sun directly overhead. Use the top position of the planet. Draw a flagpole here that points at the sun at the center of Mercury's orbit. Let 59 days pass (two-thirds of the orbital period). Mercury has moved two-thirds of the way around the circle to position 2. The flagpole will be pointing again in the same direction with respect to the stars ("down" on the paper). Look back at position 1. Which way would the flagpole point? "Up," on the paper. While moving to 1 from 0, Mercury has only made only *half* a rotation on its axis.

Continue around the circle at 59-day intervals until the flagpole points again to the sun. Total the number of days from point 0—from noon to noon. If you have done this correctly, you will come up with 176 days as the interval for Mercury to complete one rotation with respect to the sun. So the solar day equals two of Mercury's years!

7

The Jovian Planets and Moons

■

Central Concept

The Jovian planets, compared to the terrestrial ones, have greater masses and sizes but lower densities; today they pretty much resemble their early states because they preserve little history of their evolution.

■

The **Jovian planets**—Jupiter (the prototype), Saturn, Uranus, and Neptune—make up worlds completely different from those of the terrestrial planets. The Jovian satellites have surface settings that differ remarkably from those in the inner solar system. The double-planet system of Pluto and Charon stands apart from these bodies and may have little in common with them.

This chapter investigates the features of the Jovian planets that set them apart from the terrestrial ones—again, a comparative planetology approach. The major differences between the Jovian and terrestrial planets furnish additional clues about the evolution of the planets and the formation of the solar system. The key point is this: the Jovian planets are primitive worlds, looking today very much the same as when they were formed eons ago. Building their structures and histories is a very different task from seeking to understand the terrestrial worlds.

7.1 Jupiter

Jupiter (Fig. 7.1a) ranks as the largest and most massive body in the solar system (after the sun). Jupiter's total mass is about $2\frac{1}{2}$ times that of all the other planets put together (and 318 times the earth's mass). Eleven earths placed edge to edge would stretch across Jupiter's visible disk, and more than 1000 earths would be needed to fill its volume (Fig. 7.1b)!

Physical Characteristics

Jupiter is the largest planet—its diameter is almost 140,000 km (Appendix B, Table B.3). Yet, Jupiter's material is much less concentrated than the earth's, for Jupiter's bulk density is only about 1300 kg/m³. All the Jovian planets have low densities compared with the terrestrial planets—one key difference between the two classes. The terrestrial planets are basically rocks and metals, made of elements such as iron, aluminum, oxygen, and silicon. Jupiter, in contrast, is made mostly of hydrogen and helium.

The bulk density of a planet gives a key clue about its overall composition.

Jupiter At a Glance	Radius	71,492 km
	Mass	1.899 × 10²⁷ kg
	Bulk density	1330 kg/m³
	Escape speed	59.6 km/s
	Atmosphere	hydrogen, helium

The elements hydrogen and helium came together when Jupiter formed, and the giant planet has lost little, if any, since then. Jupiter's huge mass

FIGURE 7.1

Giant planets. (a) The Jovian planets, shown to scale relative to the earth. Note the banded atmospheres of Jupiter (far left) and Saturn. Both Uranus, to the right of Saturn, and Neptune have methane gas in the atmosphere, which absorbs red light and results in an overall bluish color. The rings of Jupiter, Uranus, and Neptune are not visible. (Courtesy Jet Propulsion Laboratory and NASA) (b) Jupiter and its Great Red Spot photographed by the Hubble Space Telescope in May 1991. The colors here are true colors. Note the banded structure of the clouds in the upper atmosphere caused by strong jet streams. The clouds contain small crystals of frozen ammonia and traces of colorful chemical compounds. The Great Red Spot (arrow) is a giant storm; below and to its left is a white oval. (Courtesy NASA and the European Space Agency)

(a)

EARTH

(b)

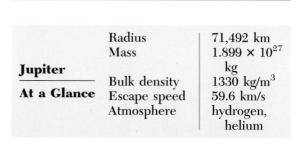

means that its escape speed is high, about 60 km/s. Jupiter's upper atmosphere is cold, only about 130 K, and hydrogen molecules there move at 1 km/s. So even low-mass hydrogen molecules do not attain escape speed. Jupiter has retained its atmosphere for eons and will hold it for eons to come. What you see now is basically the atmosphere and mass with which Jupiter was formed. The same is true for the other Jovian planets. (The earth and other terrestrial planets lost these materials early in their histories.)

Atmospheric Features and Composition

The visible disk of Jupiter (Fig. 7.1*b*) is the planet's upper atmosphere (*not* its surface!), which shows alternating strips of light and dark regions that run parallel to the equator. The light regions are called **zones**; the dark regions, **belts**. The zones mark the tops of rising regions of high pressure, and the belts mark the descending areas of low pressure (Fig. 7.2). This vigorous convective flow transports heat outward from the planet's interior, so it must be hot inside.

Markings in the clouds allow a measurement of Jupiter's counterclockwise rotation period. The period varies with latitude: Jupiter spins in 9 hours and 50 minutes at its equator and 9 hours, 55 minutes at its poles. Such a variation, called **differential rota-**

FIGURE 7.3

Close-up of a brown oval (arrow) in Jupiter's atmosphere photographed by Voyager 1. Such features last a year or two; this one has a length about equal to the earth's diameter. Such small storms appear to be clear regions that allow us a view down to lower levels. Winds in the belt above the oval flow at about 100 m/s. (Courtesy NASA)

tion, indicates that a body is fluid. (A solid body, like the earth, rotates so that each point on its surface has the same period.) This rapid rotation and Jupiter's large radius produce an equatorial speed of more than 43,000 km/h!

> Fluid (liquid or gaseous) round bodies rotate faster at the equator than at the poles.

Such an enormous rotation speed drives the circulation in Jupiter's atmosphere. Jet streams zip along at the boundaries between the belts and zones, creating atmospheric disturbances. Typical wind speeds are 100 m/s (that's about three times faster than the earth's jet streams). The visible clouds—most likely ammonia ice crystals—are blown by these winds. The Voyager missions zoomed in on these complex streams and swirls of Jupiter's upper cloud layer, showing the turbulent atmospheric flow in grand detail (Fig. 7.3).

The most famous atmospheric feature is the *Great Red Spot* (Fig. 7.4), an atmospheric storm that has been raging for centuries. The Red Spot changes in size; it gets as large as 14,000 km wide and 40,000

FIGURE 7.2

The convective circulation in Jupiter's upper atmosphere. Rising air creates high pressure regions (called *zones*) and the downflow makes low pressure areas (called *belts*). The zones generally appear lighter in color than the belts. The convection is driven by heat flowing out of the planet's hot interior.

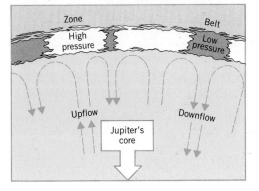

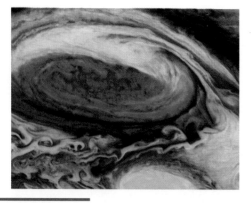

F I G U R E   7 . 4

Close-up of the Great Red Spot and the turbulent region near it. The smallest details visible are about 60 km across; the image has been computer-processed in false colors (Appendix G) to enhance these details. The Red Spot turns counterclockwise once in about a week. Its long dimension is about twice the earth's diameter. (Courtesy NASA)

km long—it could easily swallow the earth! A few degrees cooler than the surrounding zone, the Red Spot pokes above it—a rising region of high pressure that rotates counterclockwise in Jupiter's southern hemisphere. The Red Spot rotates once in 7 days. Behind it surges a region of turbulence (Fig. 7.4) from the atmosphere flowing past.

Jupiter's atmosphere is composed mainly of molecular hydrogen (82 percent by mass) and helium (18 percent by mass) and traces of other gases. Key molecules among the minor constituents are methane, ammonia, and water vapor. Jupiter has a composition similar to that of the sun.

A Model of the Interior

No distinct boundary lies between atmosphere and interior. The atmosphere gets denser and hotter farther in, gradually merging into the interior. We infer Jupiter's internal structure by constructing models that include a solar mix of materials, and the fact that Jupiter radiates into space more energy than it receives from the sun (about twice as much!). The internal thermal energy is probably left over from Jupiter's formation. Compared to the earth, Jupiter's heat flow is about 90 times greater—5.4 watts per square meter, so it must be fairly hot inside. **Power** is the term for the rate of energy transfer. It is expressed in joules per second, and 1 J/s = 1 **watt** (W). (See Appendix A.)

Large, massive planets lose heat very slowly and so remain hot in their interiors for a long time; they formed with a large store of thermal energy.

Such models come up with the following picture (Fig. 7.5). The atmosphere covers the planet like a thin skin and consists mostly of molecular hydrogen. Going down into the planet, the density, temperature, and pressure increase, and the hydrogen exists in a liquid state. At a pressure of about 2 million atm, the hydrogen is squeezed so tightly that the molecules are separated into protons and electrons, which move around freely and can conduct electricity. This state of matter, called **metallic hydrogen**, has been observed in a laboratory on the earth. The metallic hydrogen zone continues to within about 14,000 km of the planet's center. Here, perhaps, lies a core (rocky?) of heavy elements. Note that, like the terrestrial planets, Jupiter (and the other Jovian planets) have differentiated interiors.

Most of Jupiter is hydrogen, and most of that hydrogen is liquid—quite a contrast to the earth's inte-

F I G U R E   7 . 5

Model for Jupiter's interior structure. Below the atmosphere is a thick layer of liquid molecular hydrogen; below that, a region of metallic hydrogen because of the high pressure (above 3×10^6 atm). The dense core material, perhaps silicates, may be molten because it is quite hot, perhaps 40,000 K. The core may have a mass about 10 to 20 times that of the earth. (Adapted from a NASA diagram)

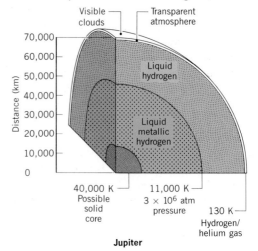

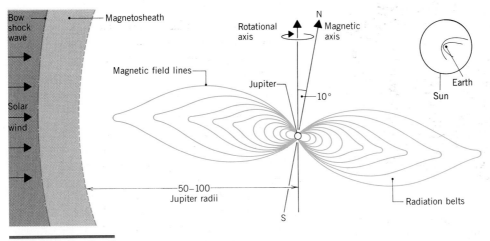

FIGURE 7.6

Model for Jupiter's magnetosphere based on spacecraft data; size and shape vary according to the strength of the solar wind. The solar wind forms a bow shock wave, where it interacts with the magnetic field; behind it lies the magnetosheath, a region of relative calm. Note that the tilt of the magnetic axis to the rotation axis (10°) is about the same as that for the earth. The size of the sun, and within it, the earth and its magnetosphere, are shown on the same scale. (Adapted from a NASA diagram)

rior (Section 4.2) and that of the other terrestrial planets. The core temperature may be about 10 times hotter than the earth's. The flow of the heat outward from the core drives the convection up to produce the atmosphere.

Magnetic Field

Jupiter has a magnetic field some 10 times as strong as the earth's. At the cloud tops, the field is about 4×10^{-4} T. The earth's magnetic field at the surface has a strength of about 0.4×10^{-4} T. (Recall that the tesla, T, is the SI unit of magnetic field strength; see Appendix A.) Jupiter's magnetic field axis is tilted about 10° with respect to the rotation axis.

Jupiter's strong magnetic field produces a magnetosphere much larger than the earth's, though the processes that make it are basically the same. The magnetic field traps plasma particles from the solar wind in belts, similar to the Van Allen belts around the earth, close to Jupiter (Fig. 7.6). The field acts as a buffer to deflect the solar wind around Jupiter. On the night side, a magnetic tail stretches out a few astronomical units in length and reaches as far as Saturn! Magnetic reconnection events here may drive the auroras that have been observed in Jupiter's atmosphere.

> Any planet with a strong magnetic field develops a magnetosphere as the field interacts with the plasma of the solar wind.

What generates this powerful magnetic field? We saw in Section 4.5 that a dynamo model of the earth's magnetic field pictures currents in the liquid metal core generating the magnetic field like an electromagnet. Liquid metallic hydrogen can conduct electric currents. So the conditions for a planetary dynamo to operate are available: a fluid able to conduct electricity, filled with convective currents driven by heat and rapid rotation. A dynamo in the large metallic hydrogen zone could produce Jupiter's intense magnetic field—the model seems to work here.

7.2 The Moons and Rings of Jupiter

Jupiter possesses at least 16 moons (Appendix B.7). The brightest and largest were observed and reported by Galileo in the seventeenth century. These huge **Galilean moons** orbit within 2 million kilometers of Jupiter; from closest outward, they are: *Io, Europa, Ganymede,* and *Callisto* (Appendix B, Table B.7). Like our moon, each is locked in synchronous rotation with the mother planet and so keeps one face toward Jupiter. Ganymede and Callisto are both larger than Mercury; Io is somewhat larger than our moon, and Europa smaller.

The Galilean moons show dramatic differences in surface features and internal structures. One key difference is the bulk densities: relative to water, Io is 3.5, Europa 3.0, Ganymede 1.9, and Callisto 1.8. The compositions of Io and Europa resemble that of our moon—mostly rock, with perhaps a little icy material. In contrast, Ganymede and Callisto contain substantial amounts of water ice or other low-density icy materials and much less rock.

Io

Io possesses a thin atmosphere. At the surface, the atmospheric pressure is about 10^{-10} atm, composed mainly of sulfur dioxide. Volcanic eruptions, at least in part, produce this atmosphere. Io has active volcanoes and fuming lava lakes on its surface (Fig. 7.7). This activity implies that the interior is hot; sulfur and sulfur compounds melt and are vented in volcanic outflows, which contain liquid sulfur. Io has a rocky interior with a sulfur-rich crust.

Io's volcanoes resemble collapsed volcanic craters. Lava simply pours out of a crater vent and spreads outward for hundreds of kilometers (Fig. 7.8). Dark-colored lava lakes, one about the size of the island of Hawaii, surround many of Io's volcanoes. The temperature of these lava lakes is about 400 K. (The melting point of sulfur is 385 K.)

A highly evolved moon or planet will have a young surface with some recently formed features, such as volcanoes and lava flows produced by volcanism.

Why is Io's interior so hot? The gravitational effects of the other Galilean moons force Io into a slightly eccentric orbit, so its distance from Jupiter changes significantly with each revolution. These variations in distance cause large and variable tidal forces to act on Io. It is because of the recurring

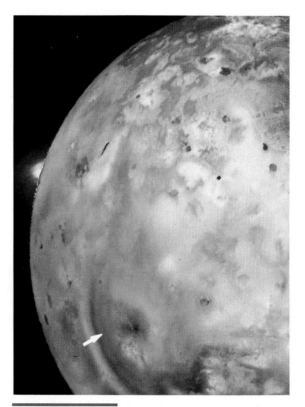

F I G U R E   7 . 7

Pele and Loki volcanic outbursts on Io. These two active volcanoes are visible in this specially processed image. The plume from Loki is visible at the upper left, against the background of space. Pele (arrow) is below it and to the right. Note that its plume spreads out like a dark, circular umbrella. (Image created by A. S. McEwen, Arizona State University and the U.S. Geological Survey. Courtesy Jet Propulsion Laboratory and NASA)

push and pull of these tidal forces that the interior of Io heats up.

Because volcanism constantly alters Io's surface, it must be very young. No impact craters appear on Io; volcanic flows have covered them up. Io's surface seems to be the youngest in the solar system, probably less than 1 million years old.

Europa

The surface features of Europa consist of bright areas of water ice among darker, orange-brown areas. Europa's surface is crisscrossed by stripes and bands that may be filled fractures in this moon's

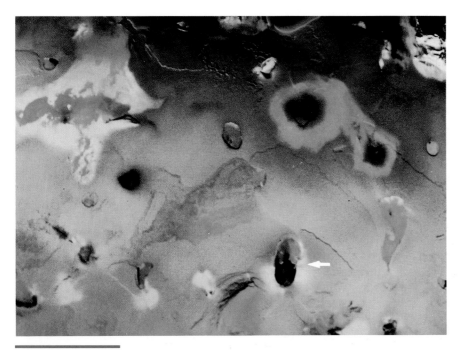

FIGURE 7.8

Wide-angle view of Io's surface in its south polar region; the width of this image is about 1000 km. Dark spots with the irregular radiating patterns are volcanic vents with lava flows (arrow). The bright white patches are made of sulfur dioxide frost. (Image created by A. S. McEwen, Arizona State University and the U.S. Geological Survey. Courtesy Jet Propulsion Laboratory and NASA)

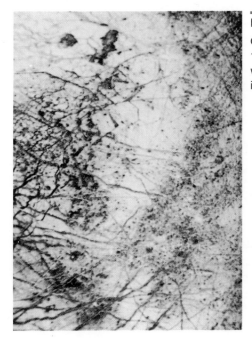

FIGURE 7.9

Close-up view of Europa's surface. The bright ridges are about 5 to 10 km wide and 100 km long. The dark bands are 20 to 40 km wide and thousands of kilometers long. This computer-enhanced image brings out different hues in the icy surface. (Courtesy NASA)

thin, icy crust (Fig. 7.9). Some of these cracks extend for thousands of kilometers, split to widths of 20 to 200 km, but reaching depths of only 100 meters or so. Europa's surface is incredibly smooth. Compared to its size, its dark markings are no deeper than the thickness of ink lines drawn on a Ping-Pong ball. Beneath the thin crust, Europa's interior is mostly rock.

Europa's surface is almost devoid of impact craters. So Europa's surface cannot be a primitive one; it has evolved since its formation. Europa's cracked surface indicates that its solid, icy crust is thin and its interior hot, perhaps molten. The crust may long ago have been a slush that wiped out early cratering. As

Europa cooled, its crust turned to smooth, glassy ice that later cracked. Beneath the ice veneer may lie a layer of liquid water.

Ganymede

Largest moon of Jupiter, Ganymede ranks overall as the largest moon in the solar system. (Titan of Saturn is second.) Ganymede has two basic styles of terrain (Fig. 7.10): *cratered* and *grooved*. Impact craters up to 150 km in size densely mark the surface of the cratered terrain. The craters are shallow for their diameter—an indication of an icy surface. Their abundance shows that the cratered terrain is some 4 billion years old. The grooved terrain separates the cratered terrain into polygonal segments. Light ridges and darker grooves, where the ground has slid, sheared, and torn apart, create a mosaic effect.

No large mountainous regions or large basins relieve the terrain on Ganymede. In fact, nowhere on the satellite is any relief greater than about 1 km. This suggests that the crust of Ganymede is somewhat pliable, probably because of the large amount of water ice in it. Ganymede's bulk density is low, only about 1900 kg/m^3, so it contains about half water ice and half rock. Occasional stresses on the water–rock crust have created the fracture patterns. Ganymede has been geologically inactive since the

waning stages of the intense bombardment of planetary surfaces early in the solar system's history.

A heavily cratered surface has been little modified and so is generally very old.

Callisto

Farthest out of the Galilean moons, Callisto has a surface riddled with craters of a wide range of sizes—and so a very old surface. Some have bright ice rays; others are filled with ice. Callisto's craters are shallow because the surface is a mixture of ice and rock. No volcanism has taken place here to wipe out any craters; the interior is cold. Overall, Callisto contains somewhat more water ice than rock.

Callisto displays a huge and beautiful multiringed feature called *Valhalla* (Fig. 7.11). Its central floor is 600 km in diameter; 20 to 30 mountainous rings that have diameters of up to 3000 km surround it like a bull's-eye. The rings look like a series of frozen waves. They were formed in the stupendous impact that melted subsurface ice, causing the water to

FIGURE 7.10

Grooved and ridged terrain on Ganymede's surface; these areas make up the newer parts of the surface. The resolution is 6 km. The impact craters with bright rays are younger than those without. (Courtesy NASA)

FIGURE 7.11

The giant ringed basin on Callisto (left of center). The bright spot at the basin's center is about 600 km across; the outer rings, 2600 km. Note the lack of ridges or mountains surrounding this impact basin, which is called Valhalla. Also note the many impact craters with bright rays. (Courtesy NASA)

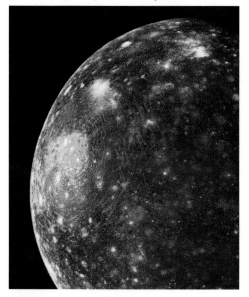

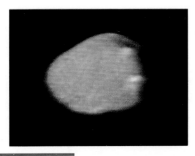

Amalthea, Jupiter's innermost moon. The indentations are craters. Note the overall reddish color. The long axis measures about 270 km. (Courtesy NASA)

spread in waves that quickly froze. The ripple marks are preserved as rings: frozen blast waves.

Asteroidal Moons

Jupiter's outer moons are asteroidlike bodies, probably captured asteroids. (This is likely, for, after all, Jupiter lies just outside the asteroid belt: Section 8.1.) We have observed one such moon, Amalthea, close up. Only 181,000 km out from Jupiter, it whizzes around once every 12 hours. It is elongated, 270 km by 150 km (Fig. 7.12); the surface is cratered and has a dark red color. This moon's irregular shape, small size, and dark, cratered surface show its asteroidlike character (Section 8.1).

Ring System

Jupiter actually has millions of moons—tiny ones that make up its **ring system**, discovered by Voyager 1

Jupiter's ring system near the edge of the planet. The sun, which is to the rear of this image, backlights the particles of the rings. Note the bright, sharp outer edge and diffuse inner region. The disk of Jupiter is to the upper left; it is dark because we are viewing the night side. (Courtesy NASA)

(Fig. 7.13). The rings are so thin (less than 30 km thick) that they are essentially transparent. They are most visible when viewed edge-on; then the particles scatter light well. To have good light-scattering properties, the particles must be small, about 10 micrometers (a *micrometer* [μm] is 10^{-6} m; see Appendix A), and probably made of silicates.

The rings have a definite structure. The outer, brightest part is 800 km wide and lies about 129,000 km from Jupiter's center. Closer to the planet is a broader ring some 6000 km wide. And within that ring lies a faint sheet of material that extends almost down to the cloud tops. A faint, outer ring surrounds the whole system and reaches out over 200,000 km, almost to the orbit of Amalthea.

Planetary rings are made of many individual, small, solid particles orbiting the planet's equator.

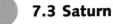 ## 7.3 Saturn

Saturn bears a marked resemblance to Jupiter, but its ring system outranks in splendor that of the larger planet (Fig. 7.14). Saturn has a slightly smaller size and less mass than Jupiter. It has the lowest density of any of the planets—only 690 kg/m^3, much less than that of water.

Saturn viewed from Voyager 2; a false-color image (Appendix G). Note that the bands in the atmosphere are less prominent than those of Jupiter. The ring system shows its overall structure. (Courtesy NASA)

	Radius	60,268 km
Saturn	Mass	5.68×10^{26} kg
	Bulk density	690 kg/m^3
At a Glance	Escape speed	35.6 km/s
	Atmosphere	hydrogen, helium

The atmospheric structure of Saturn resembles that of Jupiter: belts running parallel to the equator, driven by its rapid rotation. (Saturn's rotation period is 10 hours, 14 minutes at the equator and varies with latitude; it, too, shows *differential rotation*.) Disturbances in the belts are rarer than on Jupiter. Mighty storms do occur at intervals of about every 30 years, roughly at midsummer in the northern hemisphere (Fig. 7.15). A warming of the atmosphere may trigger these storms, which last for several weeks before fierce winds blow the great clouds apart.

The atmosphere of Saturn has roughly the same composition as that of Jupiter, with molecular hydro-

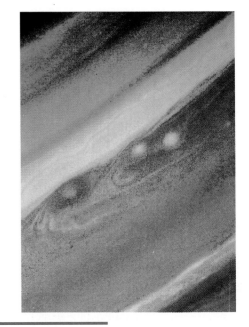

FIGURE 7.16

Voyager 2 photo of the northern hemisphere Saturn, showing the wind flow patterns. Visible here are jet streams, turbulent flows, and eddies. Winds, which can reach speeds of 100 m/s, are slower than at the equator. (Courtesy NASA)

FIGURE 7.15

Great storm in the atmosphere of Saturn, photographed by the Hubble Space Telescope in November 1990. This view is of the northern hemisphere; the north pole is in the dark region at top. The storm appears as a high, whitish cloud just above the equatorial region, where winds blow at about 500 m/s. (Courtesy NASA)

FIGURE 7.17

Model for the interior structure of Saturn. Note how much smaller the metallic hydrogen zone is compared to that of Jupiter. The inner, rocky core—if it exists!—may have a mass of about 20 times that of the earth.

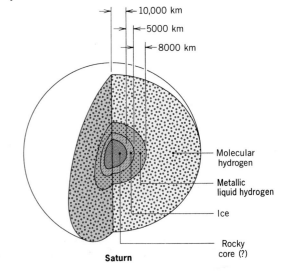

gen and helium as the most abundant constituents. However, the helium abundance (6 percent by mass) is only about one-third that found in Jupiter's atmosphere. Methane, water vapor, and ammonia make up the minority of the gases.

Saturn's clouds appear far less colorful than those of Jupiter—mostly a faint yellow and orange (Fig. 7.16). However, much of the same complexity of cloud patterns appears with wind speeds much higher than on Jupiter—up to 500 m/s near the equator. Weather patterns can change weekly, while large storm systems persist for a few years.

Saturn's interior (Fig. 7.17) probably reflects Jupiter's composition: a makeup roughly the same as that of the sun, but with a bit more in the way of heavy elements. It probably has a zone of liquid, metallic hydrogen. Saturn may also have a small, rocky core some 20,000 km in diameter and a mass of about 20 earth masses.

Saturn emits as heat radiation about twice as much power as it receives from the sun. The heat flow is some 2 watts per square meter, about 30 times greater than the earth's. As with Jupiter, this excess internal energy may be left over from the planet's formation and stored in the interior.

Any solar system body that emits more power (in the infrared) than it receives from the sun must have a source of internal energy.

Saturn also has a strong magnetic field and so a large magnetosphere. The magnetic axis aligns within one degree of Saturn's rotation axis. The magnetic field strength at the cloud tops is about one-twentieth that of Jupiter. The field probably is produced by a dynamo effect in the liquid metallic hydrogen zone of Saturn, in the same way it is presumably produced in Jupiter. The magnetic field creates regions around Saturn like the earth's Van Allen belts, which trap charged particles from the sun (Section 4.5).

7.4 The Moons and Rings of Saturn

Saturn's clique of moons totals at least 17 (Appendix B, Table B.8). The moons of Saturn fall into three

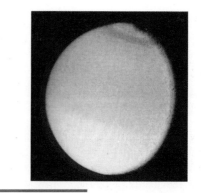

FIGURE 7.18

Clouds in the upper atmosphere of Titan. Note how they appear darker in the northern (top) hemisphere than the southern; this difference shows that the clouds in the two hemispheres differ in thickness. (Courtesy NASA)

groups: Titan by itself; the half-dozen large icy moons (Mimas, Enceladus, Tethys, Dione, Rhea, and Iapetus, in order outward from Saturn); and the 10 small moons (Phoebe, Hyperion, and the rest). Overall, their densities are less than 2000 kg/m^3, which implies that the moons are mostly water ice (60 percent) and rock (40 percent).

Most of the moons are cratered. Some cratered terrain has been modified on the larger moons, which implies internal heating to melt parts of the icy surfaces. In contrast, the small moons, which are also cratered, show no changes—they still have their original surfaces.

Titan

Titan, Saturn's largest moon (and second largest in the solar system), has a bulk density of about 1900 kg/m^3, which implies a 50:50 composition of ice and rock. Titan possesses a thin atmosphere, which consists mostly of molecular nitrogen (99 percent), with few percent methane and argon. The atmosphere's surface pressure is a surprising 1.5 atm; the surface temperature a frigid 94 K. Voyager photos (Fig. 7.18) showed a stratospheric layer of orange smog as well as a blue color along Titan's edge. This coloration suggests that the atmosphere varies in composition. No surface features were visible.

Other Moons

Saturn's four largest moons, after Titan, are Iapetus, Rhea, Dione, and Tethys. They all appear heavily

FIGURE 7.19

Iapetus, showing details as small as 20 km in size. Note the impact craters in the bright region. The dark region covers the icy crust of the hemisphere that faces in the direction in which Iapetus orbits Saturn. (Courtesy NASA)

FIGURE 7.20

Enceladus. The grooves and linear features (tens of kilometers long) on the surface imply that the crust was deformed by internal heat. The largest craters visible are about 35 km in diameter. Smallest details are about 2 km. (Courtesy NASA)

cratered. Iapetus (Fig. 7.19) has the most extremes of surface cover. One hemisphere is only one-tenth as bright as the other (like the difference between a blackboard and a field of snow). The bright hemisphere is cratered, hence ancient. No surface features are visible in the dark hemisphere, whose low albedo suggests a large carbon content.

Only Enceladus does not have a surface thick with craters (Fig. 7.20). That is a sign of recent modification of the surface. A hot interior can melt the icy surface; one Voyager photo showed a possible volcanic plume, which would clearly indicate a hot interior now. As with Io, tidal forces may heat the interior of Enceladus.

The rest of the moons are all small, heavily cratered bodies, 300 km or less in diameter. The largest is Hyperion, which has a unique shape—like a thick hamburger (Fig. 7.21). We presume that Hyperion and the other small moons are basically ice, like the larger moons.

Ring System

The stunning rings are very thin, no more than a few kilometers thick. Although thin, the rings are wide; the three main rings visible from the earth (called A, B, and C) reach from 71,000 km to 140,000 km from Saturn's center (Fig. 7.22). Voyager photographs revealed spectacular detail in the ring system. Although the A-ring is relatively smooth, the B- and C-rings break up into many small ringlets (Fig. 7.23), like grooves on a phonograph record. Many hundreds, perhaps a thousand, light and dark ringlets surround the planet, with widths as small as 2 km (and perhaps smaller).

FIGURE 7.21

Three views of Hyperion, showing the unusual shape of this moon as it rotates. Note the impact craters on the icy surface. Hyperion's diameter is about 300 km. It is one moon that is not in synchronous rotation; it may be a fragment of a larger body from a recent collision. (Courtesy NASA)

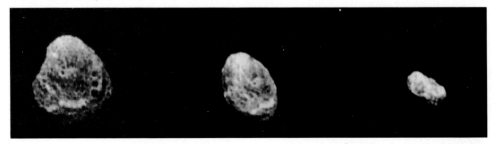

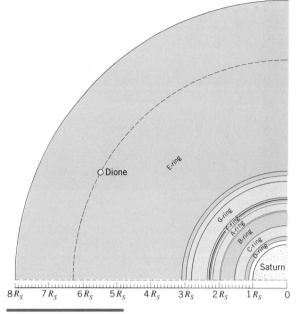

FIGURE 7.24

Saturn's B-ring, showing diffuse spokelike features. These dark markings are about 10,000 km long. (Courtesy NASA)

FIGURE 7.22

Schematic diagram of Saturn's ring system and the orbit of its moon Dione. The view is from above Saturn's north pole. Distances are given in units of the planet's radius, R_S. (Adapted from a NASA diagram)

FIGURE 7.23

Voyager 2 false-color image of Saturn's C- and B-rings, taken from a distance of 2.7×10^6 km. These colors (yellow for the B-ring, blue for the C-ring) are not the actual colors of the particles; rather, image processing was used to emphasize differences in the surface compositions and albedo of the particles in these two rings. Note the many ringlets in both rings. (Courtesy NASA)

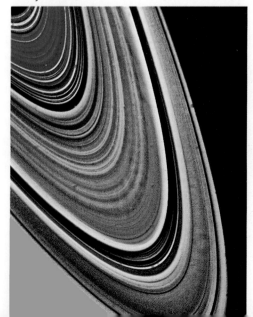

Dark, spokelike features occur in the B-ring (Fig. 7.24). Typically, the spokes are about 10,000 km long and 1000 km wide, and they revolve around the planet. The spokes consist of very small particles, much smaller than the average particle in the rings. The inner particles orbit faster than the outer ones; therefore, the spokes last only a few hours because they orbit like the particles in the rings (which each obey Kepler's laws). So the spokes may be dust-sized, charged particles held in place above the rings by electrical and magnetic forces. The spokes follow part of Saturn's magnetic field, which passes through the rings.

The revolution rate of the rings varies. The speeds range from 16 km/s at the outer boundary of the A-ring to 20 km/s at the inner boundary of the B-ring. These speeds, measured by the Doppler shift (Section 5.1), agree with those predicted by Kepler's third law for individual masses; this agreement proves that separate particles comprise the rings. (If the rings were rigid like a record, the speed of the outer edge would be higher than that of the inner edge.)

Each particle in a ring system orbits the planet following Kepler's laws.

Saturn's rings are likely made of particles of water ice, or rocky particles coated with water ice. The ice does not evaporate because the surface temperature of the particles is only about 70 K. The particles average about 1 meter in diameter, but they range in size from centimeters to tens of meters.

7.5 Uranus

On March 13, 1781, the then unknown amateur astronomer William Herschel (1738–1822) sighted an object with his telescope and "suspected it to be a comet." Observations later in March and in April proved that the orbit was not like a comet's. Herschel concluded that he had discovered a new planet—the seventh in the solar system and the first to be discovered with a telescope. Later the name of Uranus was chosen, after the father of the Titans, giants in Greek mythology.

Atmospheric and Physical Features

At an average distance of 19.2 AU, it takes Uranus 84 terrestrial years to journey around the sun. Uranus' rotation axis lies almost in the plane of its orbit—it rotates on its side. Journeying around the sun in this lopsided manner (Fig. 7.25), Uranus exposes each pole to sunlight for 42 years at a time; night at the opposite pole lasts for an equally long time.

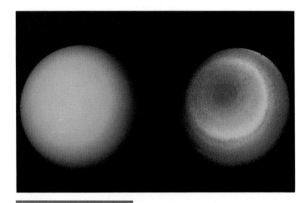

FIGURE 7.26

Uranus from Voyager 2, taken at a distance of 9×10^6 km in January 1986. Image at left has been processed to show the planet as it would appear to the eye; the bluish-green color results from the absorption of red light by methane in the atmosphere. Right-hand image uses false colors to bring out details in the structure of the upper atmosphere. (Courtesy NASA and JPL)

Far from the sun, the upper atmosphere is very cold—a mere 58 K. As on Jupiter and Saturn, the atmosphere contains mainly molecular hydrogen and helium. Uranus has a distinctive pale greenish color. This color is expected from an atmosphere that contains methane gas.

Uranus At a Glance		
	Radius	25,559 km
	Mass	8.66×10^{25} kg
	Bulk density	1270 kg/m^3
	Escape speed	21.3 km/s
	Atmosphere	hydrogen, helium

The Voyager 2 flyby in 1986 provided the first detailed views of Uranus (Fig. 7.26). They showed ammonia clouds lying below a thick layer of haze. The clouds have a delicately banded structure. Winds blow the clouds at a hundred kilometers per second in the same direction as the planet rotates. Voyager 2 found that the rotation at the cloud tops varied from 17 hours at the equator to 15 hours near the poles (opposite the pattern for Jupiter and Saturn).

The low bulk density of Uranus, 1300 kg/m^3, implies that it contains mostly lightweight elements (Fig. 7.27). Overall, Uranus is roughly 15 percent

FIGURE 7.25

The orientation of the spin axis of Uranus as the planet orbits the sun, and the seasons in the southern hemisphere. Note that the axis of rotation lies nearly in the plane of the orbit.

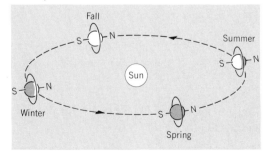

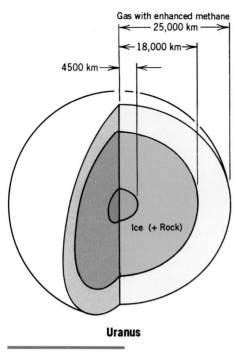

Gas with enhanced methane
25,000 km
18,000 km
4500 km
Ice (+ Rock)

Uranus

FIGURE 7.27

Model for the interior of Uranus, based on Voyager data. The interior consists of three main regions: a very small rocky core (density about 9000 kg/m³), an icy (or watery) region with perhaps some rocky material (density from 5000 to 1000 kg/m³), and a gas layer, mostly of hydrogen and helium, with enhanced concentrations of methane, ammonia, and water (density about 300 kg/m³). The rocky core, if it exists, may have a mass about that of the earth.

hydrogen and helium, 60 percent icy materials (water, methane, and ammonia) and 25 percent rocks and metals.

Moons and Rings

Five moons were known to circle Uranus before the Voyager mission: Miranda, Ariel, Umbriel, Titania, and Oberon. They all move in the planet's equatorial plane and revolve in the same direction in which Uranus rotates. Miranda, the smallest, is the closest to Uranus. The others range in diameter up to almost 1600 km (Titania). Their surfaces appear to be made of dirty ice and to have fairly high albedos—20 to 40 percent. The range of their bulk densities range—1400 to 1700 kg/m³—implies that these are bodies made of rock and ice.

Voyager 2 revealed the surfaces of these moons. Miranda displays the most complex surface, with many different types of terrain jumbled together, including craters, grooved regions, faults, and cliffs 5 km high—certainly the strangest surface seen to date in the solar system (Fig. 7.28). Oberon exhibits a dense cover of large impact craters (an ancient surface) and at least one mountain (possibly volcanic) about 5 km high. Scarps and faults also split the surface. Titania also shows a surface plastered with many impact craters, strewn with valleys 100 km wide and hundreds of kilometers long (Fig. 7.29). One valley cuts across the entire surface! Also

FIGURE 7.28

Miranda imaged by Voyager 2. (a) Overall view of the surface, showing ridges, valleys, and scarps. Smallest features visible are about 3 km in size. The largest impact craters visible are about 30 km in diameter. (b) Close-up of the surface, showing the heavily modified, grooved, and mottled terrain. Note the impact craters in both regions. (Courtesy NASA)

(a)　　　　　　　　　　　　　　*(b)*

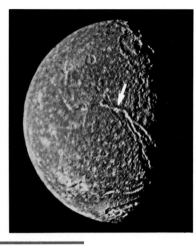

FIGURE 7.29

Titania shows many impact craters, which indicates that the surface is old and unevolved. Note the large fault valleys (arrow); they are about 75 km wide and 1500 km long. At the upper right is a basinlike structure. The overall bland, gray color is typical of the moons of Uranus. (Courtesy NASA)

FIGURE 7.30

Ariel also has valleys (arrow), probably caused by faulting of the surface. Note the light-colored impact craters. (Courtesy NASA)

visible are multiringed impact basins. Ariel has impact craters, large fractures, and valleys (Fig. 7.30). Some regions appeared to have been resurfaced by ice floes. Finally, Umbriel has the least dramatic surface, with overlapping impact craters. (The heavy cratering provides more evidence of an era of torrential impacts early in the solar system's history.)

The large moons of Uranus are made of ice and rock; most have ancient, cratered surfaces, somewhat modified.

Voyager also discovered a slew of new moons—10 that have been named after Shakespearian characters. The largest (called Puck) is only 170 km in diameter. Six moons orbit between Uranus and Puck. These inner moons have diameters between 40 and 80 km. The other moons, which are near the edge of the ring system, all have dark surfaces with low albedos (about 5 percent).

The ring system of Uranus (Fig. 7.31) is dramatically different from that of Saturn. Nine rings circle the planet in roughly three groups out to a distance of 51,000 km from the center of Uranus. The smallest rings have widths of only a few kilometers. The material that makes up Uranus' rings is extremely

FIGURE 7.31

The nine major rings of Uranus, located in the planet's equatorial plane. The view here is toward the northern hemisphere in the orbital plane. (Adapted from a NASA diagram)

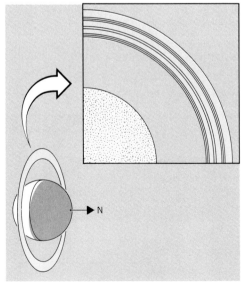

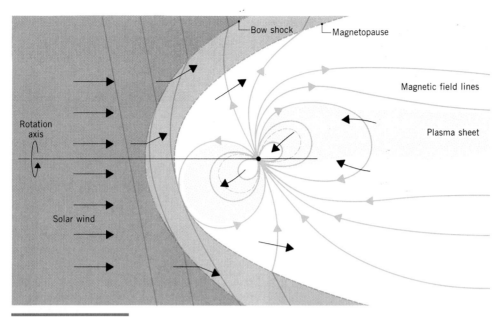

FIGURE 7.32

Model of the magnetosphere of Uranus. Note the large tilt with respect to the rotational axis. Plasma in the solar wind is picked up by the magnetosphere to form a long plasma tail downstream and also to form into Van Allen type belts of trapped, charged particles. A bow shock forms where the solar wind encounters the magnetic field; behind the shock lies the magnetopause.

dark—an albedo of 5 percent. In contrast, the particles in the rings of Saturn, because they are covered with water ice, have an albedo of more than 80 percent. The surfaces of the particles in the rings of Uranus are probably bare of ice and likely made of dark carbon materials. Most of the particles in the rings are a few centimeters across or larger.

Magnetic Field

Voyager 2 showed that Uranus' magnetic field is tilted 59° with respect to the rotational axis (Fig. 7.32), with the north magnetic pole closest to the south geographic one. The magnetic field turns once in about 17 hours, 20 minutes, and that has been taken as the planet's rotation period. (Why the magnetic field has such a large tilt is a puzzle. It also does not center on the core of Uranus!) Uranus' magnetosphere contains more particles with higher kinetic energies than Saturn's. You would expect frequent auroras here, and, indeed, they were observed on the planet's night side.

7.6 Neptune

We didn't know very much about the cold world of Neptune until the Voyager 2 flyby in 1989. The eighth planet from the sun (usually—it is ninth right now as Pluto has moved inside Neptune's orbit), Neptune was discovered in 1846 through its gravitational effects on the motions of Uranus (Section 3.4). The main constituents in the atmosphere are molecular hydrogen and helium; methane makes up a minor amount. The upper atmosphere displays distinct cloud bands.

Infrared observations show that Neptune's temperature is about 60 K, whereas if the planet were

Neptune At a Glance		
	Radius	25,269 km
	Mass	1.03×10^{26} kg
	Bulk density	1640 kg/m³
	Escape speed	23.8 km/s
	Atmosphere	hydrogen, helium

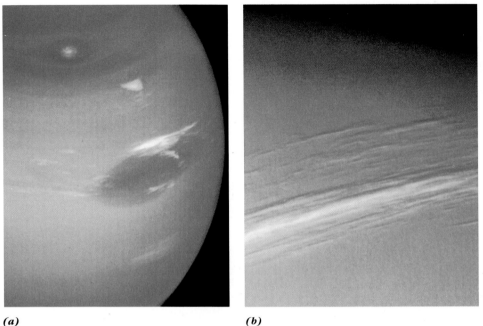

(a)   *(b)*

FIGURE 7.33

Neptune imaged by Voyager 2. (*a*) The Great Dark Spot, accompanied by white, high-altitude clouds. These clouds change their appearances in times as short as 2 hours. Note the lack of a general banded structure in the atmosphere. Relative to the planet's core, the Dark Spot moves westward. (*b*) Computer-enhanced view of white cirrus clouds in Neptune's upper atmosphere. The brighter sides of the clouds face the sun. The height of the clouds is about 50 km; their widths, from 200 to 600 km. (Courtesy the JPL and NASA)

heated by sunlight alone, we would expect 44 K. So Neptune, unlike Uranus, has internal thermal energy. It emits three times as much power as it receives from the sun; the heat flow is some 0.3 watt per square meter. This thermal energy is most likely left over from Neptune's formation.

In August 1989, Voyager 2 skirted a mere 5000 km above Neptune's clouds and imaged conspicuous markings in the upper atmosphere of Neptune. The most striking is known as the *Great Dark Spot* (Fig. 7.33*a*), which is a storm some 30,000 km across rotating counterclockwise in a few days—a region of high pressure. Bright, cirruslike clouds accompany the Dark Spot and appear in some other latitude bands. Most of these clouds change size or shape from one rotation to the next. Suspected to be condensed methane, the clouds lie about 50 km above the general cloud layer (Fig. 7.33*b*). The atmospheric activity seen on Neptune came as a surprise. As on Jupiter and Saturn, this activity is driven by the outflow of Neptune's internal heat. In general,

the atmosphere contains mostly molecular hydrogen and helium.

Voyager tracked radio signals from Neptune's magnetosphere. These signals gave an accurate measure of the rotation period: 16 hours, 3 minutes. A surprise was the discovery that the magnetic axis is tilted about 47° from Neptune's axis of rotation, almost as much as the tilt of Uranus' magnetic axis. The reason for these large tilts is not yet known. The magnetic field strength is about one-fifth that of the earth. The dipole is strangely offset toward the south pole away from the planet's center (Fig. 7.34). The magnetosphere had a very low density of trapped charged particles.

A dynamo model basically accounts for the magnetic field of Neptune, as well as those of Uranus, Saturn, and Jupiter.

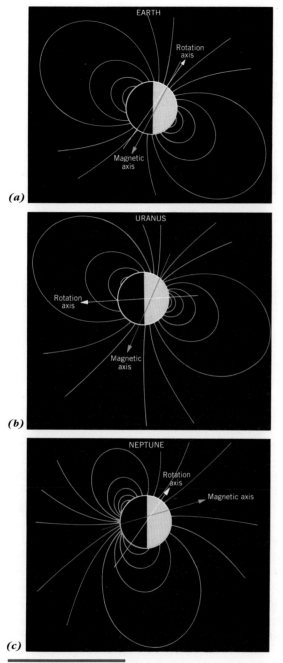

(a)

(b)

(c)

FIGURE 7.34

Comparison of the magnetic fields of the earth (a),
Uranus (b), and Neptune (c). The fields are modeled
as if a bar magnet were set in the planet's interior,
with the angle and position inferred from the
magnetosphere. Note the fields of Uranus and
Neptune, which are offset from the centers of the
planets and grossly misaligned with the rotation axes.

Voyager showed clearly for the first time Neptune's ring system, which contains five individual rings that orbit the equator (Fig. 7.35). The two brightest, outer rings have radii of 53,000 and 62,000 km with a tenuous ring spread between them. Closer in lies a wide ring some 2000 km across. The outer ring is clearly clumpy, with three brighter segments strung along a fainter but complete ring.

We knew of only two moons before Voyager: Triton and Nereid. The larger satellite, Triton, revolves with a period of 5.9 days in a retrograde (east-to-west) orbit that is inclined 23° to the plane of Neptune's equator. Its period of rotation is the same—another moon in synchronous rotation. (Almost all the moons of the Jovian planets are in synchronous rotation because of strong tidal forces: Section 6.1.) Triton has a diameter of about 2700 km, Nereid only some 340 km. Triton is one of the few moons with an atmosphere—a very thin one that contains mostly nitrogen with a trace of methane.

Voyager also captured images of six small moons. They range in size from 50 to 500 km and exhibit the usual rugged surfaces of Jovian moons (Fig. 7.36). The albedos of all Neptune's moons are low, about 5 percent or so. Two of these moons lie close to the outer and middle rings. Except for the smallest and innermost moon (called Naiad), all lie within the plane of Neptune's equator.

Most of the excitement, though, focused on Triton (Fig. 7.37), which displayed a fascinating pink and blue face. The cratering here is not too heavy, which means that the surface must be relatively young and recently modified—subject to meltings and refreezings. On parts of the surface lie frozen ice lakes. Some are stepped, which suggests a series of meltings and freezings. But in general the surface relief is quite low—less that 200 meters.

Triton's surface has been modified by volcanism.

Near Triton's south pole (which is now in a summer season), the surface ice (consisting of methane and nitrogen) appears to have evaporated in spots (Fig. 7.37). In other regions, small flows have filled valleys and fissures—slowly moving glaciers of methane and nitrogen. In other sections, the icy surface appears to have melted and collapsed. Dark,

FIGURE 7.35

The rings of Neptune; note how fuzzy the inner one appears (arrow). The black bar across the center is an artifact in the image used to block out the glare of the light from Neptune. (Courtesy the Jet Propulsion Laboratory and NASA)

FIGURE 7.36

Neptune's moon, Larissa, a mid-sized moon with a dark, icy surface. (Courtesy the Jet Propulsion Laboratory and NASA)

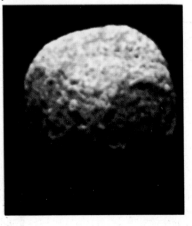

elongated streaks tens of kilometers across seem to be the trails of unusual ice volcanoes. Just below the surface, the pressure is high enough to liquefy nitrogen ice. When the surface cracks, the liquid can burst out and turn to gas, which shoots several kilometers above the surface.

Triton's bulk density is 2100 kg/m^3, which implies a rocky core surrounded by a mantle of methane and water ice. The icy surface, which is primarily molecular nitrogen ice, has an average albedo of roughly 70 percent. Triton's orbit results in a seasonal cycle of 165 years; summer in the southern hemisphere has already lasted some 30 years. Still, at this great distance from the sun, the surface remains so cold that regions of nitrogen ice are visible.

 ## 7.7 Pluto and Charon

Early in the twentieth century, Percival Lowell became fascinated with the problem of a planet beyond Neptune (called *Planet X*) and initiated a search program at Lowell Observatory. After Lowell died in

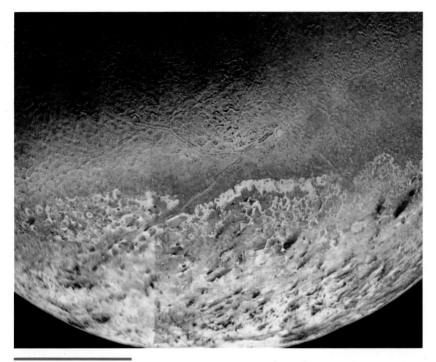

FIGURE 7.37

Triton. High-resolution mosaic of Triton's hemisphere that faces Neptune. The southern pole cap (at bottom, center) is slightly pink. From the ragged edge of the pole cap northward (up), the surface is redder and darker. (Courtesy the Jet Propulsion Laboratory and NASA)

1916, the search for Planet X was stopped until 1929, when Clyde W. Tombaugh began the work anew. Astronomers had assumed that Planet X would be similar to Neptune because of irregularities in Neptune's orbit (in analogy to the way Neptune was discovered: Section 3.4). So they had searched for a visible disk of a new planet. Instead of searching for a disk, Tombaugh looked for evidence of Planet X's motion relative to the background stars. In February 1930, he noted on different photographs two images that had shifted slightly in Gemini—there was a body orbiting the sun beyond Neptune! The discovery was announced on March 13, 1930—Lowell's birthday. The new planet was named Pluto.

Methane ice coats some of Pluto's surface. The presence of methane ice there means that the surface is bitter cold, no more than 40 K—the temperature expected from solar heating alone. Observations of Pluto's brightness have uncovered a cycle (because of patchiness of the ice) lasting about 6.4 days; this variation is generally accepted as Pluto's rotation period. The overall albedo is about 50 percent.

FIGURE 7.38

Pluto and Charon. At the time of this Hubble Space Telescope observation done with the Faint Object Camera (FOC), Charon (lower left) was about 0.9 arcsecond from Pluto. (Courtesy NASA and the European Space Agency)

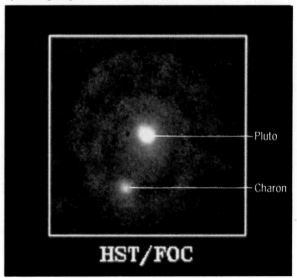

HST/FOC

On September 5, 1989, Pluto reached perihelion in its orbit. A little more than a year earlier, on June 9, 1988, Pluto passed in front of a star as observed from Australasia. Coordinated observations confirmed that Pluto possesses an atmosphere, which stretches over 600 km from the planet's surface. This atmosphere probably consists of methane gas (with a surface pressure of a mere 10^{-6} atm or so) that has been released from the surface ice as the planet is heated by its closest approach to the sun in 248 years.

In June 1978, James Christy of the U.S. Naval Observatory in Flagstaff, Arizona, noticed in several photos the faint image of Pluto's moon partially blended with the image of the planet. This moon was named *Charon*, after the mythical boatman who ferried souls across the river Styx to Pluto for judgment. The Hubble Space Telescope (Appendix H) took an image that clearly showed both Pluto and Charon (Fig. 7.38).

Recently, Pluto and Charon came into an orientation in which they eclipsed each other as seen from the earth. (Such alignments occur only twice in 248 years.) These eclipses show that Pluto has a diameter of about 2300 km and Charon roughly 1200 km. The observations of Charon show a revolution period of 6.4 days. That Charon's revolution period is the same as Pluto's rotation indicates that the two bodies are tidally locked and keep the same sides toward each other. We can use Charon to find Pluto's mass by Kepler's third law. The result: Pluto has a mass about 0.002 that of the earth.

Pluto is believed to have a bulk density of some 2100 kg/m^3. Pluto may contain as much as 75 percent rocky material. That contrasts to the Jovian moons, which have a larger percentage of water ice. Charon's density might fall as low as 1000 kg/m^3; recent infrared observations revealed water frost on its surface. So Charon may consist mostly of icy materials, like the moons of Saturn. The eclipses indicate that Charon has a dark gray surface in contrast to Pluto's reddish one.

Pluto and Charon are small bodies, orbiting a common center of mass; they made be made of different materials.

The Larger View

Our comparative planetology has focused here on the giant Jovian worlds of Jupiter, Saturn, Uranus, and Neptune. All are alien places, but we can still apply the chief concepts for building models of planets: bulk density for overall composition, internal structure from a layering of materials of different densities, dynamos for magnetic fields, and outward heat flow that spurs change. We can even apply these concepts to the Jovian satellites— worlds unto themselves, many with ancient cratered terrains, some that have been modified.

Compared to the terrestrial planets, the Jovian planets are primitive places, clinging to their pasts. They look pretty much the same now as they did 4 billion years ago. They have kept their original atmospheres. They do not have solid surfaces where volcanoes can vent and erosion can change the landscape. Infalling debris from space does not blast out any craters. The atmospheres are stormy but the changes are ethereal. These planets emit more heat than they receive from the sun, so we know that they store thermal energy. The loss of heat to space drives changes, but they do not leave a record. In that sense, the Jovian planets are forever young.

Despite the differences between terrestrial and Jovian planets, we are able to build models from our observations of these gaseous and liquid worlds with compositions like the sun. How did these planetary differences occur? We will find out in building a model for the formation of the solar system in Chapter 8.

Key Ideas and Terms

comparative
   planetology
density
differential rotation
differentiated
Jovian planets

1. As a group, the Jovian planets differ most from the terrestrial ones based on chemical composition, as well as on lower bulk densities, greater diameters, and greater masses. The Jovian planets largely consist of hydrogen and helium, with common molecules of methane, hydrogen, ammonia, and water, and are mostly fluid in their differentiated interiors.

atmospheric
   temperature
escape speed

2. The large masses and low temperatures of the Jovian planets imply that these bodies have retained their atmospheres of formation for billions of years.

belts
convection
heat
power
thermal energy
watt
zones

3. Jupiter, Saturn, and Neptune emit into space, as infrared radiation, more power than they acquire from incoming sunlight. This is because their interiors are hotter than would be possible if the sunlight were the sole source of their heating. The outflow of thermal energy causes convection in the atmospheres, with regions of strong upflows and downdrafts.

dynamo model
magnetic field
magnetosphere
metallic hydrogen
plasma
solar wind

4. Jupiter, Saturn, Uranus, and Neptune possess strong planetary magnetic fields, which implies that they have liquid, conducting interiors—but their compositions rule out metal cores. Their rapid rotation drives have organized currents in the conducting regions that generate planetary dynamos.

density
heat
impact cratering
intense
   bombardment
Galilean moons
thermal energy
volcanism

5. Jupiter's largest moons decrease in density outward from the planet. These Galilean moons have undergone different amounts of crustal evolution, as indicated by the extent of their impact cratering—Io the most evolved from volcanism, and Callisto the least. Io is now the most volcanically active body in the solar system. Its surface has been extensively modified, and so this moon is relatively young. The volcanism it shows now implies a hot interior.

albedo
Kepler's laws
ring system

6. The rings of Jupiter are thin and consist of small particles. The rings of Saturn are wide and thin; they likely consist of larger, icy particles, and contain many small ringlets. The rings of Uranus and Neptune are thin and narrow and consist of very dark, small particles. All ring particles orbit the planets around the equators and obey Kepler's laws.

albedo
density
impact cratering
intense
  bombardment

7. Except for Titan, Saturn's largest moons are basically ice and have evolved somewhat since the time of their formation, as indicated by their modified impact craters.

atmospheric
  composition
density
differentiated

8. The interior structures of Uranus and Neptune differ, while their atmospheres are pretty much the same in composition. Neptune has more violent weather than Uranus.

albedo
center of mass
Kepler's laws

9. Pluto is a small, icy world, with a methane-coated surface. Its moon, Charon, is about half Pluto's size and a fraction of its mass. The two bodies orbit a common center of mass as a double-planet system.

Review Questions

1. How does the density of Jupiter compare to that of the earth?
2. What is the major difference between the interior of Jupiter and the interior of the earth?
3. What two factors result in the Jovian planets having retained their atmospheres for billions of years?
4. Are the belts of Jupiter regions of convective upflow or downflow?
5. Compared to the earth's magnetosphere, is that of Jupiter larger or smaller?
6. What part of Jupiter's interior is electrically conducting, as required to generate a planetary dynamo?
7. Which of the Galilean moons of Jupiter has the *lowest* density?
8. Are many impact craters visible on the surface of Io?
9. Compared to a particle in a ring close to a planet, does a particle at the outer edge of the ring orbit with a shorter or a greater period?
10. What is the typical bulk density for a moon of Saturn (excluding Titan)?
11. How does the density of Neptune compare to that of Uranus?
12. What is the orbital period of Pluto and Charon around their center of mass?

Conceptual Exercises

1. In what significant respect is Jupiter most different from the other Jovian planets?

2. Suppose you flew very close by Jupiter. What outstanding features would you see in the atmosphere?

3. How do we know the bulk density of Pluto?

4. How do we know that the rings of Saturn are thin?

5. What fact makes it relatively easy to find the masses of the Jovian planets?

6. In two sentences, compare the rings of Saturn to those of Jupiter and to those of Uranus and Neptune.

7. How is Jupiter's magnetic field similar to the earth's? How is it different? Answer the same questions for Saturn's magnetic field.

8. What features does Pluto have in common with the Galilean moons?

9. In one short sentence, describe the interior compositions of the moons of the Jovian planets.

10. In general, what does a heavily cratered surface indicate about the evolution of the surface and its relative age?

11. In what ways are the magnetic fields of Uranus and Neptune very different from those of Jupiter and Saturn?

8

Solar System Debris and Origin

■

Central Concept

The planets formed from an inter-stellar cloud of gas and dust as a natural extension to the formation of the sun.

■

How did the solar system originate? Because we don't know in detail, the origin of the solar system still arouses curiosity among astronomers. Many models have been proposed. None has been completely successful, though some aspects work out well. Any model, however, influences our notion of whether other planetary systems orbit other stars.

The current approach to building models for the origin of the solar system views the formation of the planets as a natural result of the birth of the sun from an interstellar cloud of gas and dust. The general outline of this process explains the routine features of the solar system. Many details and puzzles remain to be resolved with respect to the planets and solar system debris—which is a fruitful hunting ground for physical and dynamical clues to its past.

8.1 Asteroids

Many comets, meteoroids, and asteroids orbit the sun in the space between the planets. These bodies, along with gas and dust, make up the interplanetary debris. This rubble provides fossil evidence about early times in the solar system's history. Let's look first at the asteroids, then comets and meteoroids. The emerging story is that these objects are interrelated.

An **asteroid** is an irregular, mostly rocky hunk, small both in size and in mass compared to a planet. Ceres, the largest known asteroid, has a diameter of almost 1000 km. (That's about one-third the size of our moon.) Most asteroids orbit the sun in a region between Mars and Jupiter, at an average distance of 2.8 AU. These bodies make up the **asteroid belt** within the plane of the solar system. More than 5000 asteroids have been discovered so far, and perhaps 50,000 await future sightings.

Asteroids are small, solid bodies with irregular shapes.

Most asteroids fluctuate in brightness because of their rotation—proof that they have irregular surfaces and shapes. The Galileo spacecraft, on its way to Jupiter, captured an image of the asteroid Gaspra (Fig. 8.1). It has a lumpy shape, some 20 km by 12 km by 11 km, and a cratered surface; the smallest craters are 300 meters wide. Gaspra rotates once every 7 hours. Large chunks of the asteroid have been struck off in past collisions—probably typical events in the life of an asteroid. Deimos and Phobos, the moons of Mars (Section 5.7), resemble Gaspra, strengthening the idea that they are asteroids captured by Mars. Such a capture may well have happened because Mars orbits just inside the asteroid belt.

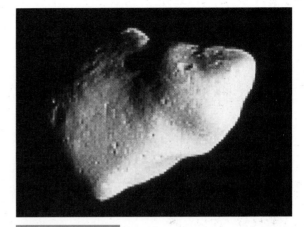

FIGURE 8.1

The asteroid Gaspra, imaged from a distance of 16,000 km by the Galileo spacecraft. The illuminated part of the asteroid is about 16 km by 12 km. The smallest craters visible here are about 300 meters across. (Courtesy the Jet Propulsion Laboratory and NASA)

The albedos of asteroids are placed in two main classes based on surface composition. Some asteroids are relatively bright (albedos of about 15 percent) and others are much darker (albedos of some 5 percent), an indication that they contain dark compounds such as carbon. Those in the lighter class, which have surfaces of silicate materials, are dubbed **S-type** asteroids; the darker carbon-rich ones, **C-type**. (Gaspra is an S-type asteroid.) A third, smaller class, called **M-type**, appear to be made of metallic substances. They have albedos of about 20 percent.

The albedo of an airless solid body provides clues to the physical composition of its surface; most asteroids are made of rocks and metals.

Compositions in the asteroid belt vary with distance from the sun. Near the orbit of Mars, almost all asteroids have S-type characteristics. Farther out, we find fewer high-albedo objects and more darker

Asteroids	Orbits	most in asteroid belt between Mars and Jupiter
	Sizes	small, less than 1000 km diameter
At a Glance	Shapes	irregular
	Surfaces	cratered
	Composition	rocks and metals

(a)

FIGURE 8.2

Spectacular comets. (*a*) Comet Ikeya–Seki in 1965, visible in the twilight sky. The tail develops from gas and dust expelled by the nucleus. (Courtesy U.S. Naval Observatory) (*b*) Comet West in March 1976, after perihelion passage. Note the separation of the tail into a gas (plasma) part and a dust one, which spreads in a wide fan. The gas tail appears bluish from its emission; the dust tail reflects sunlight and so appears yellowish. The plasma tail is blown back by the solar wind; the dust particles fall behind in their own orbits around the sun. (Courtesy Dennis DiCicco)

(b)

ones. At the outer edge of the belt, 3 AU from the sun, some 80 percent of the asteroids are C-types with surfaces like soot.

8.2 Comets

You probably associate **comets** with long, graceful tails. In fact, few comets exhibit tails—even at perihelion—and when far from the sun, comets do not have visible tails. When first sighted, a comet typically appears as a small, hazy dot. This bright head of the comet is called the **coma**; the **tail** develops from the coma (Fig. 8.2a). The coma contains the source of all comet material, called the **nucleus**. Cometary nuclei are only a few kilometers across, as confirmed by spacecraft missions to Halley's comet.

Physical Properties and Orbits

As a comet nucleus heads toward the sun, it is heated, brightens, and sprouts a tail. A comet's tail may stretch for millions of kilometers and the gaseous part of it points away from the sun. Comets have tails of two types: gas and dust (Fig. 8.2b). The gas tail is largely ionized as a plasma; the emission from the ionized gas makes the tail appear bluish to the eye. The dust tail shows up by sunlight reflected off dust expelled from the coma; visually, the dust tail has the same yellowish color as the sun. The pressure from sunlight pushes dust from the coma to form a tail. (Light can be thought of as particles—photons. Photons reflected off dust apply a pressure to the surface. If, then, a dust speck has a low enough mass, photons can push it.)

The plasma tails point away from the sun because they are blown by the solar wind, which contains ions carrying magnetic fields (Section 4.5) moving at high speeds through interplanetary space. The magnetic fields interact with the plasma of the gas tail and give a comet its stretched-out form—an interplanetary windsock!

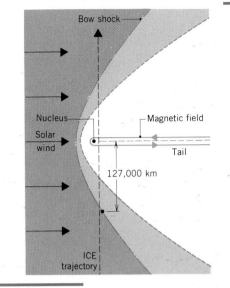

F I G U R E   8 . 3

The interaction of the solar wind and Comet Giacobini–Zinner, from observations by the ICE spacecraft. The ICE trajectory crossed the cometary region just behind the nucleus to measure the interaction of the comet with the solar wind, which produces a bow shock wave. Magnetic fields carried by the solar wind drape around the nucleus. These contain the plasma expelled from the nucleus to form the tail.

The coma, plasma tail, and dust tail of a comet originate from the material expelled from the comet's nucleus.

Spacecraft observations confirm that the magnetic field carried by the solar wind indeed drags ions from the coma. (Recall from Section 4.5 that a plasma such as the solar wind can carry along a magnetic field.) The International Cometary Explorer (ICE) mission through the tail of Comet Giacobini–Zinner in September 1985 revealed the interaction between comets and the solar wind (Fig. 8.3). The interaction of the solar wind and the coma produces energetic ions, which are carried away by the solar

Comets At a Glance	Orbits	elliptical for periodic ones
	Size of nucleus	very small, less than tens of kilometers
	Shapes	irregular
	Surfaces	craters and vents; small hills
	Composition	rocky with carbon, embedded ices (water)

wind's magnetic field to make a long ion tail. The charged particles and the magnetic fields generate a large current flow, which fixes the form of the visible comet. Once a current has been established, the magnetic field is self-sustaining. Without the magnetic field interactions, we really wouldn't have showy comets!

Comets are solar wind telltales, with magnetic fields and ionized gases playing the major role in their structure.

For all their stunning length in the sky, comets have very small masses. They lose some mass in every orbit because gas flies off at greater than escape speed. Halley's comet, one of the largest, has an estimated mass of only about 10^{16} kg and loses about 10^{11} kg during each perihelion passage, only about 0.001 percent of its total. With so little mass, a comet gets its spectacular display only by spreading itself very thin.

FIGURE 8.4

The dirty snowball cometary model (simplified!). The nucleus—about 10 km in diameter—is a pudding of ices and rocky material encrusted by a thin, rocky crust. Sunlight heats the nucleus and vaporizes the ices to produce dusty gas jets. These materials make up the coma and then are blown out into the tail; both are much larger than the scale here. Note that the gas jets do *not* directly produce the tail; rather they serve to release material into the coma. The nucleus may have a very irregular shape and rotate about its center of mass once every few days.

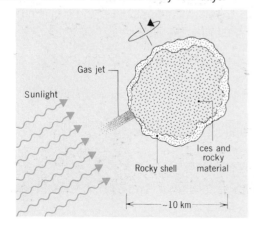

What is the nature of the nucleus? In 1950 astronomer Fred L. Whipple developed the **dirty snowball comet model** (Fig. 8.4). It pictures cometary nuclei as compact bodies made of frozen gases—ices—of water, carbon dioxide, ammonia, and methane, embedded within rocky material. The ices have a lot of dust mixed in them. When a comet nears the sun, some of the icy material vaporizes on the sunlit side and generates dusty gas jets, forming the coma—a cloud of (mostly) ionized gas and dust. More vaporizing enlarges the coma; the solar wind blows back the ions to create the plasma tail, sunlight the dust tail. As the ice evaporates, a thin coating of rocky material remains to form a solid, but fragile, crust on the nucleus. From the missions to Halley's comet, we know that this idea is basically correct, with the addition of lots of carbon material.

Comets lose some of their solid nuclear material with each passage near the sun; eventually they deteriorate into irregular asteroidlike bodies, high in carbon.

Comets that return periodically to the sun (not all do!) have elliptical orbits and are called **periodic comets**. A periodic comet eventually (in a million years or less) suffers one of three fates: it dissipates or breaks up; it collides with a planet; or, it is ejected out of the solar system into interstellar space by a close, gravitational encounter with a planet. The periodic comets we see, which are eliminated rapidly compared to the age of the solar system, must be very recent additions to the inner solar system. Hence, there has to be some reservoir, which continues to supply them. Astronomers envision a shell-like cloud of cometary nuclei, reaching as much as 50,000 AU from the sun, as this supply. This cloud, called the **Oort Cloud** (after the Dutch astronomer Jan Oort), may contain 10^{12} cometary nuclei.

Halley's Comet

When King Edward the Confessor died in 1066, he left no direct heir to the English throne, and the nobles chose Harold II as their king. In the same year, a bright comet streaked across the sky—the commonly accepted sign of a ruler's death and misfortunes to follow. Meanwhile, William of Nor-

FIGURE 8.5

Halley's comet in A.D. 1066, from a section of the Bayeux Tapestry. The comet is at the top right. (The Bettmann Archive)

mandy, who believed that Edward had designated him as the next king of England, cleverly interpreted the comet as a sign in his favor (Fig. 8.5). With this psychological edge for his men, he sailed to the Brit-

ish Isles and conquered the dispirited Saxon armies near Hastings. The comet of 1066 turned out to be one appearance of a comet that returns regularly.

Cometary orbits remained unknown until Edmond Halley calculated the orbits of the comets of 1531, 1607, and 1682 by a method devised by Newton. Halley found, to his amazement, that the orbits were almost identical: Noting that the comets appeared at intervals of approximately 75 years, Halley concluded that these several comets were in fact one. He correctly predicted it would return around 1758. This comet indeed was sighted on Christmas night in 1758 and was named in Halley's honor. Halley's comet (Fig. 8.6) was the first comet to be recognized as a permanent member of the solar system, with an elliptical, periodic orbit. The spectacular comet of 1066 was none other than Halley's comet.

Halley's comet is the granddaddy of all known *colloq.* comets. Its passage near the sun has been recorded at least 29 times, as far back as 239 B.C. Halley's comet moves in an elongated ellipse (Conceptual Activity 8). Having passed aphelion (farthest distance from the sun) beyond Neptune's orbit in 1948, it returned to the earth's neighborhood and reached perihelion on February 9, 1986, at a distance of 0.59 AU. After rounding the sun, the comet passed by closest to the earth (0.42 AU) on April 11, 1986. It returns again in the year 2061.

FIGURE 8.6

Development of the tail of Halley's comet. (*a*) In December 1985, the comet shows first signs of a tail. The background streaks are star trails in this time exposure. (Image by D. Malin; courtesy Anglo-Australian Telescope Board) (*b*) Halley's comet in March 1986; note the complex structure of streams in the plasma tail. This photo was taken shortly before the Giotto spacecraft's closest approach to the nucleus. (Image by D. Malin; courtesy Royal Observatory, Edinburgh)

(a) *(b)*

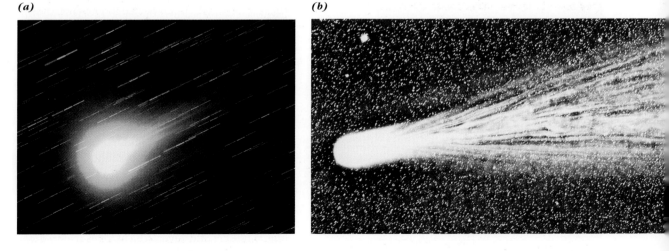

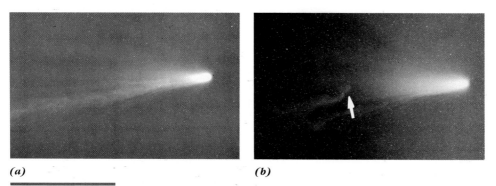

(a)   *(b)*

FIGURE 8.7

Magnetic reconnection event in the tail of Halley's comet, photographed on March 9, 1986 (a) and March 10, 1986 (b). During this time, a part of the gaseous coma (arrow) was blown down the tail when the comet crossed a change in the magnetic polarity in the solar wind. (Photos by E. Moore; courtesy Joint Observatory for Cometary Research, New Mexico Institute of Mining and Technology, and NASA)

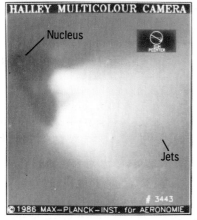

(a)

FIGURE 8.8

Giotto images of Halley's comet. (a) Image of jets and nucleus just before destruction of the camera onboard Giotto. "Sun pointer" indicates the direction to the sun. (b) Composite, detailed image of the nucleus from the Halley multicolor camera exposures on March 13, 1986. (Courtesy and copyright 1986, Max-Planck Institut für Aeronomie, Lindau/Harz, Germany)

(b)

This time round marked the first opportunity of astronomers to probe a comet and discover its true nature. Ground-based work resulted in photos that revealed details of the solar wind plasma in its interaction with the gases from the comet. Occasionally, the tail breaks off and is swept away by the solar wind. As the old tail leaves, a new plasma tail grows in its place—a snakelike shedding of gas (Fig. 8.7). Magnetic reconnection works here as it does in the earth's magnetosphere (Section 4.5).

In space, an armada of flybys penetrated Halley's comet for close-up views. Launched by the European Space Agency (ESA), the Giotto probe swooped within 600 km of the core of Halley's comet on March 14, 1986. Well before then, Giotto had sampled the environment around the comet's head, where gases are ionized. Most of the ions were related to water; in fact, most of the comet's gas was water vapor. The gas contained an unexpected amount of carbon compounds.

Giotto took its last picture from a distance of about 1700 km and revealed a strange realm, dominated by dust (Fig. 8.8). The peanut-shaped nucleus is about 10 km wide and 12 km long. And its surface is dark—black as velvet, an albedo of only a few percent. The surface also appeared rough, with at least one hill a few hundred meters high and about a kilometer wide at its base. Craterlike structures about a kilometer in size were noted in some regions. By keeping track of these structures, it was learned that the nucleus rotates once every 53 hours.

Before the Halley missions, most astronomers had imagined that a comet's nucleus was a dirty snowball of ices and dust. They were basically right, but the amount and nature of the dust surprised everyone. The sunlit side of the comet blows off heated materials—ice that vaporizes to gas, and the dust with it. These blow off in jets only a few kilometers wide—at least nine were visible during the Giotto encounter. The surface sources of the jets were less than a kilometer in diameter, and the outspray contained about 80 percent water vapor and 20 percent dust. And that dust was almost all carbon (like soot), which is why the surface of the nucleus is dark like coal.

Comets have dusty, sooty nuclei of irregular shape and containing icy materials; when heated by the sun, the nuclei expel jets of dusty gases from vaporized ices.

You can imagine that a comet exhausts its ices after many passages around the sun. What remains? An irregular, solid body containing lots of carbon, with a close resemblance to a C-type asteroid.

8.3 Meteors and Meteorites

A **meteor** is the flash of light from the entry of a solid particle into the earth's atmosphere (Fig. 8.9). As it plunges through the air, the particle is burnt up by friction and leaves a bright trail behind it. Before the particle meets its fiery doom in the upper air, it is called a **meteoroid**: a solid object traveling through interplanetary space. Of course, other objects (comets, asteroids, and planets) also travel through the interplanetary void; a meteoroid differs from those chiefly in its small size—no more than a few meters in diameter, usually much less. A mete-

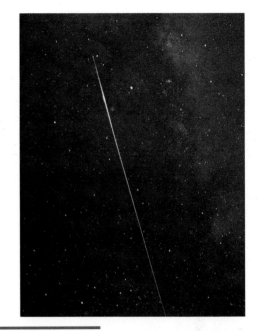

FIGURE 8.9

Bright fireball of the Perseid meteor shower in 1988. (Courtesy Steve Traudt)

oroid that survives its plunge through the atmosphere and strikes the earth's surface is then called a **meteorite**. Most meteoroids consist of fragile, delicate particles (Fig. 8.10) that crumble quickly in their contact with the air.

The dirty snowball comet model connects comets with most meteors. During a comet's passage by the sun, dust and solid particles interspersed in the ices

FIGURE 8.10

Particle of interplanetary dust, probably debris from a comet; it is about 10 μm in size. (That's about a thousandth the size of the period at the end of this sentence.) Note its open, fragile structure. (Courtesy D. Brownlee, University of Washington)

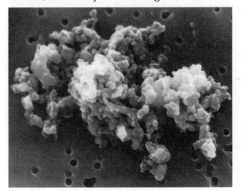

flake off and scatter in an untidy array around the comet. This solid debris is very fragile and has a low density. The older the comet and the greater its number of passages by the sun, the greater the loss of icy material and release of meteoroid material. About 99 percent of meteors (the low-density ones) are of cometary origin. (The remainder—ones of high density—are associated with asteroids; they are the ones that survive passage through the air.)

Because we intercept the orbits of old comets at the same place in the earth's orbit, showers of meteors are usually annual events. The best known is the *Perseid meteor shower*, which peaks around August 12. The debris of the *Leonid meteor shower* orbits in clumps; we pass through the densest region every 33 years. The next great Leonid show will ensue on November 17, 1999. Watch for it!

Types of Meteorites

Meteorites fall into three broad classifications: irons, stones, and stony-irons. The **iron meteorites** (irons), which are generally about 90 percent iron and 9 percent nickel, with a trace of other elements, are the most common. They are easy to identify because of their high density and obviously melted appearance. The **stony meteorites,** or stones, are composed of light silicate materials similar to the earth's crustal rocks. When examined under a microscope, many stones are seen to contain silicate spheres, called **chondrules**, embedded in a smooth matrix (Fig. 8.11). These stones are known as **chondrites**. Finally, **stony-iron meteorites** represent a cross between the irons and the stones and commonly exhibit small stone pieces set in iron.

Meteorites are samples of original solid materials of the solar system; the carbonaceous ones are probably the best samples of the most primitive materials.

F I G U R E   8 . 1 1

Chondrules, the round glassy (silicate) spheres found in chondrites. These chondrules, embedded in a finer matrix, are a few millimeters in diameter. (Courtesy O. Richard Norton, Science Graphics)

F I G U R E   8 . 1 2

Carbonaceous chondrite from Allende, Mexico. This close-up shows the many individual chondrules that make up such a meteorite. Note the very dark color. (Courtesy O. Richard Norton, Science Graphics)

A most curious kind of chondrite is the **carbonaceous chondrite** (Fig. 8.12). Their chondrules are embedded in material that contains a large amount of carbon compared with other stony chondrites— typically about a few percent carbon with respect to the total mass. This high carbon content results in a low albedo. Carbonaceous chondrites also contain about 10 percent water and volatile materials. These are chemically quite different from the earth but

Meteorites	Sizes	centimeters to meters
	Shapes	irregular
At a Glance	Composition	rocks (some with high carbon content) and metals
	Sources	comets and asteroids

very similar in composition to the sun. This suggests that carbonaceous chondrites come from out of solarlike original material and have suffered no major heating since their formation.

Origin of Meteorites

Most meteorites have too great a density to come from comet-related meteoroids. Rather, they resemble the inferred physical characteristics of asteroids. Orbits of meteoroids that have been determined are seen to be like those of asteroids rather than comets. Collisions between asteroids fragment them, and these pieces are the source of meteorites.

An important clue to the origin of meteorites is obtained when the surface of an iron meteorite is polished and etched with acid. After this treatment, large crystalline patterns, called **Widmanstätten figures,** become visible (Fig. 8.13). Terrestrial iron does not show such patterns when etched. A nickel–iron mixture that is cooled slowly under low pressures from a melting temperature of about 1600 K, will form large crystals.

The key point here is that the cooling must be very gradual. But metals conduct heat well, and in the cold of space, a molten mass of nickel and iron would cool rapidly, without forming large crystals. So for nickel–iron meteorites to grow Widmanstätten figures requires protection from the cold. It's likely that nickel–iron meteorite material solidified inside small bodies, termed **parent meteor bodies**. To allow a slow cooling, these were at least 100 km in diameter.

FIGURE 8.13

Widmanstätten figures in a nickel–iron meteorite from Toluca, Mexico. (Courtesy O. Richard Norton, Science Graphics)

Meteorite material originated in larger bodies that formed early in the solar system's history.

Parent meteor bodies may have been only a few hundred kilometers across. Once formed, they could be heated by the radioactive decay of short-lived isotopes. When melted, a parent meteorite body differentiates—the densest material falls to the center, and the least dense froths to the surface. So the object ends up with a core of metals and a cover of rocky material, which cools to form a crust. This insulates the molten metals and allows them to cool slowly and form Widmanstätten figures. Much later, the parent meteor bodies collide and fragment. Pieces from the outer crust make stony meteorites, pieces from farther down become stony-iron meteorites, and the core produces the iron meteorites.

Are any parent meteor bodies around now? Yes—as asteroids. Recall that there are three main asteroid classes: the dark C-type, containing much carbon; the lighter S-type, composed of silicate materials; and the M-type, with metallic characteristics. These probably are related to the carbonaceous chondrites, the stony meteorites, and the irons. So parent meteor bodies are changed mainly by collisions among themselves.

Since the parent meteor bodies were among the first solid objects to form in the origin of the solar system, so the ages of meteorites should provide a direct indication of the age of the solar system. Meteorites can be dated by using radioactive decay techniques; such methods give ages very close to 4.6 billion years. Meteorites provide a direct, reliable estimate of when the solar system formed.

The radioactive dating of meteorites tells us that the solar system formed about 4.6 billion years ago.

Now that we have dug into the remains of the solar system's formation, let's turn to the puzzle of how it came about, using information about debris presented here and about the planets and their moons, as developed in Chapters 4 through 7.

8.4 Nebular Models of the Formation of the Solar System

We have only one solar system to examine close up! Its regularities in chemical and dynamic properties provide patterns that result from its origin. A successful model must explain as many of these dynamic and chemical properties as possible in some internally consistent, simple fashion. Let's see what these are.

Chemically, the solar system falls into three broad categories of material: gaseous, rocky/metallic, and icy (Table 8.1). Each group is distinguished by its melting point. The gaseous and icy materials, which together are called *volatiles*, are generally gaseous under the conditions presumed during the solar system's formation. The bodies of the solar system are composed of various combinations of the three groups. Though one class of materials may dominate, any solar system body contains some of each group. For example, the sun contains mainly gaseous materials and also icy/rocky materials, but as gases (because the sun is hotter than 2000 K), not solids. The terrestrial planets and the asteroids are mainly rocky/metallic; Jupiter and Saturn mostly gaseous; Uranus, Neptune, Pluto, and Charon, and comets mostly icy.

The solar system displays a regular structure in terms of its dynamic properties, those that relate to its motions. These major patterns are as follows.

1. The planets revolve counterclockwise around the sun; the sun rotates in the same direction.

2. The major planets—except Mercury and Pluto—have orbital planes that are only slightly inclined with the plane of the earth's orbit; that is, the orbits are *coplanar*.

3. Except for Mercury and Pluto, the planets move in orbits that are very nearly *circular*.

4. Except for Venus, Uranus, and Pluto, the planets rotate *counterclockwise*, in the same direction as their orbital motion.

5. The planets' orbital distances from the sun follow a *regular spacing*; roughly, each planet lies twice as far out as the preceding one.

6. Most satellites revolve in the *same direction* as the rotation of their parent planets and lie close to the equatorial plane of their parents.

Most models today are variations of **nebular models**, in which the sun condenses from an interstellar cloud of gas and dust that also forms a disk, a **solar nebula**, out of which the planets condense. The sun forms in the center of a flattened cloud. The nebular models see the solar system as a natural outcome of the sun's formation and, perhaps, of any star's formation. If nebular models are correct, planetary systems may be very common in our Galaxy—and in other galaxies.

In nebular models, the sun and planets formed out of a cloud of gas and dust.

That's the basic picture and the nub of the problem: how to get the planets, in their present orbits, out of an originally diffuse cloud. The problem has two parts: (1) how to make a flat solar system and (2) how to get the planets to grow out of the cloud.

Table 8.1

General Classes of Solar System Materials

Class	Materials	Melting Temperature (K)
Rocky and metallic	Iron, iron sulfide, silicates, *etc.*	500–2000
Icy	Water, methane, ammonia, *etc.*	100–200
Gaseous	Hydrogen, helium, neon, argon, *etc.*	0–50

Angular Momentum and the Solar Nebula

Consider again (Section 6.5) the conservation of angular momentum. The basic point is this: once a body has started spinning, it will keep on spinning as long as no torque acts on it. If, by itself, the body changes size—for instance, if it contracts gravitationally—it will naturally spin faster to keep its angular momentum the same. (A mass contracts gravitationally when it pulls itself together by the gravitational forces between the particles that make it up—a process called **gravitational contraction**.)

A familiar example of the conservation of angular momentum is the spinup of an ice skater. She goes into the spin with her arms outstretched. As she pulls her arms in, her spin speeds up. But if she brought her head down to her chest, nothing would happen. Why not? Because angular momentum affects only the motion *around* a spin axis, not along it.

Consider now a large spherical cloud of gas and dust particles, slowly spinning. Imagine that it pulls itself inward by its gravity. What happens? It will spin faster and collapse down along the spin axis to make a flat disk with a fat center (Fig. 8.14). That's just the form of the solar system. As a natural result, we get the planets' orbits aligned in a thin disk, and the sun rotates in the same direction as the planets revolve.

The conservation of angular momentum resulted in the formation of a flattened solar nebula.

With this neat solution to most dynamic features of the solar system comes a serious objection: the present distribution of angular momentum. Although the sun holds 99.9 percent of the system's mass, it contains less than 1 percent of the angular momentum. The outer planets have the most, 99 percent of the total. If all the planets were dumped into the sun, it would spin once every few hours rather than once a month. Actually, the sun spins 400 times more slowly than this rate. The angular momentum is there, but not in the right place!

One solution involves an interaction of magnetic fields and charged particles (Section 4.5), which results in the transfer of the spin of the central part of the nebula to the outer regions. As the sun forms, it heats up the interior regions of the nebula. Here the gas is ionized to make a plasma, and the magnetic field lines trap the charged particles. As the sun rotates, it carries its magnetic field lines with it; these drag along the plasma (Fig. 8.15), which in turn unites with and drags along the rest of the gas and dust. So the magnetic field spins around the material in the nebula near the sun.

At the same time, the mass of the nebula resists the rotation. This drag on the magnetic field lines stretches them into a spiral shape. The magnetic field links the material in the nebula to the sun's rotation. So the nebular material gains rotation (and angular momentum) and in the process causes a drag on the sun's rotation, which slows it down.

Plasmas and magnetic fields may have interacted during the formation of the solar nebula, transferring spin from the inner to the outer regions, where it is now.

Here's an analogy. Imagine standing in a deep swimming pool with your arms extended. Spin around as fast as you can. Your arms will drive the water around and force it to swirl. In response, you'll

FIGURE 8.14

Gravitational contraction of a spinning cloud. (a) The collapse begins with the cloud slowly spinning. As the cloud contracts (b), material falls in along the spin axis to form a disk with a central bulge (c). Because of the conservation of angular momentum, the cloud ends up spinning faster than at the start of the collapse.

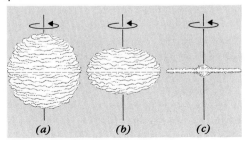

(a)   *(b)*   *(c)*

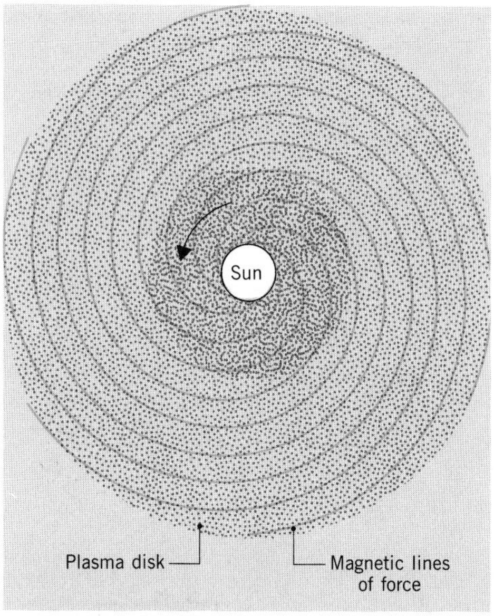
Sun
Plasma disk
Magnetic lines of force

FIGURE 8.15
Top view of a possible primeval solar magnetic field. One end of the field lines rotates with the sun; the other end is dragged by the mass of the ionized material in the disk. Hence the spin of the central regions is transferred to the outer parts, slowing down the rotation of the central regions and speeding up that of the outer ones.

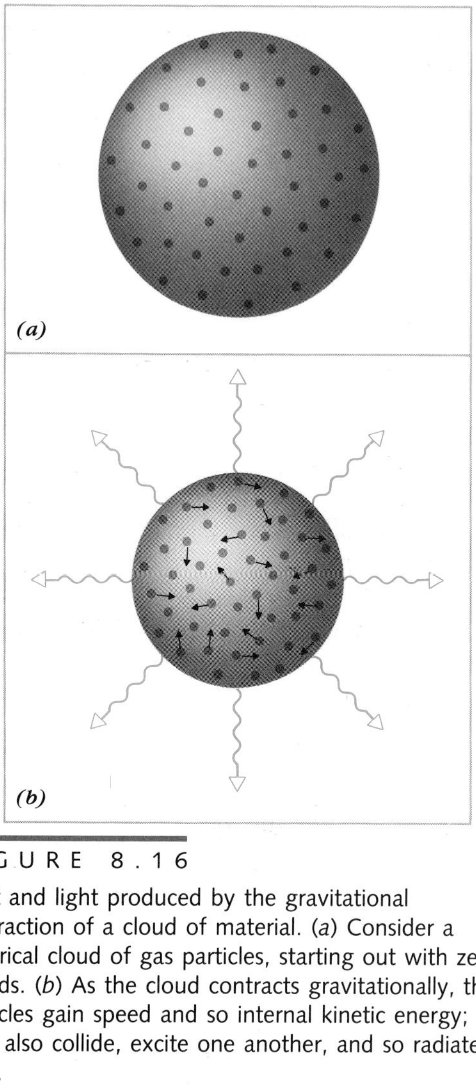
(a)
(b)

FIGURE 8.16
Heat and light produced by the gravitational contraction of a cloud of material. (*a*) Consider a spherical cloud of gas particles, starting out with zero speeds. (*b*) As the cloud contracts gravitationally, the particles gain speed and so internal kinetic energy; they also collide, excite one another, and so radiate light.

feel a drag of the water on your arms. If you didn't keep yourself spinning, you'd rapidly slow down.

Heating of the Nebula

Finally, one other basic process occurs that relates dynamics and chemistry: heating of the solar nebula from gravitational contraction. Whenever a mass pulls itself together by its own gravity, it gets hotter (Section 4.6)—a transformation of energy from one kind (gravitational potential energy) into other kinds (heat and light). Consider a ball held above the earth's surface. Its speed is zero and it lacks kinetic energy. But it does have potential energy. When you drop the ball, it falls, and its speed increases—it accelerates. So its kinetic energy increases the more it falls.

Instead of a ball, consider a cloud of gas containing a large number of particles (Fig. 8.16). Think of each as a small ball. Imagine that the cloud contracts gravitationally, from the combined attraction of all the particles. As it contracts, the particles gain speed. Though at the start the velocities are directed inward, collisions will soon distribute them in random directions, with only a slow net motion inward. The average kinetic energy of the particles increases. Since temperature is a measure of the average kinetic energy of the particles involved, the temperature of the gas likewise increases.

Meanwhile, the density also increases. So the particles collide more often, from the combined effects of the increases in speed and density. The collisions cause some atoms to emit light. So the net result of the gravitational contraction is that some initial gravitational energy converts to heat (higher temperature) and some ends up as light. In fact, half goes into raising the temperature and half into light.

When a cloud of gas contracts gravitationally, it naturally gets hotter and denser.

8.5 The Formation of the Planets

A successful nebular model must account for the important stages in the solar system's evolution: (1) the formation of the nebula out of which the planets and sun originated, (2) the formation of the original planetary bodies, (3) the subsequent evolution of the planets, and (4) the dissipation of leftover gas and dust.

Building Planets

The growth of the planets happens by three main methods: gravitational contraction, accretion, and condensation. *Gravitational contraction* works only if regions in the nebula have so much mass that they contract by their gravity to form a planet. *Accretion* occurs when small particles collide and stick together to form larger masses that eventually grow into planets. (An example: as snowflakes fall through the air, they can collide and stick to form snowflake clusters.) **Condensation** involves the growth of small particles by the sticking together of atoms and molecules. (An example: water molecules combine in clouds to form raindrops.)

Accretion probably was the main process building most of the planets (Fig. 8.17). Small particles would collide and accrete, forming larger ones. Such objects, from a few kilometers to a few hundred kilo-

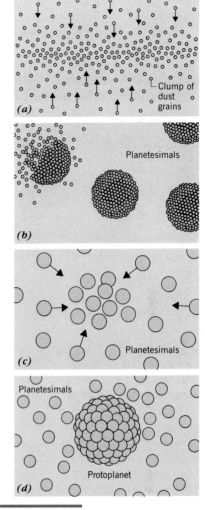

FIGURE 8.17

Possible scenario for the formation of the planets. (*a*) Dust grains collide and stick together, forming small objects. (*b*) Gravity pulls these clumps together to make asteroidal-sized bodies, the planetesimals. (*c*) The planetesimals collect in clusters to form the cores of the protoplanets (*d*), which evolve into the planets. Note that the composition of a protoplanet depends on the composition of the planetesimals that form it.

Nebular Models **At a Glance**	Origin	interstellar cloud of gas and dust
	Processes	gravitational contraction, conservation of angular momentum
	Sun	forms at center of nebula, reflects original composition
	Planets	form in disk of nebula, reflect temperature locally
	Formation processes	condensation, accretion

meters in size, are called **planetesimals**. Next, the planetesimals collided and accreted to make planet-sized masses called **protoplanets,** which evolved into the planets of today.

How to get from small dust grains to large protoplanets? Grains collide and accrete to form larger, perhaps pebble-sized, objects. These quickly fall into the plane of the nebula. These pebbles then accumulate into planetesimals by gravitational attraction. Whatever materials happen to be available at a certain distance from the center of the nebula make up a planetesimal. So the planetesimals reflect the compositions of local material.

The early planets were built in a series of stages; their composition reflects the composition of the solid materials in the region in which they formed.

Once the planetesimals have formed, they gather into larger bodies, perhaps almost as large as the moon. Their masses would be enough to help pull in other, smaller masses from a distance. So a growing planet will sweep clear a zone of the nebula to feed its mass. For the terrestrial planets to grow to their present sizes, roughly 100 million years had to pass. This clearing out of a region around a protoplanet results in the spacings of the planets that we see now.

Chemistry and Origin

How did the protoplanets acquire differences in chemical composition? Research has focused on a **condensation sequence**. The basic idea is this: the nebula's center must have been hot, a few thousand kelvins. Here solid grains, even iron compounds and silicates, could not condense. Elsewhere, the materials that would condense as new grains depended on the temperature. Just below 2000 K, grains made of rocky/metallic materials would condense (Table 8.1); below 273 K, grains could form of both rocky/metallic and icy materials.

The compositions of the planets can be explained with the condensation sequence *if* the temperature in the nebula drops *rapidly* from the center outward. Then, at different distances from the sun, different temperatures allowed different chemical compounds to condense and form grains that eventually made up the protoplanets. If a material could not condense because the temperature was too high, it would not end up in the body of the protoplanet.

For instance, the terrestrial planets lack much of the icy and gaseous materials common in the Jovian planets which, being closer to the sun, have temperatures too high for the condensation of icy and gaseous materials. It is how *low* the temperature falls that determines the chemicals that condense. In general, the condensation sequence requires that a certain *minimum* temperature be reached to account for the known chemical composition of the planets. Roughly, the temperatures are 1400 K for Mercury, 900 K for Venus, 600 K for the earth, 400 K for Mars, and 200 K for Jupiter.

The condensation sequence explains the different compositions of the planets, whose material condensed at different temperatures.

Any leftover planetesimals bombard the planetary surfaces, leaving remnant craters on the terrestrial planets—the era of heavy bombardment preserved on ancient terrains. This bombardment heats the surfaces of the planets. Radioactive decay heats the interiors and melts them. Dense elements, such as iron and nickel, sink to form a core. Less dense materials, such as silicates, float to form surface froth that will cool to become the crust. So the planets become differentiated—the first step in their evolution.

Jupiter and Saturn may not have formed according to a planetesimal accretion model. They may have contracted gravitationally from single large blobs of material in the nebula, rather than by accretion of planetesimals. The internal heat they acquired came from the conversion of gravitational potential energy into heat during gravitational contraction.

Not all planetesimals end up in larger bodies. Some rocky/metallic ones remain; they become asteroids. The icy ones hang around as the nuclei of comets. Most are tossed into the outer solar system by the gravitational influences of large masses, such as Jupiter. The Oort Cloud now contains these primordial icy bodies.

Note that the condensation sequence implies that myriad small bodies rich in volatiles and carbon formed in the cold reaches beyond Mars—the moons of the Jovian planets and even Pluto and Charon are examples. The icy-carbon objects were the original supplies of the solar system. Those that ran into one another accreted or fragmented. Those that plunged toward the sun heated and in part vaporized. Yet, they have evolved less than the planets. The survivors today make up the interplanetary debris.

The Larger View

Are there other solar systems? We have tantalizing observations but no confirmations so far—the search goes on. The nebular model developed here strongly implies that planets develop as a natural outcome of star formation (Chapter 12). Overall, this model gives a good general explanation for the formation of the solar system. Specifically, it does well in explaining dynamic properties, by gravitational contraction of the original cloud of gas and dust and the conservation of angular momentum (aided by magnetic fields). The general compositions of the planets come out right if a condensation sequence worked in a solar nebula that was hotter at the center than the edges.

We have seen that solar system debris supplies crucial insights to the story of formation. Many cold, dark bodies are out there—the original building blocks of planets. Some we see now as comets, meteoroids, and asteroids—all processed to some degree by heating and impacts. From these fossils we can infer the conditions of early times.

The nebular scenario contains key concepts that will remain in future models of planetary formation involving our sun and other stars. These include: conservation of energy, gravitational contraction and heating, conservation of angular momentum, and the influences of magnetic fields. A nebular model points to similar processes shaping other planetary systems during the birth of stars under unfamiliar skies.

Key Ideas and Terms

nebular models
scientific models

1. Any scientific model for the origin of the solar system must account for its general chemical and dynamic properties in a unified way.

angular momentum
mass

2. The key dynamic properties of the solar system include the following: the system is flat, and most of the mass is in the center (the sun); most of the angular momentum is in the planets; the planets' orbital directions are all the same; and the planets are regularly spaced.

composition
density
volatiles

3. The key chemical properties of the solar system include the general division of the terrestrial and Jovian planets, the differences among planets in each group, and the compositions of asteroids, meteoroids, and comets.

asteroid belt
asteroids
comets
meteorites
meteoroids

4. Interplanetary debris provides important clues about the origin of the solar system, especially from the physical properties of comets (dirty snowballs), asteroids (rocks and metals), and meteorites (rocks and metals).

carbonaceous
   chondrites
chondrites
chondrules
iron meteorites
meteorites
meteors
parent meteor
   bodies
stone meteorites
stony-irons

5. The crystalline patterns found in iron meteorites indicate that their material cooled slowly from a molten state, which implies that they formed in larger bodies. Other types of meteorites are different remnants of the first solids formed in the solar system. The radioactive dating of these materials provides the date of the solar system's formation: 4.6 billion years ago.

angular momentum
conservation of
   angular
   momentum
gravitational
   contraction
solar nebula

6. A spinning cloud of interstellar material (gas and dust) will flatten (from the conservation of angular momentum) and become hot and dense as it contracts gravitationally.

condensation
condensation
   sequence
volatiles

7. The condensation sequence works if the temperature in the solar nebula decreases rapidly from center to edge; which materials vaporize and condense depends on how high and then how low the temperature gets at certain distances from the sun.

angular momentum
magnetic fields
plasma

8. We can account for the distribution of angular momentum now by citing the ability of magnetic fields (exercised before the formation of the planets) to transfer the spin of the sun to the rest of the solar nebula.

accretion
condensation
gravitational
   contraction
planetesimals
protoplanets
solar nebula

9. The general process of planet formation in the solar nebula may have been as follows: condensation of grains, accretion of grains into planetesimals, and clumping of planetesimals into protoplanets, leading to the evolution of the protoplanets into the bodies we see today. Jupiter and Saturn may have formed by the gravitational contraction of large blobs of material rather than by planetesimal accretion.

asteroid belt
comets
Oort's Cloud
periodic comets
S-, C-, and M-type
   asteroids

10. Comets and asteroids are very likely leftover planetesimals formed in different regions of the solar system. Today, most asteroids are found in the belt between Mars and Jupiter, and most cometary nuclei inhabit the enormous Oort Cloud.

coma
dirty snowball
comet model
nucleus (of a comet)
tail (dust, plasma)

11. Comets interact with the solar wind to generate long, magnetized plasma tails; their nuclei are icy, irregular in shape, and very dusty. When the nucleus is heated by the sun, it vaporizes and ejects the materials that produce the coma and the tail.

Review Questions

1. What are two of the main dynamic properties of the solar system?
2. What are two of the main chemical properties of the solar system?
3. How does a comet differ in density from an asteroid?
4. Which type of meteorite has the highest density?
5. Under what conditions do large crystals form in iron meteorites?
6. As a cloud of material contracts under its own gravity, how will its density and temperature change?
7. In the condensation sequence, what kind of material condenses *last*?
8. For magnetic fields to transfer spin, the gas in the solar nebula needed to be in what form?
9. Once grains had formed in the solar nebula, what happened next in the process of the formation of the planets?
10. Where in the solar system are today's asteroids mostly found?
11. What is the source of material in a comet's tail?

Conceptual Exercises

1. Suppose you hopped in your spaceship on Saturday night and flew to an asteroid. What would you see?
2. Then you sped off to a comet. What would you see? How would it differ in appearance from the asteroid you just visited?
3. After many perihelion passages of the sun, what happens to a comet in the dirty snowball comet model?
4. How can you tell an iron meteorite from a piece of terrestrial iron and nickel?
5. What is the major weakness of the nebular models with respect to the dynamic properties of the solar system?
6. What one planet fits least well with the general chemical and dynamic properties of the solar system?
7. Use the chemical condensation sequence to explain the general chemical differences between the earth and Jupiter.

8. In what ways did the Giotto mission confirm in detail the dirty snowball model of cometary nuclei?

9. Suppose you found a dense, irregular fragment on the ground. How could you tell that it was or was not a meteorite?

10. What is the probable relationship of asteroids to meteorites?

Conceptual Activity 8   The Orbit of Halley's Comet

This activity will guide you into constructing a scale model of the orbit of Halley's comet. Because the orbit is elliptical, you will need to review the properties of ellipses. Refer to Figure F.3 for the specific parts of Halley's Comet orbit. Note that a new aspect of an ellipse will be used: the *semilatus rectum*, which is the distance from the focus (in which the sun lies) to the ellipse.

To draw the scale model, you must know the values of the important properties and decide on a scale. Here's the information you need: eccentricity is 0.97; semimajor axis, 17.9 AU; perihelion distance, 0.59 AU; aphelion distance, 35.3 AU; length of semiminor axis, 4.6 AU; length of semilatus rectum, 1.2 AU; and distance between foci, 17.4 AU.

To decide on a scale, consider the largest part of the orbit: the entire major axis. Its length is twice that of the semimajor axis, or 2 × 17.9 AU = 35.8 AU. At a scale of 1 AU = 1 cm, the orbit would fit on two pieces of paper taped

FIGURE   F.3

The orbit of Halley's comet, drawn to scale.

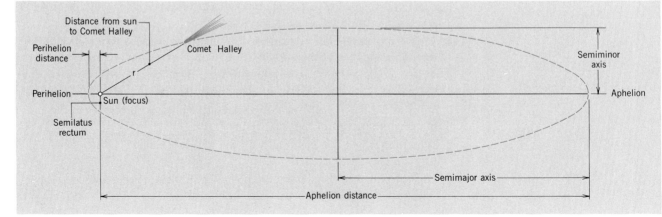

together so that their length is at least 40 cm. Cut a piece of string whose length is equal to that of the major axis on this scale.

Pin or tape the paper down on a thick piece of cardboard. Get two tacks and place them in line on the paper, separated by 17.4 AU on your scale. Set each end of the string under a tack. Place a pencil within the loop of the string and, keeping that string taut, trace out the ellipse. You may have a little trouble keeping the string tight through the perihelion and aphelion points, but be careful to do so. When the ellipse is drawn, take away the string and tacks and draw in the major and minor axes and the semilatus rectum. Check that all the dimensions are correct.

You may want to include the orbits of some planets such as Jupiter and Neptune on the same scale. Note that the comet's orbit is actually tilted with respect to the plane of the earth's orbit. You have drawn the comet's orbit as seen face-on.

The View from Miranda, Moon of Uranus

Adapted by Michael Zeilik from original material by Sheridan A. Simon

It's a strange world here—an outlandish vista compared to the environs of the earth and moon. On Miranda, the weather arrives daily clear and cold, with temperatures in the 60s—kelvin! When you exhale, your space suit emits used air, rich in nitrogen. In sunlight, the nitrogen condenses onto Miranda's surface in tiny drops; at night, it freezes to a thin frost. The gravity is no more homey, only 0.01 g. You can't jump off—Miranda's escape velocity is 160 m/s—but you feel very reduced in weight. A football offensive lineman would weigh about as much as this book.

Bluish-green Uranus monopolizes the sky. When you hold out your hand at arm's length, you can just cover the planet's disk with your fingers spread wide. Despite its size, Uranus does not shine very bright. It spans 2000 times as much sky as the full moon seen from earth but provides only 50 times as much reflected sunlight. Uranus is big but bland; no turmoil confronts the placid clouds floating in its upper atmosphere.

The other large moons of Uranus do not put on a dazzling show, either. Not much sunlight reaches the Uranian system, and the moons' gray surfaces reflect little of the light that hits them. Ariel gets close enough to us to loom twice the size of the moon as seen from the earth, but it shines only one-twentieth as bright. Distant Oberon never grows more than one-third the size of the moon and is 0.1 percent as bright. The sun gleams as a tiny, glaring dot, only one-twentieth its size as viewed from the earth—providing a scant 0.3 percent of the light. It's dark out here, near the edge of night, near the end of the solar system.

In 1986, when the Voyager spacecraft passed through the Uranian system, the south pole of the planet and all its moons pointed at the sun. Then, from the south pole of Miranda, the sun seemed to cycle in a small circle around a point directly overhead. Each complete circle took 34 hours, Miranda's orbital period.

As Uranus orbits the sun, its rotation axis points in the same direction in space, riveted to the stars. So as the year progresses, the motion of the sun as seen from Miranda slowly changes—the sun's daily circles enlarge. In A.D. 2006, as seen from the south pole, the sun will skim along the horizon, dipping slightly upward and downward during each orbit of Uranus. Later, the sun will circle below the horizon. Night will fall on the south pole of Miranda—a night that will stretch for 42 earth years, or half a Uranian year.

From Miranda's equator, the sun traces a much different motion in the sky. In 1986, the sun drew small circles on the southern horizon. It spent half the time above and half below the horizon. While above, it marched on a semicircular path. As Uranus orbits the sun, each semicircle gradually grew larger and its end points on the horizon shifted. The rising and setting points will move closer to due west and due east. In the early twenty-first century, the sun will wheel in a grand semicircle from west to east with its highest point at the zenith. Later, the semicircle will shrink, this time toward a point due north. Even later, they will fan open again and move back south. Then 84 years after the Voyager mission (in A.D. 2070), the sun will revert to the position and motions it had in 1986.

And so the year passes on Miranda—a long year marked by cycles of the sun's progress quite unlike that seen from the earth. Other places, even in the solar system, deliver different worlds and impart different astronomies.

T

2

Concepts of the Stellar Evolution

∎

Unifying Concepts

The sun is one of billions of stars making up the Milky Way Galaxy. Even the stars closest to our sun lie a few light years away; we discover their properties by analyzing the light we have received from them. The sun has produced energy—which we see as sunlight—for a few billion years and will do so for a few billion more. Fusion reactions involving hydrogen take place deep in the sun at very high temperatures and densities; these convert matter to energy. The heat supplied by these reactions supports the sun against its gravity. Other stars also generate energy by fusion reactions for most of their lives and produce starlight; as they do so, they must evolve. How a star lives and dies depends on its mass. Most stars go through a rapid, violent terminal phase before they end up as corpses.

9
Celestial Light

■

Central Concept
Matter produces light, and this light carries physical information about the sun, stars, and other celestial objects that emit it.

■

To build models of stars, we need to know their physical nature. How do we find out the physical makeup of the stars and other astronomical objects? Astronomers have in hand only the light funneled into their telescopes. Yet this light carries information, if it can be decoded.

In the twentieth century, physicists constructed models of atoms—models that explained how atoms absorb and emit light. The atoms in stars (including the sun) emit starlight (and sunlight), so an analysis of light reveals information about the physical conditions and chemical compositions of stars. This key to this approach involved displaying the light as spectra. Astronomers finally could penetrate the environments of the distant stars and the sun—and any celestial object that emits light.

9.1 Sunlight and Spectroscopy

Matter makes up the sun and stars, and matter's primary form is as atoms. Atoms can emit and absorb light; sunlight originates from atoms. The structure of atoms can be investigated by analyzing the light they emit and absorb. We also can investigate the physical environment of atoms by analyzing light.

> To understand the physical properties of the sun (and stars) requires an understanding of how atoms and light interact.

Atoms and Matter

Substances fall into two classes: chemical *elements* and chemical *compounds*. **Elements** cannot be broken by chemical reactions into simpler substances. (See Appendix F for the periodic table of the elements.) Elements are the most basic substances, such as hydrogen (H), helium (He), carbon (C), and oxygen (O). Although 92 elements occur in nature (and more have been created in the laboratory), only a few dozen are common. Most substances you encounter are not elements but **compounds**, substances made of two or more elements. Water (Fig. 9.1*a*) is a chemical compound composed of the elements hydrogen and oxygen—H_2O.

A **molecule** is the smallest unit of a compound that still has the chemical properties of the compound. Since, however, a compound can be broken down into elements, a unit of matter smaller than a molecule must exist. An **atom** is the smallest unit of an element that displays the chemical properties of the element. A compound is created when atoms join to form molecules of the compound.

The modern idea of an atom (Fig. 9.1*b*) pictures a tiny, dense **nucleus** surrounded by rapidly orbiting

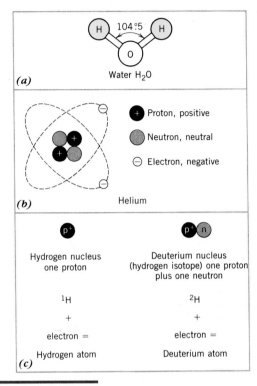

F I G U R E   9 . 1

Molecules, atoms, and isotopes. (*a*) Schematic of the chemical structure of water, H_2O, with two hydrogen atoms bonded at a certain angle to the oxygen atom. (*b*) Schematic of the structure of a helium atom. Two negatively charged electrons orbit the positively charged nucleus (two protons, two neutrons). (*c*) The difference between a hydrogen atom and deuterium, an isotope of hydrogen, lies in the additional neutron in the deuterium nucleus.

electrons. **Electrons**, which carry a negative charge, are low-mass particles (only about 10^{-30} kg; Appendix C). **Protons** and **neutrons**, which are about 2000 times more massive than electrons, make up an atom's nucleus. The protons are positively charged, and the neutrons have no electrical charge. The nucleus of an atom, because of the protons it contains,

Atoms	Nucleus	protons and neutron, positive charge from protons
	Elements	determined by number of protons
At a Glance	Isotopes	determined by number of neutrons
	Electrons	orbiting nucleus, negative charge equal to number of protons for neutral atom

is positively charged and attracts the negatively charged electrons.

Atoms consist of a dense nucleus bearing a positive charge surrounded by negatively charged electrons.

This attractive electrical force binds the electrons to the nucleus and holds the atom together. (A *strong nuclear force* binds the protons and neutrons together in the nucleus; it overwhelms the mutual repulsion of the positively charged protons.) The electrons whiz around the nucleus in orbits at very great distances in terms of the size of the nucleus. If the proton of a hydrogen atom were the size of a marble, the electron would be about a kilometer away from the nucleus.

Elements differ from each other because their atoms differ, and atoms of different elements contain different numbers of protons in their nuclei. The nucleus of a hydrogen atom, for example, contains one proton; helium, two (Fig. 9.1*b*); and carbon, six. The number of orbiting electrons equals that of the protons, so that an atom has no net charge overall. The electrons largely determine the chemical properties of an element. (Appendix F provides a summary of the elemental properties.)

Nuclei that contain the same number of protons but a *different* number of neutrons are called **isotopes** of the element. For example, one form of heavy hydrogen, called *deuterium*, has one proton and one neutron, whereas ordinary hydrogen has one proton and no neutrons (Fig. 9.1*c*). The isotopes are forms of the same element, but an atom of deuterium has more mass than an atom of hydrogen because of the extra neutron. (Appendix F gives only the most common isotope of each element.) Most elements contain approximately the same number of protons and neutrons or a few more neutrons than protons. For example, atoms of the most common isotope of carbon contain six protons and six neutrons.

The number of protons in a nucleus determines the element; different numbers of neutrons identify isotopes of an element.

Overall, an atom is electrically neutral. This condition can be changed by either adding or taking away electrons. If the number of electrons is less than or greater than the number of protons (so that the net charge is positive or negative), the particle is called an **ion**. In most astronomical cases, we find *negative* ions from the loss of one or more electrons (as in a plasma).

Simple Spectroscopy

Breaking up light into its component colors is called **spectroscopy**. You've probably seen a rainbow, with colors running continuously from red to violet. What's happening? Raindrops disperse the sunlight into a continuous band of colors. You can use a prism to disperse sunlight (Fig. 9.2). Use a slotted piece of cardboard to pass a beam of sunlight through a prism. Let the light coming out of the prism fall on white paper. The sunlight spreads into an array of colors, with red light bent the least and violet the most. This sequence of colors is called a **spectrum**. A spectrum with no breaks in it is called a **continuous spectrum**; it has a continuum of unmixed shades of color (Fig. 9.2). White light (and sunlight) has all these colors running smoothly into one an-

---

FIGURE 9.2

Experiment with a prism, a slit, and sunlight to produce a visible spectrum. The prism disperses the sunlight into its component colors.

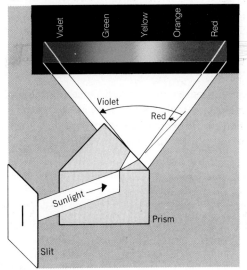

(a)

(b)

FIGURE 9.3

Observing spectra. (*a*) Photograph of a continuous spectrum from an incandescent light made through a diffraction grating. (*b*) Emission-line spectrum of a fluorescent light. (Spectra photos courtesy Dean Zollman and the American Association of Physics Teachers)

other and no colors missing. You can think of white light as a mixture of all colors.

The visible light spectrum is made by dispersing white light into its component colors from red to violet.

An inexpensive device to disperse white light into its colors is called a *diffraction grating*. If you have one, you can examine various light sources around campus or in town. An ordinary incandescent light bulb shows a *continuous spectrum* (Fig. 9.3*a* and Fig. 9.4*a*). A fluorescent light shows mostly distinct, *bright lines* (Fig. 9.3*b*).

Suppose you put another slotted piece of cardboard (or slit) *behind* your prism or grating to let just a *single color* pass through. You'd then see a bright line of that color, not a complete, continuous spectrum. This line is called a **spectral line**. Note that straight lines appear in a spectrum because the slit admitting the light to the prism is a straight-line source of light. Each line is an image of the slit in the light of one color.

If you let sunlight pass through both a prism and a slit and then magnify the spectrum with a telescope, you'll see a continuous spectrum crossed by many *dark lines*. Light is missing from some colors because something has removed, or absorbed, light in these cases. The dark lines are called **absorption lines**, and a spectrum displaying them is called an **absorption-line spectrum** (Fig. 9.4*b*). When light is emitted only at certain colors, such as from a neon sign or sodium street lamp, we speak of **emission lines** or an **emission-line spectrum** (Fig. 9.4*c–g*).

Spectra come in three basic types: continuous, emission line, and absorption line.

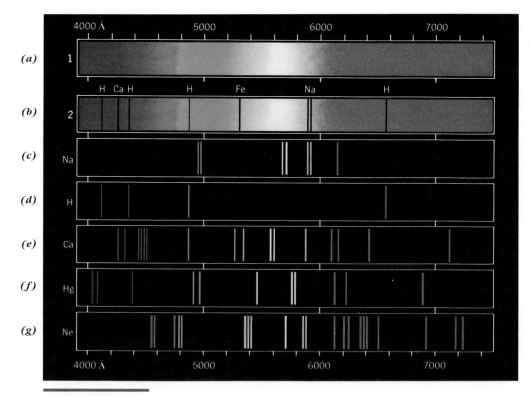

FIGURE 9.4

Major spectral types: shorter wavelengths are to the left, with wavelength (Å) increasing to the right (from blue to red). (*a*) Continuous spectrum such as that from an incandescent light bulb. (*b*) The absorption-line spectrum of the sun; only the most prominent dark lines are shown from the following elements: H, hydrogen (the Balmer series in the visible); Ca, calcium; Fe, iron; and Na, sodium. Below the sun's spectrum are the emission-line spectra of selected elements: (*c*) sodium, (*d*) hydrogen, (*e*) calcium, (*f*) mercury, and (*g*) neon. Note how the bright lines of sodium, calcium, and hydrogen line up with the dark lines in the sun's spectrum, which indicates the presence of these elements in the sun's atmosphere.

9.2 Analyzing Sunlight

To unravel the message of sunlight, let's first do a few experiments with a **spectroscope**, an instrument for observing details in a spectrum. (Conceptual Activity 9 describes how to build your own simple spectroscope.)

For the first experiment, point a spectroscope at a pure neon light, for example, in an advertising sign. You'll see an emission-line spectrum with the brightest lines in the red region (Fig. 9.4*g*). That's why such a sign appears red to our eyes. A mercury street light (which contains mercury gas) has its strongest bright lines in the yellow and green regions (Fig. 9.4*f*). These lines combined result in the greenish color of mercury street lamps. In contrast, sodium street lamps appear a ghastly yellow because the brightest lines from hot sodium gas fall in the yellow region of the spectrum (Fig. 9.4*c*).

Under the right conditions, each chemical element displays a unique arrangement of bright lines, so a spectrum can be used to identify elements in a source of light.

For the second experiment, put sodium (in the compound form of table salt) in a hot flame. Visually, you'll see the flame turn yellow. Through the spectroscope, you'll see a series of bright lines, the brightest, a pair, in the yellow region (Fig. 9.4*c*).

With the spectroscope you can find that different chemical elements (and molecules) give off different patterns of bright lines. A particular spectral pattern shows the presence of a particular element. Each pattern is unique to an element, just as fingerprints are to people.

The atoms in the hot gas produce the observed bright lines. Different gases composed of different atoms produce different patterns of spectral lines. So the bright-line spectrum of a hot gas is characteristic of the atoms in the gas. The pattern of the lines reveals which elements the gas contains.

For the third experiment, look carefully at sunlight with a spectroscope (Conceptual Activity 9). WARNING: NEVER LOOK AT THE SUN DIRECTLY! You'll see dark lines against a bright, continuous background of color (Fig. 9.4*b*). A pair of dark lines in the yellow region of the spectrum is called **D lines**. Could these dark lines be related to the bright lines of sodium observed when salt is put in a flame? Could the dark lines in the sun's spectrum also come from sodium in the sun?

To test this idea, allow light from a glowing solid (which gives off a continuous spectrum) to pass through the sodium flame. A pair of dark lines appears in the spectrum at exactly the same position in the yellow regions as the pair in the sun's spectrum.

Sodium, then, must be making these dark lines, *removing* yellow light from the solar spectrum. Now pass sunlight through a sodium flame into the spectroscope. The pair of lines becomes even darker! The sodium in the flame absorbs even more of the sun's yellow light. This proves that sodium made the lines. But why did the lines get darker?

Kirchhoff's Rules

From lab experiments such as those described above, the German physicist Gustav Kirchhoff (1824–1887) formulated empirical rules of *spectroscopic analysis*, the determination of the physical state and the composition of an unknown mixture of elements. **Kirchhoff's rules** (Fig. 9.5 and Table 9.1) are briefly stated as follows.

Rule 1. A *hot* and *opaque* solid, liquid, or highly compressed gas emits a continuous spectrum. Example: the filament of a light bulb.

Rule 2. A *hot*, *transparent* gas produces a spectrum of bright lines (emission lines). The number and colors of these lines depend on which elements are present in the gas. Example: a neon sign.

FIGURE 9.5

Basic spectral types as observed through a spectroscope and described by Kirchhoff's rules. (*a*) Continuous spectrum of colors with no lines from a hot, opaque source of light. (*b*) Emission-line spectrum with specific lines from a transparent gas. (*c*) Absorption-line spectrum with the same spectral lines as in (*b*) when observing the hotter source through a cooler, transparent gas.

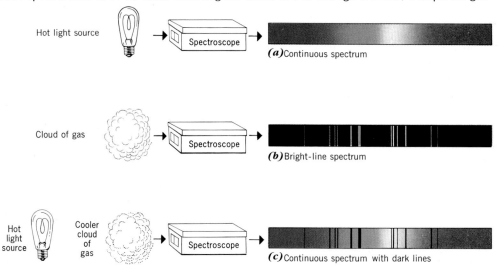

Hot light source → Spectroscope → (*a*)Continuous spectrum

Cloud of gas → Spectroscope → (*b*)Bright-line spectrum

Hot light source → Cooler cloud of gas → Spectroscope → (*c*)Continuous spectrum with dark lines

Table 9.1

Spectra of Basic Types

Spectrum Type	Appearance	Physical Conditions
Emission line	Distinct bright lines against a dark background	Hot, transparent gas
Continuous	Smooth blend of all colors	Hot, opaque gas, liquid, or solid
Absorption line	Distinct dark lines against a bright background of colors	Light from a hot, opaque gas, liquid, or solid passing through a cooler, transparent material

Rule 3. If a continuous spectrum (from a hot, opaque solid, liquid, or gas) passes through a gas at a lower temperature, the *transparent cooler* gas causes the appearance of dark lines (absorption lines). Their colors and their number depend on the elements in the cool gas. Example: light from the sun.

Essentially, the first rule says that an opaque, hot material produces a continuous spectrum. Atoms usually emit and absorb at discrete colors. However, when atoms are jammed together so densely that the material becomes opaque to light, a continuous spectrum appears. For instance, the sun is so hot that it is gaseous, but it is still opaque enough to produce a continuous spectrum.

A solar spectrum shows dark lines against a continuous background. What causes these dark lines? By Kirchhoff's third rule you expect there to be a cooler and less dense gas between the visible surface of the sun and us. And there is. The solar atmospheric layer absorbs light from the continuous spectrum passing through it to produce the dark lines. The positions of the lines in the spectrum tell which elements are present. The solar composition determined from the dark lines relates only to the region of the sun that produces the absorption spectrum (its atmosphere).

By matching the emission lines of elements to the absorption lines in the sun's spectrum, we can infer the chemical composition of the sun's atmosphere.

With these ideas in mind, you can understand what happens when sunlight passes through the so-

dium vapor to make darker D lines. The sodium in the flame is *cooler* than the sun's visible surface, so the sodium *absorbs* the light in the D lines. Though hot, the sodium gas removes some specific wavelengths from the beam of sunlight. That's why the sodium D lines got darker when sunlight passed through the sodium flame. This experiment also proves that colors are missing from the continuous solar spectrum because they have been absorbed by the atoms in the cooler gas of the solar atmosphere.

Whether the atoms in the gas emit or absorb depends on the physical conditions in the gas. Emission requires high temperatures in a transparent gas; absorption occurs when a continuous spectrum passes through a cooler transparent gas. In either case, the pattern of emission lines is the same as the pattern of absorption lines for a gas with the same chemical composition.

The lines absorbed by a gas from a continuous spectrum are the same lines emitted by that gas when energy is put into it.

The Conservation of Energy Revisited

The statement above is another way to state the *conservation of energy*, a key concept you've already encountered (Section 4.6).

Experiments have shown that within a system of objects, energy is naturally transformed from one form to another, that when all the energy is carefully tallied, the total amount remains the same with time—it is conserved. This conservation holds if the system is isolated from outside influences that could bring in or take away energy.

This chapter deals with energy in the form of light, *radiative energy*. This radiative energy comes from another form, *potential energy*, the stored energy of position. On the chemical scale, changes within and between the electrical charges of molecules result in the release of the energy; so these, too, relate to positions and are a form of potential energy. Why? Basically because electrical charges create forces between themselves. Likewise in an atom; the electrons orbiting the nucleus have specific kinetic energies from their motion. They also have specific potential energies that relate to their distances from the nucleus and their electrical attraction to the protons there. So an atom acts like an energy conversion machine, creating and absorbing photons.

Atoms convert electrical potential energy into light while conserving energy.

 ## 9.3 Spectra and Atoms

How can the physical conditions in stars be discovered from their spectra? The spectral code of the patterns of lines was cracked in a great revolution of physics in the twentieth century: the quantum theory and its explanation of the nature of the atom and of light. But to understand the power of spectroscopy, you must first know something about light.

Light and Electromagnetic Radiation

Light sometimes behaves as waves and sometimes as particles, depending on how it is observed. Both models are needed to describe the properties of light completely. In either cases, light carries energy—radiant energy. I will use both models; let me emphasize some wavelike features first.

You probably have some experience with waves. Imagine that you're at a beach with waves arriving at regular intervals. Time them with your watch. The number of waves that arrive in a given time is their **frequency**. For instance, one wave hitting the shore every minute is a frequency of one per minute. The

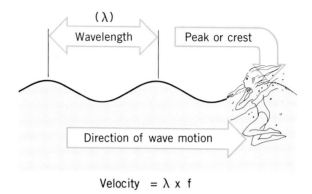

Velocity $= \lambda \times f$

F I G U R E   9 . 6

Basic properties of waves. The distance from peak to peak of a wave is its wavelength. The number of waves that pass by in a second is the frequency.

---

distance between the crests (peaks) of these waves is their **wavelength** (Fig. 9.6). For example, if the crests are 10 meters apart, the wavelength is 10 meters. The *wave velocity* indicates how fast and in what direction the waves are traveling. Here the waves move 10 meters in 1 minute, so their velocity is 10 m/min.

Note that these waves do *not* involve the mass motion of material over long distances. If you are in a boat on a lake with no currents, waves from passing motor boats cause your boat to bob up and down. But you stay in the same location.

A wave carries energy of motion from one place to another, but it does not transport material.

The velocity, wavelength, and frequency are interrelated. When you multiply frequency (number per time) and wavelength (length), the result is equal wave velocity (length per time): frequency × wavelength = velocity. For light in a vacuum, the velocity is the same for all wavelengths: 299,793 km/s. It is usually designated by a lowercase c and is easily recalled by its approximate value, 300,000 km/s (3×10^8 m/s).

So waves have three fundamental properties. The *wavelength* is the distance between two successive crests of a wave, the *frequency* is the number of waves that pass you each second, and the *velocity* is the distance covered in one second by a crest traveling in a certain direction. An observer who counts the frequency of the waves and measures their wavelength can calculate the velocity of the waves.

MATH CONCEPT

The basic wave equation is

$$f \times \lambda = v$$

where f is the wave's frequency, λ the wavelength (in meters, say), and v the velocity (say in m/s). This fundamental rule applies to waves of all kinds. For light waves only, the speed is c, and we write

$$f \times \lambda = c$$

where λ is the wavelength and f the frequency. Note that because c is constant for all kinds of light, different wavelengths must have different frequencies. The higher the frequency, the shorter the wavelength.

Visible light is one type of wave produced whenever electric charges are accelerated. Then there are radio waves, which are produced by a transmitter that uses electricity to move electrons rapidly back and forth in an antenna. The electrons are accelerated in a periodic fashion and produce radio waves at a certain wavelength. Such radiated energy, generated by accelerated electrical charges, is termed *electromagnetic radiation*. The range of all different wavelengths of electromagnetic radiation makes up the **electromagnetic spectrum** (Fig. 9.7); visible light covers only a small part of the electromagnetic spectrum.

Scientists often use special wavelength units for different regions of the electromagnetic spectrum. For example, visible light is usually measured by astronomers in *angstroms* (abbreviated Å), where 1 Å is 10^{-10} m. In the infrared region the unit of length commonly used is the *micrometer* (abbreviated μm), with 1 μm equal to 10^{-6} m. In the radio region, astronomers commonly use frequency rather than wavelength. The unit of frequency is the *hertz* (abbreviated Hz), which is one cycle (or vibration) per second.

FIGURE 9.7

The electromagnetic spectrum; size scale at bottom indicates representative objects equivalent to the wavelength scale. Wavelength in the visible region is measured in ångstroms (Å); one ångstrom is 10^{-10} m. Note that the visible part of the spectrum is only a small slice of the entire electromagnetic spectrum. In the infrared region a micrometer (10^{-6} m) is the unit of measure. Radio waves are measured in cycles per second or hertz, as explained in the text.

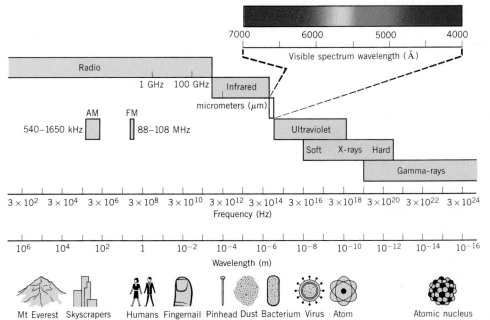

When you see a wave go by, from peak to peak, it has gone through one cycle. Even in the radio region, many cycles go by in a second. For example, the AM band covers the range from 540 to 1600 kilohertz; 1 *kilohertz* (kHz) is 1000 Hz. The wavelengths are a few hundred meters. The FM band ranges from 88 to 108 megahertz; 1 *megahertz* (*MHz*) is 1 million hertz. The wavelengths here are a few meters. Radio astronomers work at such high frequencies that they use the unit of *gigahertz* (*GHz*), 1 billion hertz. Here wavelengths are as short as millimeters.

The electromagnetic spectrum encompasses all forms of light; it can be described in terms of wavelength or frequency.

In a sense, a spectroscope works like a radio tuner. Both devices arrange a part of the electromagnetic spectrum by spreading it out from shorter to longer wavelengths. A certain wavelength can then be selected for examination. You choose a wavelength on a radio by tuning it. A spectroscope can also be tuned to look at a particular line of some region of wavelengths.

Atoms, Light, and Radiation

During the nineteenth century, astronomers were puzzled by spectral lines from stars. Why should stars emit light only at certain wavelengths and not others? The pattern of lines of hydrogen appeared in a simple series in the visible region of the electromagnetic spectrum. This series of hydrogen lines was named the **Balmer series** (Fig. 9.4*b*), after the Swiss mathematician Johannes J. Balmer (1825–1898). When hydrogen is ionized, it no longer produces the Balmer series of lines. A structured arrangement of electrons somehow determines the light-emitting properties.

In 1900 Max Planck (1858–1947) announced the revolutionary idea that radiating matter emits light in discrete chunks of energy, which he called **quanta**. Taking up this quantum idea, Albert Einstein (1879–1955) showed that each quantum, usually called a **photon** when talking about light, carries an amount of **radiative energy** that relates directly to its frequency. Light of higher frequency (shorter wavelength) transports more energy than light of lower energy. For example, a blue photon carries more energy than a red photon—almost twice as much. An x-ray photon carries almost a *million* times more energy than a visible light photon! This relationship between energy and frequency is a direct one: A photon with *twice* the frequency carries *twice* the radiative energy. Similarly, a photon with half the frequency transports half as much energy.

Photons of light carry radiative energy; the higher the frequency, the greater the energy.

MATH CONCEPT

The fundamental relation between a photon's frequency and energy is

$$E_{\text{photon}} = hf$$

where E is the energy in joules, h a constant (called *Planck's constant*) with a value of 6.63×10^{-34} J · s, and f the frequency in hertz. Since wavelength and frequency are related by the wave equation,

$$f = \frac{c}{\lambda}$$

we can rewrite the energy equation as follows:

$$E = \frac{hc}{\lambda}$$

with the wavelength in meters.

You have seen that atoms emit and absorb certain colors at specific wavelengths. Atoms must absorb and emit energy as whole photons. For each wavelength you can imagine a photon carrying off a certain amount of energy.

The Danish scientist Niels Bohr (1885–1962) meshed the quantum and atomic pictures of matter. The resulting scheme, known as the **Bohr model of the atom**, explained and predicted the absorption and emission of photons.

Bohr pictured atoms as having many possible arrangements of electrons. He imagined that the emission or absorption of light arises from a **transition** between electron arrangements. Bohr realized that only certain electron arrangements are permitted. Electron transitions can occur only between the special arrangements. For a given nuclear charge, electrons have available many possible states with specific energy values or **energy levels**. Each level corresponds to a certain orbit the electron can have—a certain total energy, the sum of its potential and kinetic energies.

In the emission or absorption of light, an atom loses or gains energy in *discrete* amounts, the energies of individual photons.

As an analogy, imagine moving a bowling ball up and down a set of stairs with uneven height rises. The first step is large, and each one after that is smaller. Each step is an incremental change in the ball's energy, and only full-step changes are permitted. (You can't climb stairs in intervals of part of a step.) If we try to add less than one step's worth of energy to the ball, it remains at its initial level. If we add exactly one step of energy, the ball moves up one step. When the ball loses energy, it descends in steps until it hits the floor at the bottom of the staircase; this is the ground level, the state of lowest energy. The key point is this: electron transitions can occur only between certain energy levels (stairs), and so an atom of a given element can produce only a particular pattern of lines (colors of light).

Consider the hydrogen atom: it has one proton for a nucleus and one electron attached by the electric force between it and the proton. (Other atoms have more protons and electrons but are held together in the same way.) The electron has a certain total energy; the essence of quantum theory is that electrons remain in stable states of specific energies, and for each state there is a particular orbit (Fig. 9.8a). For example, the electron in the lowest energy level (called the **ground state**) securely orbits the nucleus. The electron must gain energy to move out to larger orbits. The orbits, and so the energy levels, follow strict spacing rules, and the amount of energy that can be gained by an electron as it moves from one orbit to the next cannot deviate from the amount determined by the rules of quantum physics.

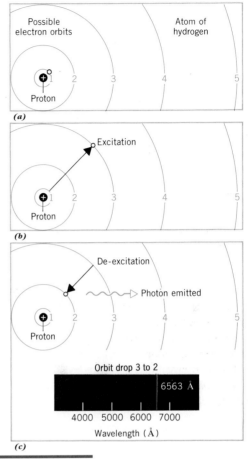

FIGURE 9.8

Energy orbits for the hydrogen atom. (*a*) The first five orbits for an electron; the energy for each increases outward from the nucleus for orbits 1 to 5. (*b*) The addition of energy can excite an electron to a higher orbit, such as from 1 to 3. This process is called *excitation*. (*c*) When an electron drops down an orbit, it loses energy, usually as a photon; here, in the drop from 3 to 2, a photon at 6563 Å is emitted. This process is called *de-excitation*.

Energy can be added to the atom either by collision with another particle or by absorption of a photon with sufficient energy. The electron jumps up one or more energy levels (Fig. 9.8b). The atom is then *excited* and the process is called **excitation**. This condition does not last long; the electron drops to a lower level in about 10^{-8} second. However, for the electron to descend to a lower energy level requires that it lose some energy (Fig. 9.8c). The electron accomplishes this by emitting a photon with an

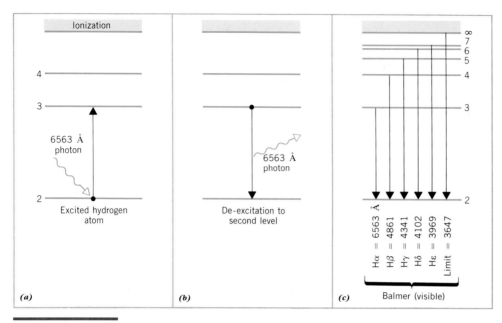

FIGURE 9.9

Energy-level diagram for the Balmer series of the hydrogen atom. The energy level differences are drawn to scale. (*a*) Absorption of a 6563-Å photon by an excited atom boosts an electron upward to level 3 from level 2. (*b*) When the electron drops from level 3 to 2, the atom emits a 6563-Å photon. This brightest line of hydrogen is called the *hydrogen-alpha* (H$_\alpha$) line of the Balmer series. (*c*) Note that the Balmer series involves transitions to and from the second energy level. Each line is given a name in the sequence of greater energy differences to and from the second level. Arrows show de-excitation that produces Balmer emission lines.

energy *equal to* the amount it needs to lose between the levels. This energy relates directly to the photon's frequency. The *larger* the energy transition of the electron, the *higher* the energy of photon emitted (or absorbed), and the shorter its wavelength.

If an atom gains enough energy, the electron flies away from the nucleus. The atom is then *ionized*, in the process called **ionization**. The loss of an electron changes the energy arrangements available and so changes the atom's spectrum also. For a given atom and electron in an atom, it takes a certain minimum energy to break the electron lose of the electrical grip of the nucleus—the *ionization energy*.

When an atom gains energy, an electron can move from a lower to higher energy level; when an electron drops from a higher to a lower level, the atom loses energy in the form of a photon.

Let's apply the energy-level concept to hydrogen. Every upward step requires the absorption of energy (Fig. 9.9*a*) and every downward one the emission of energy (Fig 9.9*b*). The greater the energy difference needed for a transition, the higher the frequency

Atoms and Light	Photons	produced and absorbed by electron transitions
	Transitions	electrons move between discrete energy levels
———	Energy levels	stable values of energies for electrons
At a Glance	Excitation	by collisions, absorption of photons

(hence the shorter the wavelength) of the photon produced from that transition. All transitions to and from the lowest energy level (level 1) involve large energy changes (short-wavelength photons), so they correspond to wavelengths in the ultraviolet range of the spectrum. This set of lines is called the **Lyman series**. The set of transitions down to and up from the second energy level (level 2) is the *Balmer series* (Fig. 9.9*c*); it lies in the visible region of the spectrum. The hydrogen absorption lines in the sun's spectrum are those in the Balmer series.

The Balmer series of hydrogen arises from transitions between the *second* energy level of hydrogen and the levels *above* the second; the transitions are *upward* from level 2 for absorption, and *downward* to level 2 for emission.

The simple Bohr quantum model needs drastic modification to work with other elements. Despite complications, one essential point remains: each atom and each ion has its own unique set of electron energy levels. So each has its unique energy-level diagram. Because spectral lines are produced by electronic transitions between energy levels, each element or ion has a unique set of spectral lines. Hence, the study of spectra reveals key information about the internal structure of atoms.

In a sense, atoms are tuned. They respond to energy input by resonating at their natural frequencies. Those are their energy levels, and as a result, the light emitted and absorbed by atoms provides information about their natural resonances, much as the playing of the open strings of a guitar tells you how the instrument is tuned.

Every atom has a unique structure of energy levels, and the transitions of electrons among those energy levels produce an atom's unique spectrum.

Caution: Do not confuse the *energy* of photons with the *brightness* of emission lines. The energy determines the wavelength (color). The total number of photons emitted at a given wavelength sets the brightness of the line.

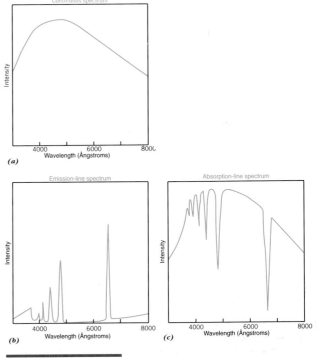

FIGURE 9.10

Schematic representations of the profiles of the main spectral types: (*a*) continuous, (*b*) emission line, (*c*) and absorption line. Since the intensity is plotted versus wavelength, emission lines rise above the level of the continuous spectrum and absorption lines fall below it. Sharp peaks correspond to emission lines; sharp dips to absorption lines. These lines are some of the Balmer series for hydrogen.

 9.4 Spectra from Atoms

Astronomers often show spectra by using a graph that plots intensity against wavelength. You will see both pictorial and graphic representations in this book, so you should know how to handle both. As a graph, continuous spectrum (Fig. 9.10*a*) lacks sharp dips or peaks, though the brightness or intensity level may vary gradually. Emission lines appear as sharp peaks (Fig. 9.10*b*); absorption lines as sharp dips (Fig. 9.10*c*).

What can excite atoms for emission or absorption? Imagine a box filled with a gas. The atoms of the gas collide with each other (and the sides of the box).

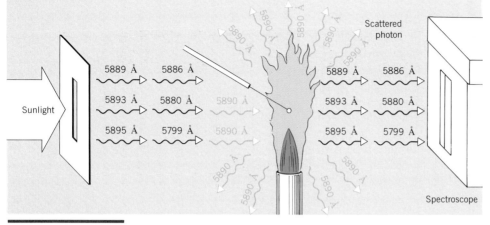

Scattered
photon

FIGURE 9.11

Paths of photons from the sun through a hot sodium flame and into a spectroscope. Photons with wavelengths of 5890 Å of one of the sodium D lines are scattered out of the beam, so that an absorption line is observed at this wavelength through a spectroscope. If you observed the flame from the side, you would see the sodium line in emission from the scattered 5890-Å photons.

Heat up the gas (add energy). The atoms then move faster and knock into each other harder. Collisions can bump electrons into higher energy levels; the harder the collision, the higher the level. This process is called **collisional excitation**. Some of an atom's kinetic energy is transferred to an electron of another atom. Photons can also excite atoms when absorbed, but only certain photons, namely those with energies corresponding to the *difference* in energies of two energy levels of the atom. This process is called **photon excitation**. Whether excited by collisions or by photons, an excited atom usually radiates quickly—in less than a millionth of a second.

Atoms can be excited by collisions (with other atoms, electrons, or ions) or by the absorption of photons.

Now return to the sodium vapor experiment with the quantum model in mind. When sodium chloride is placed in a flame, individual atoms of sodium are released from the salt. Some atoms collide with others and are excited. As the electrons return to lower levels, they emit photons of yellow light at 5890 and 5896 Å—the sodium D lines.

Let's follow one sodium D-line photon (5890 Å) on its way from the sun to the spectroscope through the sodium vapor cloud. If a photon of just this energy encounters an unexcited sodium atom, the sodium atom *absorbs* the photon. A dark line appears at 5890 Å. What happens to the sodium atom that has been excited by the photon absorption? It quickly emits a new 5890-Å photon. The original 5890-Å photon was headed directly for the spectroscope slit before it was absorbed by a sodium atom. The brand-new 5890-Å photon can be reemitted in any direction (Fig. 9.11). Very rarely will it be emitted in the same direction in which the original photon was traveling; only very few of these new photons enter the spectroscope slit. That's how the sodium line in the sun's spectrum becomes darker when it is passed through the sodium vapor.

This basic idea of how atoms in transparent gases produce line spectra applies to the sun and to stars, because their spectra resemble the spectrum of the sun (Fig. 9.12). The absorption lines are made as follows. Deep down in the sun's atmosphere, the solar gas is compressed and opaque. Here the continuous spectrum is produced. Above this region, the gas is thinner and cooler; these layers of the sun's atmosphere are the last significant barrier through which the continuous radiation must pass before it

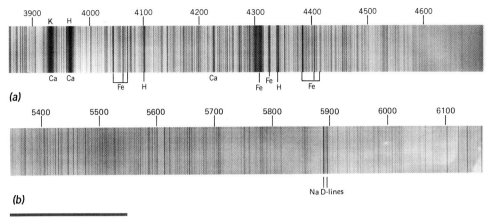

(a)

(b)

FIGURE 9.12

The sun's spectrum. (a) The short-wavelength end of the spectrum, showing absorption lines of calcium, iron, and hydrogen. The wavelength scale is at the top in ångstroms (Å). Note the very strong lines of calcium between 3900 and 4000 Å; we call these the K and H lines of singly ionized calcium. (Courtesy of Mt. Wilson Observatory, Carnegie Institution of Washington.) (b) The yellow-to-red region of the spectrum: Kirchhoff showed that the two especially strong lines in the yellow region, called the D lines, are made by sodium. (Courtesy Mt. Wilson Observatory, Carnegie Institution of Washington)

escapes. Atoms and ions at this level in the atmosphere absorb light at their characteristic wavelengths from the continuous spectrum and create the absorption lines in the sun's spectrum.

Chemical composition, temperature, and density of the gas all influence the spectrum that is produced by the sun or a star.

This approach applies generally to the spectra of stars. By similar atomic processes, the atmospheres of most stars produce absorption-line spectra that can be observed with a spectroscope attached to a telescope. (See Appendix H about telescopes.) Here is one for a star that has a temperature a little greater than the sun at its surface (Fig. 9.13). The overall

trend is the star's continuous spectrum. The spectrum contains many absorption lines; the strongest ones here are those from calcium and the Balmer series of hydrogen, which are produced in stars as they are in the sun. By matching the spectral lines of stars to those made in labs from known elements (Fig. 9.14), astronomers inferred that stars contained almost all elements found on the earth. But one important difference emerged: in contrast to the earth, stars and the sun contained mostly hydrogen and helium.

The similarity of spectra and chemical composition indicates that most stars have atmospheres like our sun's.

FIGURE 9.13

Spectrum of a star having a surface temperature like that of the sun, showing many absorption lines. The strongest of these are from the ionized calcium and the hydrogen Balmer series; many are from iron (Fe). (From *An Atlas of Objective-Prism Spectra*, by N. Houk, N.J. Irvine, and D. Rosenbush).

Ca   Fe   Hδ   Hγ   Hβ

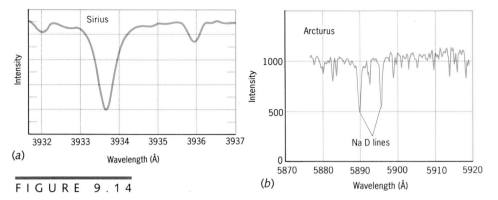

(a)

(b)

FIGURE 9.14

Profiles of stellar spectra. (a) A profile of the absorption K line of singly-ionized calcium in the spectrum of the bright star Sirius, a line also prominent in the sun's spectrum. (Adapted from a diagram in *Sample Spectral Atlas for Sirius*, by R. Kurucz and I. Furenld, Smithsonian Special Report No. 387, 1979.) (b) Section of the spectrum of Arcturus, showing the sodium D lines. (Courtesy J. E. Gaustad, Swarthmore College Observatory)

9.5 The Doppler Shift Revisited

Now that you are familiar with light and its wave properties, let's expand a bit on the Doppler shift (Section 5.1) as used in astronomy. I will focus on

See p. 84

the observations of the lines in spectra as markers for the cosmic speedometer.

The Doppler shift occurs with waves of all kinds, including light. When you are moving *toward* a wave source, the waves appear more frequent and shorter; in contrast, when you move *away* from a wave

FIGURE 9.15

The Doppler shift for light waves. In the direction of the motion, the waves appear compressed, so a blueshift is seen in the positions of the spectral lines of the source. If the source is moving away, the waves appear to be stretched out, and a redshift is seen. At right angles to the motion, no shift is seen in the spectral lines.

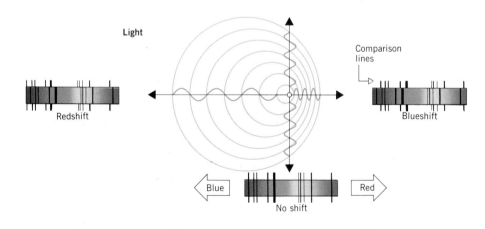

source, the waves appear less frequent and the wavelength longer. It's only the *relative velocity along the line of sight*—called the **radial velocity**—that causes the Doppler shift (for speeds much slower than that of light). For light waves, we can measure the Doppler shift by the *change in the wavelength* of lines in a spectrum. The emission or absorption lines are shifted toward the blue end of the spectrum for an object approaching you, and toward the red for one that is receding (Fig. 9.15). These shifts are called the **blueshift** and the **redshift**. (Keep in mind that red light has a longer wavelength than blue.)

How large is the shift? First, we measure the shift in the line compared to a standard at rest. We take that shift and divide it by the rest wavelength of the line. That then equals the radial velocity of the source divided by the speed of light.

The Doppler shift is an apparent change in wavelength from the relative motion of a source and receiver. It allows astronomers to find the line-of-sight velocities of luminous objects without having to know their distance. For stars, the astronomer takes advantage of the Doppler shift to determine the radial velocities relative to the earth by using spectral lines to measure wavelength shifts. The astronomer takes a spectrogram of the star and at the same time superimposes a comparison spectrum from a local laboratory source (Fig. 9.16). The comparison source is at rest with respect to the telescope and provides the normal (zero relative velocity) placement of the lines with respect to which the astronomer measures the shift. The larger the radial velocity, the greater the shift. The direction of the motion—approach or recession—determines the direction of the shift of the lines in the spectrum. So the amount of the shift and its direction give the radial velocity of the star.

MATH CONCEPT

With the measured shift and the value of c, the astronomer calculates the relative velocity between the source and earth by the expression

$$\frac{\Delta\lambda}{\lambda_0} = \frac{V_r}{c}$$

where V_r is the relative radial velocity, λ_0 the rest wavelength of the observed line, $\Delta\lambda$ the observed shift in the line ($\Delta\lambda = \lambda_{observed} - \lambda_0$), and c the speed of light. This equation shows that the ratio of the shift to the original wavelength is directly proportional to the ratio of the radial velocity to the speed of light. If we want to solve for the radial velocity, we can put the Doppler shift equation in the form

$$V_r = c\left(\frac{\Delta\lambda}{\lambda_0}\right)$$

which again highlights that the key property to measure is $\Delta\lambda/\lambda_0$.

FIGURE 9.16

Doppler shifts in the spectrum of a star, observed at different times. Above and below the spectra of the star, which contain absorption lines, are the emission-line spectra of a source at the telescope; these serve as the at-rest comparison spectra—the lines have zero shift. Spectrum (a) shows a blueshift and spectrum (b) a redshift. The star is in a binary system; at times it orbits toward the earth, at times away from it. (Courtesy Lick Observatory)

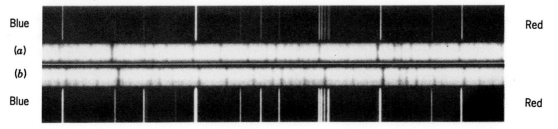

The Larger View

Celestial objects emit electromagnetic radiation over a wide range of wavelengths. Their emissions span the electromagnetic spectrum; visible light makes up only a small portion. On earth, we receive mostly visible light and radio waves; the rest of the spectrum is absorbed by the atmosphere—a natural blinder to the full power and glory of the cosmos. To gather information from the full range of the electromagnetic spectrum, we use telescopes on satellites and space missions. (See Appendix H on telescopes.)

Light reveals its origins through spectra. Each wavelength represents a different energy. Atoms emit and absorb light in discrete chunks—photons—from the transitions of electrons between energy levels. These jumps produce or absorb photons with a specific energies and wavelengths that show up as absorption and emission lines in spectra. Such photons convey information across the grand distances of space.

Astronomers try to observe the spectra of astronomical objects over as wide a range of wavelengths as possible. Why? So we can extract the maximum information about them—information about physical properties such as temperatures and chemical compositions. (You'll see how we do so for the sun in Chapter 10 and other stars in Chapter 11.) We can even use the Doppler shift to find out how fast they are moving. We need these essential facts to build models of the sun, stars, and other celestial objects.

Key Ideas and Terms

absorption-line spectrum
continuous spectrum
emission-line spectrum
spectroscope
spectrum

1. A spectroscope spreads light into its component colors, which is called a spectrum. Three general types of spectra exist: continuous, with a smooth spread of colors; emission, which shows discrete, bright lines; and absorption, in which dark lines appear in a continuous spectrum.

composition
element
absorption line
absorption-line spectrum
spectral line

2. The sun (and most other stars) have absorption-line spectra; by matching the dark lines in a star's spectrum with those produced by elements in the lab, the chemical composition of the star can be inferred. This procedure works because each element has a set of spectral lines unique to itself.

D lines (of sodium)
emission line
emission-line spectrum
Kirchhoff's rules
spectroscopy

3. Kirchhoff's rules of spectral analysis describe the physical conditions under which each type of spectrum is produced: continuous spectrum from a hot opaque gas, liquid, or solid; emission spectrum from a hot transparent gas; absorption-line spectrum from a continuous spectrum passing through a cooler, transparent material (usually a gas).

electromagnetic spectrum
energy
frequency
photon
quanta
radiative energy
wavelength

4. Light has both particle (quantum) and wave properties; red light has longer wavelengths than blue light. Light also carries radiative energy: the shorter the wavelength, the greater the energy, so photons of blue light carry more energy than those of red light. The complete range of light over all possible energies makes up the electromagnetic spectrum.

atom
Balmer series
Bohr model of the atom
excitation
ground state
Lyman series
transition

5. In an energy-level model for the hydrogen atom, Balmer lines arise from excitation of an electron, starting in the second level and jumping to any higher level (for absorption lines) or dropping from a higher level to the second level (for emission lines). The electron makes transitions between energy levels.

collisional excitation
energy level
electron
ground state
photon excitation
nucleus
transition

6. The quantum energy-level model for the emission and absorption of light by atoms views the electrons as playing a critical role. Electrons can inhabit only stable orbits called energy levels. Electrons can move to higher levels only if they gain energy (by collisions or absorption of photons); when they drop to lower levels, they give off energy (usually as photons). The greater the difference between levels, the greater the energy in the emitted photons.

blueshift
Doppler shift
radial velocity
redshift
spectroscopy
spectrum

7. Spectroscopic analysis reveals that stars have physical environments (temperatures and densities) and chemical compositions in their atmospheres basically like the sun. The same elements exist in stars as on the earth (though their relative proportions differ). By the Doppler shift, we can also find the relative speeds of the stars, either approaching us (blueshift) or receding (redshift).

conservation of energy
heat
kinetic energy
potential energy
radiative energy
temperature

8. A key rule in the universe is that energy is conserved, though it can mutate into many forms; atoms can transform potential and kinetic energy into radiative energy, and also do the reverse.

atom
ion
ionization
isotopes
neutron
nucleus
proton

9. Atoms are the basic units of matter. An atom has a nucleus of protons and neutrons, with the number of protons determining the element, the number of neutrons plus protons the isotope. Electrons in number equal to that of the protons orbit the nucleus; if electrons are lost or gained, the atom is ionized.

Review Questions

1. What type of spectrum shows a smooth blend of colors?
2. What is the basic function of a spectroscope?
3. What kind of spectrum do we observe from the sun?
4. By Kirchhoff's rules, what two properties are necessary if a gas is to produce a continuous spectrum?
5. Does ultraviolet light have greater or less energy per photon than blue light?
6. If light has a high energy, does it have a high or low frequency?
7. What atom produces Balmer lines?
8. If an electron drops from a higher to a lower energy level, what happens to the energy?
9. What element produces the D lines in the solar spectrum? What element would produce D lines in the spectra of other stars?
10. When an atom produces light, in what sense is energy conserved?
11. What is the electrical charge of an ionized atom that has lost two electrons?
12. How do the various isotopes of any one element differ?

Conceptual Exercises

1. You throw ordinary table salt, which contains sodium, into the flame of a Bunsen burner. What kind of spectrum do you see when you look at the vaporized salt with a spectroscope?
2. How do we know that sodium exists in the sun?
3. The spectra of most stars look like the spectrum of the sun: absorption spectra. How can particular elements be identified in these spectra?
4. Suppose you had a box containing hydrogen *ions* only. What kind of line spectrum would the ions produce?
5. The Balmer lines in the sun's spectrum are absorption lines. Do electrons jump up or fall down energy levels to produce them?
6. Arrange the following kinds of electromagnetic radiation in order from the least to the most energetic: x-rays, radio, ultraviolet, infrared.
7. When you examine the spectrum of the moon, it resembles that of the sun. Explain by Kirchhoff's rules.
8. What happens to a hydrogen atom when it absorbs light with enough energy to knock off its electron?
9. Use Kirchhoff's rules to predict what would happen if the emission line spectrum from a hot, transparent gas passed through another transparent gas that was hotter and had a different chemical composition.

Conceptual Activity 9   Using a Spectroscope

You can examine spectra yourself by building a simple spectroscope and using it to observe light produced under different physical conditions.

You will need a shoebox, some extra cardboard, two single-edged razor blades, a diffraction grating, and black paint or suitable fabric (see below). The grating will be the hardest piece to obtain; your instructor may have one that you can borrow. Or your instructor might suggest a source where you can buy these inexpensive items. A grating is inscribed with very closely spaced parallel lines, which use the wave nature of light to produce spectra.

To build the spectroscope, first cut a hole 1 inch (2.5 cm) square in the center of each end of the shoebox (Fig. F.4). Both openings should be the same distance from the upper edge. Tape a grating inside the box over one of the holes. It must be positioned so that the grating lines run up and down. Next, cut out a piece of cardboard 2 inches (5 cm) square. With a razor blade, make a vertical slot in it that is a quarter-inch (10 mm) wide and 1 inch (2.5 cm) long. Tape a razor blade on one side of the slit so that its edge is parallel to the slit. Place a small piece of cardboard tight against the taped blade, and tape the other blade on the other side as close to the cardboard spacer as possible. The opening between the blades should be narrow and parallel.

Remove the cardboard spacer between the blades and mount them on the inside of the box on the end opposite the one at which the diffraction grating is mounted. Paint the inside of the box a flat black, or cover the inside with a dark, dull fabric. Test the spectroscope by aiming the end with the slit at a light bulb. You should see a continuous spectrum. Note that you'll have to look through the grating at an angle to see the spectrum completely.

Now you can put your spectroscope to work. Go out one night and observe different street lights (such as mercury and sodium ones; you might also find xenon) and illuminated display signs. You'll find that not all signs are made with neon! Compare what you see to the color plate of spectra in the book.

FIGURE F.4

Assembling the pieces for a homemade spectroscope.

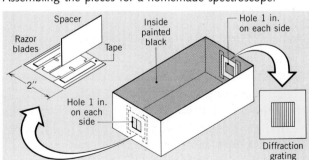

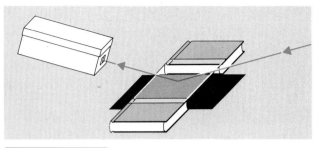

FIGURE F.5

Using a homemade spectroscope to observe the sun's spectrum safely.

NEVER USE A SPECTROSCOPE TO OBSERVE THE SUN DIRECTLY. SERIOUS EYE DAMAGE COULD RESULT.

To observe the sun's spectrum safely, place a pane of glass, say 8 by 10 inches in size, on top of two books at least 4 inches apart (Fig. F.5). Put a sheet of black paper beneath the glass. Set this arrangement so that sunlight hits the glass and reflects off it. Tape a piece of wax paper in front of the slit of your spectroscope; this will act as a filter. Now look at the sun's reflection with your spectroscope, pointing the slit at the glass. What do you see? Look for dark, vertical lines. Compare them to the photos of the solar spectra in the book. Can you match the lines of any elements? You should be able to do so for sodium.

10
Our Sun: A Model Star

■

Central Concept
The sun produces its life-giving energy by nuclear fusion reactions in its core; the outward flow of this energy determines the sun's physical structure.

■

The sun is basically a hot, huge ball of gas powered by fusion reactions in its core. There, deep in the heart of the sun, the energy bound up in nuclear matter is unleashed. Slowly, over millions of years, that energy (as electromagnetic radiation) slowly flows to the sun's surface. Free of the sun, the light flies into space. Some 8.3 minutes later, a very small part of that light strikes the earth—light that astronomers can analyze, light that energizes life.

The sun serves as our local link to the stars. It has the same basic structure as the other stars in the sky, which we can see directly only as pinpoints of light. To find out the physical characteristics of these distant lights, we use our sun as a model. The sun serves as the local laboratory for testing our ideas about stars in general; our understanding of stars hinges on that for the sun. To build plausible models of stars, we need models of the sun.

10.1 The Sun's Physical Properties

How large is the sun? How massive is it? Before we can tackle such questions, we need to know how far the earth is from the sun. Let's see what we know more or less directly about the sun.

The earth's orbit about the sun is elliptical (Section 2.5). The semimajor axis of its orbit, which is the average earth–sun distance, has a length called one *astronomical unit* (AU). So the earth is 1 AU from the sun. (Remember this; it will be used often.) How large is the AU in basic physical units, such as kilometers? To find out, we use the accurately known speed of light. Radar signals, which travel at light speed, are bounced off Venus, and the time between transmission and reception is accurately measured and converted into kilometers. Kepler's laws give the earth–Venus distance in AU. So we know a fraction of an AU in kilometers, hence the AU in kilometers: 1.496×10^8 km. (Remember this by rounding off, to 150,000,000 kilometers: 150×10^6 km).

Viewed from the earth (Fig. 10.1), the average angular diameter of the sun is 32 arcminutes. (That's about $\frac{1}{2}°$, or half the tip of your finger at arm's length.) Because we know the AU in kilometers, we get the sun's actual size from its angular size. The sun's diameter is roughly 1.4 million kilometers, or some 110 times the earth's diameter. Imagine the earth the size of a dime. The sun would be about 2 meters in size and 200 meters from the coin-sized earth. (See Conceptual Activity 10.)

To find the sun's mass, you need the earth–sun distance, the earth's orbital period, and Newton's version of Kepler's third law (Section 3.4). The sun's mass comes out to be about 2×10^{30} kg, or more than 300,000 times the earth's mass. The sun is a huge, massive object. Yet its bulk density (the mass divided by the volume) is low: about 1400 kg/m^3, only 40 percent denser than water! In fact, the sun is a hot gas throughout, even in its deep, dense core.

FIGURE 10.1

The visible face of the sun, its *photosphere,* in visible light: the photosphere fades out at the edge. The dark regions are sunspots; note that those in the bottom hemisphere appear to be lined up at about the same latitude. (Courtesy William Sterne, Jr., Sterne Photography)

10.2 Ordinary Gases

To grasp the physical conditions inside the sun (and other stars), you need some idea of how gases behave. Let me describe a simple model for ordinary gases. (You will come across extraordinary gases in Chapter 14.)

A gas consists of particles—atoms, molecules, electrons, or ions, or perhaps all of these. Simplify the situation in a real gas by imagining all the particles to be the same and like hard spheres. Picture these spheres trapped in a small box. Once set into motion, the spheres keep moving, bouncing off the walls of the box, and colliding with one another (Fig. 10.2).

Over time, after many collisions, all the spheres will have some average speed. This average speed relates to the average kinetic energy of particles in a gas as a measure of its *temperature.* If all the particles were motionless, a gas's temperature would be

The Sun's Properties At a Glance		
	Radius	6.966×10^8 m
	Mass	1.989×10^{30} kg
	Luminosity	3.863×10^{26} W
	Surface temperature	5780 K
	Average density	1410 kg/m^3
	Age	4.5 to 5 billion years
	Composition (surface) by mass	hydrogen, 72%; helium, 26%; other elements, 2%

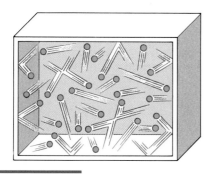

FIGURE 10.2

Hard-ball model of a gas in a box. The temperature of the gas is a measure of the average kinetic energy of the particles. The internal pressure is created by collisions among the particles.

zero—absolute zero on the Kelvin temperature scale (Appendix A). At room temperature, about 300 K, the average speed in a gas of oxygen molecules is about 0.5 km/s! Higher temperatures mean higher average speeds of the particles, and so a higher average kinetic energy.

This book uses the *Kelvin*, or absolute, temperature scale. One kelvin degree is the same amount of a temperature difference as one degree in the centigrade or Celsius scale. The zero points, however, are different. The Kelvin scale uses absolute zero; the Celsius scale uses the freezing point of water. On the Kelvin temperature scale, the freezing point of water is 273.15 kelvins, abbreviated *K*.

The temperature of a gas relates to the average kinetic energy of the particles that make it up.

Imagine putting a partition in the gas container. The spheres ram and bounce off both sides of this partition; each collision exerts a small force on it. The combined force of all collisions is the *pressure* of the gas, the force on a unit of area. Imagine that you increased the temperature of the gas. Then the average speed of the particles increases, so they collide with one another and into the partition more frequently and with greater force. The pressure increases. How, in relation to the increase in temperature? For ordinary gases, the increase is a direct one.

So if you double the temperature (in kelvins), the pressure doubles.

What happens to the pressure if you increase the number of particles in the box? (You've done this if you've pumped up a bicycle tube, by forcing more air molecules into it.) Adding more particles to the same space increases the density of particles; on the average, each cubic meter contains more particles. For a gas, the number of particles in a unit volume is called the *number density*. Suppose you increased the number four times without changing the temperature. Each cubic meter now contains four times as many particles. So four times as many collisions occur, on the average, against the partition, and each collision has the same average force as before. The pressure increases; it is four times greater. Increasing the number density directly increases the pressure of an ordinary gas.

In an ordinary gas, the gas pressure depends directly on *both* the number density and the temperature.

With these basics in mind, let's turn to that huge hot ball of gas that makes up the sun and infer some of its important physical properties.

 ## 10.3 The Sun's Spectrum

Most of what we know about the sun and other stars comes by way of light. Decoding the message of sunlight and starlight takes up much of the time, energy, and ingenuity of astronomers. How is the message read?

Solar Luminosity and Temperature

The sun's **luminosity** (or *power*) is its total output in radiative energy each second. By the time the light reaches the earth, it has spread out over a large region of space—over a sphere whose radius is 1 AU. We will need to know the area of this imaginary sphere to measure the sun's luminosity.

To find the luminosity with precision, we place a special detector in a satellite orbiting the earth.

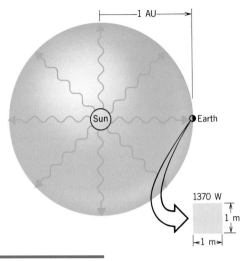

FIGURE 10.3

An imaginary sphere with a radius of 1 AU would capture all the sun's radiative energy output. At the earth's distance, one square meter captures only a small part of this total energy output, or about 1370 watts. This is the sun's flux at the earth.

(Why not a ground-based detector? Because the earth's atmosphere absorbs some sunlight.) Point the detector directly at the sun. Measure the radiant energy absorbed by the detector: It amounts to 1370 watts (abbreviated W; see Appendix A) for each square meter of the detector's area (Fig. 10.3); or about energy input of a hand-held electric hair dryer. So on a surface 1 AU in radius (an imaginary sphere surrounding the sun), every square meter catches this amount of energy, which is almost enough to light 14 hundred-watt light bulbs. Totaled up over the entire surface of the imaginary sphere, the energy amounts to roughly 3.8×10^{26} W—the same amount that the sun emits.

Astronomers use a special name for the amount of energy that passes each second through one given unit of area (such as a square meter): **flux**. The sun's flux, 1370 W/m², will serve as the standard of comparison for the flux from other celestial objects. A *spectrum* can be thought of as a plot of flux over a range of wavelengths.

We can find the sun's luminosity from its flux and distance.

Geological evidence indicates that the sun's luminosity (and its flux) has not varied more than a few tens of percent over the past 3.5 billion years. Satellite measurements show that it does not vary now more than a few tenths of one percent over decades. Careful, longer-term observations indicate a variation of only 0.25 percent in the past 50 years. Such seemingly small variations are important to us because they may affect the earth's climate.

The sun's color (yellowish white) provides a clue to its surface temperature. We can assign a temperature to the sun's surface by examining its continuous spectrum, ignoring the lines for the moment. (Recall, from Section 9.3, that the sun shows an *absorption-line spectrum*.) To see how, consider heating a piece of metal in a very hot flame. From Kirchhoff's first rule (Section 9.2), you know that this opaque solid emits a continuous spectrum. To your eye, the metal emits a dull red light, as do the coils of an electric stove at its highest setting. Then, as the metal gets hotter, it glows more brightly reddish, then orange, yellowish, yellowish green, white, and, finally, bluish white. The overall color change in the visible region of the spectrum relates to the temperature of the metal. So a solid's continuous spectrum changes with temperature in a special way (Fig. 10.4). Note three features in these spectra: (1) the emission *peaks* at some wavelength, (2) the peak shifts to *shorter* wavelengths as the metal gets hotter, and (3) the metal emits *more* flux at *all* wavelengths at *higher* temperatures.

An object whose continuous spectrum has this characteristic shape (Fig. 10.4) and the variation with temperature just described is called a **blackbody radiator**, or simply a **blackbody**. The spectrum is also known as a **Planck curve**, in honor of the German physicist Max Planck. The energy output as electromagnetic radiation, and the spectrum of a blackbody, depend only on its temperature and *not* on other properties (such as composition). Perfect blackbodies do not actually exist, but the sun (and many other stars) emit radiation somewhat like an ideal blackbody. The wavelength at which a star's continuous spectrum peaks relates to its temperature: the *higher* the temperature, the *shorter* (and thus more energetic) the peak wavelength.

The spectrum of a blackbody depends only on its temperature.

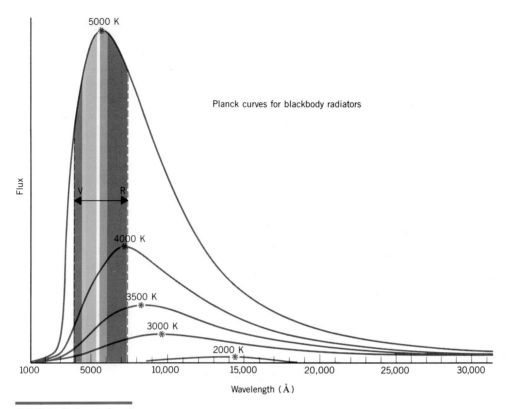

FIGURE 10.4

The spectrum of a hot metal (or opaque solid, liquid, or gas) as it varies with increasing temperature (2000–5000 K). The visible region (from violet, V, to red, R) is indicated; the long-wavelength end of the curve goes out into the infrared. An asterisk marks the peak of the emission for each temperature; note that the peak is at a shorter wavelength for higher temperatures. The shapes of these continuous spectra, called *Planck curves*, are typical of the emission from blackbody radiators.

A blackbody has a number of special characteristics. First, a hotter blackbody emits *more* energy at *all* wavelengths than does a cooler one. Second, a hotter blackbody emits a *greater proportion* of its radiation at *shorter* wavelengths than a cooler one. (That's why a hot object appears bluish.) And third, the amount of radiation emitted by the surface of a blackbody depends on the *fourth power* of its temperature. So if one blackbody is twice as hot as another of the same size, it emits $2^4 = 2 \times 2 \times 2 \times 2$, or 16 times as much energy in total. The hotter body also emits 16 times as much energy from each square meter of surface as the cooler one. Loosely speaking, for blackbodies: the hotter, the bluer, the brighter.

A common question about blackbodies is why they are called *black* when they give off light? Blackbodies are so named for their light-absorbing abilities: they absorb light at all wavelengths completely and reflect none. An object that absorbs all wavelengths of visible light appears black to a viewer; hence the name. When a blackbody absorbs radiative energy, it heats up and emits at all wavelengths, even though the peak of emission may not be visible to our eyes. Physically, a good absorber, when heated, is a good radiator.

Because the sun and many other stars emit radiation almost like a blackbody does, we can apply this concept to their continuous radiation. The sun's spectrum peaks at about 5000 Å. Without earth's atmospheric absorption, the sun's spectrum follows a more or less continuous curve with a peak in the yellowish-green part of the spectrum. To produce this peak requires a **surface temperature** of about 5800 K. Note that any opaque and hot solid, liquid, or gas will produce a blackbody spectrum, as you expect from Kirchhoff's rules (Section 9.2). The key word here is *opaque*.

The sun (like most other stars) emits a continuous spectrum that has the characteristics of a blackbody of a specific temperature.

Opacity

If the sun's continuous spectrum has a blackbody shape, the emitting region must also be a good absorber of light. What does the absorbing? To see this point, let me expand on the concept of **opacity**. On a clear day, you can see to great distances (many kilometers) through the air. The air is transparent to light; its opacity is low. On a very foggy day, you can't see far at all (only a few meters!). The air is opaque to light.

What makes a gas opaque? Interactions of light with atoms and electrons. When a photon is absorbed, it no longer exists and so can't carry energy any farther. The photon's energy is not destroyed; it has just been transferred to an electron. When this electron loses that energy, it emits a photon. But— and this is the key point—the photon can be emitted in *any* direction, including back in the direction from which it originated. It heads off and moves only a short distance before it is again absorbed. When another photon is reemitted, it usually zips off in a different direction. Photons in an opaque gas do not travel very far before they are reabsorbed; the lower the opacity of the gas, however, the greater the distance traveled. Photons bounce around in an opaque gas and so travel slowly through it, whereas they fly straight through a transparent gas.

Photons scatter many times before leaving a gas with a high opacity; a gas with low opacity lets photons fly straight through—it is transparent.

So the opacity of a gas relates to how far photons can travel, on the average, between absorptions. The opacity depends on how much of the absorber is there (its density) and how effectively it absorbs (what it is). If we define the visible surface of the sun as the first layer at which the gases become visibly transparent, the layer we see is only a few hundred kilometers thick. It has an average temperature of 5800 K, which we take as the *surface temperature* of the sun (even though the sun does not have a solid surface). This region defines the sun's **photosphere** (Fig. 10.1). It marks the place where visible light photons can escape straight out from rather than bounce around within the sun's atmosphere.

The Absorption-Line Spectrum

A spectroscope shows that the solar spectrum is an absorption-line spectrum. How is it produced? We use the concept of opacity again. Consider the 4383-Å line of iron (Fig. 10.5). We find when we measure the absorption of this line that it is not perfectly sharp. It has a finite width in terms of wavelength, centered on 4383 Å, where the iron atom absorbs very well. At somewhat shorter or longer wavelengths, the iron atom can still absorb light, but not as well. In other words, a gas containing iron has a much higher opacity for 4383-Å photons (at the

---

FIGURE   10.5

The absorption profile of the 4383-Å line of iron in the photosphere of the sun. This profile results from the ability of the iron atom to absorb photons at wavelengths near 4383 Å: that is, the opacity. The opacity is the greatest at the center of the line and drops off rapidly on either side.

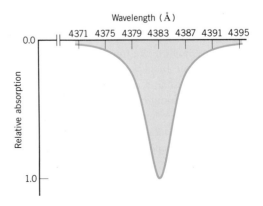

FIGURE 10.6

The sun's dark-line spectrum, from the red (top left; lower energy) to the violet (bottom right; higher energy). The spectrum is arranged so that the first row turns at the top right; the next at the left, and so on, with the result that the wavelength range is continuous. Note the many absorption lines; each appears at a specific color (wavelength). (Courtesy Sacramento Peak–NSO/NOAO)

line's center) than for those with wavelengths somewhat shorter or longer than the central wavelength.

In the photosphere, the high opacity at the centers of absorption lines traps photons. They travel very short distances, and so have little chance of escaping into space. The density of the photosphere decreases rapidly from bottom to top. So the opacity decreases as fewer atoms per cubic meter are available to absorb. Eventually the opacity falls enough to permit the escape of photons, even at the wavelengths of centers of lines.

As a result, the absorption lines form at *different levels* in the photosphere. At their centers, where the opacity is high, the light emerges from higher up in the photosphere, where the gas is cooler and emits less intensely. Off the line centers, where the gas is more transparent, the light emerges from lower, hotter layers, which emit more intensely. So the lines are brighter (less dark) away from their centers. Note again that absorption lines are *not* perfectly black and they do contain *some* energy. They are dark only relative to continuous emission at neighboring wavelengths.

In the photosphere, the temperature gets hotter with depth: a few thousand kelvins in a few hundred kilometers. It is this sharp temperature change that results in the sun's dark-line spectrum. In effect, you can consider the atmosphere as lying over a hot surface—just the situation needed, according to Kirchhoff's third rule (Section 9.2), to produce an absorption-line spectrum.

Astronomers have analyzed more than 20,000 lines in the solar spectrum to find the chemical composition of the sun's atmosphere. The intensity of the dark lines from a particular chemical element relates to how much of that element is there; it also depends on the temperature and density in the photosphere. Iron produces most of the absorption lines; other strong lines come from hydrogen, calcium, and sodium (Fig. 10.6). Identifying particular elements simply requires matching their "fingerprint" patterns. Most of the atmosphere's mass is hydrogen (72 percent) and helium (26 percent). All elements with more than two protons (loosely called *metals* or *heavy elements* by astronomers) make up a mere 2 percent.

The composition and chemical abundance revealed in analyzing the sun's absorption-line spectrum applies only to the *photosphere*.

If an element's absorption lines don't show up in the visible spectrum, can we assume that the element is not present in the sun? Not necessarily. Perhaps so little of the element exists that it does not produce detectable absorption lines. Or the strongest absorption lines may be in a region of the sun's spectrum that is unobservable through the earth's atmosphere. Another possibility is that the temperature and density of the sun's atmosphere may inhibit the formation of the element's spectral lines. If the conditions are too hot or too cool, the lines cannot form. So which lines form tells us about the temperature and density of the photosphere.

 ## 10.4 Energy Flow in the Sun

The thermal energy flow, from the hotter core to cooler surface and beyond the atmosphere, controls the internal structure and environment of the sun (Table 10.1). The sun that we see directly consists of the sun's outer layers, together known as the atmosphere. Because the sun's atmosphere is a gas, it does not have a distinct layer with sharp boundaries. But we have discovered that the atmosphere has three different zones: the *photosphere*, the *chromosphere*, and the *corona*. These regions are the visible

Table 10.1

The Sun's Structure

Region	Property
Core	Site of fusion reactions; temperature 8×10^6 to 16×10^6 K; energy transport by radiation.
Convection zone	Energy transport by convection; temperature below 500,000 K; outer 0.3 of the sun's radius, below photosphere.
Photosphere	Origin of continuous and absorption-line spectra; temperature 6400 to 4200 K; energy transport by radiation.
Chromosphere	Energy transport by radiation and magnetic fields; temperature 4200 to 10^6 K.
Corona	Energy transport by radiation and magnetic fields; temperature 1 to 2×10^6 K; source of the solar wind.

expression of the sun's energy outflow. Before dealing with each, let's look at how heat flows from one place to another.

In general, energy is carried by the processes of *conduction, convection,* or *radiation.* If you've ever relaxed in front of a roaring fire, you have experienced all these. You were directly warmed by the fire's infrared radiation, which traveled directly from the fire to you, to be absorbed by your skin. That's energy transported by **radiation**.

Much of the energy from a fire in a fireplace, however, is wasted up the chimney. The fire heats the air just around it. Air is gas, and the air expands and becomes less dense. Cooler, denser air flows in and pushes the hotter air up and out of the house, where it cools off. Recall that this transport of energy by mass motions of a gas (or liquid) is *convection*, which you've already encountered in planets (Sections 4.3 and 4.4).

A hot-air balloon provides a visualization of convection. On the ground, the crew heats the air in the balloon with a gas burner while they hold down the balloon. The air trapped in the balloon becomes hotter and less dense than the surrounding air. When released, the balloon rises, and the hotter air releases heat to the air. When the air has cooled and become more dense, the balloon sinks (unless the burner is used to reheat the air), generally landing at a different location than its launch site. The rising and falling of the balloon mimics the motions of blobs of air in convective cells that mark the main feature of energy transport by convection.

Finally, you may have by mistake left the poker in the fire. If you grabbed the handle without thinking, you learned about **conduction**: The poker's electrons that were actually in the fire were heated and so moved around at high speeds. They banged into their neighbors, agitating them. These collided with their neighbors, and so on throughout the poker from one end to the other. The kinetic energy was transferred by direct collisions along the poker from regions of higher to lower temperature. The poker feels hot to your touch because heat is flowing from the poker to your hand.

In ordinary stars like the sun, radiation or convection can carry energy along. (Conduction does not play an important role because the sun's material is a gas and not very dense. Conduction does become important in extremely dense stars where matter is in a different state than the ordinary: Chapter 14.)

Which process is at work? That depends on the local conditions in the gas. The general rule is this: the transport process that works most efficiently prevails.

Convection, conduction, and radiation are the basic processes of heat flow; which one dominates depends on which is more efficient, given local conditions.

What determines this efficiency? How well radiation transports energy through a material depends on *opacity*. Recall that opacity measures how efficiently a material absorbs photons. In the sun's core, photons travel about 1 mm between absorptions. The gas here is fairly opaque, but radiation is still more efficient than convection for transporting the energy. As photons journey toward the surface, they run into a region where the temperature is low enough to permit the formation of hydrogen atoms, which suddenly make the gas extremely opaque. The photons are bouncing around so much that they make little progress outward, and so do not transport energy very well. Energy transport by circulation of the gas can transport energy more efficiently, and a convection zone forms.

The sun's *photosphere* has a bubbly look, like the surface of a pan of boiling water (Fig. 10.7). Each

FIGURE 10.7

The granular structure of the photosphere at the top of the sun's convective zone. An individual granule is about 1500 km across; it marks the top of hot, rising bubbles of gas from convective upflow. The darker regions between are the places of downflow of the cooler, denser gas. (Courtesy Pic du Midi Observatory)

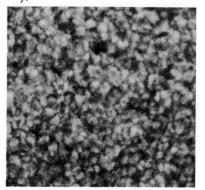

bubble has an irregular shape about 2000 km across (half the size of the continental United States!) and lasts for about 5 minutes. This phenomenon is called *photospheric granulation*. You can visualize the photospheric granulation as the top layer of a seething zone where hot blobs of gas spurt to the surface, radiate energy, cool, and flow downward. Just below the seething photosphere is a region of convection, where the outward flow of energy heats the gas and makes it rise. The base of the convection zone lies at about 0.7 to 0.8 solar radius. For most of the interior, though, radiation transports energy outward from hotter to cooler regions.

Just before and after totality in a solar eclipse, a bright, pink flash appears above the edge of the photosphere. This bright region marks the **chromosphere**, the solar atmosphere just above the photosphere. The pink color comes from the emission of the first line of the Balmer series of hydrogen (Section 9.3), in the red region of the spectrum. A spectroscope directed at the chromosphere during its fleeting appearance shows a bright-line spectrum (Kirchhoff's second rule, Section 9.2).

The chromosphere begins just above the photosphere and extends only a few thousand kilometers, where upon it merges into the corona. The chromosphere is about a thousand times less dense than the photosphere but, surprisingly, gets much hotter. The temperature rises from about 4000 K to above 400,000 K in the first two thousand kilometers above the photosphere. This sharp rise to high temperatures produces the emission lines from this region. Why the chromosphere generally is hotter than the photosphere is not yet known for certain.

The sun's splendid **corona** is the outermost layer of the atmosphere (Fig. 10.8). The corona emits bright lines from highly ionized atoms of iron, nickel, neon, and calcium in addition to hydrogen. Because it takes large amounts of energy to rip many electrons from an atom, the corona must be very hot—1 to 2 million K. At the base of the corona, the temperature rises rapidly, roughly 500,000 K in only a few hundred kilometers in a thin transition region between the chromosphere and corona (Fig. 10.9). This sharp rise may result from the transport of energy by magnetic fields.

The **solar wind** originates from the sun's hot corona, which expands from the sun into space. The solar wind is a swift plasma flow, carrying trapped magnetic fields along with it (Section 4.5). At the

FIGURE 10.8

The sun's corona as seen in the Philippines during the March 1988 solar eclipse. A special camera was used to accent details. Note the streaming structure in the outer regions (arrow). (Courtesy NCAR)

earth's orbit, the solar wind whips by at roughly 700 km/s. The speed varies from 300 to 800 km/s; the solar wind blows in gusts. The particles in the solar wind travel the distance from the sun to the earth in about 5 days. The earth catches some of the solar wind particles in its magnetic field to create its magnetosphere (Section 4.5). The wind sculpts comets' tails and blows well beyond the orbit of Pluto.

FIGURE 10.9

The sharp rise in temperature from the chromosphere to the corona. Note that the corona's temperature increases rapidly above the chromosphere and levels out to about 2 million kelvins. The horizontal axis gives the height in kilometers about the photosphere.

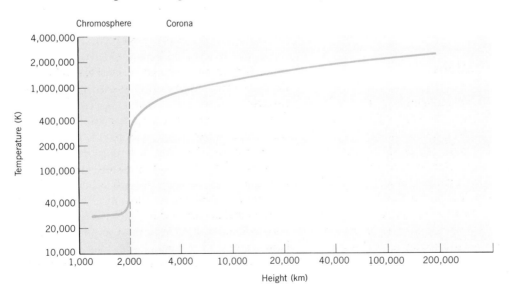

10.5 Energy Production in the Solar Interior

The interior of the sun contains the furnace that generates its energy. Most features of the sun, from photosphere to solar wind, result from the energy flow out from the interior. How does the sun yield so much energy? The problem is not the rate of energy production but its longevity. The sun has been at roughly the same luminosity for at least 3.5 billion years, according to present geological evidence. Energy from ordinary chemical reactions, such as burning, could not provide the amount necessary for so long. If the sun were composed entirely of oxygen and coal, it would have burned to a dark cinder in a few thousand years. Even energy from gravitational contraction of the sun's large mass could power the sun for only tens of millions of years.

Albert Einstein provided the key idea about the sun's energy at the beginning of this century. In his special theory of relativity, mass and energy are related by the equation

$$E = mc^2$$

where E is the energy (in joules) released in the conversion of mass m (in kilograms), and c is the speed of light (m/s). Because c^2 is a large number (about 9×10^{16}), a minute mass stores enormous energy. For example, the conversion into energy of 1000 kg of matter (the mass of a small car!) unleashes about 10^{13} kWh, roughly equal to the total energy consumption of the United States in one year!

Nuclear Transformations

How can matter be changed into energy? Two operations in nature unleash the energy frozen in the nucleus of an atom: *nuclear fission* and *nuclear fusion*. In the process of **nuclear fission**, the nucleus of an atom of a heavy element (such as uranium or plutonium) splits into two lighter nuclei. The mass of the remnants adds up to less than that of the original nucleus. The deficit in mass is released as energy. In **nuclear fusion**, the nuclei of lighter elements are fused together to create a heavier nucleus. However, the product has less mass than the original particles that were put in. The missing mass has been converted to energy.

> Nuclear fission and fusion processes convert matter into energy.

The sun produces its energy by fusion. Hydrogen is the sun's most abundant element and also has the smallest nuclear charge (one proton). The fusion of four hydrogen nuclei produces one helium nucleus. To make a helium nucleus from hydrogen nuclei releases 4.2×10^{-12} J of energy. That minuscule amount would raise the temperature of one gram of water only 10^{-13} K. Many hydrogen nuclei must react each second to supply the sun's luminosity.

Two sets of fusion reactions can transform hydrogen to helium: the **proton–proton chain** (*PP chain* for short) and the **carbon–nitrogen–oxygen cycle** (*CNO cycle*). The CNO cycle contributes a minor amount to the energy of the sun, but it acts as a key source in more massive stars (Chapter 13).

The Proton–Proton Chain

Let's look at the steps in the proton–proton chain. A collision between two protons starts it off (Fig. 10.10); if these nuclei collide with enough energy (a temperature of at least 8 million kelvins), the protons stick together. A deuterium nucleus (^{2}H) forms, consisting of a proton and a neutron. The other positive charge breaks away as a positron. (A **positron** is the antiparticle of the electron and carries a positive instead of a negative charge.) Almost lost in the shuffle is a neutral, supposedly massless particle called a **neutrino**, which zips away at the speed of light. The dense solar interior is transparent to the neutrino's travel, and, in about 2 seconds, the neutrino escapes into space, carrying away some energy.

In the solar interior, the positron quickly collides with an electron, and the two antiparticles annihilate to form two gamma rays, high-energy electromagnetic radiation. Meanwhile, the deuterium crashes into another proton and forms a light isotope of helium (^{3}He) and a gamma ray. This chain occurs again. Another ^{3}He nucleus is created; it collides

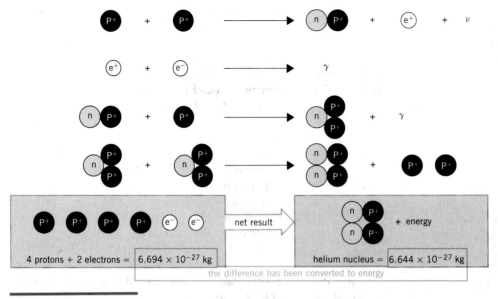

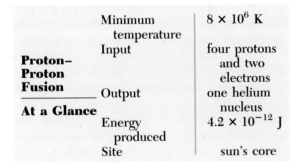

FIGURE 10.10

The primary proton–proton fusion reaction that occurs in the core of the sun. Net result is the formation of one helium nucleus that has a mass of about 5×10^{-29} kg less than the unbound particles that go into the reaction. The missing mass has been released as energy in this sequence of fusion reactions.

with one formed earlier. In this final reaction of the PP chain, the usual result is ordinary helium (^{4}He) plus two protons and another gamma ray.

Proton–Proton Fusion At a Glance	Minimum temperature	8×10^6 K
	Input	four protons and two electrons
	Output	one helium nucleus
	Energy produced	4.2×10^{-12} J
	Site	sun's core

The proton–proton chain is a fusion process that converts protons into a helium nucleus and transmutes some matter into energy.

Keep in mind that very high temperatures (at least 8 million kelvins) and high densities are needed for protons to collide with enough energy to fuse and frequently enough to generate the sun's energy. This requirement restricts the energy production to the sun's **core**, about the inner 25 percent of its radius. Only here is the temperature high enough.

In summary, the input for the PP chain is six protons and two electrons, and the usual output is one helium nucleus, two protons, two neutrinos, and assorted gamma rays. Note that the net result of this nuclear cooking is the creation of helium, and mass converted to energy. Each completed PP chain unlocks about 4.2×10^{-12} J. To account for the sun's luminosity, 1.4×10^{17} kg of matter must be converted to energy each year. That amounts to 4.5×10^9 kg of matter transformed each second, or about the mass of 2 million automobiles!

Fusion reactions, mostly the proton–proton chain, generate the sun's energy in its hot core.

The neutrinos fly off into space with about 2 percent of the sun's energy, but the gamma rays, which carry off the bulk of the energy, find the sun's interior opaque. The photons bounce along random paths as they are absorbed and reemitted; photons transport the energy. Slowly, the photons wander out toward the sun's surface, from regions of higher temperature to cooler regions. With this temperature decline, the average photon's energy declines. The original gamma rays are degraded, by interaction with the sun's material, into many lower-energy photons. In a few tens of thousands of years, the photons break out of the photosphere in the form of less energetic, visible radiation. The energy you see now was produced in the core many years ago.

The sun is a fusion furnace, forging helium from hydrogen in its core. Slowly, the core's helium abundance increases, and its hydrogen abundance decreases. Right now, the core is about 40 percent hydrogen and 60 percent helium. How long can the sun survive at this rate? The sun has enormous hydrogen supplies. Without convection, only the hydrogen in and close to the core can burn. This amounts to about 10 percent of the total. Also, the PP chain transforms only 0.7 percent of the mass into radiant energy. Even with these restrictions, the sun's fusion energy can last about 10 billion years. Given the age of the solar system—about 5 billion years (Section 8.3)—that means the sun has about 5 billion years left to live.

The sun is a middle-aged star; it has fused hydrogen for about 5 billion years and will continue for about 5 billion more.

Solar Neutrino Experiments

How to probe the sun's interior when we can't see below its surface? We have some idea of the interior conditions because we can make theoretical models of the sun. Computers calculate those models, which are matched to the observed characteristics of the sun. The concepts to build a model of the sun are as follows: energy source (fusion reactions) to account for the luminosity, the properties of ordinary gases, the opacity of the solar gas (which determines the form of energy transport), and the chemical composition (starting with that of the photosphere). The resulting model must have the same mass, luminosity, radius, and surface temperature as our sun does now, at an age of 5 billion years.

These models of the sun's interior have been constructed as carefully as possible. However, ongoing experiments cast some doubt on them. The problem: the neutrinos from fusion reactions. These particles fly directly out of the sun's core and reach the earth in about 8.3 minutes. So if we could detect these neutrinos, we could see into the sun's core.

How can we do so? Raymond Davis and his colleagues have developed a strange telescope to catch the solar neutrinos. This telescope consists of about 378,000 liters of a chlorine compound (cleaning fluid!) placed in a huge tank located about 1.6 km below the earth's surface in a gold mine in Lead, South Dakota. How does this telescope work? A chlorine atom can absorb a neutrino and be converted to argon. By a very delicate procedure, the argon gas can be flushed out of the tank and the amount measured, which allows the number of neutrinos captured during a given time span to be determined.

The experimental results? To date, the detected solar neutrinos amount to only *one-third* the number predicted from standard models of the sun's interior and our current knowledge of the nuclear reactions involved! To put the result another way, the experimental result implies that fusion reactions take place now in the core at one-third the rate needed to account for the sun's current luminosity.

What's wrong? First, the experimental equipment may contain unknown problems (though this is unlikely because the experiment ran for 7 years). The chlorine capture experiment is the only one to have run for a long time. Its uniqueness demands results from another, independent experiment. About 10 other experiments are now in preparation or progress. One of them, the Kamiokande II experiment in Japan, has reported a neutrino flux half that predicted by standard models of the interior. So the deficit of neutrinos is real, though the exact amount is still uncertain.

Second, if the results are accurate, they are telling us that the standard solar model is incorrect or that something happens to the neutrinos on the way to the earth; that is, the experiment may provide new information on the properties of the sun's interior or on the properties of neutrinos. Perhaps solar models

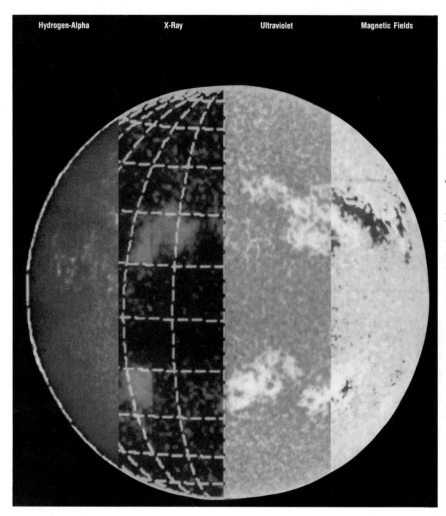

FIGURE 10.11

Active regions on the sun. The sun viewed in four different regimes of its emission at the same time; from the left to right, as it appears in the hydrogen alpha line, x-rays, ultraviolet, and regions of strong magnetic fields. Each image relates to a different level in the sun: magnetic fields in the photosphere; hydrogen-alpha marks the upper photosphere and lower chromosphere; ultraviolet, the upper chromosphere; x-ray, the corona. The effects of active regions are visible at each of these levels. (Courtesy *Astronomy* magazine)

made with much lower abundances of heavy elements produce far fewer neutrinos—but they contradict our general picture of the manufacture of elements in stars. Solar models can be rigged to agree with the experiment, but then they contradict other astrophysical data or what we know about the earth's climatic history. (The neutrinos observed from Supernova 1987A, Section 14.4, confirmed that some of the expected fusion reactions *do* produce neutrinos.) Or, if neutrinos have even a little mass, they would behave differently from expectations.

We do not detect enough neutrinos to account for the sun's luminosity; no really satisfactory explanation for the discrepancy between theory and results has been worked out.

10.6 The Active Sun

Events of the active sun involve occasions of localized, short-lived turmoil on or near the solar surface. The most important of these are sunspots and flares, which are associated with the locally intense magnetic fields. In general, these areas of solar activity are called **active regions** (Fig. 10.11).

One other solar property contributes to solar activity: the sun's nonuniform rotation. Because the sun is a gas, it can and does spin faster at the equator (one rotation every 25 days) than at the poles (one rotation every 31 days). So as you travel from a pole toward the equator on the sun, the photosphere's rotation rate increases; this effect is called *differential rotation*. (You saw that the Jovian planets rotated this way because they are fluids.) The differential rotation distorts solar magnetic field lines and drives the breeding of active regions.

Because the sun is a fluid, it rotates differentially, at least at its surface.

Sunspots and Flares

Sunspots appear as dark blotches on the solar disk (Fig. 10.12). With a temperature of about 4000 K, a sunspot is relatively cooler than the photosphere and so appears dark in contrast. Sunspots have a strong tendency to form in groups. A large, single sunspot group may contain 100 individual spots, which persist for months. Intense magnetic fields are associated with sunspots—the strongest have field strengths exceeding 0.4 T. (That's about 8000 times stronger than the earth's field at its surface.)

Sunspot numbers vary periodically. About 11 years pass between sunspot *maxima* (times of greatest numbers of sunspots) through sunspot *minima* (times of least numbers of sunspots) to the next maximum. During a cycle, the sunspots' locations also change. Sunspots usually appear only in the zone between the solar equator and 35° north or south latitude. At the start of a sunspot cycle, a few spots emerge at high latitudes. As the cycle progresses, the

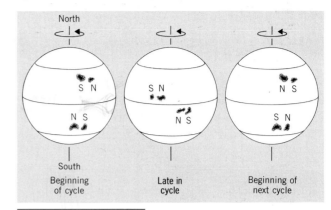

FIGURE 10.13

The magnetic cycle for sunspots. At the beginning of a sunspot cycle, if spots with north polarity are leading groups in the northern hemisphere, then south polarity leads groups in the southern hemisphere. (Here, "leading" means going in the same direction as the sun's rotation.) Late in the same cycle, the spots appear to be closer to the equator but the polarities are the same. At the beginning of the next cycle, the polarity of the leading spots is reversed for each hemisphere.

sunspot zone migrates toward the equator. As the survivors of one cycle expire near the equator (about 5° north or south latitude), new spots from the next cycle form at the higher latitudes.

Sunspot groups contain spots of opposite magnetic polarity. For example, the east spot of a twin group might have north polarity, the west spot south. If this order holds at a given time in the northern hemisphere of the sun, the situation at the same time in the southern hemisphere is reversed (Fig. 10.13). The magnetic polarities for the west and east spots have a 22-year cycle, just double the 11-year cycle of sunspot numbers. That is, if a spot group has the west spot with a south polarity, in the next sunspot cycle it will have a north polarity, and in the following cycle, south polarity again.

Sunspots have a roughly 11-year cycle in numbers and a 22-year cycle in magnetic polarity.

Magnetic fields drive the sun's activity over a 22-year cycle. We believe, although all the details are

FIGURE 10.12

Unusual sunspot group, visible in 1982, with a spiral shape in the large, main spot. The large spot below the center has a diameter about that of the earth. (Courtesy NSO/NOAO)

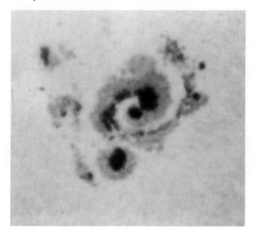

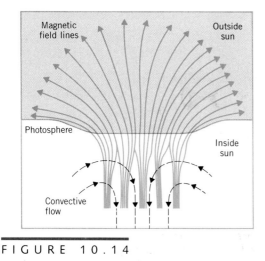

F I G U R E   1 0 . 1 4

One model of the distribution of magnetic field lines in a sunspot driven by the convective upflow and downflow below the photosphere. The convective upflow of the sun's plasma pushes the magnetic field lines together below the photosphere. (Adapted from a figure by E. N. Parker)

not worked out, that the cycle occurs from a coupling of the sun's overall magnetic field, which is driven by convection, with its differential rotation. In other words, the sun has an electromagnetic *dynamo* in its convection zone. The convection and the dif-

ferential rotation drive this dynamo to generate magnetic fields, and they interact to create a cycle. We believe that similar processes create magnetic fields and activity cycles in other stars.

Sunspot magnetic fields are generated by enormous electric currents, much as the field of an electromagnet is produced. The hurricane of currents and fields in turn creates sunspots. How this happens is unclear. One possibility: the sunspots' magnetic fields suppress the hot gas rising from the convective zone. The hot gas runs into a magnetic thicket and has trouble breaking through the surface in the sunspot's center (Fig. 10.14). A convective downflow draws away some of the hot gas, removing heat from the region. That may explain why a sunspot is cooler than the surrounding photosphere.

Sunspots mark regions of high concentration of magnetic field lines in the photosphere.

Sunspots are floating islands of electromagnetic storms, which generate short-lived, violent discharges of energy called **solar flares** (Fig. 10.15). These energetic bursts appear in active regions. The elapsed time between the birth of a flare and its rise to peak is only a few minutes, even for a large flare, and the decay time is about an hour. Emitting myr-

F I G U R E   1 0 . 1 5

The enormous solar flare of March 1989. This sequence of photos taken in hydrogen-alpha light of the Balmer series shows the early stages, followed by the ejection of charged particles, visible as the dark, spraylike structure at the lower left (arrow). When the flare was at maximum (last photo), it had become the strongest optical flare ever recorded. (Courtesy Sacramento Peak–NSO/NOAO)

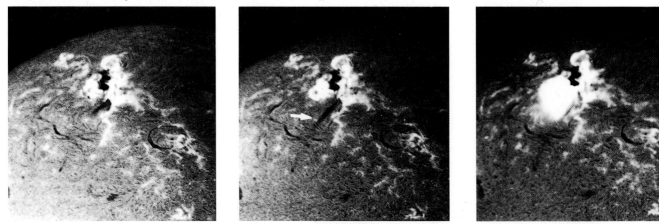

1516 UT   **1523 UT**   **1532 UT**

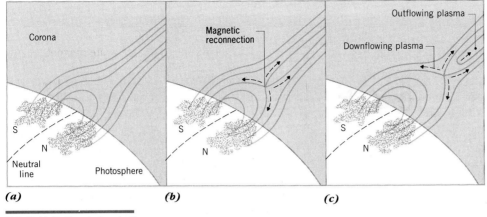

(a)   **(b)**   **(c)**

FIGURE 10.16

One model for the physical process that creates a solar flare. (a) Magnetic loops extend into the corona from active regions of north and south polarity. A *neutral line* between the north and south magnetic regions indicates where the field has zero strength. (b) Magnetic reconnection occurs above the photosphere and connects field lines of opposite polarity. (c) The point of reconnection moves up the field lines into the corona, driving plasma outward and downward as the field lines, like stretched rubber bands, release energy. (Adapted from a figure by R. Noyes)

FIGURE 10.17

The sun imaged in x-ray light in February 1991. Very hot regions (a million kelvins or so) appear bright in this picture, so most of these regions lie in the sun's corona. The looped appearance of streamers in the corona arises from strong magnetic fields. This image of the sun shows fine structure in the coronal loops, where the plasma temperature is 1 million K. Note the active regions strung out parallel to the equator. (Courtesy L. Golub, IBM and Harvard–Smithsonian Center for Astrophysics)

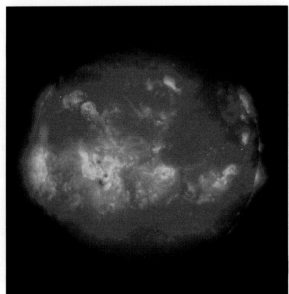

iad forms of energy—x-rays, ultraviolet and visible radiation, high-speed protons, and electrons—a large flare blows off about 10^{25} J, the equivalent of the energy released by a bomb of 2 billion megatons!

One model for how a flare works involves magnetic reconnection (Section 4.5) of magnetic loops in the corona (Fig. 10.16). Such loops connect the north and south polarities of an active region; at the top of the loops, field lines extend away from the sun. Stresses on the magnetic field prompt reconnection at the top of the loop. This reconnection forces some charged particles down into the chromosphere and some out into space—a burst that we see as a flare. When these energetic particles are captured in the earth's magnetosphere, they can generate auroras (Section 4.5).

Flares are produced by the sudden release of magnetic energy in an active region.

Coronal Loops and Holes

Because the coronal gas is so hot, it emits low-energy x-rays and shows up in x-ray images of the sun (Fig. 10.17), which reveal that the coronal gas has an ir-

FIGURE 10.18

Streamers in the sun's corona during a total eclipse in March 1988. This is a composite of a white-light and x-ray image (look back at Fig. 10.8). The corona photographed in white light shows the streamers representing the extension of the sun's magnetic field into space and is the source of the solar wind. The x-rays (red) show the corona a few hours before the eclipse. (Courtesy NCAR)

regular distribution above and around the sun. Large loop structures indicate where the ionized gas flows along magnetic fields that arch high above the sun's surface and return to it. The hot gas is trapped in these magnetic loops. The corona consists primarily of such loops.

Some regions of the corona appear dark where the coronal gas is much less dense and less hot than usual; these regions are called **coronal holes**. The coronal holes at the poles do not appear to change very much, but those above other regions seem to be somehow related to solar activity. Coronal holes mark areas where fields from the sun continue outward into space rather than bending back to the sun in loops. So the coronal gas, not tied down in these regions, can flow away from the sun out of the coronal holes; this flow makes the solar wind (Fig. 10.18).

The coronal gas does *not* follow the differential rotation of the photosphere. Rather, like the earth, it rotates at the same angular speed at all latitudes. This property implies that the bottoms of the magnetic loops are anchored deep below the photosphere, perhaps at the very bottom of the convective zone, where such fields might be generated by a solar dynamo. The magnetic loops, rising from the photosphere up to 400,000 km into the corona, play the key role in heating the corona, although we don't yet know how in detail.

Magnetic fields play a prominent role in most forms of solar activity.

10.7 Solar Eclipses and Relativity

At the start of this century, Einstein created an insight about gravitation that replaced the one developed by Newton. This new picture emerged in two stages: the **special theory of relativity**, then the *general theory*. The special theory dealt with the laws of physics as seen by *unaccelerated* observers—those who experience uniform motion. The general theory copes with the nature of gravitation and so deals with *accelerated* observers. The sun during solar eclipses played a key role in the verification of this theory.

Einstein showed in his special theory that mass and energy are two aspects of the same essential stuff. He was able to derive the famous formula used in Section 10.5 for understanding the source of the sun's energy. This relationship means that you can think of all matter as a form of energy and all kinds of energy as possessing mass. This mass produces gravity and also shows up as inertia.

The special theory resulted in the uniting of the concepts of space and time. Scientists had looked at these as separate aspects of the universe. In Einstein's view, all events in the universe involve space and time together: **spacetime**, which has four dimensions (three of space and one of time). Einstein made special note of what seems so obvious: when anything happens in the world, it takes place in *both* space and time.

After working out the special theory, Einstein tackled the problem of dealing with *accelerated* observers in spacetime. This investigation led to the **general theory of relativity**, which is Einstein's theory about the nature of gravitation. Einstein rethought gravitation by noting that the acceleration of gravity is *independent* of the mass of an object. Consider the moon: it accelerates toward the earth. A spacecraft of much different mass placed at the moon's distance has the *same* acceleration! Einstein felt that this was a fundamental fact about the universe and gave it a special place in the general theory.

Note that the moon and a spacecraft are free-falling. To Einstein, free-falling objects are simply following their *natural motion in spacetime*. Consider the moon again. A scale placed under it would read zero—it is weightless as it orbits the earth. So is the spacecraft. What determines their paths? That depends on the local geometry of spacetime. You probably assume that it is flat, but it could be curved. In Einstein's general theory of relativity, the distribution of mass and energy determines the local geometry of spacetime. All objects produce a curvature of nearby spacetime. And that curvature shows itself by accelerated motions, which Newton would have attributed to gravitational forces.

The general theory of relativity views gravity as an indicator of the curvature of spacetime.

Einstein's theory predicts that light rays will be bent by the curvature of spacetime near massive objects. Remember, light has energy, and so mass—but not much! Testing this prediction requires a strongly curved region of spacetime. In the solar system, that region is found where the most mass is located: the sun. Relativity predicts that the sun's mass is deflecting light rays away from Euclidean, straight-line paths.

How to test this? Take a picture of the stars and sun during a total solar eclipse. Later, when the sun is not in the same part of the sky, take a picture of the stars. Compare the angular separation of two stars close to but on opposite sides of the sun. With the sun present, the angular separation of the stars is *greater* than when the sun is not there (Fig. 10.19). Is this shift visible? Yes! Such observations have been made through the toil of many people over many years. They come out very close to Einstein's prediction from general relativity of 1.75 arcseconds. (That's about $\frac{1}{1000}$ the angular diameter of the sun.)

What's happening? In Einstein's view, light pursues a straight line in spacetime. But the geometry in which it travels is curved by the sun's mass. You can picture the sun's mass as creating a warp in the geometry of spacetime (Fig. 10.20), like a bowling ball placed on a trampoline. Light crossing the warp appears to us to take a curved path, although it is actually traveling on the shortest distance between two points through a curved region of spacetime.

Einstein's general theory makes testable predictions, such as the bending of the paths of light, which have been confirmed.

Einstein considered the orbits of the planets to be straight-line paths in curved spacetime. We view these paths as elliptical orbits in three dimensions. The paths appear curved to us because spacetime is warped, not flat. The amount of warping decreases as the distance from the sun increases. That's why the earth, for example, follows an orbit more strongly curved than that of Jupiter.

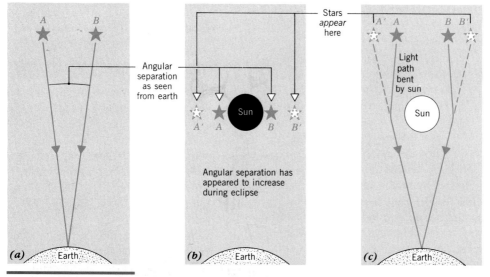

FIGURE 10.19

The effect of the sun's mass on the apparent angular separation of two stars. Consider two stars (*A* and *B*) that have a measured angular separation when the sun is not nearby. During an eclipse, when the sun lies between the two stars, they appear to be farther apart (at *A'* and *B'*) because of the bending of the light paths by the curvature of spacetime near the sun. The mass of the sun acts as an imperfect lens in changing the paths of the light rays.

FIGURE 10.20

Einstein's explanation for the change in two stars' angular separation during an eclipse. The mass of the sun curves the spacetime around it: the closer to the sun, the greater the curvature. The light from the stars follows this curved spacetime when the sun lies between them as seen from the earth (*A'B'*). When the sun is not there, the light paths travel through a flatter region of spacetime, and so the stars appear closer together (*AB*). (Adapted from a diagram by C. Misner, K. Thorne, and J. Wheeler)

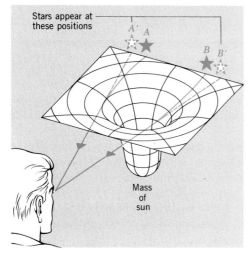

The Larger View

The sun is a star—an unremarkable one among the billions around us. But as our only resident stellar specimen, it provides the key to understanding stars in general. As for any star, we first analyze its light to find, among other facts, its luminosity, temperature, and chemical composition. Then we apply what we know about ordinary gases, nuclear fusion, and heat flow to build a reasonable model for the sun—and so the general structure of stars.

Our model sees the sun as a huge, gaseous, thermonuclear reactor, powered by fusion reactions, which convert hydrogen to helium in its hot core. The photons spawned in the fusion reactions in the core bounce outward and take many years to reach the surface. Eventually they emerge through the photosphere and produce the sun's spectrum. All the features of the sun result from the outward heat flow interacting with matter (Fig. 10.21). For us the most essential is the sunlight we receive at the earth—energy first formed in the inferno deep in our sun, energy that powers life on our planet. But this sunlight does not last forever. In about 5 billion years, the sun will run out of hydrogen fuel and die.

In Chapter 11, we'll look beyond the sun to other stars. Many are ordinary, but some are fantastic! Yet, we can still apply the basic model of a sphere of gas, contained by gravity, energized by a fusion

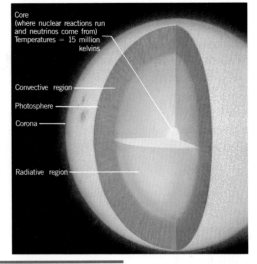

FIGURE 10.21

Model of the general interior structure of the sun. The fusion reactions take place in the core; the energy released here flows as radiation through the radiative zone. When the opacity becomes high enough, energy transport occurs by convection. Above the convective zone lies the photosphere. The corona marks the outer limit of the sun's atmosphere. (Courtesy Paul DiMare)

furnace, with heat flowing from the core to the surface—then to space and to us. Starlight teaches us about the stars.

Key Ideas and Terms

astronomical unit
density
gravitation
mass

1. The astronomical unit (AU), the average earth–sun distance, is about 150 million km. It is found accurately by bouncing radar signals off Venus. With the distance known, the sun's radius can be calculated from its angular size, its mass from the earth's orbital period and Newton's law of gravitation, and its low bulk density from its volume and mass.

flux
luminosity
power

2. To find the sun's luminosity or power (total radiative energy output per second) requires a knowledge of the earth–sun distance in kilometers and a measurement of the sun's flux at the earth.

blackbody **blackbody radiator** **continuous** ** spectrum** **Planck curve** **surface temperature**	3. The sun's surface temperature can be found from its color or from the wavelength at which its continuous spectrum hits a peak intensity. The continuous spectrum of the sun approximates that of an ideal blackbody radiator made of opaque material.
absorption-line ** spectrum** **absorption lines** **composition** **opacity** **photosphere**	4. The sun's surface—its photosphere—is the region in which the gases become transparent to light. The continuous spectrum forms in this region. The absorption lines form in higher, cooler layers, making the overall absorption-line solar spectrum. The sun's absorption lines indicate what elements are in the photosphere. The sun contains by mass about 72 percent hydrogen, 26 percent helium, and 2 percent all other elements (generally called by astronomers *metals*).
chromosphere **conduction** **convection** **corona** **heat** **photosphere** **radiation** **temperature**	5. Heat flows from the hotter to the cooler regions inside the sun. The photosphere has a bubbly structure, indicating that a zone of convection exists beneath it. This convection carries energy to the surface to be radiated into space. The chromosphere above the photosphere is hotter and is visible by its emission lines during an eclipse; the corona is even hotter than the chromosphere, as indicated by emission lines of highly ionized atoms. It is not yet clear why the chromosphere and corona are so hot.
CNO cycle **core (of the sun)** **neutrino** **nuclear fission** **nuclear fusion** **positron** **PP chain**	6. The sun's energy comes from fusion reactions. In them, four protons (hydrogen nuclei) are fused into a helium nucleus with the loss of mass and release of energy. Only in the sun's core is the temperature (and density) high enough for these fusion reactions to occur; the sun has enough hydrogen to fuel these reactions for billions of years. Gradually, the sun's core increases in its percentage of helium as its hydrogen decreases.
core (of the sun) **neutrino** **nuclear fusion**	7. Solar neutrino detectors to date have found far fewer neutrinos than are predicted by models of the sun and nuclear reactions. A good reason for this discrepancy has not yet been worked out.
active region **convection** **differential rotation** **dynamo** **magnetic field** **solar flares** **sunspots**	8. The sun undergoes cycles of activity in which active regions—which tend to be sites of sunspots and flares—become more common and then less so. This cycle lasts 22 years, as indicated by the polarity of sunspots; the magnetic field and activity cycle are driven by a solar dynamo, established by the sun's convection and differential rotation.
active regions **auroras** **magnetic field** **solar flares** **sunspots**	9. Active regions relate to the development of strong local magnetic fields, which produce sunspots (dense concentrations of magnetic field lines) and solar flares. These emit electromagnetic radiation (x-ray, ultraviolet, and radio) and high-speed particles that cause auroras when they reach the earth.

corona
coronal holes
magnetic field
solar wind

10. The solar wind (a plasma) leaves the sun through coronal holes, cool regions in the corona from which the sun's magnetic field streams into space.

general theory of
   relativity
spacetime
special theory of
   relativity

11. Observations of stars near the sun during total solar eclipses verify the general theory of relativity—Einstein's concept of gravitation. Space and time are unified in this theory. What we call gravity is understood as a curvature of four-dimensional spacetime. Objects move along the shortest paths in curved spacetime; for the planets, such motion appears as elliptical orbits.

Review Questions

1. What is the earth–sun distance in light travel time?
2. How does the density of the sun compare to that of water?
3. How does the mass of the sun compare to that of the earth?
4. How does the radius of the sun compare to that of the earth?
5. What is the rough value of the sun's flux at the earth?
6. What is the rough value of the sun's surface temperature?
7. What region of the sun's atmosphere produces the absorption lines in its spectrum?
8. What is the *second* most abundant element in the sun's photosphere?
9. How is energy transported to the sun's photosphere?
10. What kind of reaction produces the sun's energy?
11. In what part of the sun do the energy-producing reactions take place?
12. How does the observed number of solar neutrinos compare to the expected number?
13. What is the main feature of a solar active region?
14. Where in the sun does the solar wind originate?
15. In Einstein's view, how many dimensions does spacetime have?

Conceptual Exercises

1. Why do astronomers bounce radar signals off Venus to find the distance to the sun?
2. What measurements must be made to find out the sun's luminosity? Which of these is *least* accurate?
3. Suppose you examined sunlight with a spectroscope. Describe, in general, what you would see.
4. Explain the sun's spectrum by simple atomic processes.

5. Why do you *not* see helium absorption lines in the sun's visible spectrum, even though helium is the sun's second most abundant element?

6. Do you expect the chemical composition of the sun's core to differ from that of its photosphere? Why or why not?

7. Describe how to estimate the sun's surface temperature from its continuous spectrum.

8. Use the concept of opacity to explain why it takes photons millions of years to arrive at the surface from the sun's core.

9. In one short sentence, describe the source of the sun's energy.

10. Describe a method by which you could infer the temperature of a sunspot from its continuous spectrum.

11. We can see spectral lines of helium from the sun's chromosphere but not the photosphere. Why?

12. How do we know that the photosphere marks the top of the sun's convective zone?

13. As the sun evolves, how do you expect the chemical composition of its core to change?

14. In one sentence, describe Einstein's concept of gravitation.

Conceptual Activity 10   The Size of the Sun

You can actually measure the size of the sun by hand! Here's how. Get a piece of cardboard about the size of a playing card (use an old one), a piece of white paper, a ruler, a pencil, a pin, and a sunny day. With the pin, make a hole in the center of the card. Line up the card and the white paper so that an image of the sun is projected through the pinhole onto the paper (Fig. F.6). Have a friend mark with the pencil around the edge of the sun's image. Also have your friend measure the distance from the pinhole to the image in centimeters. Then measure the diameter of the sun's image, also in centimeters.

F I G U R E   F . 6

Making an image of the sun with a pinhole.

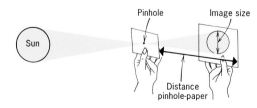

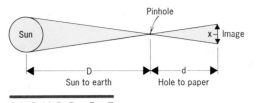

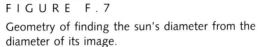

FIGURE F.7

Geometry of finding the sun's diameter from the diameter of its image.

From the position of the pinhole, the angular sizes of the sun's image on the paper and the sun in the sky are the same (Fig. F.7). So the ratio of the sun's actual diameter to its distance from the earth is the same as the ratio of the sun's image to the distance between the paper and pinhole:

$$\frac{\text{diameter of sun}}{\text{distance earth-sun}} = \frac{\text{diameter of image}}{\text{distance hole–paper}}$$

Let S be the sun's diameter, D the earth–sun distance, x the image diameter, and d the distance from the pinhole to the paper; then

$$\frac{S}{D} = \frac{x}{d}$$

and solving for the sun's diameter, we have

$$S = \frac{D \times x}{d}$$

You have measured x and d; now you just need D. Use the AU: 150×10^6 km; the result will then come out in kilometers for the sun's diameter. Compare with the value given in Appendix C.

WARNING: NEVER LOOK AT THE SUN DIRECTLY!

11

The Stars as Suns

■

Central Concept

Astronomers determine the physical properties of stars by finding their distances and analyzing the light received from them.

■

The stars are enormous fusion reactors in space, like our sun. That view is recent, developed in full in this century, when astronomers finally acquired the conceptual tools to analyze starlight and compare the stars to the sun. By finding distances to nearby stars, we are able to infer basic physical properties that are applied to more distant stars.

These revelations reinforced the connection between the sun and the stars, *the* essential link to the rest of the cosmos. The stars are suns; the sun is the nearest star. Using the sun as a guide to stellar physical properties, we can build up our models of stars.

FIGURE 11.1

In the southern Milky Way, the star cluster (to left of center) called New General Catalog 6520. (NGC 6520). The stars have various colors: white, reddish, yellowish, and bluish. To the right of middle is a cloud of dust blocking out light from stars behind it. (Image by D. Malin; Courtesy Anglo-Australian Observatory)

11.1 Messages of Starlight

Go out on a clear January night to view the constellations. Face south around midnight. Orion (refer back to Fig. 1.1) immediately catches your eye. Two stars in Orion shine the brightest: Betelgeuse, which looks reddish, and Rigel, blazing bluish white. To the south and east of Orion lies Canis Major, the Great Dog, Orion's hunting companion. Sirius, the jewel of Canis Major, is the brightest star in the sky. (See the winter star map in Appendix I for these constellations; Appendix E lists the physical properties of Betelgeuse and Sirius.)

The differences in these stars typify the astronomers' dilemma: What can we know about these distant stars? How do these stars compare with our sun? Are they larger? More luminous? Hotter? How to find out these physical properties? In what ways do these stars resemble the sun? How do they differ?

Starlight carries information about the physical properties of stars (Fig. 11.1). By deciphering starlight and comparing it to sunlight, we can infer the nature of stars. Their physical properties include chemical composition, surface temperature, radius, luminosity, and mass. For some of these properties, we do not need to know the distance to the stars; for

Ordinary Stars	Composition	hydrogen, helium
	Energy sources	fusion reactions, gravitational contraction
	Core	site of fusion reactions
At a Glance	Luminosity	total amount of energy produced each second
	Photosphere	source of stellar spectra

others, we do. In all cases, we observe, measure, and analyze the light from the stars as was done for the sun. We need this information to be able to make reasonable models of stars.

Luminosity and Flux

Stars do not have the same brightness. You make this judgment with your eyes, which work somewhat like a telescope (Appendix H): the lens of each eye focuses the light onto the retina, which detects the light (Fig. 11.2). How bright a star appears to you depends on how much energy, each second, strikes your retinas. The more energy received, the greater brightness you perceive.

A telescope works like your eye. It gathers light to a focus, and a detector there senses the radiative energy. How bright a star appears in a telescope depends on how much energy, each second, arrives at the telescope from the star and how large the telescope is. Recall that *flux* is the name given to the amount of energy striking each unit of area in a second (Section 10.3). The flux at the earth from the sun is 1370 watts per square meter. Compared to the sun, Sirius, the brightest star, sends a small flux to the earth—a mere 10^{-7} watt per square meter. To put it another way, we received about 10 *billion* times more energy per square meter from the sun than we do from Sirius!

Flux is a measure of the amount of energy per second received over a certain area.

Remember how to find the sun's *luminosity* (Section 10.3)? Measure the sun's flux, find the earth–sun distance in kilometers, construct an imaginary sphere with a radius of 1 AU, and total up all the energy hitting that sphere. We can find a star's luminosity by the same procedure (Fig. 11.3). Measure the star's flux at the earth. Then, knowing its distance, find the area of a sphere surrounding the star, and add up the flux over the total area. Simple enough—but we need to know the distance to the star, and that may be hard to come by (Section 11.2).

A star's luminosity is the amount of radiative energy the star emits each second, and that's fundamental to understanding stars as suns. To find a star's total luminosity, we measure its flux over the *complete* range of the electromagnetic spectrum. If the flux is measured over a limited range of the spectrum, say the visual, astronomers report this as the *visual flux*. Then the luminosity calculated from it is

FIGURE 11.2

Comparison of the optics of the human eye (top) and a refracting telescope (bottom). Both bring light to a focus to make an image on a detector. In the case of the eye, the retina serves as the light detector.

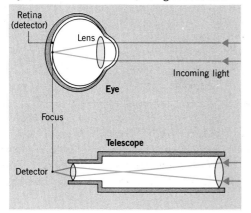

FIGURE 11.3

Determining the luminosity of a star from its distance and the flux at the earth. When it reaches the earth, a star's light has spread out over a sphere whose radius is the distance between the earth and the star, and whose area is $4\pi d^2$. The flux received at the earth is a very small fraction of the star's luminosity.

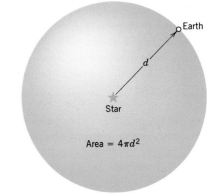

only the *visual luminosity* of the star; most luminosities given in this chapter are visual ones. Fortunately, stars like the sun emit most of their light in the visual part of the spectrum, so their visual luminosity is almost their total luminosity.

A star's luminosity is the total amount of radiative energy it emits each second.

The Inverse-Square Law for Light

You may have noted that *distance* plays a key role in relating flux to luminosity, because the brightness of a light source relates in a very specific way to your distance from it.

Consider this experiment. Put a bare light bulb in a socket and turn it on. Take a light meter—a device for measuring the flux of light, commonly used with automatic exposure cameras—and place it 1 meter from the bulb (Fig. 11.4). Note the reading on the light meter, and call it one unit. Now move the light meter to 2 meters from the bulb; its reading will be one-fourth that at the 1-meter position. Move the light meter to a distance of 3 meters; the reading will

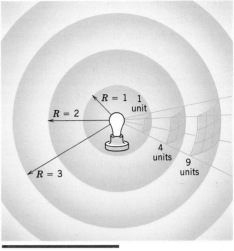

FIGURE 11.5

Geometry for the inverse-square law for light. Place a light bulb in a series of concentric, transparent shells. Since each larger shell has a greater area, light that covers one unit of area in the first shell covers 4 units in the second and 9 in the third. So the flux at the second shell is $\frac{1}{4}$ the first; at the third, $\frac{1}{9}$. The spreading out of the light dilutes the flux.

now be one-ninth that at 1 meter. Note that the flux decreases as the *inverse square* of the distance.

How does this inverse-square relation happen? Imagine the bulb placed in the center of concentric transparent spherical shells (Fig. 11.5). As the light moves away from the bulb, it expands in a sphere in all directions. The total amount of energy in the light remains the same, no matter how large the sphere. As the light passes through the first sphere, it covers a certain area. As it goes through the second (with twice the radius of the first), it covers a larger area. The area of a sphere is proportional to the radius squared; so the larger sphere has four times the area of the smaller one, and the light spreads out four times as much.

Now pick any square meter of surface for both spheres. Because of the radiation's dilution over a larger area, the small patch you select on the larger sphere has only one-fourth as much light striking it as the patch on the smaller sphere. For the next sphere, with a radius of 3, the decrease in brightness would be by 9; if increased to 4, the decrease would be by 16. This is the **inverse-square law for light**

FIGURE 11.4

Flux and distance. As a light meter is moved away from a light bulb, the measured flux decreases. The greater the distance, the lower the reading, so that at 2 meters, the flux is only $\frac{1}{4}$ that at 1 meter, at 3 meters, $\frac{1}{9}$ as much.

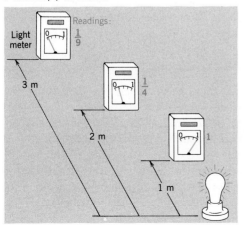

(or any electromagnetic radiation): the flux is inversely proportional to the square of the distance.

Suppose you observe that the flux of one star is 100 times that of another. If you assume that both stars have the same luminosity, how do their distances compare? The brighter one must be 10 times as close as the fainter one, according to the inverse-square law.

We can find a star's luminosity from its flux and distance, using the inverse-square law for light.

You have already encountered an inverse-square law in a different guise—Newton's law of gravitation (Section 3.3) involves another physical property, but the way in which it changes with distance is the same.

11.2 Stellar Distances: Parallaxes

The key to finding a star's luminosity from its flux is its *distance*. For nearby stars, we have a direct method to find distances: triangulation, similar to the procedure used by surveyors on the earth. Stellar triangulation is called **heliocentric** or **trigonometric parallax**.

For example, put your hand out at arm's length with a pencil held upright (Fig. 11.6). Now alternately open and close each eye. The pencil will appear to jump back and forth relative to the distant background. This angular shift in the pencil's position is the parallax of the pencil. Now if you use angular measure to determine the amount of shift the pencil appears to have and also the distance between your eyes, you can calculate how far the pencil is from your head.

Imagine that your pencil is a nearby star, the background is more distant stars, and your eyes are the sighting positions of the earth in orbit around the sun separated by a time of 6 months. From these two positions in the earth's orbit (separated by 2 AU), the nearby star appears to shift in position (its parallax) compared with the more distant stars (Fig. 11.7). Measure the angular size of that shift. Half of this angular shift is called the parallax of the star. Be-

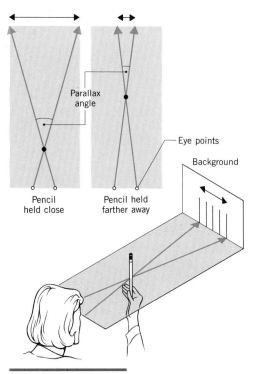

FIGURE 11.6

Observing the parallax of a pencil held at arm's length. As you alternately open and close each eye, the pencil appears to jump in position relative to the background markers. The farther away you hold the pencil, the smaller the parallax angle. The baseline for the shift is the distance between your eyes, and the angle shown is the parallax angle.

cause you know the diameter of the earth's orbit, you can calculate the distance to the star.

The farther away a star is, the smaller its parallax angle will be; the closer, the larger the angle.

Parallax, then, occurs when you view a relatively close object from each of two ends of a baseline. As the earth moves from one place in its orbit to another, a nearby star's position seems to change relative to the more distant stars. The maximum shift occurs when you view the star on nights 6 months apart, so you are sighting from opposite sides of the earth's orbit. The angular shift (actually half the shift) is the star's *parallax*. Measure the amount of

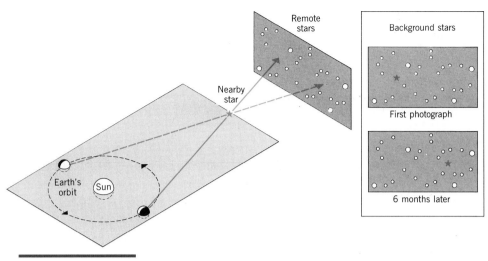

Remote stars

Background stars

First photograph

6 months later

Nearby star

Earth's orbit

Sun

FIGURE 11.7

Trigonometric (heliocentric) stellar parallax. As the earth goes around the sun, the positions of nearby stars shift relative to the positions of more distant stars in 6 months. The nearby star returns to its original relative position in a year. The closer the star, the greater the observable shift. The shift has a one-year cycle, the earth's period of revolution. The baseline for the shift is the diameter of the earth's orbit. (Technically, half of this total shift is what astronomers call the parallax angle.) Note that a larger baseline results in a larger shift, so that from Mars, for instance, the same star would have a larger parallax (but the cycle is longer, about 1.5 years).

that shift. Then, because you know the radius of the earth's orbit (1 AU), you can calculate the distance to the star.

MATH CONCEPT

When a star is 1 *parsec* (pc) from the sun, then it is at the distance at which its *parallax* is one *second* of arc, 1 arcsecond, a distance roughly equal to 3.26 light years (ly). Suppose the star were twice as far away; it would have half the parallax, $\frac{1}{2}$ arcsecond. If at half the distance, its parallax would double and be 2 arcseconds. Note this inverse relationship of parallax and distance. Algebraically, the distance d is related to the parallax p (arcsec) by

$$d \text{ (pc)} = \frac{1}{p}$$

or using light years as the unit of distance,

$$d \text{ (ly)} = \frac{3.26}{p}$$

The parallax of Sirius is 0.38 arcsecond, and so its distance is 8.6 ly. The distance to Sirius is about 546,000 times the distance to the sun (546,000 AU). From this information, we use Sirius' flux to calculate its luminosity compared to the sun's luminosity from its flux. We measure the flux from a star. With a star's distance known, the inverse-square law gives us the flux from a star, *if* it were 1 AU from us. We compare that figure to the flux from the sun. We then know how luminous a star is with respect to the sun. For example, if you imagined that we moved Sirius next to the sun, it would appear visually 23 times brighter than the sun. So Sirius is 23 times more luminous a the sun.

Parallax works accurately only for close stars. Note that the parallax of Sirius, a very close star, is less than 1 arcsecond, about $\frac{1}{2000}$ the angular diameter of the moon! That's the size of a United States quarter at a distance of a little more than 5 km. Because parallaxes are so small, it should not be surprising that it was only in the nineteenth century that astronomers succeeded in measuring the first stellar parallaxes. Current ground-based techniques have an accuracy of about 0.001 arcsecond, so we can get accurate distances out to about 300 ly (and some 10^4 stars). In the near future, space telescopes will measure parallaxes to an accuracy of about 10 times better than can be achieved today—out to a few thousand light years.

11.3 Stellar Colors, Temperatures, and Sizes

Let's return to the stars in the winter sky. One observable property is their fluxes. Another property you can observe is **stellar color**. Betelgeuse looks as if it were tinged reddish, Rigel and Sirius appear bluish white, and Capella shines yellowish white. What can we learn from these differences? The different stellar colors suggest that these bodies have different **surface temperatures**. Rigel is the hottest of the four; it is so hot that it emits more blue light than any of the other visible colors. In contrast, Betelgeuse is the coolest, for it radiates mostly red light.

The colors of stars relate to their surface temperatures: redder is cooler; bluer is hotter.

When we look at starlight with our eyes, we see only the visible part of its entire spectrum. Stars radiate somewhat like blackbodies (Section 10.3), so their continuous spectra, from shortest to longest

F I G U R E   1 1 . 8

Colors and the continuous emission from stars. (*a*) The fluxes of blackbodies with temperatures equivalent to (top to bottom) those of Rigel, Sirius, Capella, and Betelgeuse compared in the color bands violet-blue (v-b), green (g), yellow-orange (y-o), and red (r). These are compared to continuous spectra of different temperatures, which match closely that of each star. The blackbody spectra have the distinctive shape known as a *Planck curve*. Note that the visible range is only a small part of the total spectrum; the color of a star depends on the eye's perception of a star's continuous spectrum in the visible range. (*b*) For the 10,000 K curve, the flux in the blue range is greater than that in the red. The ratio of these fluxes, measured accurately, defines the color of the star.

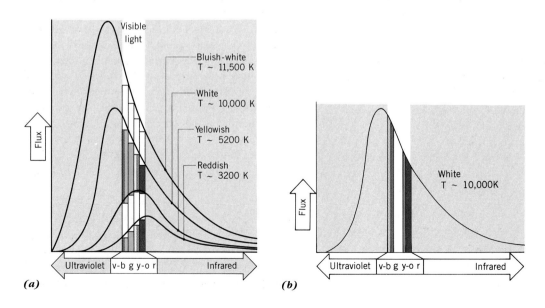

wavelengths, have the characteristic blackbody shape, called a *Planck curve*. The visible range covers but a small part of a blackbody's spectrum (Fig. 11.8*a*).

One way to measure a star's temperature is by measuring the *relative* fluxes at any two wavelengths—that's the meaning of *color* (Fig. 11.8*b*). A reddish star emits more than red light alone, but the relative amount of the longer (red) wavelengths is greater than that of the shorter (blue) wavelengths. All stars emit light over the entire visible range, but each shows relatively different amounts of red and blue light.

The stellar temperatures obtained from the colors help us to infer the sizes of stars, if they radiate like blackbodies. An example: in the constellation Scorpius lies the bright reddish star Antares, which is a binary star (Section 11.6). Antares and its companion revolve around each other, bound by gravity. The reddish star is called Antares A, its bluish-white companion Antares B. Antares A has a surface temperature of roughly 3000 K; its companion, about 15,000 K.

We receive about 40 times more flux from Antares A than from Antares B. Now both stars lie at the *same* distance from the earth, so the difference in flux *cannot* arise from different distances. It comes from differences in the *luminosities*: Antares A is 40 times more luminous than Antares B. How can a cooler star be so much more luminous than a hotter one?

The amount of energy emitted by *each unit of area* of a blackbody's surface depends only on its temperature—in fact, on the fourth power of the temperature. For example, Antares B is about five times hotter than Antares A; every second, therefore, each square meter of Antares B emits 5^4, or $5 \times 5 \times 5 \times 5 = 625$ times the energy of Antares A. If Antares A had 625 times the surface area of Antares B, both stars would have the same luminosity. Because Antares A has 40 times the luminosity of Antares B, it must have $40 \times 625 = 25,000$ times the surface area. Once we know the surface area, we can calculate a star's radius—Antares A is 160 times bigger than Antares B.

A star's luminosity is related to its surface temperature *and* to its surface area.

A key property of blackbodies is the amount of energy, E, emitted for every square meter of a blackbody at temperature T (kelvins):

$$E = 5.67 \times 10^{-8} \, T^4$$

in units of watts per square meter. This relation is called the *Stefan–Boltzmann law*. For example, every square meter of the sun's photosphere (5800 K) emits about 6.3×10^7 W:

$$E = 5.67 \times 10^{-8} \, (5800)^4$$
$$= 5.67 \times 10^{-8} \, (1.1 \times 10^{15})$$
$$= 6.3 \times 10^7 \text{ W/m}^2$$

If a star radiates as a spherical blackbody, then its power or luminosity is its surface area times its emission per unit area. Since the area of a sphere is $4\pi R^2$, then

$$L = 4\pi R^2 \, (5.67 \times 10^{-8} \, T^4)$$

11.4 Spectral Classification of Stars

The absorption spectra of many stars resemble the spectrum of the sun. From the sun's spectrum, astronomers infer its photospheric composition. The same procedure can be applied to stars. But there's one twist: a star's spectrum is affected by its *surface temperature* as well as by its *chemical composition*.

Most stars have pretty much the same chemical composition, but their spectra are *not* all alike. Consider the Balmer lines of hydrogen (Fig. 11.9). Stars cooler than the sun have spectra with weaker Balmer lines. Stars somewhat hotter than the sun show stronger (darker) Balmer lines. But stars *much* hotter than the sun again have spectra with weak Balmer lines.

Recall (Section 9.3) that the Balmer series arises from transitions between the *second energy level* of hydrogen and any level above it, upward for absorption and downward for emission. To absorb energy and produce a Balmer line, a hydrogen atom must be excited to level 2. In the sun, only one of every

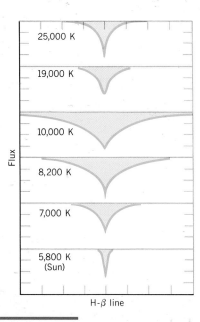

25,000 K

19,000 K

10,000 K

8,200 K

7,000 K

5,800 K
(Sun)

Flux

H-β line

FIGURE 11.9

The variation of the intensities (darkness) of one Balmer line of hydrogen (the second line in the series, the hydrogen-beta line) in stars with different temperatures. The flux profile of the line in the sun is at the bottom. Note how the line increases in intensity and width from higher temperatures until it reaches a maximum at 10,000 K stars; it then decreases.

10^8 atoms is excited; not enough energetic collisions occur to kick up many electrons from the ground level. So the Balmer lines in the sun's spectrum are not very strong. In a cooler star, fewer energetic collisions occur, and fewer hydrogen atoms are excited. The Balmer lines are much weaker. So the Balmer lines act as an indirect thermometer, signaling the temperature of a gas.

In a hotter star with more energetic collisions, more atoms are excited, about one out of every million. The Balmer lines are darker in such a star. If the star is sufficiently hot, collisions are so violent that many electrons are knocked out of the atom entirely, leaving hydrogen *ions* (protons) behind. There are not many hydrogen atoms with electrons in the second level from which Balmer absorptions can occur. So with fewer excited atoms, the Balmer lines are even weaker.

Balmer lines of hydrogen, which are produced by atoms with an electron in the second energy level, show the temperature in the atmosphere of a star.

If for some reason there aren't very many hydrogen atoms excited to the second level in a star, the Balmer lines in that star will be weak. For example, if the star has a very high temperature, virtually all its hydrogen will be ionized, or if the star has a relatively low temperature, even though there is much neutral hydrogen, there will be very few excited atoms in the second level.

At the turn of this century, workers at Harvard College Observatory classified stellar spectra by using absorption lines, especially hydrogen Balmer lines. The original classification scheme was set up strictly on the basis of the strength of various lines (Fig. 11.10), well before there was any understanding of the effects produced by different temperatures. The Balmer lines played a vital role in this scheme.

The **stellar spectra sequence** from hotter to cooler in the Harvard classification now runs **O-B-A-F-G-K-M**. [A mnemonic for the sequence is: *Oh, Be A Fine Girl* (or *Guy*), *Kiss Me!*] Almost all stellar spectra fit into this sequence of **spectral types**. The O stars have spectra with weak Balmer lines of hydrogen and lines of ionized helium, whereas A stars have the strongest Balmer lines. In F stars the Balmer lines are weaker, and many other lines appear, mostly of metals. The sequence from types O to M, looking at the continuous spectra from the stars, is also a color sequence. The O stars appear bluish white, G stars yellowish, and M stars reddish. Astronomers further divide each type into subtypes from hotter to cooler, usually labeled from 0 to 9. For example: G0, G1, G2, G3, G4, G5, G6, G7, G8, G9; each subtype is distinguished by slightly different intensities of specific absorption lines.

The strengths of the Balmer lines suggest that the differences in stellar spectra reflect primarily differences in *temperature*, not in the abundance of elements. These temperature differences result in different degrees of ionization and excitation of the atoms in the star. How many atoms are excited and

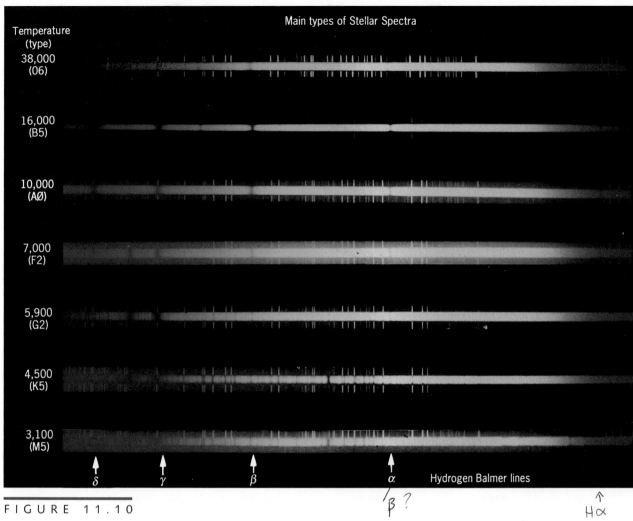

Main types of Stellar Spectra

Temperature
(type)

38,000 (O6)

16,000 (B5)

10,000 (AØ)

7,000 (F2)

5,900 (G2)

4,500 (K5)

3,100 (M5)

δ   γ   β   α   Hydrogen Balmer lines

/β ?   Hα

FIGURE 11.10

Stellar spectra from type O (hottest) to M (coolest) classified on the Harvard system. Note the intensities of the hydrogen lines, which appear strongly in some spectra and not at all in others. The positions of the Balmer line series are indicated by the arrows. The spectral types at left also have their surface temperatures given. Each stellar spectrum (absorption lines) has a comparison spectrum (emission lines) above and below it to provide a wavelength reference. (Courtesy Yerkes Observatory)

how many are ionized determine the strength of the atom's spectral lines. The variation in Balmer line intensities arises from collisions that excite and ionize atoms. How much the collisions ionize or excite depends on temperature. So each spectral type corresponds to a restricted range of surface temperatures, which are listed in Table 11.1.

A star's spectrum is determined mostly by the temperature of its photosphere, as indicated by the strengths of various absorption lines.

Other lines from other elements can be analyzed in a fashion similar to that for the Balmer lines (Table 11.1). For example, the line from once-ionized calcium is strongest in G-type stars and weaker in hotter or cooler ones. The observation of stellar spectra coupled with an understanding of the atom gives astronomers information about the physical conditions (such as the temperature) in the atmosphere of a star. An analysis of spectral lines based on atomic theory also provides information about the abundance of elements in stars in the same way as for the sun. Astronomers have found that, like the sun, ordinary stars consist mostly of hydrogen and helium.

Table 11.1

Features of the Stellar Spectral Classes

Spectral Class	Color	Surface Temperature (K)	Principal Features	Examples
O	Bluish-white	30,000	Relatively few absorption lines. Lines of ionized helium and other lines of highly ionized atoms. Hydrogen lines appear only weakly.	Naos
B	Bluish-white	11,000–30,000	Lines of neutral helium. Hydrogen lines more pronounced than in O-type stars.	Rigel, Spica
A	Bluish-white	7,500–11,000	Strong lines of hydrogen. Also lines of singly ionized magnesium, silicon, iron, titanium, calcium, and others. Lines of some neutral metals show weakly.	Sirius, Vega
F	Bluish-white to white	6,000–7,500	Hydrogen lines are weaker than in A-type stars but are still conspicuous. Lines of singly ionized metals present, as are lines of other neutral metals.	Canopus, Procyon
G	White to yellowish-white	5,000–6,000	Lines of ionized calcium are the most conspicuous spectral features. Many lines of ionized and neutral metals are present. Hydrogen lines are weaker even than in F-type stars.	Sun, Capella
K	Yellowish orange	3,500–5,000	Lines of neutral metals predominate.	Arcturus, Aldebaran
M	Reddish orange	3,500	Strong lines of neutral metals and molecules.	Betelgeuse, Antares

11.5 The Hertzsprung–Russell Diagram

How do the surface temperatures and luminosities relate to the internal anatomy of the stars? Again we face the astronomer's dilemma: how to extract vital information from points of light in the sky. The solution lies in the spectroscopy of these stars.

Sorting Stars by Luminosity and Temperature

Let's look at a graph: a temperature–luminosity diagram (Fig. 11.11) based on the spectral types of the stars. Such diagrams were independently set up early this century by Ejnar Hertzsprung (1873–1967) and Henry N. Russell (1877–1957). In their honor, such a plot is called a **Hertzsprung–Russell diagram** (commonly abbreviated to *H–R diagram*). Note that astronomers follow the strange convention of plotting temperature so that it increases to the *left* on the horizontal (temperature) axis of the H–R diagram. (Conceptual Activity 11 gives you the information to make an H–R diagram for selected stars.)

Examine Figure 11.11. Notice Sirius B in a corner by itself and Betelgeuse alone in the upper right-hand region. Regulus, Sirius A, Procyon, the sun, and Epsilon Eridani fall along a sloping line close to each other. No obvious patterns, though, appear in this plot.

Let's now inspect a different H–R diagram, one for the nearest stars (Appendix D), all within 20 ly of

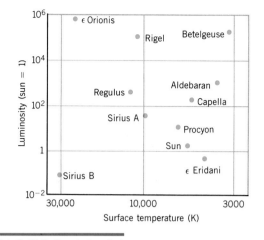

Hertzsprung–Russell (temperature–luminosity) diagram for important stars in the winter sky (Appendix E). The sun is included for comparison. The vertical axis is luminosity (increasing upward) in solar units; the horizontal axis is surface temperature, which increases to the left rather than to the right as you might expect. This is a historic convention that astronomers have not changed. Overall, no pattern is obvious in the plotted points, in part because this is a small sample of stars.

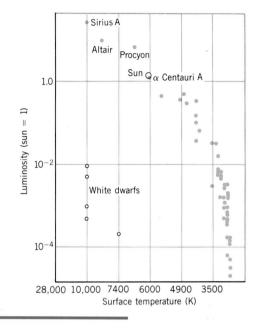

F I G U R E   1 1 . 1 2

Hertzsprung–Russell (luminosity–temperature) diagram for the selected stars nearest to the sun (Appendix D). The vertical axis is luminosity in solar units, sun's luminosity = 1.0. The horizontal axis is surface temperature in kelvins. The sun and Alpha Centauri A have essentially the same values of the luminosity and temperature, so their points overlap. Except for the white dwarfs (marked by the open circles), the stars' points fall along a narrow band.

the sun (Fig. 11.12). Notice that most of the stars are less luminous than the sun, and cooler. (We can see them only because they are so close to us.) The star Alpha Centauri A has almost the same luminosity and temperature as the sun. This star is also the star nearest to us. (Alpha Centauri A is one member of a triple-star system with Alpha Centauri B and Proxima Centauri.)

Finally, a pattern can be seen! Note that the stars' properties in Figure 11.12 clearly do *not* fall in a random scatter. Rather, there is a trend; if you draw a line through the points from luminous, hot, Sirius A to the coolest, faintest star in the lower right-hand corner, you have identified the **main sequence**. Most nearby stars fall on the narrow strip of the main sequence in the H–R diagram. Note the few stars in the lower left, Sirius B included. These stars have very high surface temperatures but low luminosities, so they must be very small. Such a peculiar body is called a **white dwarf** star.

Consider another H–R diagram (Fig. 11.13), one for the brightest stars you can see in the sky (Appen-

dix E). Compare it with the preceding plot. What a difference! Almost all these stars have a much higher luminosity than the sun. And many of them are also much hotter (stars in spectral classes O and B). The main sequence no longer appears so obvious. Only a handful of stars show up in both diagrams.

What are the physical differences among these stars? Take the star Betelgeuse, whose properties put it in the upper right of the H–R diagram. Here is a star whose surface is much cooler than the sun's. So if Betelgeuse were the same size as the sun, it would be much less luminous. But Betelgeuse has a visual luminosity some 20,000 times that of the sun. To be so much cooler and more luminous than the sun, Betelgeuse must be very much larger (Section 11.3), at least 400 times the size of the sun. Astronomers call Betelgeuse a **supergiant** star. If the sun were replaced by Betelgeuse, the earth would orbit

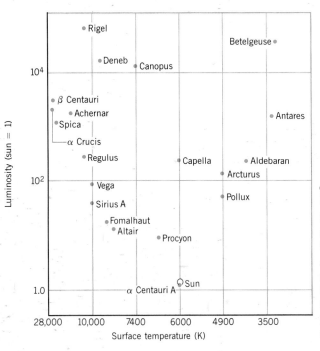

FIGURE 11.13

Hertzsprung–Russell (luminosity–temperature) diagram for the brightest stars in the sky. Axes are the same as for Figure 11.12. Again, the sun and Alpha Centauri A have almost the same luminosity and surface temperature, and so they overlap on this diagram. The stars split into two groups, one that rises up to the left and the other to the right. Rigel and Betelgeuse, both supergiant stars in Orion (Fig. 1.1), fall at the tops of these ranges.

within its atmosphere! Capella, which has a surface temperature similar to the sun's, also is larger than the sun—it is a **giant star**.

Now we can see why the H–R diagram for the nearest stars (Fig. 11.12) differs from that for the brightest stars (Fig. 11.13): the first diagram contains ordinary stars with sizes like that of the sun; no giants are among the nearest stars, for they are very rare. The sun is a main-sequence star. Most of the stars in the sun's vicinity are those cool, main-sequence stars, of spectral class M, which are not very luminous. The second diagram contains many giant and supergiant stars, still visible among the brighter stars because of their high luminosity, even though they are scattered widely through space.

Now piece these two diagrams together and add more stars (Fig. 11.14). Most stars fall on the gentle curve of the main sequence. A scattering of stars cuts across the tip of the diagram; these are the very luminous supergiants. A group of luminous stars extends off the main sequence; these are the giants. Finally, note the white dwarf stars in the lower left-hand corner of the diagram.

An H–R diagram plots the *luminosities* of the stars against their *surface temperatures*; it graphically sorts different classes of stars by some of the important physical properties.

Luminosity Classes

Once we have an H–R diagram for many stars, we can use it to infer approximate distances to stars. First, find out the spectral type of a star. Suppose it's an M star. Look at the H–R diagram to find the luminosity. Then measure the flux and find the distance from the luminosity, the flux, and the inverse-square law for light (Section 11.2).

But a problem arises (refer to Fig. 11.14): you can see that M stars have a wide *range* of luminosities, from 10^{-5} the sun's luminosity for main-sequence M stars to about 10^5 solar luminosities for supergiant ones. How do we decide what luminosity an M star has over a range of 10^{10}?

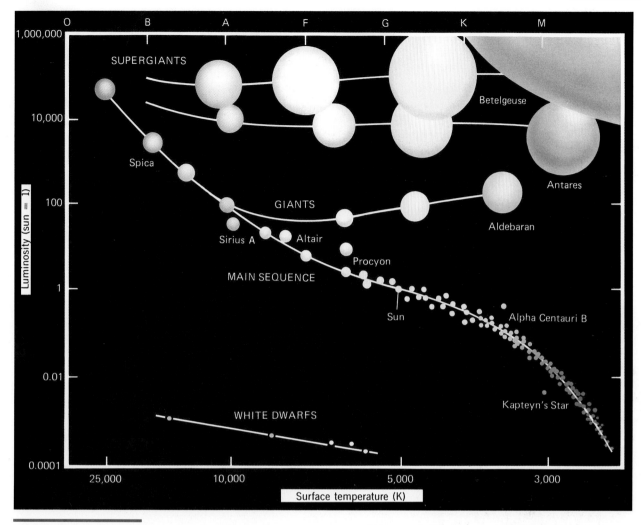

F I G U R E   1 1 . 1 4

True-color representation of the Hertzsprung–Russell diagram. As usual, the horizontal axis is temperature; the vertical, luminosity. The solid white lines show where stars of different luminosity classes fall on the diagram: Ia supergiants at the very top; Ib supergiants below, III giants just below them; and finally V main-sequence stars. At the bottom lie the white dwarfs. The relative sizes of the stars are shown correctly within each luminosity class, but not between them. The colors are those as perceived by the eye looking at these stars through our atmosphere with a telescope.

Fortunately, we can tell from a star's spectrum. Recall that the strengths of dark lines relate to a star's temperature. How energetic the collisions are depends on the temperature in a star's atmosphere. But for gases at the same temperature, the *rate* of collisions depends on the density of the gases. So in a denser gas, collisions are more frequent than in a less dense gas at the same temperature.

Giant stars, because they are so huge, typically have lower density atmospheres than main-sequence stars. The more frequent collisions in main-sequence stars make certain absorption lines in their spectra appear broader than the same lines in spectra of giant or supergiant stars (Fig. 11.15). So a star's size, hence its luminosity, is given indirectly by the *widths* of certain absorption lines when compar-

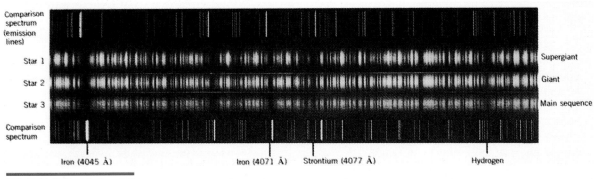

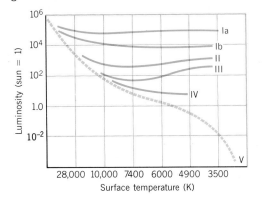

FIGURE 11.15

Luminosity class and stellar spectra: the spectra of three stars that have the same surface temperature (spectral class G8). The top spectrum (Star 1) is that of a supergiant star, the middle (Star 2) for a giant, and the bottom (Star 3) for a main-sequence star. The different intensities of the dark lines allow astronomers to infer the luminosity class of the stars (Star 1 is I; Star 2 is III; and Star 3 is V). Above and below the stellar spectra are the bright lines of comparison spectra that allow the stellar lines to be identified and a wavelength scale to be established. (Lick Observatory photo)

ing the spectra of stars of the same spectral type. In general, main sequence stars have sharper absorption line than giant stars.

Such an analysis reveals that stars fall into a **luminosity class**. The recognized luminosity classes (Fig. 11.16) are: *Ia*, *most luminous supergiants*; *Ib*, *less luminous supergiants*; *II*, *luminous giants*; *III*, *normal giants*; *IV*, *subgiants*, and *V*, *main-sequence stars*. The sun falls into luminosity class V. (White dwarfs do not have luminosity classes because they all have essentially the same size.)

The strengths of certain absorption lines in the spectra of stars with the same surface temperature provide a means to sort them by luminosity class and therefore by size.

So a star's spectrum allows it to be classified by spectral type *and* luminosity class. For a given spectral type, you can estimate the luminosity (Fig. 11.17). If you know a star's luminosity and its flux, you can calculate its distance. The procedure for working out distances from spectra is called taking **spectroscopic distances**.

FIGURE 11.16

Luminosity classes of stars on the H–R diagram. The classes run from I, the largest supergiants, to V, stars like the sun. Classes Ia and Ib are supergiants; class II, luminous giants; class III, normal giants; class IV, subgiants; and class V, main-sequence stars. The lines in this diagram represent the center of the actual range for each class.

11.6 Weighing and Sizing Stars: Binary Systems

Some stars are larger than the sun, some smaller. But sizes do not tell us directly whether a star is more or less massive than the sun. How to find a star's mass?

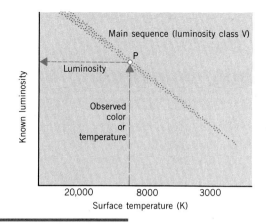

FIGURE 11.17

Spectroscopic distances, which use spectral class and luminosity class (in this case V, main sequence) to infer the luminosity and so the distance to a star, if its flux is known. From a star's color (or spectrum), we can estimate its temperature on the horizontal axis. We move straight up until we hit the main sequence and then move horizontally to the left to find the star's luminosity. Comparing the luminosity to the measured flux allows an inference of the distance, because the flux varies in accordance with the inverse-square law.

Binary Stars

To find masses, we examine the gravitational effects of one object on another. Recall how we find the sun's mass (Section 10.1): we look at the acceleration of the earth as it orbits the sun. Similarly, we use the accelerations of two stars orbiting each other to find their masses. Two stars bound by their mutual gravity revolving around a common center of mass (Section 3.3) are called a *binary star*.

If both stars are visible, we can trace out their orbital motion by observing them over a long time, which gives us the angular size of the orbit and the orbital period. Then we need to find the distance to the binary system so that we can convert their angu-

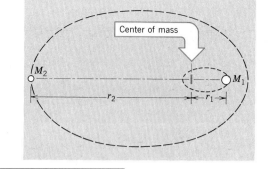

FIGURE 11.18

Center of mass in a binary star system. Both stars (M_1 and M_2) move in elliptical orbits (r_1 and r_2) around the center of mass. The more massive star is the one closer to the center of mass: in this case, star M_1.

lar separation into a physical one. We also need to account for any tilt of the plane of a star's orbit from a direct face-on view. Then we can find the *sum* of the masses from Newton's revised form of Kepler's third law (Section 3.4). To find the *individual* masses, we must have one more piece of information: how far each star is from the center of mass of the system. (The *center of mass* is the balancing point between the stars, as if one were on each end of a seesaw.)

With sufficient information, we can use the laws of Newton and Kepler to measure the masses of stars in binary systems.

In a binary system each star orbits the center of mass, with the more massive star lying closer to the center of mass and the less massive one farther away. For instance, in Figure 11.18, M_2 lies 4.5 times farther from the center of mass than M_1. That means that M_1 has 4.5 times the mass of M_2. As the system travels through space, and so across our line of sight, the center of mass traces a straight line (Fig. 11.19),

Binary Stars	Binary system	two stars orbiting common center of mass
	Visual	both stars visible through telescope
	Spectroscopic	periodic cycle of Doppler shifts in spectra
At a Glance	Eclipsing	stars pass in front of each other
	Source of	stellar masses and diameters

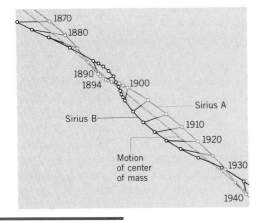

FIGURE 11.19

Motions of the binary star system, Sirius A and its companion, Sirius B, between 1870 and 1940. This figure shows two motions: that of Sirius A and its companion about their center of mass, and that of the center of mass of the two stars relative to background stars, as the center of mass moves through space. So the two stars make a corkscrew motion against the sky.

while the two stars appear to spiral around it. This corkscrew motion identifies the stars as binary and locates the center of mass.

Remarkably, *most* stars are in binary or multiple star systems. For nearby sunlike stars, no more than 45 percent are probably single, while at least 55 percent are known to be members of double, triple, and even quadruple star systems! For example, Alpha Centauri is a triple system. Two of the stars, Alpha Centauri A and B, orbit each other with a separation of about 20 AU. The third star, called Alpha Centauri C (or Proxima Centauri) orbits a few thousand AU from them.

Suppose two stars are very close together: their orbital periods are just a few days, with the stars moving quickly in their orbits. We can identify this binary by looking for two sets of lines in the spectrum (one from each star) and measuring the Doppler shifts (Section 9.5) produced by the orbital motion. This is called a **spectroscopic binary**.

Let's apply the Doppler shift to a binary star. Imagine the more massive star to be stationary, with

FIGURE 11.20

The cycle of Doppler shifts from a spectroscopic binary system. Consider the primary star as if fixed (not revolving around the center of mass). When the secondary is at position 1, it is moving toward the earth (negative radial velocity); so its spectral lines appear to be blueshifted. At 2, the secondary moves across our line of sight, so we see no shift. The same is true at 4. At position 3, the secondary moves away from us (positive radial velocity), so we see a redshift. So from 1 to 2, we see a decreasing blueshift, from 2 to 3 an increasing redshift, from 3 to 4 a decreasing redshift, and so on.

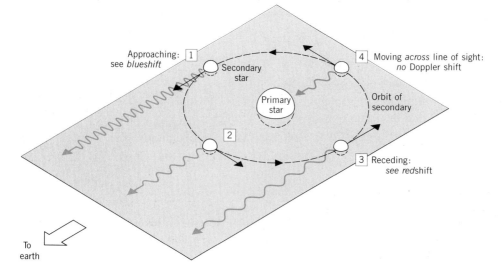

the secondary revolving about it. As the secondary recedes from the earth, you see its spectral lines redshifted compared with those of the primary; as the secondary approaches, you see its lines blueshifted (Fig. 11.20). At the intermediate points, when the secondary travels across the line of sight, you see no shift. If the two stars do not differ greatly in luminosity, both spectra can be observed, especially the cycle of shifts of the secondary with respect to the primary. (The smaller-mass star will have the higher velocity, because it is farther from the center of mass.) Note that the lines of both stars undergo a cyclic shift, but the more massive primary moves slower than the secondary.

These wavelength shifts can be turned directly into shifts in relative velocity by using the Doppler effect (Section 9.5). We then use the speeds and the period to get the circumference of the orbit. Then we work out the radius of the orbit and so the separation of the two stars. So a spectroscopic binary gives us direct information on the system's orbit and the stellar masses. Note that we can measure the actual velocity of the stars in kilometers per second; we get the actual radius of the orbit, in kilometers, for each star, relative to the center of mass. We can use this information, along with Kepler's third law, to determine the stars' masses.

In some cases, the orbits are tilted just so that one star passes in front of the other, producing an eclipse. That's an **eclipsing binary system** (Fig. 11.21a). Algol, the Arabic name for "demon star" in the constellation Perseus, is the prototype of eclipsing binaries. It has a period of 2.87 days and, in mideclipse, plummets sharply in brightness—easy to see by eye in about 6 hours. Algol A (Fig. 11.21b) the brighter star (260 solar luminosities) is B8 V (surface temperature 12,500 K); its companion, Algol B, is a fainter (5 solar luminosities) K star (surface temperature, 4500 K). From the measured speeds and the period, Algol A has a mass 4.6 times that of Algol B. From Kepler's third law, the total (combined) mass of the system is 4.51 solar masses. So Algol A is 3.7 solar masses and Algol B 0.81 solar mass.

Eclipsing binaries give us directly one other property of the stars: their diameters. When the less luminous star Algol B passes in front of the more luminous Algol A (primary eclipse), the duration of the eclipse depends on the diameter of B relative to A, and the relative orbital speeds. When Algol A swings

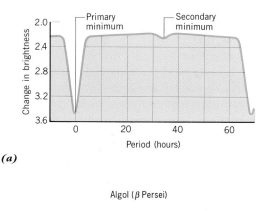

(a)

Algol (β Persei)

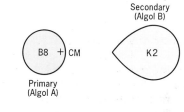

(b)

FIGURE 11.21

Algol, an eclipsing binary system. (a) The orbits of the two main stars lie almost edge-on as seen from the earth, so eclipses can occur. The eclipse of the primary star by its cooler companion produces the largest dip (*primary minimum*) in the light curve, a plot of how the observed brightness changes with time. When the companion circles behind Algol A, less light is lost, so a smaller dip (*secondary minimum*) occurs in the light curve. (b) Top view of the Algol system. The center of mass (CM) is located just inside the primary star. The shape of the secondary is distorted by the gravitational force of the primary.

in front of Algol B (secondary eclipse, during which the more luminous star crosses in front of the less luminous one), the duration depends on the diameter of A relative to B and the relative orbital speeds. So we can find the radius of each star: 2.9 solar radii for Algol A and 3.5 solar radii for Algol B.

Eclipsing binary systems allow us to determine both the masses and the radii of the two stars.

Luminosities, Masses, and Lifetimes

Binary systems act like scales that allow us to weigh the two stars. In most cases, the luminosities of the two stars also can be determined. When the luminosities of main-sequence stars are plotted against the stars' masses, the points fall into a definite pattern (Fig. 11.22). For main-sequence stars, the mass determines the luminosity, and the resulting correlation is called the **mass–luminosity relation** (sometimes abbreviated *M–L relation*).

Basically, the mass–luminosity relation shows that a star's luminosity is *roughly* proportional to the third power of its mass. For example, a star with a mass 10 times that of the sun has about 10^3 times the sun's luminosity. Main-sequence stars follow the mass–luminosity relation fairly well; hence the upward swing in luminosity of the main sequence from M to O stars reflects an increase in the stars' masses (Fig. 11.23). So main-sequence O stars are more massive than main-sequence M stars. Astrophysicists had predicted the mass–luminosity relation theoretically, and its confirmation came from investigations

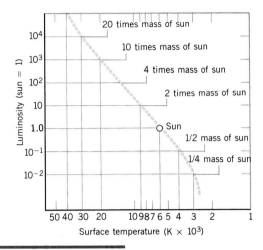

FIGURE 11.23

Approximate masses of the stars along the main sequence. The more massive stars lie on the upper end of the main sequence; the less massive ones at the lower end.

FIGURE 11.22

The mass–luminosity relation for main-sequence stars, as determined from binary systems, in which the individual masses can be determined. The luminosities of these stars can be calculated because their distances can be measured. (Adapted from a diagram by W. D. Heintz)

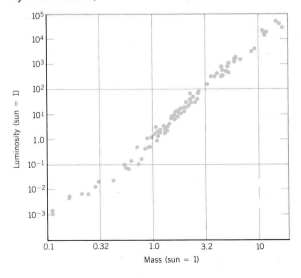

of binary stars. Giants and supergiant stars also follow a mass–luminosity law, but it differs from that for main-sequence stars.

A star's mass determines its luminosity; the more massive a main-sequence star, the greater its luminosity.

From the mass–luminosity diagram, you see that the masses of stars do not differ widely from the sun's mass: roughly from less than about 100 solar masses to greater than roughly 0.1 solar mass.

The mass–luminosity law provides a way of comparing **stellar lifetimes**. The argument goes like this. The total amount of energy available to a star from the conversion of hydrogen to helium is directly proportional to its mass. So a star with more mass can produce energy for a longer time than one with less mass, *if* the two bodies give off energy at the *same* rate. But the rate at which a star loses energy is given by its luminosity. A greater luminosity means that a star is producing energy faster, using up its mass more quickly. Because luminosity increases roughly as the cube of the mass, more massive stars use up their mass faster and have shorter

Table 11.2

Typical Properties of Main-Sequence Stars

Spectral Class	Surface temperature (K)	Mass (M_{sun})	Luminosity (L_{sun})	Radius (R_{sun})	Lifetime (years)
O	40,000	40.0	500,000	20.0	1×10^6
B	15,000	7.0	800	4.0	8×10^7
A	8,200	2.0	20	2.0	2×10^9
F	6,600	1.3	2.5	1.2	5×10^9
G	5,800	1.0	1.0	1.0	10×10^9
K	4,300	0.78	0.16	0.7	20×10^9
M	3,300	0.21	0.008	0.3	50×10^9

lifetimes. For example, a star of 30 solar masses has a lifetime of a few million years, in contrast to the sun's 10 billion years (Table 11.2).

The more luminous (and massive) a star, the shorter its lifetime.

Do not confuse the age of a star with its lifetime! The *lifetime* is the total span of active life from fusion reactions. The *age* is how much time has passed since fusion reactions began in the core. The sun has a lifetime of 10 billion years, but its age is about 5 billion years. When we say a star is "young" or "old," we are comparing it to stars with *similar lifetimes*.

11.7 Stellar Activity

If stars are suns, do they also show the aspects of the active sun (Section 10.6)? Do they have coronas, starspots, and flares? Remember that these phenomena occur in active regions, which result from strong magnetic fields.

Because stars are far away, we cannot directly see their coronas (if they have them). How to observe them? The hot plasma in the sun's corona shows up in x-ray photos as bright regions; in contrast, the photosphere and chromosphere appear dark. So x-ray observations of stars provide an indirect view of their coronas—if stellar coronas resemble the sun's in general ways.

From these observations, we find that stars of *all* spectral classes have x-ray emission and, by deduction, coronas. The sun when most active puts out some 10^{22} W. Other G stars emit upward of 10^{24} W—about 100 times more than the active sun. The implication is that such stars have more extensive and hotter coronas (temperatures up to 10 million K), connected with extensive active regions. Coronas appear common in other stars.

The sun's corona is the source of the solar wind. With coronas, many stars have outflows as stellar winds. Most O and B stars have hot, rapid winds that blow off 10^{-6} solar mass per year at speeds of thousands of kilometers per second. Cool giant and supergiant stars also have strong winds, but they are cooler and less speedy that those from hot stars.

How could we observe **starspots**? During times of sunspot maxima, sunspots cover only 0.2 percent of the sun's total surface. If you observed the sun as a star from light years away, you'd need extremely sensitive equipment to note any variation in the sun's light as it rotated, with sunspots coming into and passing out of view. So you might expect that it would be useless to try to search for starspots by observing the same kinds of change in light output as stars rotate.

Luckily, this solar analogy proves right in concept but wrong in scale. Some stars, even G stars, turn out to be hyperactive, with enormous concentrations of starspots compared to the sun—starspots covering such a large fraction (up to 50 percent) of the surface that we can actually see them indirectly! As the star rotates, we alternately view the spotted and clean sides. When few spots face us, we see the most light from the star. When the spotted hemisphere turns our way, we see a decrease in the amount of light.

We know that flares occur in active regions on the sun. The same appears to be the case for some other stars, but the **stellar flares** burst forth with much more energy than on the sun.

Clearly, many stars must be observed over long times (decades) to confirm activity cycles like the sun's. Such a long series of observations of sunlike stars was begun in 1976 at Mt. Wilson Observatory and continues today. The project so far shows that some sunlike stars exhibit activity cycles. A few last longer than decades; others on the order of decades; and some show no sign of activity. Basically, it appears that all stars of spectral type F or cooler undergo some type of magnetic activity at some time in their lives.

Other stars have characteristics of the active sun, so strong magnetic fields, probably twisted and tangled, lie at the roots of this activity.

The Larger View

Stars appear as pinpricks of light in the sky. What are they, really? Astronomers made a crucial breakthrough in this century when they realized that stars are in essence other suns. We reached this insight by analyzing starlight in the same way that we analyze sunlight. We also measure the distances to stars (by parallax) and their motions (by the Doppler shift). We then can infer important physical properties, such as mass, luminosity, surface temperature, size, and chemical composition. Stars show a range of values for these properties; yet, most are counterparts to the sun, peers in the realm of space. Stars are naturally controlled fusion reactors, pretty much like our sun.

We order the properties of stars with the Hertzsprung–Russell diagram of stellar temperatures plotted against their luminosities. This graph is a fundamental sorting tool for the astronomer. The luminosities are found from fluxes and distances; the temperatures from the spectral types or colors. Stellar spectra are classified by which dark lines are visible and how intense they appear. The spectra also allow us to sort stars by sizes, in terms of their luminosity classes. The patterns that emerge tell us that the physical structure of most stars follows a common model.

Stellar luminosities are related to stellar masses. More luminous stars have short lifetimes compared to less massive ones. Like the sun, stars evolve. How? We'll begin to answer these questions in Chapter 12 by looking at how stars are born.

Key Ideas and Terms

flux
inverse-square law

1. Viewed from the earth, stars display different brightnesses. Astronomers measure the brightness of stars as seen in the sky by their flux, or how much energy from a star reaches the earth each second over a given area (such as a

square meter). Light intensity changes with distance; the flux varies as the inverse square of the distance. This law applies to electromagnetic waves of all kinds.

luminosity
parallax
   (trigonometric)

2. Only the distances of the closest stars can be measured directly by a triangulation technique called *trigonometric (heliocentric) parallax*: the closer the star, the larger its parallax. If a star's distance is known, we can find its luminosity from its flux.

blackbody
stellar colors
continuous
   spectrum
Planck curve
stellar surface
   temperatures

3. Assuming that stars radiate like blackbodies, their colors indicate their surface temperatures. We can infer their sizes from their surface temperatures and luminosities. Many stars do have continuous spectra that resemble Planck curves pretty closely.

Balmer lines
absorption-line
   spectra
spectral type
stellar spectral
   sequence
stellar surface
   temperatures

4. Based on observations of certain absorption lines in their spectra (especially hydrogen Balmer lines), stars fall into a spectral sequence of spectral types from higher to lower surface temperatures (and so over a range of colors). A star's surface temperature mainly determines which absorption lines appear in its spectrum; chemical composition plays a secondary role.

giant
Hertzsprung–
   Russell diagram
luminosity class
main sequence
spectroscopic
   distances
supergiant
white dwarf

5. The Hertzsprung–Russell diagram is a graph of the surface temperatures and luminosities of stars. On it, stars fall into distinct groups: main sequence, giants, supergiants, and white dwarfs. Each star's position on the diagram is established by its spectral type (temperature) and luminosity and luminosity class (size). Once an H–R diagram has been made for a large number of stars, we can use it to estimate distances from spectra.

binary star system
eclipsing binary
   system
spectroscopic binary

6. Binary stars are two stars, bound by gravity, orbiting a common center of mass. Binaries are catalogued by astronomers into categories that depend on the way they are observed. If spectroscopic observations are available of the Doppler shifts of the two stars in an eclipsing binary system, we can find both the individual masses and radii of the stars.

mass–luminosity
   relation
stellar lifetimes

7. Binary stars provide the only means of finding directly the masses of stars. For main-sequence stars, we find that the masses and luminosities are related (the more massive stars are more luminous); we infer that more massive stars have much shorter lives than less massive ones.

active regions
magnetic fields
starspots

8. We are beginning to recognize that stars have magnetic activity, both erratic and cyclic, similar to that of the active sun. This activity includes both starspots and flares that are driven by magnetic fields localized in active

stellar coronas
stellar flares

regions. We also infer that many other stars have coronas like the sun and lose mass from the outflow of stellar winds.

Review Questions

1. What property of a star do you sense when you observe it in the sky?
2. If you could move closer to a star, what would seem to happen to its flux?
3. If one star has a *smaller* parallax than another, is it closer or farther in distance?
4. How large is one parsec in light years?
5. If you observe two stars, and one appears yellowish and another reddish, which star is hotter?
6. A star's luminosity depends on its surface temperature and what other physical property?
7. Which stellar spectral type shows the strongest Balmer lines?
8. If one star falls into spectral class B and another into class F, which is hotter?
9. Stars that fall into the upper right-hand corner of the Hertzsprung–Russell diagram are of what kind?
10. Most stars fall into what region of the Hertzsprung–Russell diagram?
11. What key stellar property can we find out from binary stars?
12. If main-sequence star A is more massive than star B, which is the more luminous? Which has the longer lifetime?
13. What is one observational clue that stars have active regions like the sun?

Conceptual Exercises

1. In the winter sky, you see the following stars with their colors: Capella, yellowish; Betelgeuse, reddish; and Sirius, bluish. List these stars in order of increasing surface temperature. Estimate the surface temperature of Betelgeuse and of Sirius. Do they differ much?
2. Consider the following stellar types: M I, G III, and A V. Which star is the largest? Which the most luminous?
3. Can you think of considerations that limit the accuracy of heliocentric parallax measurements? (*Hint*: How would these measurements differ if you were on Mars?)
4. Refer to the H–R diagram in Figure 11.13 to answer the following questions: (a) Capella and the sun have roughly the same surface

temperature. Which star is larger? (b) Regulus and Capella have about the same luminosity. Which star is larger? (c) Vega and Sirius have about the same surface temperature. Which star is more luminous? (d) Which star would appear redder, Vega or Pollux?

5. Since we can't see the disks of other stars directly, how do we know that some of them have coronas like the sun?

6. What procedure does an astronomer follow to find out a star's density? (*Hint*: Divide mass by volume to get density.)

7. Consider a noneclipsing binary star system in which one star is much brighter than the other. Then the absorption lines from the fainter star do not appear in the spectrum, but those of the brighter one do. Describe how the Doppler shift would appear from the orbital motion of the stars. Assume that we view the orbit pretty close to edge-on.

8. The main-sequence star Regulus has a mass about five times that of the sun. Use the mass–luminosity relation (Fig. 11.22) to estimate the luminosity of Regulus. Compared to the sun, what is Regulus' potential lifetime?

9. Consider two binary systems (A and B) with the same periods of revolution. In B, the two stars have a greater separation than in A. How do the orbital speeds of the stars in B compare to those of A? How does the total mass of system B compare to that of system A?

10. Imagine you have two stars with the same luminosity. One is located at 10 ly from the earth; the other at 30 ly. How do the fluxes received at the earth compare?

Conceptual Activity 11   Classifying Stars

A glance at the stars in a dark sky may strike you as overwhelming. Yet, astronomers can measure their traits well. By focusing on just two of these stellar characteristics—surface temperature and luminosity—we can get an inkling of how stars are different and how they are alike.

Table F.2 provides a list of selected stars. It contains both some nearby stars and some of the brightest stars in the sky. Most of the nearby stars will have unfamiliar names. Given for each star are its approximate luminosity (relative to the sun) and surface temperature (in kelvins). Get yourself a piece of graph paper and make the horizontal axis temperature, starting with the high end to the left and going lower to the right. Start the temperature axis label at 25,000 K and run it in equal steps down to 2500 K. Set the luminosities on the vertical axis. Plot each star at the correct combination of luminosity and temperature. Do the sun first, it's easy! Plot each star's point using its number, so you can tell which stars are where on the graph.

Table F.2

Data on Selected Stars

Number	Star Name	Approximate Visual Luminosity	Approximate Surface Temperature (K)
1	Sun	1.0	5,800
2	Luyten 726-8A	6.0×10^{-5}	2,600
3	Tau Ceti	0.45	5,200
4	Alpha Hydri	6.0	7,400
5	Theta Eridani	18	9,400
6	Epsilon Eridani	0.30	4,600
7	Omicron-Two Eridani	0.30	4,900
8	Aldebaran	690	3,800
9	Pi-Three Orionis	3.0	6,400
10	Eta Aurigae	580	16,000
11	Rigel	89,000	12,000
12	Capella	130	5,200
13	Betelgeuse	20,000	3,300
14	Beta Aurigae	110	9,800
15	Zeta Canis Majoris	780	17,000
16	Mu Camelopardalis	150	3,000
17	Canopus	9,100	7,400
18	Sirius A	23	10,000
19	Sirius B	3.0×10^{-3}	30,000
20	BD + 5° 1668	1.5×10^{-3}	3,000
21	Procyon A	7.6	6,500
22	Procyon B	5.5×10^{-4}	8,000
23	Pollux	36	4,100
24	Iota Ursae Majoris	11	7,800
25	Zeta Leonis	50	8,800
26	Wolf 359	2.0×10^{-5}	2,600
27	Lalande 21185	5.5×10^{-3}	3,300
28	Ross 128	3.6×10^{-4}	2,800
29	Spica	1900	20,000
30	Arcturus	76	3,900
31	Alpha Centauri A	1.3	5,800
32	Alpha Centauri B	0.52	4,200
33	Proxima Centauri	6×10^{-5}	2,800
34	Beta Canis Minoris	240	12,000
35	Antares	3,600	3,000
36	Zeta Ophiuchi	4,500	23,000
37	Barnard's Star	4.5×10^{-4}	2,800
38	70 Ophiuchi A	0.45	4,800
39	Vega	52	11,000
40	Ross 154	3.6×10^{-4}	2,800

With a completed graph in hand, answer the following questions.

1. How are the stars arranged overall on this temperature–luminosity graph? Can you divide them into two or three large groups?

2. The low-luminosity stars tend to be the nearby stars (otherwise, we wouldn't see them!). Where on the diagram do these stars fall? Do their surface temperatures show any similarity?

3. Pick out any star hotter than the sun. Then choose another. Do you expect stars hotter than the sun to be generally more or less luminous than the sun?

4. Does your graph resemble any of those in the book? Which ones? Why are they similar?

12

Starbirth and Interstellar Matter

■

Central Concept

Stars are born out of short-lived clouds of material in the space between the stars.

■

How are stars born? In clouds of gas and dust between the stars. Interstellar space contains gas and dust, both thinly spread out and in clumps. Hydrogen makes up most of the gas, which outnumbers the solid dust particles enormously. A large fraction of the interstellar gas is locked up in short-lived clouds. From these interstellar clouds, gravitational contraction forms stars. We need to investigate the births of stars before we can assemble stories of their lives. You will see that a great deal of the matter in the universe is not tied up in stars, planets, and their moons.

The gas and dust between the stars marks the place where star-forming action is occurring in the Milky Way. Here, stars form in vast stellar nurseries. A few will become massive stars, some sunlike stars, a majority binaries, and some will develop with planetary systems.

12.1 Interstellar Gas

The **interstellar medium** in the Milky Way consists of all the gas and dust between the stars. (The **Milky Way** is a faint, broad band of light in the sky. It consists, when viewed with an optical telescope or binoculars, of a multitude of faint stars, all belonging to our Galaxy, of which we see only a small part.) The **interstellar gas**—neutral atoms, ions, and molecules—dominates, yet it is *very* tenuous. The distance between interstellar atoms is roughly 100 million times the size of the atoms themselves. If two people were separated by a proportional distance relative to their size, they would be about 100 million meters apart, about the distance between the earth and moon! Sparse as it is, the interstellar gas occasionally clumps in dense clouds and forms stars.

The interstellar medium contains both gas and dust; the gas, which dominates, includes atoms, ions, and molecules.

FIGURE 12.1

Close-up of the Orion Nebula, a typical emission nebula of ionized gas. A smaller nebula appears above it. The colors of the emission depend on the elements that have been excited by collisions. The dark regions within the nebula are those with concentrations of dust. This image has been processed to bring out the fine detail. (Image by D. Malin; courtesy Anglo-Australian Observatory)

Emission Nebulas

On a winter night, you can easily spot the constellation Orion (look back at Fig. 1.1). Dangling from Orion's belt is a short sword; if you look closely, the middle star appears fuzzy. A small telescope pointed at this fuzzy patch shows you a diffuse, convoluted cloud surrounding a small cluster of stars. This bright cloud is called the *Orion Nebula*. (The use of *nebula*, the Latin word for "cloud," is an astronomical holdover from the nineteenth century.) Only 1500 ly from the earth, the Orion Nebula is roughly 20 ly in diameter (Fig. 12.1).

Spectroscopic analysis demonstrates that these nebulas consist of hot gases because they show spectra of emission lines. Hence, they are called **emission nebulas**. Recall from Kirchhoff's rules for spectra (Section 9.2) that an emission-line spectrum indicates a hot, transparent gas. The Orion Nebula displays strong emission lines of hydrogen, helium, and oxygen in its spectrum, so it is hot and at a low density.

Nebulas such as Orion shine by the energy from hot (up to 30,000 K) OB stars located within them.

The Interstellar Medium — At a Glance	Contents	gas and dust
	Dust	small grains made of silicates and ices
	Gas	atomic. molecular, and ionized; mostly hydrogen
	Structure	clumpy, with a thin gas between clouds
	Source of	stars and planets

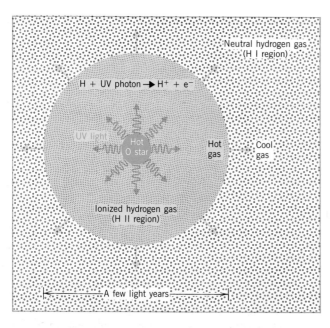

FIGURE   12.2
Schematic diagram of an idealized H II region. A hot O star emits ultraviolet light that can ionize hydrogen for a few light years around the star. The absorption of the ultraviolet heats the ionized gas to about 10,000 K. The hot, ionized gas expands into the cooler, neutral gas surrounding it. Unlike this drawing, real H II regions are not perfectly spherical.

The essential physical process is this: the gas absorbs high-energy ultraviolet photons given off by the central star (or stars) and in turn gives off photons in emission lines at lower energies. For a hydrogen gas, the visible light comes out mainly in the red region

FIGURE   12.3
Flux contour map (Appendix G) of the radio emission at 8 GHz from Messier 8, the Triffid Nebula. The contour lines are overlaid on a photo of the optical nebula; note that the peak coincides with the central part of the nebula. Just below and to the left of M8 is the peak of millimeter-wave emission from the surrounding molecular cloud; the molecule indicated is formaldehyde (H_2CO). At the lower right is the angular scale for 2 arcminutes. (Observations by E. Chaisson at Haystack Observatory)

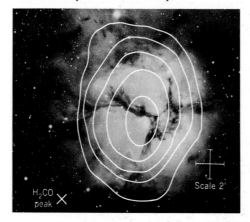

of the spectrum as the hydrogen-alpha line of the Balmer series (Section 9.3).

The star or stars in the nebula emit many ultraviolet photons with enough energy to ionize hydrogen. These photons ionize the gas surrounding such a star out to a few tens of light years. This zone of ionized hydrogen is called an **H II region** (Fig. 12.2). (*H I* stands for neutral hydrogen and *H II* for ionized hydrogen. In general, a neutral atom is labeled I, an atom with one electron removed is labeled II, with two electrons removed, III, and so on.).

Radio astronomers can see H II regions by their continuous radio emission, produced when electrons speed past protons. As the opposite charges attract, the electrons are bent from their straight-line paths. The electrons are accelerated, their energy changes, and they emit electromagnetic radiation, mostly at millimeter and centimeter wavelengths. Since many different electrons undergo different energy changes at the same time, a continuous spectrum results. Radio telescopes map this continuous radio emission (Fig. 12.3). From such maps, astronomers can infer how much ionized gas an H II region contains (the Orion Nebula, for example, is about 300 solar masses).

H II regions are created when interstellar gas is ionized by hot (O- and B-type) stars.

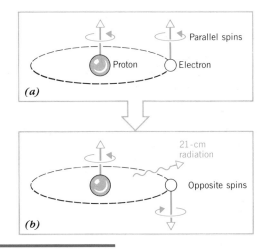

FIGURE 12.4

The 21-cm emission from hydrogen atoms. (*a*) Collisions can line up the spins of the proton and electron so that they are parallel. (*b*) If after some time the atom does not hit another atom, the electron flips so that the spins are opposed, a lower energy state. This loss of energy can be detected because the atom emits a 21-cm photon.

Interstellar Atoms and Clouds

Investigations of emission nebulas show that they are composed almost entirely of hydrogen. So the neu-

tral gas ionized to form them contains mostly hydrogen, as neutral atoms (designated an **H I region**).

In interstellar space, hydrogen atoms can emit radio waves. The hydrogen atom has one proton in the nucleus and one electron in orbit around it. Both the proton and the electron have angular momentum; you can imagine each spinning. The electron and the proton can be oriented in the atom so that the two spins either align or oppose each other. If the spins are opposed, the total energy of the atom is just a bit less than if the spins are aligned. As usual, the atom prefers to be in the lower energy state. Suppose the spins are aligned; eventually the proton flips over and emits a low-energy photon—energy that corresponds to a wavelength of 21 cm (Fig. 12.4).

The protons and electrons in hydrogen atoms are aligned by collisions with electrons and other atoms. The gas in interstellar space is very sparse, and collisions between two atoms occur only once every few million years. Once the spins in a hydrogen atom have become aligned, about 10 million years on the average passes before the electron flips and the atom drops to its lowest energy state—a rare event for any one atom. But because so many hydrogen atoms exist in interstellar space, enough are emitting 21-cm radiation at any given time that the interstellar gas radiates strongly at this wavelength, and the **21-cm line** can be detected with radio telescopes (Fig. 12.5).

FIGURE 12.5

Observations of the 21-cm-line emission from atomic hydrogen in interstellar space. This false-color image is an all-sky view with the plane of the Milky Way across the center. The colors indicate the amount of neutral hydrogen gas: black and blue the least, red and white the most. (Image assembled by C. Jones and W. Forman)

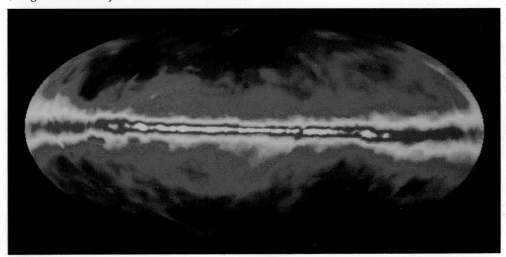

Sky surveys at 21 cm find that most neutral hydrogen is concentrated in the plane of the Milky Way. On the average, the hydrogen atoms have a temperature of 70 K and a number density of 10^6 atoms in a cubic meter. (For comparison, the density of air at the earth's surface is about 10^{25} particles per cubic meter.) Such observations show that the neutral hydrogen gas is very patchy, with H I clouds that range in diameters from tenths up to tens of light years.

The radio emission line at 21 cm indicates the presence of many hydrogen atoms in the interstellar gas, usually in clumpy H I regions.

The space between the interstellar clouds also contains gas. This intercloud gas consists mostly of hydrogen. Is it ionized or neutral? Well, it seems to be both! Observations at 21 cm indicate a thin, neutral gas. Other radio observations point to an even thinner, hotter (about 10,000 K) ionized gas also between the clouds.

Interstellar Molecules and Molecular Clouds

A molecule consists of atoms linked together in particular arrangements by electron bonds. It can have different energy states if the atoms vibrate or the molecule spins in various ways. As with changes in electronic states in an atom, when a molecule changes its vibrational or rotational state, it can emit or absorb a photon. For changes in vibrational states, the photons are infrared ones; for rotational states, radio. In the cold regions where molecules can exist in interstellar space, occasional collisions between molecules (or perhaps with atoms) kick the molecules and get them rotating. Eventually, the interstellar molecules emit radio photons that can be observed as a radio emission line, generally at millimeter wavelengths.

The radio search for molecules with many atoms began in earnest in the 1960s. More than 80 molecules have been found so far. Table 12.1 lists some key ones. Note that carbon monoxide (CO) is one of the most common molecules in space; it is easily excited by collisions in places where the densities are a mere 10^8 molecules per cubic meter. Many of the

Table 12.1

Some Important Interstellar Molecules

Complexity	Molecule Name and Symbol
Diatomic	Hydrogen, H_2 Carbon monoxide, CO
Triatomic	Water, H_2O Hydrogen sulfide, H_2S
4-Atomic	Ammonia, NH_3 Formaldehyde, H_2CO
5-Atomic	Methane, CH_4 Cyanamide, H_2NCN
6-Atomic	Methyl alcohol, CH_3OH Methyl cyanide, CH_3CN
7-Atomic	Methylamine, CH_3NH_2 Methylacetylene, CH_3C_2H
8-Atomic	Methyl formate, $HCOOCH_3$ Methyl cyanoacetylene, CH_3C_3N
9-Atomic	Ethyl alcohol, CH_3CH_2OH Ethyl cyanide, CH_3CH_2CN

molecules are *organic*: that is, compounds in which carbon plays a central role. The most abundant atoms in these molecules—carbon, hydrogen, nitrogen, and oxygen—are also the most abundant in living creatures on the earth.

By far the most abundant molecule is molecular hydrogen (H_2). But even though radio telescopes can observe atomic and ionized hydrogen, they *cannot* detect molecular hydrogen, because it does not emit or absorb at radio wavelengths. The hydrogen molecule does absorb and emit ultraviolet and infrared wavelengths, however. Infrared emission lines of molecular hydrogen have been observed from heated interstellar clouds (at a temperature of about 2000 K, the hydrogen molecules are excited and emit infrared lines), and ultraviolet absorption lines from cool clouds in front of hot stars have been detected in the spectra of these stars. Overall, about half the interstellar hydrogen in the Milky Way appears to be in the form of molecules.

Interstellar molecules of many kinds are visible from their radio and infrared emission; molecular hydrogen is the most abundant.

Most interstellar molecules are concentrated in dark, dense, cold conglomerates called **molecular**

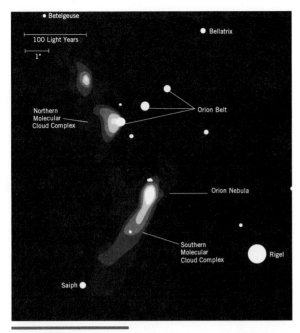

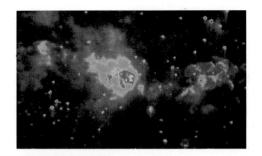

FIGURE 12.6

Molecular clouds and regions of starbirth in Orion. A carbon monoxide flux map of the emission from giant molecular cloud associated with the Orion Nebula is superimposed on the stars. This map covers about 15° on the sky. The hottest, densest part of the cloud lies near the Orion Nebula; it is called OMC1 for Orion Molecular Cloud 1. It is part of the Southern Molecular Cloud Complex. Another complex lies just to the north and east of the last star in Orion's belt. This map was made by a millimeter-wave antenna at Columbia University. The scale for one degree and 100 ly on this map is shown in the upper left.

clouds (Fig. 12.6). These clouds often lie near H II regions; one of the closest molecular clouds sits behind the Orion Nebula, as we view it from earth. The molecular cloud here is at least 30 ly across, has a density of 10^9 hydrogen molecules per cubic meter, and contains at least 10^4 solar masses of material. Its core, which is only 0.5 ly in size, has a peak density of 10^{11} hydrogen molecules per cubic meter, and a mass of only 5 solar masses.

The Orion region presents an excellent example of a **giant molecular cloud** (sometimes called *GMC* for short). Observations so far indicate that the bulk of the material of the interstellar medium is bound up in complexes of giant molecular clouds (Fig. 12.7). These immense globs of molecules, held together by gravity, consist mostly of molecular hydrogen; many other molecules are present, but these make up only a small fraction of the mass. GMCs

FIGURE 12.7

False-color infrared map of emission from hydrogen molecules in the constellation of Cepheus: dark blue for the weakest emission, red for the strongest. The emission comes from hydrogen molecules at a temperature of 2000 K; the molecules have been heated by shock waves generated by newborn stars. The peak near the center comes from a buried cluster of young stars. These molecular clouds are about 2400 ly away; the area covered is 3.3 ly by 1.7 ly. (Image by A. P. Lane and J. Bally; courtesy KPNO/NOAO)

often come in groups called *cloud complexes* having average densities of a few hundred million molecules per cubic meter; the individual clouds are slightly denser, with a few billion molecules every cubic meter. A typical cloud has a size of a few tens of light years. The total masses of the complexes range from 10,000 to 10 million solar masses; 100,000 solar masses is typical. Masses of individual clouds are about 1000 solar masses.

The cores of these clouds have temperatures of some 10 K, and densities get as high as 10^{12} mole-

FIGURE 12.8

Messier 17 and its adjacent molecular cloud. This image combines an optical photo of the H II region (black and white) with infrared observations (false color). This infrared emission comes from the dust in the molecular cloud to the right of M17; the color changes from violet to red to illustrate the energy flow here. (Courtesy A. Myers, Ames Research Center, NASA)

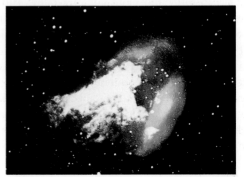

(a)

(b)

FIGURE 12.9

Dark clouds. (*a*) The Horsehead Nebula in Orion in visible light. The red emission nebula contrasts with the famous dark cloud called the Horsehead from its silhouette. (Image by D. Malin; courtesy Anglo-Australian Observatory) (*b*) Gas and dust clouds in the region of the star Rho Ophiuchi (star at top middle). The bluish cloud around this star is dust scattering the starlight. The dark clouds below and to the left are denser regions in which stars have recently formed. (Image by D. Malin; courtesy Anglo-Australian Observatory)

cules per cubic meter. Giant molecular clouds are so massive, though, that they contain an enormous number of molecules in total. And, even though densities are low by terrestrial standards, they are high enough to permit simple ions and molecules in the gas to form into more complex ones. And the internal temperatures and pressures are high enough in most clouds to prevent them from collapsing by their own gravity.

Giant H II regions, which surround small clusters of young, massive stars, are always found near molecular cloud complexes (Fig. 12.8). This proximity suggests that giant molecular clouds play a key role in the process of star formation.

Interstellar molecules most commonly occur in large, cool molecular clouds, which are often found in large molecular cloud complexes.

12.2 Interstellar Dust

Dust as microscopic particles also occupies interstellar space. There's not much out there; on the average, there's one dust particle in every million cubic meters—that's roughly a cube with sides one football field long! The **interstellar dust** amounts to about one percent of the total mass of interstellar matter, but it can cut out light from distant objects or from those shrouded in dense clouds. Even at low densities, dust particles are very effective at absorbing visible light. Piercing the dust veil has been an important goal of astronomers to reveal the process of starbirth.

Evidence for Cosmic Dust

Visual observations hint at dust between the stars. Dark clouds, such as the famed Horsehead Nebula in Orion (Fig. 12.9a), display dramatic cutoffs of light due to dust. A **dark cloud** is an interstellar cloud that contains so much dust that it blots out the light of stars within it and behind it (Fig. 12.9b). The dark rifts and lanes in the Milky Way, once attributed to the lack of stars, are actually regions heavily obscured by dust. In some regions of the sky, small dusty, isolated dark clouds blot out the light of the stars behind them.

Some bright clouds are dust, reflecting light from nearby stars. One of the best examples of such a nebula is that around the Pleiades (Fig. 12.10a). The spectrum of this nebula does *not* exhibit the bright emission lines characteristic of an H II region. It shows the absorption-line spectrum of the Pleiades stars—light reflected by dust. A bright nebula that arises from the reflection of starlight by dust is called a **reflection nebula**; note that it has a bluish color (Fig. 12.10b). (This is the same blue cast that the sky has: Section 4.4.)

Generally, interstellar dust makes itself known in two ways: by **extinction**, the dimming of starlight, and by **reddening**, the scattering of the blue wavelengths more than the longer wavelengths. Let's look at each of these processes (which also occur in the earth's atmosphere).

Imagine starlight traveling through a dust cloud. The dust particles can absorb some of this light as it comes through. The dust particles can also scatter the starlight, so it goes off in a different direction from the original one. In either case, less light exits the dust cloud than entered it. Astronomers call this dimming of starlight *extinction*, and when it hap-

FIGURE 12.10

Reflection nebulas. (a) The Pleiades star cluster immersed in a reflection nebula. Dust surrounding the stars reflects their light preferentially in the blue. (Image by D. Malin; courtesy Royal Observatory, Edinburgh) (b) Reflection nebula surrounding a cluster of stars in Orion (NGC 1977). (Image by D. Malin; courtesy Anglo-Australian Observatory)

(a)

(b)

pens, blue light is more strongly affected than red. Similarly, when you observe a star through the dust cloud, red light penetrates the dust cloud more readily than blue because more of its red light than its blue light reaches your eye. Thus the star appears redder than it actually is: astronomers call this process *reddening*. The blue light that is scattered bounces around the dust cloud until it finally exits. So the cloud, a reflection nebula, appears bluish. The sun appears redder when near the horizon because of the preferential scattering of blue light through much more air than when the sun is overhead (Section 4.4). In a similar way, interstellar dust makes stars appear redder than you would expect from their spectral class.

We can measure the reddening of starlight and estimate the quantity of the dust from the amount of reddening. From the H–R diagram, we know that a certain spectral classification of a star corresponds to a certain color. If we obtain a spectrum of a star, its spectral type can be determined (from the strength of absorption lines), even if its light is reddened. We measure the star's color and compare that to the color expected for its spectral type. The difference in color, the reddening, tells how much dust lies along the line of sight to the star.

At visual wavelengths, interstellar dust reveals itself by its extinction and reddening effects on starlight.

FIGURE 12.11

Photo in red light of the core of the Orion Nebula. The small group of stars in the center is the Trapezium cluster (arrow). (Courtesy J. Riffle, copyright 1990 by Astro Works Corporation)

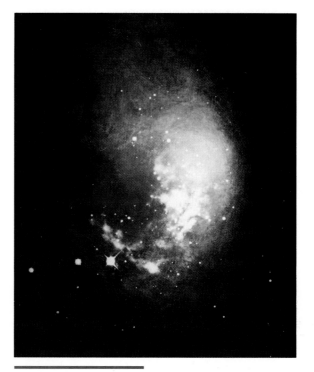

FIGURE 12.12

Infrared false-color map of the central part of the Orion Nebula. The colors represent infrared emission from 30 to 100 μm and are superimposed on a black and white photo that shows the visible stars. The strongest part of the infrared emission comes from the core of the molecular cloud that lies behind the H II region. (Courtesy NASA)

The interstellar dust is made of small, solid particles called *grains*. Basically, dust grains act roughly like (very small) blackbody radiators. If the grain has a temperature of around 100 K, its emission will roughly peak in the infrared, which we can observe with infrared telescopes.

The Orion Nebula marks a region studded with strong infrared sources. Optically, the core of the nebula (Fig. 12.11) is the densest part: hot gas (mostly hydrogen) is ionized and excited by the O stars there. The brightest stars form a trapezoid figure (easily seen in a small telescope) called the *Trapezium cluster*.

Now let's look at an infrared map (Fig. 12.12) of the core region. Quite a difference! What we are seeing is the infrared emission from cool dust (about 70 K) located at or near the center of the molecular cloud—dust heated by something capable of putting out 70,000 solar luminosities. The visible Orion

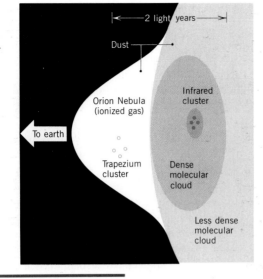

FIGURE   12.13

One model of the association of the Orion Nebula, its molecular cloud, and infrared sources. The Orion Nebula is a hot gas bubble, expanding outward, on the front of the molecular cloud. The infrared sources are embedded in the molecular cloud.

Nebula, illuminated and sustained by the Trapezium stars, lies in front of the molecular cloud like a hot bubble (Fig. 12.13). Enough dust floats around one of the Trapezium stars to be a weak infrared source.

Interstellar dust reveals itself directly by its infrared emission.

The Nature of Interstellar Dust

What is the composition of interstellar dust? Elements that make up an appreciable fraction of the interstellar material (hydrogen, oxygen, carbon, and nitrogen) can contribute in a large part to the dust grains. The most abundant elements make up rather common substances: hydrogen and oxygen for water, carbon and hydrogen for methane, carbon and oxygen for carbon dioxide, nitrogen and hydrogen for ammonia, silicon and oxygen plus metals for silicates (compounds commonly found in earth rocks). Compounds like water, methane, and carbon dioxide are

loosely called *icy materials* because they are solid at temperatures below about 100 K. (Recall that these substances make up the bulk of the nucleus of a comet: Section 8.2.)

To account for the properties of interstellar extinction, astronomers have developed models of **core–mantle grain** (Fig. 12.14). The small core, about 0.05 μm in radius, can consist of silicates, iron, or graphite; silicates are most likely. The mantles, about 0.5 μm in radius, are made of icy materials, likely some mixture of them all. When grains drift into hot regions, such as an H II region, their mantles evaporate, leaving behind a bare core. Infrared observations bolster the idea that silicates and ices make up a lot of the interstellar grains. Note that these grains are microscopic; they have a typical size of 10^{-3} that of the period at the end of this sentence.

Interstellar dust contains elements common to the interstellar gas in the general form of ices, silicates, graphite, and metals (such as iron); some of the mantles of grains contain organic compounds.

The icy materials in the mantles may be processed into organic compounds. Laboratory simulations of the conditions in interstellar space show that dirty ice mantles can absorb ultraviolet light, which has enough energy to break up chemical compounds and

---

FIGURE   12.14

Simplified model of an interstellar dust grain, which may consist of any of the materials listed or some combination of them. The main feature is that a small core is coated with a much larger mantle. The surface may be coated with a tarry substance.

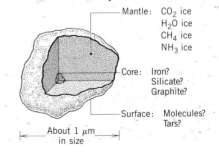

promote the formation of new ones. The experiments have also created tarry residues, whose compositions are unclear, but which can survive much higher temperatures than ordinary ices—up to 500 K. So they can survive in relatively hot environments, such as solar nebula from which planets formed (Section 8.5).

The Formation of Molecules and Dust

Dust grains play a key role in the formation of molecules. Consider the basic problem of forming an interstellar *molecule*. You have to get widely separated atoms together and chemically bound—no easy task in the dilute gas of interstellar clouds. Solution? Use cold dust grains as the sticking and forming surfaces. If a hydrogen atom hits a cold grain, it will stick. Add another hydrogen atom to the same grain and a hydrogen molecule forms. The larger particle does not stick to the dust grain as well as atomic hydrogen, and eventually it pops off into space. Then chemical reactions in the gas alone seem to explain the formation of molecules up to those containing four atoms.

What about more complex molecules? They may form in the icy mantles of grains, where ultraviolet light provides the energy for chemical processing of dirty ice to more complex molecules. Some of these compounds are unstable and can actually explode, sending the molecules into the interstellar gas.

Finally, dust also appears to protect molecules once formed. Ultraviolet light can break up molecules. Dust absorbs some ultraviolet, and enough dust can act as an effective shield (say, in the centers of molecular clouds) to prevent stellar ultraviolet light from breaking up recently formed molecules.

Interstellar dust aids in the formation and preservation of interstellar molecules.

How and where are grains made? The denser grains probably form in the atmospheres of cool supergiant stars. We know that such stars blow mass into space at rates of about 10^{-5} solar mass per year. The surfaces of these M stars have temperatures of only 2500 K or less. As gaseous material streams outward from them, the temperature drops, and solids can condense out of the vapor. In fact, spectra of some supergiant stars show silicate features and others carbon, indicating that such dust exists around them.

Cool giant stars, which are more common than supergiants, also lose mass at about 10^{-6} solar mass per year and contribute to the influx of dust to the interstellar medium. For both giants and supergiants, some of spectral class M have the highest mass loss rates of all, as great as 10^{-4} solar mass a year. They are large contributors to interstellar dust.

The ices that make up grain mantles likely condense on cores in the deep interiors of dense molecular clouds, where they are protected from ultraviolet radiation. Here the temperatures are low and the gas densities high, so bare grains can grow crusts of ices.

The cores of interstellar dust grains condense from gas ejected by stellar winds; the mantles form on the grains when they are embedded in a molecular cloud.

12.3 Starbirth: Basic Concepts

Here's the big picture: stars are born out of interstellar molecular clouds. Because these clouds contain many times the mass of a single star, they fragment

	Found in	interstellar gas clouds
Interstellar	Detected by	extinction, reddening
Dust	Structure	small core with larger mantle
	Radius	0.1 to 1.0 μm
At a Glance	Composition	silicate core with icy mantle; tarry crust
	Formation site	stellar winds of cool stars

The basic process of star formation is that of *gravitational contraction*, as you saw for the origin of the solar system. A cloud with enough mass and a low temperature will naturally contract from its own gravity. As gravitational potential energy becomes kinetic energy, the material in the cloud heats up (Section 8.5). This process eventually has two results: (1) the temperature (and density) build up enough that the outward pressure halts the contraction and balances gravity, and (2) the kindling temperature of fusion reactions is reached, at which moment, a star is born. Prior to the ignition of fusion reactions, while the cloud is contracting and heating, it is a **protostar**, which shines by the conversion of gravitational energy to light.

Protostars of different masses evolve differently; generally, there is one history for solar-mass clouds and stars and one for massive clouds and stars (which have 10 solar masses or more). Despite differences, the models have common features: (1) the collapse starts out in free-fall, that is, controlled only by gravity; (2) the central regions collapse more rapidly than the outer parts, and a small condensation forms at the center—this core will become a star; (3) once the core has formed, it accretes material from the in-falling envelope of material surrounding the core; and (4) the star becomes visible to us, either by accreting all the surrounding material onto itself or by somehow dissipating the shroud of dust (Fig. 12.15).

Solar-Mass Protostellar Formation

Imagine an interstellar cloud of gas, mostly molecular hydrogen, and dust, with an initial diameter of a few light years. The contraction begins. Material at the cloud's center increases in density faster than at the edge. Because of the density increase, the contraction at the center speeds up. It collapses faster, grows denser, and so collapses still faster. The core heats up. Meanwhile, the envelope falls inward, showering mass on the core. As the matter piles up, it increases the core's mass and temperature, making a protostar. Its size is twice that of the sun, its luminosity a few times the sun's. Fusion has not yet begun. Models suggest that the total time from the start of contraction to this stage is about one million years.

The dusty envelope, still falling in, blocks the birth from visual view. However, the light absorbed by the

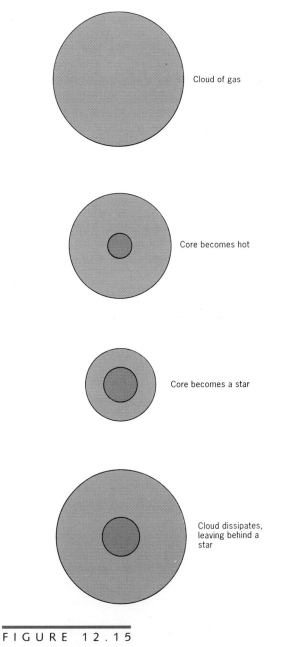

Cloud of gas

Core becomes hot

Core becomes a star

Cloud dissipates, leaving behind a star

FIGURE 12.15

Schematic sequence of gravitational contraction and star formation from a nonrotating cloud of gas and dust.

into much smaller pieces during the process of star formation. I have divided the topic into two broad parts: basic concepts (this section) and observational clues (Section 12.4).

dust heats these particles so that they give off infra-red radiation; an observational sign of protostars should be small, intense sources of infrared radiation in or near interstellar clouds.

Eventually, the star rids itself of its cloaking cloud, which may be blown away by a strong stellar wind. As the cloud dissipates, we see a **pre-main-sequence star**, one that is larger and cooler than it will be later on the main sequence. The total time elapsed from the onset of contraction to reaching the main sequence: about 50 million years.

Stars form by the gravitational contraction of interstellar clouds.

Contraction with Rotation

Interstellar clouds rotate, and any rotating mass has *angular momentum* (Section 6.5). An isolated, rotating mass must conserve angular momentum. So as a spinning, spherical mass contracts gravitationally, it must spin faster. It will eventually collapse into a disk along its rotation axis, as described in Section 8.5 for the formation of the solar nebula.

The addition of spin to theoretical models of protostar formation makes the calculations much tougher and the results less conclusive. In some cases, a ring or bar of material results. These rings and bars turn out to be unstable in some instances; they break up into two or three blobs. Sometimes these blobs coalesce and become fewer. The smallest mass of a fragment is about 0.01 solar mass; even these small fragments end up rotating rapidly, even if the cloud originally rotated very slowly.

Angular momentum results in a flattened disk during the gravitational contraction of a rotating interstellar cloud; a star or stars form within it.

Recent three-dimensional computer models (Fig. 12.16) have probed this process more extensively. The cloud begins with a smooth structure, but its rapid rotation causes it to fragment into smaller, distinct lumps. These can then develop into protostars. The important point to remember is this: with spin

FIGURE   12.16

Computer model of the birth of protostars from a spinning cloud of gas. The calculation is designed to conserve angular momentum. These false-color images show the density of the gas, with blue and black the densest regions. The sequence in time from (a) to (c) spans about a thousand years. Note that the cloud fragments into blobs whose masses are appropriate for protostar cores. (Courtesy A. P. Boss, Carnegie Institution of Washington)

(a)   *(b)*   *(c)*

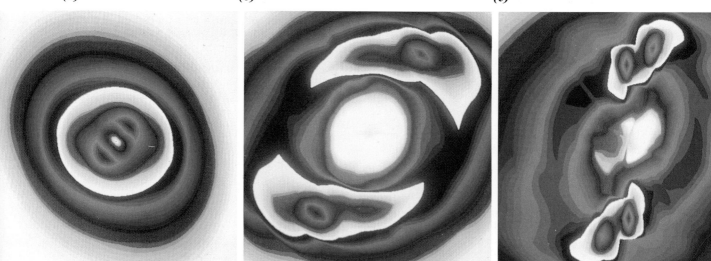

added to the clouds, computer models show that there is fragmentation into a few blobs. If each blob eventually becomes a star, we then have a natural explanation for the fact that most stars in the Galaxy are in binary or multiple systems. Or, if the masses are much less and end up in rings, they can result in planetary systems.

12.4 Starbirth: Observations

Observations have uncovered more information about massive starbirth than about the birth of solar-mass stars. There's a good reason: massive protostars have greater luminosities than solar-mass ones, and, once they have reached "stardom," massive stars ionize the gas around them. Radio telescopes can then detect the ionized gas. Since all this action takes place cloaked by dust, infrared and radio observations are necessary if we are to inspect stellar wombs.

Let's return to the Orion Nebula. The H II region around the Trapezium marks the oldest (*most evolved* in an evolutionary sense) part of the region. The Trapezium cluster consists of a few hundred stars. The massive OB stars of this group, which ionize the gas, are no more than one million years old. The molecular cloud core that lies behind the Orion Nebula is the youngest (least evolved) part of the region. Where is starbirth happening now? Probably at infrared sources in the molecular cloud; their observed characteristics match those expected from massive protostars.

Infrared images reveal that the core of the Orion Nebula contains a dense cluster of young stars (Fig. 12.17). More than 500 stars appear, most of them visible only in the infrared; these make up the densest young cluster known. Infrared observations of hot hydrogen molecules in this region show that the gas moves along at high speeds—some 100 to 150 km/s—probably powered by a wind from one or more protostars here. So star formation in the Orion Nebula now occurs in the densest regions, stirred by the outflows from the newborn stars.

FIGURE 12.17

Young stars in the Orion region. This infrared image shows at least 500 stars, most of which are invisible at optical wavelengths. The false-color coding has the hottest sources blue and the cooler ones red. The angular area covered is about 5 arcminutes on a side, centered on the Trapezium cluster. (Courtesy Goddard Space Flight Center, NASA)

Messier 17 (or M17, the seventeenth object in the catalog compiled by Charles Messier)—another bright H II region—presents a clearer observational view of one sequence of massive star formation (look back at Fig. 12.8). The optical H II region marks a site of star formation some 10 million years old; contained within it is a small cluster of O and B stars. To the west of the nebula, nothing much appears optically. But millimeter radio observations reveal two pieces of a molecular cloud. Both fragments have temperatures above those usually found in molecular clouds, so something there is heating up the gas.

In the south fragment lurks a starlike infrared source that has some characteristics of a protostar. To the west, connected to the two fragments next to M17, lies a gigantic molecular cloud, some 70 ly by 300 ly in size and containing more than a million solar masses of material. Infrared photos of M17 show a cluster of newly born stars in the depths of the molecular cloud (Fig. 12.18).

Starbirth	Where	molecular clouds
	When	now, at a rate of about 10 stars per year in the Milky Way
At a Glance	How	gravitational contraction, fragmentation
	What	young stellar objects, visible in the infrared, with bipolar outflows

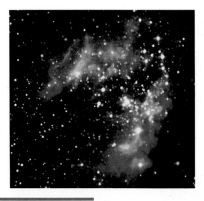

FIGURE 12.18

Infrared image of the molecular cloud region near M17, revealing a cluster of young stars embedded within it. This false-color map combines emission at 1.2 μm (blue), 1.65 μm (green), and 2.2 μm (red). (Courtesy KPNO/NOAO)

Infrared and radio observations together show regions of recent starbirth in molecular clouds.

The Birth of Massive Stars

We can develop a sequential model of massive star formation from giant molecular clouds. On one side of the molecular cloud lies the oldest region of starbirth. Here hot stars in a small cluster have heated the gas around them, ionizing it. The hot gas of the H II region slowly expands and runs into the cold, dense molecular cloud, and a shock wave forms. The shock wave, moving at about 10 km/s, prompts gravitational collapse and star formation in the molecular cloud (Fig. 12.19).

Moving farther into the fragmented molecular cloud, the shock wave will probably trigger additional star formation, and each group of OB stars will develop an H II region and another shock wave. In this model, small groups of massive stars are born in a sequence of bursts across the molecular cloud. Whatever material doesn't make it into stars eventually is dissipated (by stellar winds, for instance) by the stars that have formed.

Massive stars are born out of the fragments of giant molecular clouds.

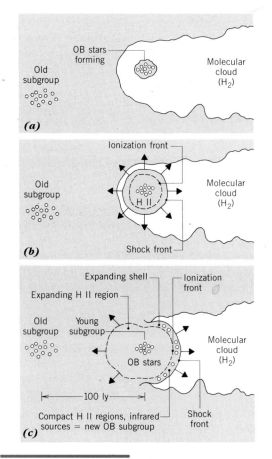

FIGURE 12.19

Schematic for a sequential model of massive star formation from a giant molecular cloud. (*a*) The formation of a small cluster of hot stars in a molecular cloud creates (*b*), an expanding H II region and shock wave that drives the collapse of more of the molecular cloud to make another small cluster of stars (*c*). Each episode of star formation leaves behind an OB subgroup as the formation process moves through the elongated shape of a giant molecular cloud.

What starts the first burst of star formation? The answer is not yet clear. Perhaps the collision of molecular clouds is the point of origin. Or, more likely, it is the blast wave of a supernova explosion plowing into an end of a molecular cloud. If so, a supernova—the violent death of a massive star—may ignite the birth of other massive stars (Section 14.4).

The Birth of Solar-Mass Stars

The observational evidence about forming solar-mass stars is skimpy and doesn't yet hold together in

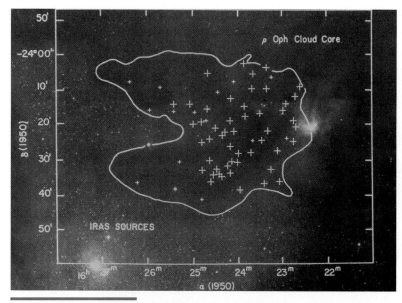

FIGURE 12.20

Observations of young stellar objects (YSOs) in the rho Ophiuchi dark cloud. Infrared observations at 2 μm show YSOs (marked by crosses) in the darkest, densest parts of the clouds. This dark cloud region lies near the center of Figure 12.9*b*. (Courtesy C. J. Lada and L. Wilking)

a complete way. One point is clear: like massive stars, solar-mass stars are born from molecular clouds. The questions are, In which clouds, and how?

Some clusters of stars are thought to be young because of their close association with gas and dust in the form of dark clouds, which constitute one type of small molecular cloud. Hidden among these dark clouds are *T-Tauri stars* (named after their prototype, T Tauri). Studies indicate that most T-Tauri stars are low-mass (0.2 to 2.0 solar masses), premain-sequence stars. They gather in dark clouds where star formation is most active. Some T-Tauri stars have clear evidence of thin disks of circumstellar material with diameters of a few hundred AU. These might mark future sites of planetary formation.

A prime example of a dark cloud with active star formation is called *rho Ophiuchi* (in the constellation of Ophiuchus; see the star map for summer in Appendix I). The central region of this cloud emits far-infrared radiation. Near-infrared observation (Fig. 12.20) show many possible **young stellar objects** (*YSOs*), the generic name for all the stellar objects in early stages of formation. At least 100 YSOs appear in the central 2°, the densest part of the dark cloud.

One view of the birth of solar-mass stars is this: they are born in fairly massive dark clouds, perhaps in giant molecular clouds along with massive stars, forming from fragments throughout the cloud, rather than at the edges as massive stars do. The birth of the massive stars—or perhaps the death of one in a supernova—sweeps away the gas and dust to reveal the stars. In this picture, most starbirth takes place in dark, massive clouds, out of which OB stars form.

We suspect that, in general, stars are born from molecular clouds; massive stars from massive clouds, and less massive stars (the majority of those in the Galaxy) from less massive molecular clouds. Observations by the Infrared Astronomical Satellite (IRAS, Appendix H) support this general notion. Its survey of the sky found many warm (temperatures from 70 to 200 K) sources of infrared emission in the cores of dark, molecular clouds (Fig. 12.21). These may be protostars, heating up the surrounding environment.

Solar-mass stars are born in dark clouds, perhaps in conjunction with the birth of more massive stars.

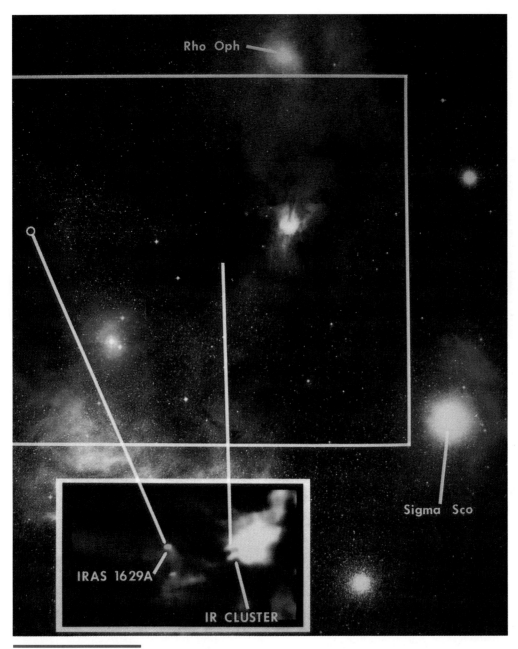

FIGURE 12.21

Protostar formation in the rho Ophiuchi dark cloud. The image shows an optical view of the region near rho Ophiuchi, just north of Antares. The dark cloud lies near the center; it contains the YSOs shown in Figure 12.20. The small insert is a false-color map from IRAS, in which 12 μm emission is blue, 60 μm green, and 100 μm red. It reveals a group of young stars called the *IR cluster*. The object called IRAS 1629A appears to be a collapsing cloud forming a protostar. (Courtesy of the University of Arizona; optical photograph by Anglo-Australian Observatory)

Molecular Outflows and Starbirth

Observations of molecules around YSOs have discovered high-speed (typically 50 km/s) flows of gas. Doppler shift measurements show that these flows tend to be *bipolar*—two streams moving in opposite directions (one has blueshifts, the other redshifts.) Such **bipolar outflows** carry considerable mass

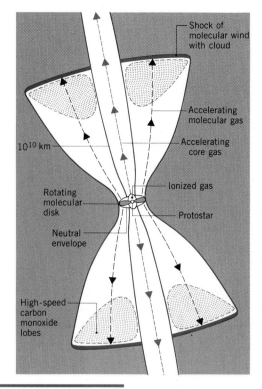

FIGURE 12.22

Model for the source of the bipolar outflows from a young stellar object. The key feature is the formation, by accretion, of a disk of material around the star. This disk is hot enough to produce ultraviolet photons to ionize some of the gas nearby. If the star has formed with a magnetic field, the ionized outflow follows the magnetic field lines. The outflow creates a shock front that backs up the material in two lobes. (Adapted from a diagram by R. E. Pudritz and C. A. Norman)

process are buried in gas and dust, but later stages are visible to optical and radio telescopes.

Models for these bipolar flows basically involve a disk of material around the protostar, threaded with a magnetic field (Fig. 12.22). The disk forms from gravitational contraction with spin. The magnetic field resides in the surrounding molecular cloud. Ionized gas that falls into the disk drags the magnetic field with it. The hot disk generates both a molecular flow (from the outer regions) and an ionized flow from its inner parts. The ionized outflow is channeled by the magnetic field. Essentially, the rotating disk acts like a flywheel, which stores energy that is released in the outflow. The outflow rams into the surrounding molecular cloud, forming a shock front.

Keep in mind that the bipolar flow stage marks a *very* brief act in the process of starbirth and the evolution of YSOs. The sequence goes like this: a star forms in the gravitational collapse of part of a rotating molecular cloud; the central part gathers into the protostar, which appears as an infrared source surrounded by a dense disk. Next, the star develops a powerful wind that breaks through the disk in oppo-

FIGURE 12.23

Circumstellar disk around the star Beta Pictoris. This computer-enhanced image shows a disk of material on both sides of the star, whose image is artificially blocked; the dark lines across the image are the supports for the circle of the blocking material. The disk around the star appears nearly edge-on and extends out some 60×10^9 km from the star, or about 400 AU. The black spots are artifacts of the technique. (Courtesy B. A. Smith and R. J. Terrile; observations at Las Campanas Observatory, operated by the Carnegie Institute of Washington; image processing at the University of Arizona and Jet Propulsion Laboratory)

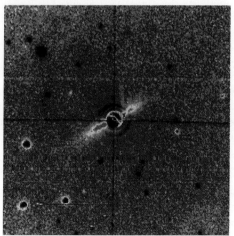

(many times that of the sun) and can span a few light years; so enormous amounts of energy push them along (some 10^{40} J—or about a million years of sunlight). Such bipolar outflows do appear associated with the birth of a massive star, and they probably last only a short time—no more than 10,000 years.

A simple model to explain the outflows envisions a young, massive star putting out a strong stellar wind. Surrounding the star is a dense disk of gas and dust. This disk would naturally channel the flow of the stellar wind, inducing it to run out along the thin axis of the disk, making two streams. When these two streams push enough material outward, two opposing lobes of gas should form. The early stages of this

site directions. The wind carries along clumps of molecular material and produces shock waves where it strikes the cloud. Cavities carved by the outward flow enlarge and push away more of the dark cloud, eventually revealing the star. For massive stars, the remaining gas ends up as an H II region.

The presence of molecular bipolar outflows from young stellar objects implies that the YSOs are formed with a surrounding disk of material.

Planetary Systems?

The discovery of the bipolar outflows strongly hints that disks of material typically form around massive stars during their formation. From such disks planetary systems might form. So we have a clue that the nebular model for planetary formation might actually operate elsewhere in the Milky Way—other stars may well have planets.

In fact, we are just finding observational clues that planetary formation may be happening now. A specially processed image of a nearby star called Beta Pictoris (Fig. 12.23) shows a possible planetary disk. The disk of dusty gas extends 60×10^9 km from the star—roughly five times the diameter of our solar system. It shows up edge-on, and calculations indicate that planets may already have formed here.

IRAS (Appendix H) has also gathered some strong, indirect evidence for planetary systems. It found a cool cloud of solid particles around the nearby star Vega. The cloud has a diameter of roughly 170 AU, about twice the diameter of our solar system—a possible site of planetary formation.

Finally, observations of bipolar outflows from T-Tauri stars, which have about the same mass as the sun, suggest that many of them have disks of materials with sizes on the order of 100 AU—just what you would expect for nebulas from which planetary systems can form.

Planetary systems may commonly form out of disks surrounding newborn stars.

The Larger View

Stars evolve and so must be born—some right now! Our current model sees them as embarking on their lives in dense clouds of the interstellar medium, which contains both gas and dust. The gas comes in a variety of forms in both cool and hot clouds. By far the largest clouds in the Milky Way are the giant molecular clouds. Many of the solid dust particles are found gathered in giant molecular clouds and in smaller dark clouds.

These molecular clouds are the stellar nurseries in the Milky Way. For any one star (like the sun), our model for birth basically follows a common route: A spinning clump of gas and dust becomes dense enough to contract by its own gravity. A star's formation releases gravitational energy, which powers a protostar until it gets hot enough

in the core to ignite fusion reactions. Even stars of different masses follow this general model for starbirth. Once formed, how do stars progress in their lives? Chapter 13 will probe the lifetimes of stars, some long and serene, some short and brutal, all necessary in the flow of stellar evolution.

During starbirth, a pancake-shaped disk of material develops around the star. Here planetary systems may be born. The gaseous material can collect in Jovianlike planets (and in the star). The dust may gather in earthlike planets, making globes of rocks and metals with a bit of carbon needed for life. Starbirth may well spur the building of planetary systems. If so, a multitude of planets may exist now.

Key Ideas and Terms

interstellar dust
interstellar gas
interstellar medium
Milky Way

1. The interstellar medium of the Milky Way contains both gas and dust. The gas, mostly hydrogen, comes in a variety of forms: molecules, atoms, and ions—each observed in different ways.

dark clouds
emission nebula
giant molecular clouds
H I region
H II region
molecular cloud

2. The gas clumps in clouds of various sizes, ranging from small clouds of atoms to the giant molecular clouds and complexes of GMCs. The hydrogen in the clouds comes in the form of molecules, atoms, and ions. Between the patchwork of clouds is a hotter, thin intercloud gas. A wide variety of molecules has been found in molecular clouds; the most common is molecular hydrogen. Some of these molecules are organic compounds.

extinction
interstellar dust
molecular clouds
reddening
reflection nebula

3. Interstellar dust is far less abundant than the gas but does contain about 1 percent of the total mass of the interstellar medium. Dust makes itself known by the reddening, scattering, and extinction of light and also by infrared emission when it heats up. Dust grains are associated with molecular clouds and aid the formation of simple interstellar molecules.

core–mantle grains
interstellar dust
molecular clouds

4. The dust is made of grains about a micrometer in size. The cores of these grains contain silicates, graphite, or iron. The mantles are made of icy materials and organic compounds. The grains are formed in the outflow of material from cool, supergiant and giant stars.

conservation of energy
gravitational contraction
gravitational potential energy
kinetic energy
pre-main-sequence star
protostar

5. Stars are born out of molecular clouds by the process of gravitational contraction. A protostar forming in a cloud gets its energy from the conversion of gravitational potential energy to kinetic energy. The process of starbirth is hidden from our direct view, but we can infer its operation by infrared and radio observations. Somehow the surrounding material must be dissipated to reveal the star.

binary stars
conservation of angular momentum

6. The gravitational contraction of rotating clouds naturally results in the formation of planetary systems with a single star or with multiple star systems. The clouds fragment as they contract; the fragments can form stars or planets.

giant molecular clouds
pre-main-sequence star

7. Infrared and radio observations imply that massive stars are formed in small groups out of giant molecular clouds in a chain-reaction sequence in which one groups triggers the birth of the next. The process may be started by a supernova remnant hitting the giant molecular cloud or by collisions with

young stellar objects (YSOs)

other giant molecular clouds. Solar-mass stars may be formed out of small molecular clouds or as a spin-off of the birth of massive stars from giant molecular clouds.

bipolar outflows conservation of angular momentum gravitational contraction

8. The formation of stars involves a bipolar outflow of gas (both molecular and ionized), collimated in part by a magnetized disk or ring around the young stellar object. The disk of material could form a planetary system.

Review Questions

1. What are the two main forms of material in the interstellar medium?
2. What is the astronomical name given to a cloud of neutral hydrogen gas in the interstellar medium?
3. What is the astronomical name given to a cloud of ionized hydrogen gas in the interstellar medium?
4. What is the most common molecule in the interstellar medium?
5. How does the amount of mass as gas in the interstellar medium compare to the amount of mass in the form of dust?
6. How does dust make a reflection nebula? What color does such a nebula have?
7. What effect will interstellar dust have on the light of a distant star?
8. What is the approximate diameter of an interstellar dust grain?
9. What is the source of a protostar's energy?
10. For what physical reason do we expect a disk of material to form around a young star?
11. What observational evidence do we have for bipolar outflows from young stellar objects?

Conceptual Exercises

1. Describe one way in which astronomers observe each of the following: interstellar H I, interstellar H II, and interstellar molecules.
2. Outline *two* ways in which astronomers "see" interstellar dust.
3. What evidence do we have, if any, that dense materials make up part of the interstellar dust? And that icy materials do?
4. List the observational evidence that leads to the model for the formation of massive stars in small groups from giant molecular clouds.

5. In a theoretical picture for the formation of a massive star, how is it that we never see a massive star directly (optically) until it reaches the main sequence?

6. Once a massive star reaches the main sequence, how does it influence its parent cloud?

7. What *observational* evidence do we have that solar-mass stars form from dark clouds?

8. In one sentence, argue that starbirth must be occurring in the Milky Way now.

9. In the sequential model for massive star formation from giant molecular clouds, what mechanism promotes the formation of a new group of stars, once the process has begun?

10. What is one possible source of interstellar dust grains?

13

Star Lives

Central Concept

Stars evolve; their physical properties change as they go through their normal lives.

Gravity controls the history of newborn stars. A star survives as long as it can counteract the relentless gravitational crunch. The story of this battle against gravity runs like the aging of a person from birth to death, but it plays out over millions to billions of years. How a star lives depends mostly on how much mass it has at birth. Stars that are much more massive than the sun have short, frenetic lives.

This chapter presents, first, theoretical ideas about stellar evolution, contrasting the lives of stars like the sun to those of more massive ones. Second, it offers some observational evidence in support of these theoretical models. You will see how stars burn, falter, and fade out. Our sun, too, will eventually die—and the earth with it.

13.1 Stellar Evolution and the Hertzsprung-Russell Diagram

As a star loses energy to space, it must change. Nuclear fusion reactions generally furnish the energy for this heat flow. When one fuel runs out, another may ignite, at a temperature made higher by the jamming of matter by gravity. Only when a star can no longer fuse matter will it finally cool off. Then the star has died (Chapter 14).

This chapter presents mostly theoretical ideas, because observations of one star's evolution are impossible. The sun's anticipated lifetime is more than 100 million human lifetimes. So there is no way you could watch a single star—like the sun—evolve. But you can see many different stars at one time. You can organize these stars on the Hertzsprung-Russell (H–R) diagram if you know their luminosities and spectral types. Then you can use the H–R diagram containing information about many stars to glean the evolution of one star.

Stellar lifetimes are too long compared to the human lifespan to permit us to observe the evolution of any single star.

Suppose you went around to ask some 18- and 19-year-old males in the United States for their weight and height. Plot the data as a graph of weight

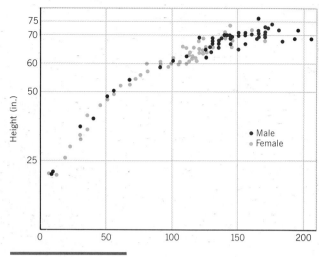

FIGURE 13.2

Height–weight diagram for a sample of U.S. males (black dots) and females (blue dots), ranging in age from 1 to 50 years. Units are inches and pounds. (Data collected by an introductory astronomy class at the University of New Mexico)

versus height (Fig. 13.1). Note the trend: the points tend to fall along a line that shows that weight generally increases with height.

Now suppose you recorded the height and weight of *every* person you encountered randomly. Plot the data again (Fig. 13.2) and compare with the other graph. What's the difference? You still have a general trend, rather than a scatter of points. But you also have other groups that don't follow the first set. Why the difference exists is clear: the first graph represents people of the *same* age; the second, people at *different* ages.

A third graph (Fig. 13.3), which plots weight versus height for an average U.S. male at different times in his life from birth to 20 years of age, shows how a *single* person's height and weight change as he ages. Note that this evolutionary graph for one person follows the trend in the height–weight graph of many different people.

If we know how a representative individual changes with time, we can infer the evolution of the group that is represented.

FIGURE 13.1

Height–weight diagram for a sample of 18- and 19-year-old U.S. males from an introductory astronomy class at the University of California, Berkeley. Units are inches and pounds.

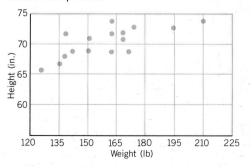

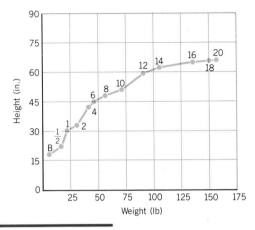

FIGURE 13.3

Height–weight–time diagram for a typical U.S. male from birth (B) to 20 years of age. The age is indicated by dots along the line. Units are inches and pounds. (Data from the National Center for Health Statistics Growth Chart)

You can correctly interpret the graph for many people in evolutionary terms *if*, and *only* if, you know how one person ages. Time was *implicit* in that graph. In the same way, time and age implicitly play a role in an H–R diagram, in which are plotted two essential properties of stars: surface temperature and luminosity.

How does the H–R diagram tell you about an evolutionary sequence of a star? Imagine a large family. They have gathered for a snapshot of the whole crew, from the newest-born to the great-grandparents. Most of the people in the picture are in their middle age, with a few infants, some children and teenagers, and a few old people. Middle-aged people are the majority because most human life spans these years; relatively less time is lived as an infant, teenager, or old person. Note that although the ages

differ, the expected lifetime is the same for all these people.

Now suppose you have a collection of objects that evolve in a specific sequence. You can estimate the relative time spent in any evolutionary stage by the relative numbers you find at that stage compared with others. (This argument holds true only if births and deaths occur continuously; if no more people are born, eventually you will see only old people—and then none.)

The H–R diagram shows the luminosity and surface temperature of many stars. Most of the stars fall on the main sequence, so stars found here are going through the longest, most stable stage in their evolution. The main sequence represents a sequence from higher to lower masses. Here's the evolutionary meaning of the main sequence: *it marks stars at the stage of converting hydrogen to helium in their cores*. Stars remain at this stage for the greatest part of their lives. Any star fusing hydrogen to helium in its core is a main-sequence star; it falls on the main sequence of the H–R diagram. That's why we see so many main-sequence stars now. Other stages, such as becoming a red giant, must be shorter because we see far fewer red giants now than main-sequence stars.

Time is implicit in an H–R diagram of many stars; the main sequence marks the longest stage, that of fusion of hydrogen in a star's core.

A star's mass (and so the pressure and temperature in its core) determines how the star will evolve. So the H–R diagram reveals how stars of different masses evolve—if we know how to read it. What is the correct interpretation? We need some hints from the physical nature of stars.

Stellar Lives **At a Glance**	Mass	mainly determines the life of a star
	Chemical composition	secondary effect on star lives
	Energy sources	alternation of gravitational contraction and fusion
	Lifetimes	from one million to tens of billions of years
	Evolution	changes in radius, luminosity, surface temperature, and chemical composition

13.2 Stellar Anatomy

A star is a huge, hot ball of gas (mostly hydrogen) heated by thermonuclear reactions in its core—a naturally controlled hydrogen bomb! A star does not explode, because gravity persistently pulls it together. All its life, a star must withstand the inward squeeze of gravity by producing an outward pressure force.

For most of its life, this pressure comes from the star's high internal temperature. A star consists of gas; a hot gas has a high pressure, and this outward force balances the inward gravitational force. This balance holds true at every level throughout the star (Fig. 13.4); otherwise it would be unstable.

Most of its life, a star maintains a balance between self-gravity and internal pressure, neither expanding nor contracting.

FIGURE 13.4

The balance of gas pressure and gravity in a star. The upward pressure from the internal heat (generated by the fusion reactions in the core) just balances the inward pull of gravity for the star to be stable. This balance must hold at every level in the star. If it does not, the star will either expand (if pressure is higher) or contract (if gravity is greater).

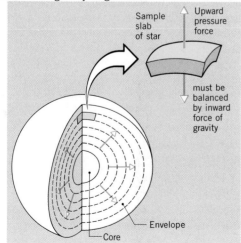

A star must generate internally energy that flows from the hotter core to the cooler surface. Mainly, thermonuclear fusion reactions operate as the internal furnace (Section 13.4). The total energy produced inside the star every second must equal the rate at which the energy radiates away at its surface. This balance holds not only overall but also at each layer within the star. As an analogy, consider a car assembly line, on which cars pass through different stages in the plant. At each stage, the rate at which the cars come in and go out must be equal. Otherwise the cars will pile up at one location. If the heat flow through a star were uneven, the temperature of various layers would change. These temperature changes would result in pressure changes that could cause the star to expand or contract.

A star produces energy at a rate equal to the rate of energy it loses.

A star must transport energy from its core to the surface. The heat flows from hotter to cooler regions. Basically, three methods of heat flow are available: *conduction*, *convection*, and *radiation*. Which of these occurs depends on the star's opacity (Section 10.3). The *opacity* of a star's material directly affects its radiative energy transport. The more opaque a star's material, the slower the flow of radiation through it.

Opacity acts like insulation in a house in winter. The furnace is generating energy. The greater the house's insulation, the slower the flow of the heat to the cold outside. (The resistance of insulation to heat flow is called its R-value. The higher the R-value, the slower the heat flow.) A slow flow keeps the exterior of the house cold and the inside hot. In a poorly insulated house, the outside is warmer and the inside colder than in a well-insulated house. Likewise, if the opacity of a star were suddenly lowered, radiation would escape more easily, and the star would become more luminous, but its inside would become cooler. So the internal pressure would drop, and the star would contract until it regained its balance. The opposite occurs if the opacity increases: the radiation is dammed up, the star becomes hotter, and it expands.

If the stellar material is transparent, energy will flow more easily by radiation. If it is so opaque that the radiative energy flow gets bottled up, convective transport will take over and operate instead. Generally, a star's opacity depends on its chemical composition, density, and temperature. A greater density or abundance of metals increases the opacity; a greater temperature decreases it.

A star transports energy from its center to its surface by radiation or convection, depending on the opacity.

A star loses energy by heat flow to space. Thermonuclear fusion reactions supply the energy to keep a star hot. Eventually the star runs out of fuel, and the pressure can no longer balance its gravity. Then the star contracts. A star can survive as long as it can find a means to produce energy, to stay hot, hence to withstand gravity. When it fails to do this, the star dies.

13.3 Star Models

To build a theoretical model of a star, we incorporate all the physical conditions just described. In addition, we must know how the matter in a star behaves; luckily, for most stars for most of their lives, that behavior is the same as for an ordinary gas (Section 10.2). We also need to know exactly what thermonuclear processes produce energy, as well as the mass and chemical composition of the model star. We can then build models as was done for the sun (Section 10.5). These models must match up with the masses, temperatures, luminosities, and chemical compositions of actual stars.

Models are calculated with computers to find such physical conditions as temperature, pressure, and density in the star, from the center to the surface. We know, for instance, that the temperature of a star increases inward from the surface to the center. The models say exactly how these conditions change within the interior at different depths. This catalog of values for important physical properties for a specified mass and composition is called a **star model**. One pair of numbers of this catalog, the

luminosity and the temperature at the surface, specifies a point on the H–R diagram—the *theoretical* point for a model star of given mass and chemical composition.

The study of stellar evolution rests on the proper construction of physically reasonable models of stellar interiors. We rely on the models because we cannot directly observe the interiors of stars, and we cannot wait millions to billions of years to watch individual stars evolve.

Stellar models, developed on computers and based on physics, reveal how time enters into the H–R diagram for real stars.

Almost everything said here rides on the validity of stellar models. But the solar neutrino experiment casts some doubt on our models for the sun (Section 10.5). And even the best models to date are but simplifications of the actual conditions of within stars, especially in the later stages of evolution.

Star models tell us that as a star evolves, its radius, temperature, and luminosity change in complicated

FIGURE 13.5

Movement of a star's point on an H–R diagram as its size changes. A star that stays at a constant luminosity but increases in temperature must be decreasing in size (arrow *A*). If a star stays at a constant surface temperature but increases in luminosity, its size must increase (arrow *B*). Generally, a star's changes in luminosity, temperature, and radius are more complex than shown here for these two cases.

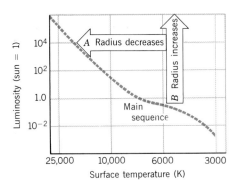

ways. Such changes result in the change in the position of a star's representative point on the H–R diagram (Fig. 13.5). For example, if a star's surface temperature (but not its luminosity) increases, the star's position on the H–R diagram moves horizontally from right to left, with no vertical change. If the luminosity increases (but not the surface temperature), the star's point on the H–R diagram moves vertically from bottom to top, with no horizontal change. These motions on the H–R diagram represent changes in the physical properties of the star; the path is called a star's *evolutionary track*.

Note in the first case that to keep the luminosity constant at a higher temperature requires that the star's surface area, and so its radius, decrease. In the second case, with no change in temperature, the star's luminosity can go up only if its surface area—and so its radius—increases. A star's surface temperature, luminosity, and radius are all interrelated.

A star's physical evolution emerges as the changes in size, luminosity, and surface temperature that show up as a star's evolutionary track on the H–R diagram.

13.4 Energy Generation and Chemical Compositions

The core of the star spurs its evolution. Here thermonuclear reactions cook lightweight elements into more complex ones by fusion. This change in chemical composition and its effects on a star's structure mark another major theme of stellar evolution. *Note:* Astronomers use the word "burning" to denote fusion reactions; keep in mind that we are *not* talking about *chemical* reactions in this context but *nuclear* ones.

Hydrogen Burning

The sun generates energy by the proton–proton (PP) reaction (Section 10.5). Four hydrogen nuclei combine to form one helium nucleus and release a certain amount of energy. A minor part of the sun's

energy comes from another reaction sequence, called the carbon–nitrogen–oxygen (CNO) cycle. In more massive stars, this cycle becomes more important than the proton-proton cycle. If a star's central temperature is greater than about 20 million kelvins, the CNO cycle produces more energy than the PP reaction. (Both can go on at the same time.) The net

---

FIGURE 13.6

The carbon–nitrogen–oxygen (CNO) cycle, a hydrogen fusion process. Note that the net result is the conversion of four hydrogen nuclei to one helium nucleus: the carbon entering the first step, which returns in the last, serves as a catalyst for the reaction. This process generates most of the energy for stars more massive than the sun, and a minor amount in the sun itself.

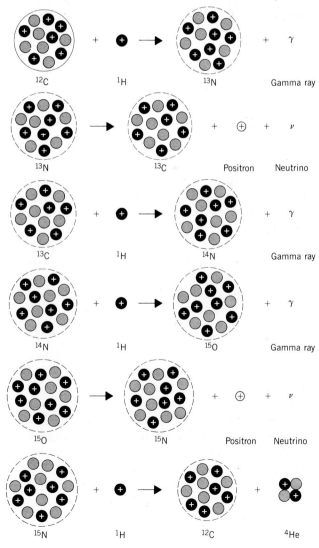

result of the CNO cycle is the same as that of the PP reaction: four hydrogen nuclei are converted to one helium nucleus, with the conversion of a small amount of matter into energy (Fig. 13.6). The CNO cycle predominates in stars greater than about 1.5 solar masses.

These reactions take place only in the star's core, where temperatures are high enough to keep them going. Because both involve the fusion of hydrogen as a fuel, they are given the generic name of *hydrogen burning*. Because a star has only a finite amount of hydrogen to burn, the core is eventually converted to helium, and the CNO and PP reactions cease.

The proton–proton reactions and the carbon–nitrogen–oxygen cycle are the two main fusion processes that convert hydrogen to helium in stars.

Helium and Carbon Burning

What next? The core contracts and therefore heats up. When the core's temperature gets up to roughly 100 million kelvins, another reaction can take place: the **triple-alpha reaction**, so named because three helium nuclei (also known as alpha particles) fuse to form one carbon nucleus, with the release of energy (Fig. 13.7). This process is generally called *helium burning*.

What happens when the helium runs out? The core contracts and heats up again. If the temperature increases enough, carbon can be fused into heavier elements in the processes of *carbon burning*. Such processes require extreme temperatures, at least 600 million kelvins. Iron, the most stable of all nuclei, ends the sequence of nuclear fusion. The steady climb from hydrogen to iron in fusion reactions in stellar cores is one type of cosmic **nucleosynthesis**.

You may be wondering why a star's temperature goes *up* when fusion reactions *stop*. The cause is *gravitational contraction*. If energy is not produced by nuclear reactions, it comes from gravitational potential energy. As the star contracts, some of the gravitational potential energy of its particles is transformed into kinetic energy and some is radiated away (Section 8.5). So the temperature goes up until

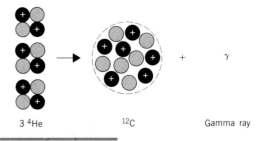

FIGURE 13.7

The triple-alpha process, a helium fusion reaction, converts three helium nuclei into one carbon nucleus and produces energy.

3 ^{4}He ^{12}C   Gamma ray

the ignition temperature of the next set of fusion reactions is reached. Fusion turns on again, and the core contraction stops. (Just the *core* needs to contract, not the star as a whole.) More massive stars have greater gravity and so can build heavier elements than less massive stars.

During a star's life, short periods of gravitational contraction alternate with long spells of fusion burning, which occur in stages from lighter to heavier elements.

13.5 Theoretical Evolution of a One-Solar-Mass Star

The details of a star's evolution are controlled by its mass. Let's look at the evolution of a star like the sun, namely a 1-solar-mass star, having the same chemical composition as the sun.

Main Stages of Evolution

Let me briefly review a 1-solar-mass star's pre-main-sequence evolution (Section 12.3). The key point is that a protostar gets its energy from gravitational contraction, *not* fusion reactions.

A protostar forms by the gravitational contraction of an interstellar cloud. Once the dense core of the cloud has formed (point 1 in Fig. 13.8), the rest of the cloud accretes upon it (points 1 through 4). For a

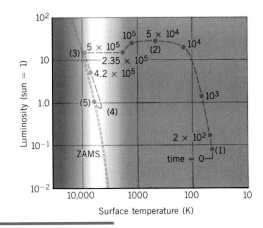

FIGURE 13.8

The evolutionary track (from 1 to 5) on an H–R diagram of a pre-main-sequence, solar-mass star. The labels at the dots indicate the approximate time in years since the formation of a core of an interstellar cloud at point 1. The zero-age main sequence (ZAMS) is indicated at the left. (Adapted from a diagram by K.-H. Winkler and M. J. Norman)

while, a protostar has a larger radius than it will have as a main-sequence star, and the surface temperature is lower (point 2 in Fig. 13.8). And the protostar has a higher luminosity than it will have when it reaches the main sequence—about 30 times larger after 100,000 years has elapsed. At this stage, the star's temperature is so low that its opacity is relatively high (even given its low density). Convection transports the energy outward. So a protostar is a huge, bubbling ball of gas. The efficient transport of energy makes the star very luminous (points 2 to 3 in Fig. 13.8).

The star shrinks in size, and its luminosity decreases (points 3 to 4 in Fig. 13.8) as it becomes a *pre-main-sequence star*. Eventually the core temperature hits a few million kelvins, high enough to start thermonuclear reactions. When the star begins getting most of its energy from thermonuclear reactions (PP reactions in the case of the sun) rather than gravitational contraction, it has achieved full-fledged stardom (point 5 in Fig. 13.8). The star is now a **zero-age main-sequence (ZAMS)** star. It settles down to the longest stage in its life, calmly converting hydrogen to helium in its core. The total time elapsed from initial collapse to arrival as a star on the main sequence is only 50 million years (from point 1 to point 5 in Fig. 13.8).

A star has reached the main sequence when most of its energy output comes from hydrogen fusion reactions.

Where the star ends up on the main sequence depends chiefly on its mass. The more massive the star, the hotter and more luminous it is; the less massive the star, the cooler and less luminous it is. The main sequence, on the H–R diagram, is a series of stars of decreasing mass but similar chemical composition, from the upper left-hand corner (O stars with high mass) to the lower right-hand corner (M stars with low mass). A sunlike star spends about 80 percent of its total lifetime on the main sequence as it slowly transforms its hydrogen core to helium.

How the star evolves further also depends on its mass. Because massive stars have higher luminosities, the hydrogen in them burns faster than in low-mass stars. As a result, massive stars spend less time on the main sequence, even though they have more fuel to burn, because they consume it at faster rates, as expected from the mass–luminosity relation (Section 11.6). Such stars are spendthrifts compared with sunlike stars and their miserly generation of energy.

The main-sequence phase ends when almost all the hydrogen in the core has been converted to helium. During this phase, the temperature in the core increases gradually, and the star's luminosity increases (Fig. 13.9). The star is now poised to become a *red giant*.

When the hydrogen in the core is used up, the thermonuclear reactions cease there. However, they keep going in a shell around the core, where fresh hydrogen still exists. At the end of fusion reactions in the core, gravity takes over and the core contracts. This heats up the shell of burning hydrogen, so the reactions go faster and the luminosity increases. The shell of burning hydrogen heats the surrounding layers and causes them to expand. The star increases in size, and its surface temperature decreases. Convection carries the energy outward in the star's envelope. In about 500 million years, a 1-solar-mass star ends up at about 1000 solar luminosities, a surface temperature of about 3000 K, and a size of about

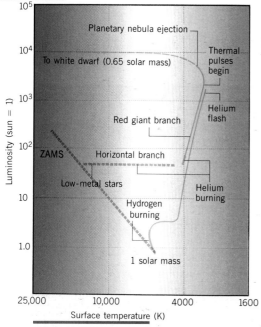

FIGURE 13.9

Theoretical evolutionary track on an H–R diagram for a 1-solar-mass star. The evolution begins at the ZAMS and ends with the formation of a white dwarf from the core of a red giant star. As hydrogen is depleted in the core, the star's luminosity increases. When the core runs out of hydrogen, the star burns hydrogen in a shell around the core; it becomes a red giant. Gravitational contraction heats the core until the temperature becomes high enough to start the ignition of helium in the core; for a short time, helium fusion occurs in the core. When the star's core is depleted of helium, the star burns both hydrogen and helium in a shell and becomes a red giant again. Finally, the star throws off its outer layers to expose its core and becomes a white dwarf. (Based on a diagram by I. Iben, Jr.)

100 solar radii—roughly the size of the earth's orbit! The bloated star now is a **red giant**.

Degenerate Gases

Gravity has compressed the red giant core to such a high density that it no longer behaves like an ordinary gas. The core becomes a **degenerate electron gas**. In this state, the electrons produce a **degenerate gas pressure**, which depends only on density, not temperature, and this enables the core to attain and preserve a balance even though no fusion reac-

tions are going on. (You will encounter degenerate gases again in Chapter 14.)

In an ordinary gas, the particles are widely separated; they rush helter-skelter into one another and rebound away (Section 10.2). In highly compressed material, little space exists between particles. The matter is so jammed together that the electrons on the outside of the atoms are, in a sense, touching each other. The nuclei can no longer hold electrons in their usual energy levels, and the electrons move among the nuclei. But there's not much space for moving about, and the electrons get along at high speeds. They exert a great pressure. A gas in this state is called a *degenerate electron gas* and the pressure is called the *degenerate gas pressure*. Unlike an ordinary gas, degenerate gas pressure is nearly independent of the temperature.

Electrons become degenerate at densities of about 10^8 kg/m^3. When electrons are degenerate, they conduct heat very efficiently, and temperature variations are quickly smoothed out. (A degenerate gas acts like a metal; it conducts both current and heat well.) So degenerate cores have the same temperature throughout. A high enough temperature can relieve the electrons of their degenerate condition. This requires a temperature of at least a few hundred million kelvins for electrons at the densities found in stellar cores.

> In a degenerate gas, which occurs at high densities in the cores of stars, conduction is the main process of heat flow.

Evolution to the End

In the red giant, the rising core temperature eventually hits the minimum to start helium burning by the triple-alpha process. This helium core is dense and degenerate. Once part of it has ignited, the energy spreads rapidly throughout the core because a degenerate gas is a good conductor of heat. The rest of the core quickly ignites. The increased temperature runs up the rate of the triple-alpha process, generating more energy, further increasing the temperature, and so on. This runaway process is called the **helium flash** (Fig. 13.9). The whole process is completed in a very short time—perhaps only a few minutes!

The star now burns helium in the core and hydrogen in a layer around the core (Fig. 13.9). This phase is the helium-core-burning analog to the star's main-sequence phase (that of hydrogen core burning). Eventually the triple-alpha process converts the core to carbon. The reaction stops in the core but continues in a shell around it. This situation—core shut down but thermonuclear reactions going on in a shell—resembles that of the star when it first evolves off the main sequence. The burning shell makes the star expand. The star again becomes a red giant. The electrons in this carbon-rich core become degenerate.

The helium-burning shell becomes unstable. Bursts of triple-alpha energy cause thermonuclear explosions in the shell; they have the prosaic name of **thermal pulses** (Fig. 13.9). The explosions occur about every few thousand years and cause the luminosity of the star to rise and dip rapidly by 20 to 50 percent. "Rapidly" here means in a few years! The explosions cause the star to pulsate in size and to vary in luminosity. Each blast generates a rush of energy; to move it out efficiently, the region becomes convective and the bubbling gases carry outward the elements fused in each explosion.

Meanwhile, the star has developed a very strong outflow of mass from its surface, sometimes called a *superwind* to distinguish it from the normal *stellar wind* of a red giant. The superwind is triggered by the pulsations of the star and blows in gusts that quickly (in about a thousand years) rip the envelope from the star. A hot core is left behind. The expelled material forms an expanding shell of gas heating by the hot core. Astronomers call this a **planetary nebula** for historic reasons. (It looks like a Jovian planet viewed with a small telescope.) The hot core appears as the central star (with surface temperatures up to 250,000 K!) of a planetary nebula (Fig. 13.10). The nebula keeps expanding until it dissipates in the interstellar medium. Such events repeat several times, each throwing off a shell. You can imagine the planetary nebula stages as a stellar strip tease.

What happens to the core? For a star of roughly 1 solar mass or less, the core never reaches the ignition temperature of carbon burning. Why not? Because the core has become degenerate and cannot contract and heat up to ignite carbon burning. In about 75,000 years it forms a white dwarf star, composed mostly of carbon. Without energy sources, the white dwarf cools to a **black dwarf**, the dark culmination of a 10-billion-year biography. The degenerate electron gas pressure supports it against gravity even as it cools down.

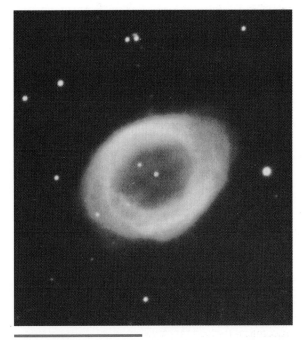

FIGURE 13.10

The central star of this planetary nebula (Messier 57 in Lyra) was once the core of a red giant. The nebula, which forms a spherical shell, was the envelope of the red giant. (Courtesy Palomar Observatory, California Institute of Technology)

What will happen to the earth when the sun dies? The sun will expand to a size of about 1 AU, with the result that the earth will orbit in its outer atmosphere. The earth's air will be ripped off, and the mantle vaporized. The drag will cause the earth's orbital radius to decrease, and in less than a few hundred years later, our planet will fall into the sun's core to vaporize completely.

Lower-Mass Stars

The evolution of stars of lower mass than the sun resembles that for the sun, with two exceptions. First, few stars less massive than the sun have had time to evolve off the main sequence. The universe simply isn't old enough. A star of 0.74 solar mass, for example, will have a luminosity about 37 percent that of the sun, according to the mass–luminosity law (Section 11.6), and so a main-sequence lifetime of 20 billion years—longer than most estimates for the age of the universe.

Second, if the mass of a star is less than about 0.08 solar mass, it will not even reach the main sequence. Gravitational contraction does not heat it enough to turn on the PP chain. Before it gets hot enough to start nuclear reactions, the density has risen so high that the matter becomes degenerate. Then the pressure of the degenerate electrons supports the star and keeps it from contracting any further. If gravitational contraction is prevented from heating the star, the nuclear fires can never be lit, and the star becomes a *brown dwarf* (Section 14.1).

Chemical Composition and Evolution

A star's mass directly affects its evolution. What about its chemical composition? If a star has a much smaller percentage of heavy elements than the sun, say 0.01 percent instead of 1 percent, would its evolutionary track make much different gyrations? The answer is *no*. Computer models indicate that the general trends stay the same: the star becomes a red giant, undergoes a helium flash, sits for a while during helium core burning, returns to red giant status, undergoes a series of thermal pulses, and finally expels its outer layers to make a planetary nebula encircling a white dwarf.

However, low-metal stars do differ significantly with respect to where they bide their time on the H–R diagram while they are burning helium in the core. Basically, stars with smaller percentages of heavy elements and less mass than the sun make a horizontal line on the H–R diagram, called the **horizontal branch**, during their phase of burning helium in their cores (Fig. 13.9).

Stars with lower percentage of metals than the sun follow the same general sequence of evolution on the H–R diagram.

13.6 Theoretical Evolution of a Five-Solar-Mass Star

Now to examine the history of a star much more massive than the sun: a 5-solar-mass star. Massive stars differ in their evolution from less massive stars

because they can reach higher temperatures in their cores. The greater temperatures have important consequences: (1) while on the main sequence, the star burns hydrogen mostly by the CNO cycle; (2) the main-sequence lifetime is roughly 25 times shorter; (3) the higher temperatures kindle carbon and heavier-element fusion in the core; and (4) the helium-rich core does *not* become degenerate. All this happens much more rapidly than for a solar-mass star. For a solar-mass star, the total lifetime, from main sequence onward, is about 10 billion years; for a 5-solar-mass star, it is only 400 million years.

A 5-solar-mass protostar evolves to the main sequence at roughly constant luminosity. The star is so hot that its opacity is low, and radiative transport is more effective than convection. The star's point on the H–R diagram moves horizontally to the left.

The PP reaction first ignites in the core (Fig. 13.11). When the core's temperature rises to about 20 million kelvins, the CNO cycle produces more energy each second than the PP reaction. Stoked by the high core temperature, the CNO cycle uses up the core's hydrogen in about 60 million years, while the star's luminosity becomes twice as large.

The main hydrogen fusion process in massive stars is the CNO cycle.

Note that a 5-solar-mass star on the main sequence is more luminous—500 times more—and hotter than a 1-solar-mass star. Less massive stars on the main sequence are cooler and less luminous.

When the central hydrogen fusion fires are exhausted, the core contracts. New hydrogen falls to the inner regions and ignites in a shell around the burnt-out core. The luminosity increases, but the radius expands, so the surface temperature drops. The star becomes a red giant (Fig. 13.11).

Meanwhile the core contracts until it gets hot enough to ignite the triple-alpha process. In contrast to a 1-solar-mass star, the core has *not* become degenerate. So no helium flash occurs in the core, just a relatively gentle triple-alpha ignition. (Stars with greater than about 2 solar masses do not develop degenerate helium cores because they do not become dense enough before helium ignition takes place.)

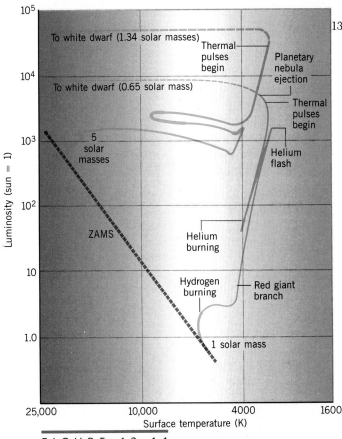

FIGURE 13.11

Theoretical evolutionary tracks off the main sequence (ZAMS) for stars of 1 and 5 solar masses. The thicker parts of the evolutionary tracks indicate phases of slow evolution with long-term fusion reactions in the cores. The dashed sections show where the evolution is rapid and not well known. This diagram shows both stars going through a planetary nebula phase before they become white dwarfs; note the large percentage of mass lost by each. Note also the overall similarity of the sequence of evolutionary stages. The thermal pulses occur in a helium-burning shell after helium burning has ceased in the core. (Based on a diagram by I. Iben, Jr.)

The star burns helium in its core for about 10 million years. When the helium runs out, the core again contracts, and the star becomes a red giant again. During the second time as a red giant, thermal pulses ignite in a shell around the core (Fig.

13.11), driven by the triple-alpha process. The star flashes and pulsates.

What next? We're really not sure; here are two possibilities. One, a superwind blows off the outer layers to leave behind a hot core surrounded by a planetary nebula. Or, two, a 5-solar-mass star can develop a degenerate carbon core of about 1.4 solar masses. When the core gets hot enough to ignite carbon burning, it should do so in a flash. This reaction may generate enough energy to blow the star apart. Such a cataclysmic explosion may result in a **supernova:** the outer layers of the star blast into space, and the core is crushed to immense densities. Whatever the physical reasons for a supernova, this explosion marks the end of the life of a massive star. (More in Section 14.4.)

> Massive stars end their lives violently after one or more phases as red giants.

13.7 Observational Evidence for Stellar Evolution

The description of evolution given in the preceding sections relies on theoretical calculations by computers based on the physics of stars. How well do these ideas connect with the real astronomical world? To find out, we see if we can find actual stars that are like the stars predicted by our models of stars at various stages in their life cycles with different masses and compositions. The key problem lies in determining whether any star is young or old relative to its expected lifetime. Luckily, we can look at many stars to find a clue.

Most of the stars in the Milky Way are in multiple systems, which are isolated from other stars by great distances. But we do find that some stars come in larger groups held together by their own gravity, at least for some time. These groups are called *star*

Red Giants	Mass	less than one and up to 8 solar masses
	Size	tens of solar radii
At a Glance	Energy transport	convection
	Phase	hydrogen shell burning
	Signatures	thermal pulses, stellar winds

clusters. The Milky Way contains two main types of star system: **galactic clusters** (or **open clusters**) and **globular clusters**. How do they differ?

Open Clusters

The Pleiades in the constellation Taurus (see the constellation charts for winter) is a good example of an open cluster, with a loose array of stars. The Pleiades lie about 400 ly from the sun. The cluster contains about 100 stars within a diameter of 10 ly for an average density of about 0.1 star in a cubic light year. These statistics are pretty typical of open clusters: they contain from fewer than 100 up to about 1000 stars in a space a few tens of light years in size; so their star densities are not more than a few stars per cubic light year—roughly 30 times the density of stars near the sun. Astronomers have catalogued some 1200 open clusters to date and perhaps 20,000 inhabit the Milky Way (Fig. 13.12). The stars in most clusters are moving apart and will eventually disperse (in billions of years) because their collective self-gravity is not enough to hold them together.

One key characteristic of open clusters is their H–R diagrams. The one for the Pleiades is pretty typical (Fig. 13.13). Note that the stars below a surface temperature of 10,000 K fall squarely on the main sequence. Above 10,000 K, the stars lie above and to the right of the main sequence. Most open clusters show the same properties: the lower-mass stars fall on the main sequence, but at some point the higher-mass stars turn off it (Fig. 13.14).

Globular Clusters

Globular clusters contrast dramatically with open clusters (Table 13.1). As the name implies, globular clusters have a distinct spherical shape (Fig. 13.15). Binoculars show them as fuzzy spheres, like cotton balls; a small telescope reveals some of the brightest stars.

The stars at the center of a globular cluster are packed as high as 100 per cubic light year! If the earth orbited a star in the core of a globular cluster, its nearest neighbors would be a few light *months* away. You would see thousands of bright stars scattered evenly over the sky. In all, a globular cluster contains 10,000 to one million stars of roughly 1 solar mass jammed into a space only a hundred light years or so in diameter.

FIGURE   13.12

New General Catalog 188: an open cluster in the Milky Way. Note how spread out the stars appear. (Courtesy Lee C. Coombs)

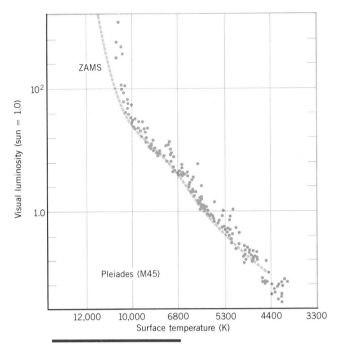

FIGURE 13.13

H–R diagram for the Pleiades. The blue line indicates the zero-age main sequence (ZAMS); the blue dots, the observed points for the stars. The luminosities are given for the visual range of the spectrum, with the sun's visual luminosity equal to 1.0. (From *An Atlas of Open Cluster Colour-Magnitude Diagrams*, by G. L. Hagen, David Dunlop Observatory, 1970)

FIGURE 13.14

Schematic H–R diagram for selected open clusters whose distances are known. Each cluster turns off the main sequence at a different position, called the *turnoff point*. In the older clusters, the massive stars above the turnoff point have evolved off the main sequence to become red giants. All clusters have stars that occupy the lower part of the main sequence below their respective turnoff points.

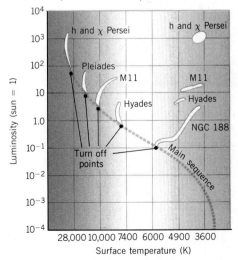

Table 13.1

A Comparison of Star Groups

Characteristic	Open Clusters	Globular Clusters
Mass (solar masses)	10^2–10^3	10^4–10^5
Diameter (ly)	5–50	50–300
Color of brightest stars	Reddish to bluish-white	Reddish
Density of stars (M_{sun}/ly^3)	0.1–10	1.0–100
Examples	Pleiades (Messier 45)	Hercules (Messier 13)

FIGURE 13.15

Globular cluster (Messier 13 in Hercules). Compare the distribution of stars to the open cluster NGC 188 (Fig. 13.12); note the great concentration of stars in the center. (Courtesy NASA)

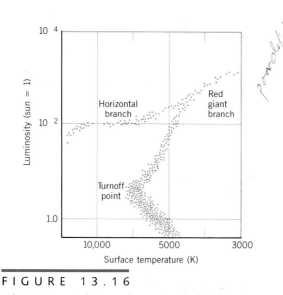

FIGURE 13.16

Schematic H–R diagram for a typical globular cluster. The stars along the lower part of the main sequence have masses less than 1 solar mass. The *horizontal branch* indicates the points for stars that have metal abundances less than that of the sun; to the left in this region, the stellar masses are smaller. Note the well-defined red giant branch and turnoff point to it.

An H–R diagram of a globular cluster (Fig. 13.16) clearly shows how it differs from an open cluster in terms of stellar type. The main sequence turns off to the red giant branch, and the upper end of the main sequence has disappeared. A horizontal branch of stars returns from the red giant region to the region of the absent upper main sequence. This slash across the H–R diagram, the *horizontal branch*, is the special signature of a globular cluster.

The horizontal branch contains low-mass stars, burning helium in their cores. They have chemical compositions with a lower percentage of metals than the sun. The horizontal branch observed for globular clusters strongly implies that the stars there are less massive and have fewer heavy elements than the sun. They are poor in metals because the gas and dust from which they were born contained only a small percentage of heavy elements.

Stellar Populations

The striking difference between the H–R diagrams of open and globular clusters implies that the stellar types in the two kinds of cluster are quite different. Astronomers call the stars formed in open clusters **Population I stars** and those in globulars **Population II stars**. The brightest Population I stars are bluish white; the brightest Population II stars, reddish. Also the brightest Population I stars (O stars) have about 100 times the luminosity of the brightest Population II stars (red giants).

Chemical composition, a really crucial distinction, cannot be seen directly in the H–R diagram. Spectroscopic observations show that typical Population I stars have essentially the same chemical composition as the sun—1 to 2 percent, by mass, of heavy elements (which are all the elements more massive than hydrogen and helium). Population II stars typically contain less, perhaps as little as 1 percent of this amount, 0.01 to 0.02 percent of the mass.

As a group, Population I stars tend to be *younger* than Population II stars, in the sense that they were born *later*. When a massive star dies, it spews a lot of material back into the interstellar medium. This blown-off material has been enriched in heavy elements; from it, new stars will be born. So Population II stars are *older* (formed *earlier*) than Population I stars because the former have fewer heavy elements. Population I stars have been formed out of enriched, recycled material.

Stars fall roughly into two populations, most distinguishable by their abundances of metals: Population II stars have less metals than Population I stars.

Comparison with the H–R Diagrams of Clusters

When a star cluster forms, its stars are born at essentially the same time with the same chemical composition, but the masses vary. So the lifetimes of these stars are different. The more massive stars, OB stars, evolve more rapidly than the less massive ones. So the more luminous stars evolve more quickly to red giants. As a cluster ages, stars of lower and lower mass will evolve off the main sequence. At the beginning of its life, a cluster's H–R diagram will resemble that of a young open cluster (Fig. 13.17*a*); a little later, that of a middle-aged open cluster, such as the Pleiades (Fig. 13.17*b*). Much later, the H–R

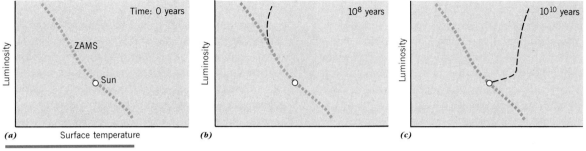

FIGURE 13.17

Theoretical H–R diagrams of one cluster of stars (with the same chemical composition, born at the same time) shown at different ages. "Sun" indicates the position of a solar-mass star. The ZAMS (blue line) is the zero-age main sequence of stars that have just begun hydrogen burning in their cores. The black dashed line shows the evolution of stars away from the ZAMS.

diagram is similar to that of an old open cluster (Fig. 13.17c). This is the observable effect of the aging of a cluster of stars with different lifetimes.

Note that the **turnoff point** away from the main sequence moves down to lower-mass stars as the cluster ages; *a cluster's turnoff point indicates its age*. An example: the old open cluster M67. Figure 13.18 shows an H–R diagram for the stars in M67. A

FIGURE 13.18

Comparison of theoretical models (solid line) and observations of the stars in the cluster Messier 67 (dots) to find the turnoff point from the ZAMS and therefore the age of the cluster (roughly 5 billion years, which means that it is a very old open cluster). (Based on calculations by D. A. VandenBerg)

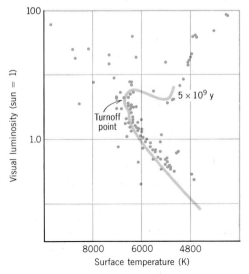

line corresponding to stars with masses ranging from 0.7 to 3 solar masses after an evolution of 5×10^9 years fits the turnoff point, so we infer that the cluster has this age, 5 billion years.

Open clusters have a large range of ages as estimated by this technique of matching their turnoff points to stellar models. For example, Praesepe (the Beehive) has an approximate age of 900 million years; and the Pleiades, 100 million years. In the Pleiades, only a few stars at the upper end of the main sequence have evolved away from it, so this cluster is rather young.

In contrast to galactic clusters, globular clusters have remarkably similar H–R diagrams, with roughly the same turnoff points. This implies that the ages of globular clusters are roughly the same; calibrated with theoretical models, the turnoff points for globular clusters indicate that they range from 12 to 16 billion years old, give or take about 2 billion years. The stars in the globular clusters are the oldest stars known.

The age of a cluster of stars can be determined by its turnoff point from the main sequence, based on stellar models.

As a cluster ages, the most massive (and luminous) main-sequence stars will evolve into red giants first, then the next most massive (next most luminous) ones will evolve, and so on. The main sequence will gradually shorten as the stars peel off, in order of mass, and evolve over into the red giant region. The turnoff point will move down the main sequence.

The stars last a relatively short time in this red giant phase; then they die and disappear from that part of the H–R diagram. By comparing theoretical evolutionary models with actual H–R diagrams for clusters, we can establish their ages. Such comparisons also confirm the general validity of the models for main-sequence evolution.

13.8 The Synthesis of Elements in Stars

Gravitational contraction provides the heat required to initiate fusion reactions. To survive, a star must fuse lighter elements into heavier ones to generate energy. The more mass a star has, the greater the central temperature produced by gravitational contraction and the heavier the elements it can eventually fuse. From knowledge of the ignition temperatures needed for fusion reactions, we can set limits on the heaviest elements that a star of a certain mass can fuse (Table 13.2). For example, our sun can burn helium to carbon in its core but will never get hot enough to fuse carbon.

Table 13.2 summarizes the principal stages of nuclear energy generation and nucleosynthesis in stars. Note that the products (or ashes) of one set of reactions usually become the fuel for the next set of reactions. What a beautiful scheme for energy production in the universe!

Also note that only very massive stars (those with masses greater than about 5 solar masses) can produce elements heavier than oxygen, neon, sodium, and magnesium in their cores. Few stars have this much mass, so most stars come to the end of their nuclear evolution having manufactured only a few heavy elements.

Massive stars create heavy elements *and* throw some back into the interstellar medium.

Red giant stars also play a vital role in the scheme of cosmic nucleosynthesis. That view, based on theoretical models, sees the thermal pulses in a helium-burning shell as the site for producing certain isotopes, especially those that are rich in neutrons. (A neutron-rich isotope is one that has more neutrons than protons in the nucleus.) This process can occur in two stages for stars of low and middle mass.

One stage takes place when any star first becomes a red giant. Thermal pulses then aid in the development of a convective zone that reaches down to the star's core and pulls up elements that have been made with hydrogen burning. At the base of the convection zone, carbon can be converted to nitrogen. The convection brings this processed material up to the surface.

For a medium-mass star, such as one of 5 solar masses, a second phase of nucleosynthesis takes

Table 13.2

Stages of Thermonuclear Energy Generation in Stars

Process	Fuel	Major Products	Approximate Temperature (K)	Approximate Minimum Mass (solar masses)
Hydrogen burning	Hydrogen	Helium	2×10^7	0.1
Helium burning	Helium	Carbon, oxygen	2×10^8	1
Carbon burning	Carbon	Oxygen, neon, sodium, magnesium	8×10^8	1.4
Neon burning	Neon	Oxygen, magnesium	1.5×10^9	5
Oxygen burning	Oxygen	Magnesium to sulfur	2×10^9	10
Silicon burning	Magnesium to sulfur	Elements near iron	3×10^9	20

place after the star has burned the helium in its core. Thermal pulses then can convert helium to carbon, carbon to oxygen, nitrogen to magnesium, and iron to certain neutron-rich isotopes of heavier elements. Convective brings these materials to the surface.

All these processes would have no effect on the rest of the cosmos except for one crucial fact: red giant stars have strong **stellar winds,** which blow off material from the surfaces of the stars, so that the processed material gets sent out into the interstellar medium.

Red giants fuse heavy elements, bring them to their surfaces by convection, and blow them off by stellar winds.

The Larger View

Once they are born, stars lose energy to space. So they must evolve. During its lifetime, a star constantly struggles against gravity. It resists gravitational collapse by internal pressure from interior energy. Fusion reactions in the core provide the energy. A star must fuse heavier and heavier elements to withstand gravity; these ideas allow us to make up models of stars to track their evolution. These models rely on the basic physical processes we have inferred for the sun and other stars: ordinary gases (occasionally extraordinary!), fusion, and heat flow.

We discover that the Hertzsprung–Russell diagram reveals more about stars than simply their luminosities and temperatures. The patterns in the diagram reflect stars of different masses and compositions at different stages of their lifetimes. Time lies hidden here; only by comparing models to clusters of stars can we illuminate how any one star evolves. The key is that a star's history is determined by its mass.

Stars spend most of their lives on the main sequence of the H–R diagram as they burn hydrogen to helium in their cores. As they age, stars become red giants. In this stage, they generate some of the key elements in the universe—including those necessary for life. In this and later stages of evolution, stars lose mass and so recycle some of their material, enriched with heavy elements, back to the interstellar medium. Some material is locked in the star's corpse, never again to partake in the recycling scheme. We'll explore the deaths of stars in Chapter 14.

Key Ideas and Terms

Hertzsprung–Russell diagram

mass

stellar lifetimes

1. A Hertzsprung–Russell diagram provides a picture of stars at different stages of their lives. A star's mass determines how it lives; more massive stars have lives that are shorter than and different from those of less massive ones.

energy

fusion reactions

pressure

star model

2. Stars maintain a balance between gravity and internal pressure generated by fusion reactions. To strike this balance, a star must produce energy inside; the loss of this energy to space requires that the star evolve. Stellar models guide our understanding of this evolution.

CNO cycle
gravitational
    contraction
nucleosynthesis
PP chain
triple-alpha reaction

3. Fusion reactions normally generate a star's energy. When fusion reactions stop, gravitational contraction can produce the energy to ignite their next stage. As fusion reactions use up fuel, the ashes produced can become the next fuel, if the temperature gets high enough. Stellar fusion reactions start with hydrogen and end with iron.

CNO cycle
main sequence
PP chain
zero-age main
    sequence (ZAMS)

4. A newly–born star shines from energy produced by gravitational contraction. When fusion fires ignite, the star achieves the main-sequence stage of its life. Stars on the main sequence in the H–R diagram are fusing hydrogen to helium in their cores; more massive stars (upper part of the main sequence) evolve faster than less massive stars (lower part of the main sequence).

black dwarf
degenerate electron
    gas
degenerate gas
    pressure
helium flash
planetary nebula
red giant

5. When a star has fusion reactions occurring in a shell (or shells) around the core, not in the core, it expands in size as a red giant and evolves away from the main sequence. Our sun (and other solar-mass stars) will evolve to a red giant, undergo a helium flash, and finally blow off its outer layers to make a planetary nebula. The former red giant core (made of a degenerate gas) will become a white dwarf, then slowly cool to a black dwarf.

red giant
supergiant
supernova

6. Medium-mass stars (5–10 solar masses) become red giants a few times in their later lives, then die in supernova explosions. High-mass stars (greater than 10–20 solar masses) become supergiants; they also die in supernovas.

horizontal branch
Population I stars
Population II stars

7. Stars with the same chemical composition as the sun undergo the same evolutionary scheme, its rate being determined by their mass. Stars with a lower percentage of heavy elements than the sun follow the similar overall trends but form a horizontal branch as red giants.

galactic (open)
    clusters
globular clusters
turnoff point

8. A comparison of the H–R diagrams for clusters of stars confirms our basic ideas about stellar evolution. The turnoff point on the main sequence indicates the age of a star cluster, in which all stars were formed at the same time.

nucleosynthesis
stellar winds
thermal pulses

9. Massive stars can fuse elements up to iron in their cores in their normal lives. Red giants can fuse some in convective helium-burning shells. These are brought to the surface by convection and blown off by stellar winds.

Review Questions

1. What two properties of stars are plotted on an H–R diagram?
2. Gravity pulls a star together. What keeps it from collapsing during its normal life?

3. How does the interior of a star become hot enough to ignite fusion reactions?

4. What is the first type of fusion reaction to operate in a star?

5. What kind of fusion reactions are main-sequence stars undergoing?

6. What happens to a star when fusion reactions take place in shells around the core rather than in the core?

7. After their main-sequence phase, most low- and middle-mass stars evolve away from the main sequence. What kind of star do they become?

8. High-mass stars, too, evolve off the main sequence. What kind of star do they become?

9. Do stars with chemical compositions that have an abundance of heavy elements much lower than the sun undergo different evolutionary paths on the H–R diagram?

10. If a star cluster's turnoff point from the main sequence is high up, is that cluster relatively young or old?

11. How do massive stars end their lives?

Conceptual Exercises

1. A star like the sun consists completely of an ordinary gas. Why doesn't it suddenly collapse gravitationally? When the sun becomes a white dwarf, what prevents it from collapsing?

2. Present calculations indicate that a solar-mass protostar is much more luminous than the sun. Yet it's much cooler at the surface. How could the protosun be much cooler and yet more luminous than the present sun?

3. How can you tell from an H–R diagram that the stars in the Pleiades cluster are younger than those in the Hyades? (*Hint*: See Fig. 13.14.)

4. Why are massive stars able to fuse heavier elements than less massive stars?

5. What observational evidence do we have that red giant stars become white dwarfs?

6. What is a main difference between the evolution of a 1-solar-mass star and a 5-solar-mass star?

7. Compare and contrast the stellar *types* in open and globular clusters.

8. Outline the evolution of a cluster of stars containing half 5-solar-mass stars and half 1-solar-mass stars.

9. What kind of fusion reaction can produce carbon in a star? What minimum mass is needed for this reaction to occur?

10. In what sense is time *implicit* in an H–R diagram? How do astronomers make it *explicit*?

11. Many stars become red giants as they evolve. Their luminosities increase, their sizes increase, and their surface temperatures decrease. In very general terms, what happens to the fusion reactions as a star becomes a red giant? To the means of energy transport?

12. What ingredients are essential in constructing a model of a star? (*Hint*: What observed properties must the model match?)

14
Stardeaths

■

Central Concept

Most stars die violently and leave behind strange corpses: white dwarfs, brown dwarfs, neutron stars, and black holes.

■

At the end of its life, a dead star's remnant core is locked tight by gravity. For some, the burned-out core becomes a white dwarf star, a solid carbon crystal. In others, the core develops into a neutron star, a smooth, spinning sphere of nuclear matter. Or the core may disappear through a warp in spacetime as a black hole.

Almost all stars throw off mass before they meet their ends. That loss is mild compared to a supernova—the most destructive agent of mass loss for massive stars. But a supernova is constructive too; in its immense explosion, many heavy elements of the universe are made and thrown to the currents of space. Supernovas spice the interstellar medium with some of the elements needed for life.

14.1 White and Brown Dwarf Stars

Because gravity never lapses, the final state of any star depends on the physical properties of matter at high densities and the total mass of the star. When all thermonuclear reactions cease, what pressure can support the star?

White Dwarfs

The conventional model of the atom surrounds the nucleus with a cloud of electrons. The distance from the nucleus to the first electron is about a thousand times the diameter of the nucleus, so an atom is mostly empty space. At the high temperatures inside stars, atoms are ionized, and the electrons run around free of the nuclei. As a star is crushed to higher densities in its evolution, the electrons form a *degenerate electron gas* (Section 13.5).

A degenerate star of 1 solar mass has an average density of about 10^9 kg/m^3. If the sun were compressed to this density, it would be approximately 7000 km in radius, about the size of the earth (Fig. 14.1). In such a degenerate mass, the pressure exerted by the electrons can resist the force of gravity only for stars of less than 1.4 solar masses. Such a star, at the end point of its thermonuclear history, is a **white dwarf**.

Surprisingly, the more massive the white dwarf, the *smaller* its radius. (This contrasts to main-sequence stars, which are larger, the more mass they have.) More mass means more gravity. To balance gravity requires internal pressure. In a white dwarf, the pressure does *not* come from the internal heat of thermonuclear reactions. Instead, it arises from the nature of a degenerate gas, where greater pressures are a response to greater density. So the more massive a white dwarf, the smaller its size.

A crucial point arrives when the mass of the white dwarf is about 1.4 solar masses; such a star has the

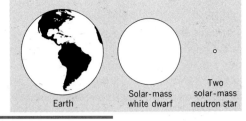

FIGURE 14.1

Comparison of the relative sizes of the earth, a solar-mass white dwarf, and a 2-solar-mass neutron star drawn to the same scale.

for a w.D.

highest density and smallest radius possible. Add a bit more mass and the gravitational forces overwhelm the degenerate electron gas pressure. The star collapses; it cannot be a stable white dwarf. This amount of mass, 1.4 solar masses, signals the point at which degenerate electron matter is crushed by gravity. This maximum mass for a white dwarf is called the **Chandrasekhar limit**, after the Indian astronomer Subrahmanyan Chandrasekhar, who first applied the physics of a degenerate electron gas to the model of a star.

A white dwarf is a stellar corpse with roughly the mass of the sun and the size of the earth; a degenerate electron gas supports it against the crush of gravity.

Sirius B is an example of a white dwarf. It has a mass of about 1 solar mass, a luminosity of about 3×10^{-3} times the sun's luminosity, a surface temperature of about 30,000 K, and a radius around 7×10^{-3} the sun's radius. Given its size and mass, Sirius B has an average density of about 3 *billion* kilograms per cubic meter! On the earth, a spoonful of white dwarf stuff would weigh about as much as a car.

White Dwarfs	Mass	up to 1.4 solar masses (Chandrasekhar limit)
	Size	about 7000 km in radius
	Density	roughly 10^9 kg/m^3
At a Glance	Internal pressure	degenerate electron gas
	Energy	stored internal heat; no fusion reactions

Very slowly (over billions of years), a white dwarf's stored internal heat (left over from past nuclear fusion) radiates into space. Eventually, the white dwarf becomes a *black dwarf* (not to be confused with a black hole!). Our sun will become a white dwarf, then a black dwarf—a cold corpse in space.

Brown Dwarfs *See also Top p.282*

White dwarfs are the corpses of stars that start their main-sequence lives with masses ranging from 0.08 to (perhaps) 5 solar masses. What becomes of objects born with less mass?

Small masses do not reach high enough temperatures for hydrogen fusion. Gravitational energy released by contraction provides their power; they have the promise of protostars but never reach stardom. Astronomers call such a mass a **brown dwarf**, a misnomer because it is not really brown (and has never really been a star because it is never fully supported by nuclear energy generation). In physical makeup, brown dwarfs are more like very large Jovian planets.

Brown dwarfs probably have low surface temperatures—less than 3000 K—so that most of their emission lies in the red and infrared part of the spectrum. These cool temperatures and expected low luminosities (roughly a few times 10^{-5} the sun's luminosity) make brown dwarfs hard to detect. But infrared observations could succeed in this task, and searches are now under way in regions near young stellar objects and dark clouds.

A brown dwarf is a low-mass object that has never reached the stage of hydrogen fusion burning.

 14.2 Neutron Stars

What happens to stars that have more than 1.4 solar masses of core material at the end of their evolution? Degenerate electron gas pressure cannot support them, and gravity crushes them to higher and higher densities.

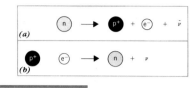

FIGURE 14.2

Beta decay and its reverse process, inverse beta decay. (An electron is sometimes called a *beta particle* for historic reasons.) (*a*) In beta decay, a neutron decomposes into a proton (p), an electron (e), and an antineutrino ($\bar{\nu}$, an antimatter neutrino). (*b*) In inverse beta decay, a proton plus an electron is transformed into a neutron (n) and a neutrino (ν).

At about 10^{13} kg/m³, **inverse beta decay** occurs (Fig. 14.2). In this process, an electron and a proton combine to form a neutron and a neutrino. (An electron is sometimes called a *beta particle*, and so the name of the process.) Both free protons and protons in the nucleus of a heavy element are subject to inverse beta decay. At 10^{17} kg/m³, the nuclei suddenly fall apart into a gas with 80 percent neutrons. A spoonful of this material on the earth would weigh about a *billion* tons—as much as all the cars ever produced.

At this density, the neutrons become degenerate in the same manner that electrons do at white dwarf densities. The neutrons provide a degenerate gas pressure, and so balance the inward pull of gravity. This pressure allows the formation of a stable **neutron star**, a star composed mainly of neutrons. Depending on its mass, its diameter will be about 10 to 20 km (about the size of a city!), As with white dwarfs, the *greater* the mass of a neutron star, the *smaller* its radius, because it is made of a degenerate gas.

A neutron star is a weird beast compared with an ordinary star. In a typical model (Fig. 14.3) with a diameter of about 15 km, the inner 12 km consists of a neutron gas at such high densities that it is a fluid. The next 3 km out from the center is a mixture of the neutron fluid and neutron-rich nuclei, arranged in a solid crystalline lattice. In the outer few meters, where the density falls quickly, the neutron star has an atmosphere of atoms, electrons, and protons. The atoms here are mostly iron.

An ordinary star that *finishes* its evolution with a core mass greater than 1.4 solar masses probably

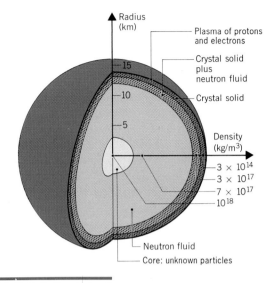

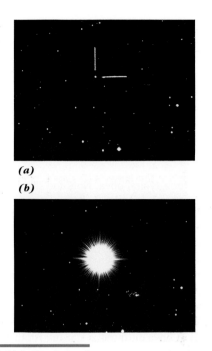

(a)

(b)

FIGURE 14.3

Theoretical model for the cross section of a solar-mass neutron star. The solid crust is mostly iron, topped off by an atmosphere of protons and electrons only a few meters thick. The interior is mostly a fluid of neutrons; the composition of the core, where densities are greater than those of the nuclei of atoms, is unknown.

FIGURE 14.4

Nova Herculis 1934. These photos show the star (a) before (marked by white lines) and (b) during its outburst in 1934. (Lick Observatory Photograph)

ends up as a neutron star. In analogy to the Chandrasekhar limit, a mass limit for neutron stars is reached when the gravitational forces overwhelm the neutron degeneracy gas pressure. This limit—not known exactly, but about 3 solar masses—signals the next crushing point of matter by gravity.

A star with a mass between 1.4 and (roughly) 3 solar masses at the time of its death naturally forms into a degenerate neutron gas to support a neutron star, which is some ten kilometers in radius and lacks any fusion reactions.

14.3 Novas

Occasionally, a new star is spotted in the sky where none was visible before; such a star is called a *nova stella* (a Latin phrase usually contracted to simply *nova*). With the advent of photography and large telescopes in the nineteenth century, astronomers discovered many novas. By the beginning of this century, a **nova** was no longer considered a new star but rather the sudden eruption of light from an existing star (Fig. 14.4). Also in this century, astronomers recognized that some of these outbursts take place with extraordinary violence. These special flare-ups are now called *supernovas* to distinguish them from ordinary novas.

Neutron Stars	Mass	up to 3 solar masses
	Size	about 10 km in radius
	Density	roughly 10^{17} kg/m^3
At a Glance	Internal pressure	degenerate neutron gas
	Energy source	stored internal heat; rapid rotation

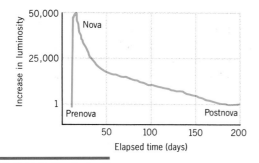

FIGURE 14.5

Light curve for a typical nova. Note the sharp rise from the prenova luminosity to the peak and then the more gradual decline over a span of a few hundred days. (Adapted from a diagram by C. Payne-Gaposhkin)

Both novas and supernovas represent explosions of stars. For ordinary novas, only the outer layers participate in the explosion. For supernovas, the interior regions of the star are also involved.

In a typical nova outburst, a star in just a few days increases up to 50,000 times in brightness (sometimes up to a million). It stays near peak luminosity (some 100,000 solar luminosities) for several hours. Then the nova's light slowly declines to an inconspicuous level—usually brighter, however, than the star's prenova level. A plot of a nova's rise and fall in flux (or luminosity) is called its *light curve* (Fig. 14.5). Most novas have the same general shape for

FIGURE 14.6

Nova Herculis 1934, its radius about 0.05 ly, photographed 40 years after the outburst. Note the shell of blown-off material, which is expanding into the interstellar medium. Many years elapsed before the shell had expanded enough to be visible. The star that supplied the fresh fuel for the outburst is at the center; the white dwarf in this binary system is too faint to show up here. (Courtesy R. E. Williams, University of Arizona)

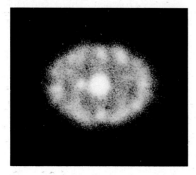

their light curves: a sharp rise and a gradual decline. All told, a nova emits during its flare-up and demise—the space of a few hundred days—about as much energy as the sun generates in some 100,000 years!

During the outburst, a shell of material is blown off the star and rapidly expands away from it (Fig. 14.6). Measured Doppler shifts of the shells' spectra indicate speeds from a few hundred kilometers per second in some novas to a few thousand kilometers per second in others. Overall, a nova spurts off about 10^{-4} solar mass. Since the prenova star had about 1 solar mass, the ejected material makes up only a small fraction of the total.

What prompts a nova to explode? One major clue: most novas occur in close binary systems, that is, binary stars with short orbital periods (typically about 4 hours). In such systems the two stars are so close (about 1 solar diameter) that matter may flow between them. We believe that a nova occurs on the white dwarf companion of about solar mass in such binary systems.

In a close binary star system, hydrogen can flow from the companion star to the white dwarf. A stellar wind causes the companion star to lose mass, and the material falls toward the white dwarf, forming a disk around it. The material gathers in this new structure, called an **accretion disk** (Fig. 14.7), because of the conservation of angular momentum (Section 6.5). The matter spirals onto the white dwarf's surface

FIGURE 14.7

Artist's conception of a red giant–white dwarf binary system with matter flowing from the red giant into an accretion disk around the white dwarf. (Courtesy Steward Observatory, University of Arizona)

from the accretion disk. Fresh hydrogen gradually accumulates on the white dwarf's surface, forming a new layer of fuel, which is compressed and heated as additional material piles on.

When the temperature at the bottom of the accreted layer reaches a few million kelvins, hydrogen fusion reactions ignite explosively and heat the entire layer to some 100 million kelvins. This material blows into space at speeds of hundreds of kilometers per second. The ejected material will have more

FIGURE 14.8

Supernova explosion in the giant elliptical galaxy Centaurus A: (*a*) the galaxy before the explosion and (*b*) the galaxy with the supernova at its peak brightness (arrowhead). (Courtesy CTIO, NOAO)

(*a*)

(*b*)

heavy elements from the thermonuclear processing. Observations of the expanding nebula around novas do show enhanced (compared to the sun) abundances of elements such as carbon, nitrogen, oxygen, and neon.

Novas occur when hydrogen accretes onto a white dwarf, igniting runaway thermonuclear reactions that blow off the outer layers; most novas are members of short-period binary systems.

14.4 Supernovas

Supernovas spew out energy in extraordinary amounts, about 10 billion times the sun's luminosity at their peak (more than 100,000 times that of a nova). A **supernova** usually signals the death of a massive star. More than 300 supernovas have been found in other galaxies (Fig. 14.8). One was seen in the Large Magellanic Cloud in February 1987. One occurs perhaps every 50 years in the Milky Way Galaxy. (Over the history of the Galaxy, about 15 billion years, hundreds of millions of supernova explosions have likely occurred.)

FIGURE 14.9

The Crab Nebula, a supernova remnant, in 1973. (See the winter star chart in Appendix I for the location of the Crab Nebula in Taurus.) This true-color photograph emphasizes the filamentary structure in the expanding material. (Image by W. Schoening and N. Sharp; courtesy KPNO/NOAO)

Chinese astronomers observed a supernova that flared in the sky on July 4, 1054. Close study of Chinese and Japanese accounts of this supernova confirms that the star remained visible to the unaided eye for more than 650 days in the night sky and for 23 days in daylight! The position noted by the Oriental astronomers placed the event in the constellation Taurus. In the twentieth century, the Crab Nebula became the first firmly identified supernova remnant in our Galaxy (Fig. 14.9).

Classifying Supernovas

Astronomers classify supernovas into two general categories (Table 14.1) by the shape of their light curves: **Type I**, which exhibit a sharp maximum (reaching about 10 billion solar luminosities) and die off gradually (Fig. 14.10*a*), and **Type II**, which have a broader peak at maximum (emitting about 1 billion solar luminosities) and die away more rapidly (Fig. 14.10*b*). Studies of other galaxies have revealed that Type II supernovas occur in association with Population I stars. In confusing contrast, Type I supernovas occur in association with *both* Population I and II stars.

The spectra of Type I and II supernovas can contain both emission and absorption lines. The basic difference is that Type II spectra show strong hydrogen lines and Type I do not. This indicates that Type I explosions involve stars that lack hydrogen—stars that are highly evolved.

The total energy output from any supernova in a few days is about as much energy as the sun will have

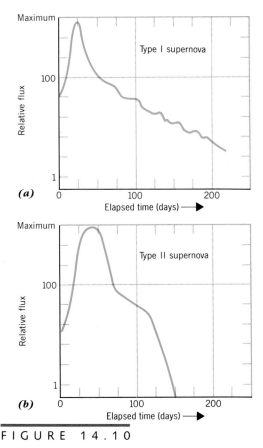

(a)

(b)

FIGURE 14.10

Typical light curves for (*a*) Type I and (*b*) Type II supernovas, showing flux over time. Both have a sharp rise to maximum, but their declines in flux differ in character. (Based on diagrams by W. Straka)

produced in its entire lifetime of 10 billion years. At its brightest, a supernova shines with a light of *10 billion* suns—about that of a typical galaxy! Both types of supernova violently eject a large fraction of the original star's mass at speeds of up to 10,000 km/s. At maximum brightness, a supernova attains a size about that of the solar system!

The Origin of Supernovas

What kinds of star become supernovas? Because Type I supernovas often occur in Population II stars, these stars may have a mass about that of the sun. In contrast, Type II supernovas are thought to include stars much more massive than the sun—stars that live their normal lives as OB-type stars. These stars probably explode once they have evolved off the main sequence and formed an iron core.

Table 14.1

Properties of Supernovas

Property	Type I	Type II
Ejected mass (solar masses)	0.5	5
Velocity of ejected mass (km/s)	10,000	5,000
Total kinetic energy (J)	5×10^{43}	10^{44}
Visual radiated energy (J)	4×10^{42}	10^{42}
Frequency	1 in 60 years	1 in 40 years

Type I supernovas are more puzzling, for it is hard to see how a 1-solar-mass star can detonate as violently as a supernova. A widely held idea resembles that for binary novas (Section 14.3). Imagine a binary system containing a white dwarf and a red giant star. Assume that the white dwarf has a mass close to the Chandrasekhar limit. As material flows onto the white dwarf, it builds up enough to push the star's mass over the Chandrasekhar limit. The star collapses because it cannot support the increased gravitational forces. The collapse heats the material and ignites the carbon of the white dwarf. The carbon fuses swiftly into nickel, cobalt, and iron—elements that are seen in the spectra of Type I supernovas.

Type I supernovas involve a low-mass stellar corpse in a binary system; Type II supernovas result from the explosion of a massive star.

Supernova 1987A

In February 1987, a supernova appeared in the Large Magellanic Cloud (LMC; see Section 16.6). The supernova (Fig. 14.11) was given the name SN 1987A ("A" for the first supernova discovered in 1987). The star that blew up appears to have been a blue supergiant, which was located in a region of active star formation in the LMC. The star, of spectral type B3, is estimated to have had a mass of 20 solar masses.

By astronomical standards, the LMC is close by (170,000 ly), so we now have the *first* opportunity to study a supernova in detail from modern observatories. That means we finally have confirmed key aspects of theoretical models about supernovas. First and foremost has been the detection of neutrinos from the explosion. In Kamioka, Japan, a joint project of the United States and Japan, called the Kamiokande II, detected a burst of about a dozen neutrinos one day before the supernova burst into

FIGURE 14.11

Supernova 1987A (arrow) in the Large Magellanic Cloud. To the upper left of the supernova is the Tarantula Nebula, a large, complex H II region. (Courtesy European Southern Observatory)

visibility. Neutrino events were also observed from detectors in a salt mine in Mentor, Ohio. The detection of neutrinos (the first for a source outside of the

Supernova 1987A	Type	variant of Type II
	Location	Large Magellanic Cloud
	Progenitor star	B3 I supergiant
At a Glance	Mass	20 solar masses
	Total energy	10^{46} J

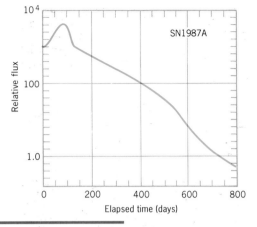

FIGURE 14.12

The light curve of SN 1987A from the first day of the visible outburst until May 16, 1988, some 800 days later. These observations were made at visual wavelengths. Compare to Figure 14.10. (Courtesy CTIO, NOAO)

FIGURE 14.13

Supernova remnant in Cygnus. Note how the bright, reddish wisps almost make a circle, as if blown out from a central point. (Courtesy J. Riffle, Astro Works Corporation; photographed with an Astromak™ telescope)

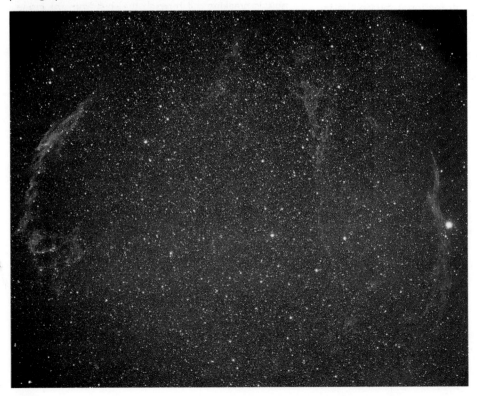

solar system) strongly implies a Type II supernova—the death of a massive star.

In the models for Type II supernovas, a star with a mass of 10 or more solar masses develops an iron core. When the core has grown to a mass greater than 1.4 solar masses, it suddenly collapses. In less than one second, the electrons in the core are forced to merge with protons to form neutrons. The infalling material bounces off this neutron core, sending out a shock wave that creates the optical supernova explosion, which takes a day or so to become visible. In contrast, the neutrinos that are produced (one for each combination of an electron and proton) zip right out of the core. These neutrinos may, in fact, carry away most of the supernova's energy.

Most observations support the idea that SN 1987A was a Type II explosion, but with a twist. Initial spectra showed strong, broad hydrogen lines, and Doppler shift measurements indicated an expansion rate of about 17,000 km/s—as expected for a Type II event. But the supernova did not get as bright as expected; it was about 100 times fainter than a Type II at maximum. In fact, it stayed pretty much constant before dropping off (Fig. 14.12), which is *not* typical behavior for a Type II.

SN 1987A provided a number of crucial tests for supernova models; neutrinos from the blast were detected—the first time we have seen the core collapse of a massive star.

Supernova Remnants

A supernova bangs out a material into the interstellar medium. Traveling at supersonic velocities, the shell of material creates a shock wave that plows through the interstellar gas and dust. The shock wave's collisions with the cool clouds of the interstellar medium excites the interstellar material so it glows. This luminous material marks a **supernova remnant**. The Loop Nebula in Cygnus (Fig. 14.13) is such a remnant. Note that its outline looks roughly spherical—a shell produced by the interaction between the interstellar medium and a supernova shock wave. These shocks stir up the interstellar medium, creating turbulence in the gas there. Radio astronomers can observe low-density excited gas that has no detectable optical emission. They were the first to detect Cassiopeia A, another supernova remnant (Fig. 14.14).

Supernova shock waves deposit enormous amounts of energy into the interstellar medium and so create supernova remnants.

Radio astronomers recognize supernova remnants by a special property of their radio spectrum—the radio flux plotted versus frequency displays a *nonthermal spectrum*. (Fig. 14.15). The special shape of the spectrum indicates that the radio emission comes from electrons moving at nearly the speed of light, which are accelerated in magnetic fields producing **synchrotron radiation**. The display of nonthermal spectra by supernova remnants means that the synchrotron process generates their radio emis-

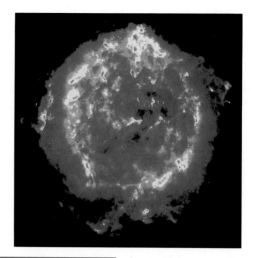

FIGURE 14.14

False-color radio image of Cassiopeia A, a supernova remnant. This Very Large Array (VLA) map, made at a wavelength of 6 cm, has red and yellow representing the strongest emission and blue the weakest. The supernova that created this remnant exploded in the year 1680. The outer layers of the star were ejected at a high velocity into the interstellar medium. The expanding shell has decelerated enough that the more slowly expanding material from deeper within the star now is breaking out through the shell from the inside. (Courtesy NRAO; VLA observations by P. E. Angerhofer, R. Braun, S. F. Gull, R. A. Perley, and R. J. Tuffs)

FIGURE 14.15

Radio spectra of two supernova remnants. Plotted here are the overall spectra at frequencies from 10 to 10,000 MHz of the Crab Nebula and Tycho's supernova. Note how the flux decreases at higher frequencies (shorter wavelengths)—a typical trend for the nonthermal spectra of supernova remnants.

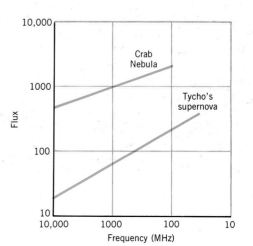

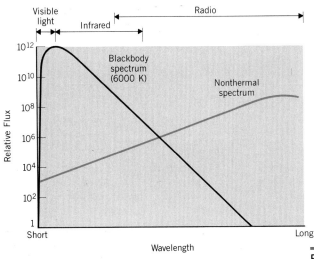

FIGURE   14.16

Schematic comparison of the spectra of a thermal (blackbody) source at 6000 K and a nonthermal (synchrotron) source. Note that the wavelengths run from short to long on the horizontal axis.

sion, and that means that the remnants contain magnetic fields and high-energy particles.

Recall that the spectrum of blackbody radiation has a characteristic shape—a Planck curve. The spectrum of blackbody emission is the archetype of *thermal emission*, which arises basically from the motions of the particles involved. The faster the motions, the higher the output of radiation (and the hotter the source). Nonthermal emission has a different spectrum. In general, a nonthermal spectrum increases in flux at longer wavelengths (Fig. 14.16).

Synchrotron emission is a common example of nonthermal emission. Moving charged particles spiral around the magnetic field lines rather than traveling across them. The particles continually change direction, accelerate, and so emit electromagnetic radiation. The frequency of emission is directly related to how fast the particle spirals: the faster the spiral, the higher the frequency. Increasing the magnetic field strength tightens the spiral and also increases the frequency. As the particles radiate, they lose energy and generate lower-energy (longer-wavelength) radiation. So a synchrotron source needs a continually replenished supply of electrons to keep emitting at relatively short wavelengths (high-energy photons).

Supernova remnants are visible from synchrotron emission, which has a distinctive shape to its spectrum.

FIGURE   14.17

The Crab Nebula viewed through polarizing filters. The arrows in the lower right-hand corners indicate the orientation of the polarization of the filters. The differences in brightness and structure in these images show that the light is polarized because different amounts are passed through the polarizing filters in their two different orientations. (Courtesy Palomar Observatory, California Institute of Technology)

(a)

(b)

The Crab Nebula

The material blown off in the explosion that produced the Crab Nebula is still expanding today. What has powered the nebula since the explosion expelled the outer layers of the star?

The source of the Crab Nebula's radio and optical emission is the synchrotron process. Synchrotron-emitting electrons, when viewed side-on in their spiral motion, appear to be moving back and forth along almost straight lines. Their synchrotron emission has its waves more or less aligned in the same plane. So synchrotron radiation can be polarized and, at visible wavelengths, it can be observed—as has been done for the Crab Nebula (Fig. 14.17).

What does it mean for electromagnetic radiation to be *polarized*? Light has wave properties and vibrates up and down. Waves are said to undergo **polarization** if the planes of their vibrating motion tend to be oriented in some direction—for instance, the plane of this page of paper. Light that is *unpolarized* has no preferred orientation: the planes of wave vibration occur in all directions in equal amounts.

As the electrons in the Crab Nebula emit synchrotron radiation, they lose energy rapidly. For the electrons producing the optical emission, half their energy would be drained off in about 100 years. Those that emit x-rays decline by half in only 7 years (Fig. 14.18). What replenishes them? The answer came in 1968 with the discovery of a pulsar in the Crab Nebula (Section 14.6).

FIGURE 14.18

X-ray image of the Crab Nebula, taken by the Einstein X-Ray Observatory. The bright circular patch near the center is the pulsar (arrow); the diffuse emission around it is synchrotron emission from the expanding material. (Courtesy Einstein Data Bank, Harvard–Smithsonian Center for Astrophysics)

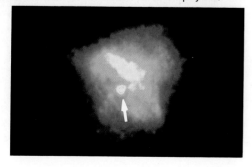

14.5 The Manufacture of Heavy Elements

A massive star dies in a supernova. The blast can leave behind a neutron star or a black hole. So a massive star digs its own grave. But it also blasts newly synthesized elements into interstellar space.

Massive stars (those of 8 solar masses or greater) burn heavier elements in the sequence of thermonuclear energy generation. These fusion reactions—carbon to oxygen, neon, sodium, and magnesium, and onward to silicon burning—can fire only in the cores of massive stars. They have very sort durations (compared to the main sequence hydrogen burning)—a thousand years or less. The end to the fusion chain comes with iron. To split iron into lighter elements takes an input of energy. To fuse it to heavier ones also requires additional energy. So nuclear reactions naturally stop at iron; no other rearrangement of nuclei can generate any more energy.

A supernova explosion acts as nature's special workshop for forging some elements heavier than iron. The most important process involves high-energy neutrons bombarding various nuclei. Because they have no charge, the neutrons are not repelled by the protons in a nucleus. This nuclear capture of neutrons leads to a buildup of heavier nuclei.

Two processes take part in this buildup. A neutron under normal conditions of low density will disintegrate (in about 1000 s) into a proton, an electron, and an antineutrino. This is called **beta decay**. The rate of beta decay naturally divides neutron nucleosynthesis into two processes. In one process, neutrons are captured at a rate *faster* than the beta-decay rate, so that neutron-rich nuclei are formed. This process is called the **rapid process** (*r-process* for short). In the other, nuclei capture neutrons at a rate *slower* than the beta-decay rate, so proton-rich nuclei are made. This process is called the **slow process** (or *s-process*). A combination of both these processes leads to the manufacture of most of the elements and isotopes heavier than iron. In a supernova explosion, the times are short, and it is mainly the r-process that is effective.

Here's one model of how a Type II supernova happens (Fig. 14.19). Imagine a very massive star with a core of iron. Its interior temperature decreases from the core outward. Because of this de-

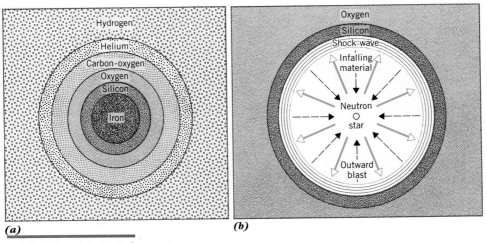

(a)　　　　　　　　　　　　　　　　　　　　*(b)*

FIGURE 14.19

One idealized model of the interior of a massive star undergoing a supernova explosion. We are looking at the core of a Population I star at the end of its life. Fusion reactions in shells give layers of different elements (*a*). At high enough temperatures, the iron core disintegrates into helium. This process soaks up heat from the core, and it collapses rapidly (*b*), sparking explosive ignition of fusion reactions. This explosion produces a burst of neutrinos (not shown). (Adapted from a diagram by J. C. Wheeler)

cline, the star's interior is layered like an onion. Around the iron core is a silicon layer; here temperatures do not get high enough to fuse silicon to iron. Around that layer is one of oxygen; here temperatures are too low to fuse oxygen to silicon. These shells are still burning—in the silicon core, oxygen is being formed; in the oxygen shell, carbon; and so on.

Once the core has ended up as iron, its fusion reactions stop and no more energy is generated. Relentlessly, gravity squeezes the core to higher temperatures and densities. When the core's temperature gets to about 5 billion kelvins, the photons there have so much energy that they break the iron nuclei down into helium nuclei. The disintegration of iron into helium uses large amounts of energy. Photons that normally would provide the radiation pressure to support the star are splitting iron nuclei instead. The pressure in the core no longer supports the star, and it collapses suddenly. The gravitational collapse rapidly pumps heat into the material. This collapse is probably *fast*—speeds in the outer part of the core hit 70,000 km/s. In one second, a core the size of the earth drops to a radius of only 50 km and its central density rises to several times that of an atom's nucleus!

Crucial events occur in this collapse. First, protons and neutrons released by the disintegration of nuclei in the core pelt and penetrate remaining nu-

clei. These can capture neutrons and be transformed to heavier elements. Second, the layers above the core plummet inward toward the core and heat up. Suddenly, ignition temperatures of many fusion reactions are reached, leading to explosive ignitions in which enormous numbers of neutrinos are produced. Third, the inner core's collapse creates a neutron star; its degenerate neutron pressure halts the collapse, and the material rebounds outward. As the infalling matter from the outer core crashes into the rebounding inner core, a shock wave forms. The shock bullies its way outward from the core in just tens of milliseconds—about the time it takes to blink your eyes—carrying with it into interstellar space material enriched with heavy elements. Meanwhile, about 99 percent of the energy released in this explosion comes out in the form of neutrinos.

Supernovas synthesize many elements heavier than lead, such as uranium and thorium. If you wear gold or silver, those materials were synthesized in the fierce death of a massive star.

A Type II supernova creates heavy elements by rapid neutron capture in the early stages of the core collapse.

14.6 Pulsars

How would a neutron star be visible? In a way not anticipated by astronomers: as a *pulsar*, accidentally discovered in the summer of 1967. A **pulsar** emits radio bursts at very regular, short intervals.

For a given pulsar, the period between pulses repeats with very high accuracy, better than 1 part in 10^8. The amount of energy in a pulse, however, varies considerably; sometimes complete pulses are missing from the sequence. Although the flux and shape vary from pulse to pulse, the average of many pulses from a pulsar defines a unique shape at a given frequency of observation (Fig. 14.20). The average pulse typically lasts for a few tens of milliseconds. (A *millisecond* is 10^{-3} second; the safety airbag in a car expands in a few milliseconds.) For the well-studied pulsars, periods range from 0.002 to 4.0 seconds, with an average value of about 0.6 seconds. When accurate radio observations have been made, these pulse periods have steadily increased.

FIGURE 14.20

Average pulse profiles from the Crab Nebula pulsar (called PSR 0532+21) observed at a variety of frequencies, from radio (300 and 600 MHz) to optical and x-rays. Each pulse from a pulsar has a special shape at a given frequency of observation. Averaged over many pulses at that frequency, every pulsar has pulses of a characteristic shape. Note how these differ at different frequencies. (Adapted from a diagram by F. G. Smith)

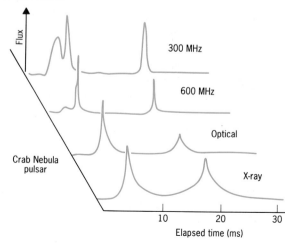

Physical Properties

Pulsar models must account for the precise clock mechanism of pulsars, that is, the extremely regular repetition of pulses. Basically, one clock mechanism works the best: *rotation*.

Consider one rotation period to be equal to the period between pulses. Then what kinds of object are rotating? Here's the basic idea: if a spherical mass rotates too rapidly, its gravity will not be able to hold it together, and mass will fly off tangent to its equator. Neutron stars have average densities of about 10^{17} kg/m^3. So they can rotate up to once every millisecond without losing mass—fast enough for even the fastest pulsar. Of course, they can rotate more slowly, too.

A rotating neutron star provides the clock mechanism for pulsars.

The Crab Nebula pulsar (Fig. 14.21) is called PSR 0532 + 21. (*PSR* stands for pulsar, and the numbers give the pulsar's position in the sky.) PSR 0532 + 21 has a fast period: 0.033 second, or 30 pulses per second! The power emitted in just the pulses is about 10^{28} W, which is about 100 times more than the total luminosity of the sun. At all wavelengths, the Crab Nebula emits about 10^{31} W, or about 10^5 times the sun's luminosity.

If the pulsar is a rotating neutron star, its slow-down in rotation period gives a change in rotational energy of about 5×10^{31} W. That's enough to power the nebula—if the rotational kinetic energy of the neutron star can somehow be converted to the kinetic and radiative energy of the nebula. In other words, the light we see now from the nebula ultimately derives from the pulsar—the rotational kinetic energy of the neutron star.

The Lighthouse Model

Now to tie these observations together in the accepted basic model for pulsars—a rotating, magnetic neutron star—otherwise known as the **lighthouse model**. The model has two key components: the neutron star, whose great density and fast rotation ensure a large amount of rotational energy, and

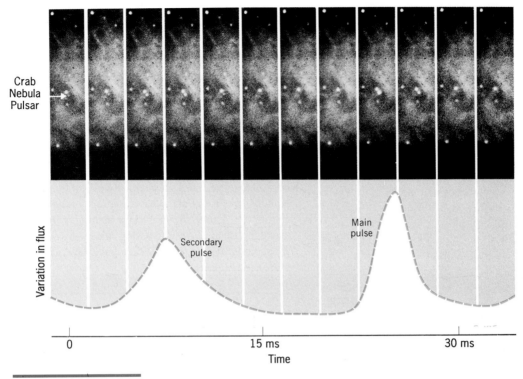

Crab Nebula Pulsar

Variation in flux

Secondary pulse

Main pulse

0   15 ms   30 ms

Time

FIGURE 14.21

The Crab Nebula pulsar at visual wavelengths: a complete sequence in the pulse cycle. The series of photographs, with the pulsar marked, shows the visual light variations over a time of just some 30 ms. The graph indicates in an average way how the flux changes with time. Note the large main pulse and a smaller secondary pulse. (Courtesy KPNO, NOAO)

a dipole magnetic field that transforms the rotational energy to electromagnetic energy. Needed is a very strong magnetic field—10^8 T, or about 100 million times the strength of the earth's field.

The region close to the neutron star, where the magnetic field directly and strongly affects the motions of charged particles, is called the pulsar's *magnetosphere* (in analogy to planetary magnetospheres). Here all the energy conversion action takes place. One model pictures the magnetic axis as tilted with respect to the rotational axis (Fig. 14.22). (The magnetic field of Uranus is tilted in such a fashion.) As the pulsar spins, its strong magnetic field spins with it, and this spin induces an enormous electric field at its surface. This electric field pulls charged particles—mostly electrons—off the solid crust of iron nuclei and electrons. The electrons flow into the magnetosphere, where they are accelerated by the rotating magnetic field lines. The accelerated electrons emit synchrotron radiation in a tight beam more or less along the field lines.

If the magnetic axis falls within our line of sight,

we see a burst of synchrotron emission each time a pole swings around to our view (like the spinning light of a lighthouse). The time between pulses is the rotation period. The duration of the pulses depends on the size of the radiating region. As the pulsar generates electromagnetic radiation, the torque from accelerating particles in its magnetic field slows down its rotation. This slowdown is observed.

Note that to see a pulsar in the lighthouse model, the neutron star's magnetic axis must be oriented just right for the pulses to beam at the earth and give a blinking effect. Many pulsars, which we cannot see, probably exist in the Milky Way with unfavorable orientations.

A rapidly rotating neutron star with a strong magnetic field can generate electromagnetic radiation in its magnetosphere and be visible as a pulsar.

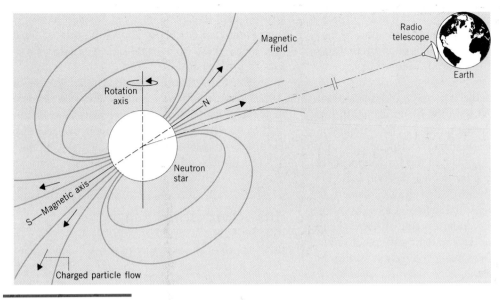

The lighthouse model of a pulsar. The magnetic axis of the neutron star is tilted with respect to the spin axis. Electrons from the neutron star's surface flow out along the magnetic field lines and so escape mostly at the north and south magnetic poles. These electrons emit synchrotron radiation, which we see as pulses when a pole spins across our line of sight. The rotation of the neutron star provides the clock mechanism for the pulsar.

Very Fast Pulsars

PSR 1937+214 is an extremely fast pulsar with a period of 1.558 *milliseconds*. Applying the lighthouse model to this pulsar requires that it spin 642 times per second (20 times faster than the Crab pulsar), which means that its surface rotates at roughly one-tenth the speed of light! In addition, the neutron star must lie very close to its breakup speed. Such pulsars with periods less than 25 ms are called **millisecond pulsars**. (Some 20 pulsars so far are known to be millisecond pulsars.)

One of the curious features of millisecond pulsars is that their rotation rates are very stable. PSR 1937 + 214, for instance, slows down at only 3.2×10^{-12} ms in a year. One explanation is that the millisecond pulsars have very weak magnetic fields, perhaps a thousand times weaker than typical. Then how did such a pulsar become observable? The scenario proposed is that of pulsars resurrected in binary systems. Millisecond pulsars may have once been formed in a supernova and aged gracefully. Then billions of years later, perhaps their low-mass companions evolved finally to red giants and matter flowed from them into a disk around the dead pulsar. This material might have made a rapidly spinning accretion disk around the neutron star. If the magnetic field of the pulsar were entwined with plasma in the disk, this linkage would spin up the pulsar so that it lives again.

Pulsars with Planets?

Millisecond pulsars are thought to be old (a billion years or so) and companion to some kind of star in a binary system. Radio astronomers time their pulses with great precision and can detect even small changes. One millisecond pulsar, called PSR 1257 + 12, showed two periods in the small variations of its timing compiled over more than a year: one around 67 days and the other around 98 days. (These changes were so small that they amounted to a 0.7 *meter* per second wobble in the pulsar.) What could cause such anomalies? At least two (and possibly three) planet-sized masses orbiting a 1.4-solar-mass neutron star! One may have a mass of about 3.9 earth masses, swinging around the neutron star in 67 days at a distance of 0.36 AU. The other may have a mass of 3.2 earth masses, orbiting in 98 days at a distance of 0.47 AU. The third and much less massive (and less likely!) body would orbit at 1 AU with a period of about one year. Though these objects have masses like the Jovian planets, it is doubtful that they have a physical makeup like any of the solar system's planets.

How might such objects form? A millisecond pulsar is likely to be surrounded by an accretion disk made of matter from the pulsar's stellar companion. Somehow planet-sized masses form in this disk, maybe in a manner similar to the formation of the protoplanets in the solar nebula (Section 8.5). For now, this is a speculative idea. But, if planets do form under conditions that involve a mass as different from the sun as a neutron star, it may well be plausible that planetary systems are much more common than is now believed. They would be strange places, indeed!

14.7 Black Holes

Many stars in the Milky Way will have masses much greater than the neutron star limit of about 3 solar masses when they die. For such masses, the crush of gravity overwhelms all outward forces, including the repulsive forces between particles with the same charge. No material can withstand this final crushing point of matter. The collapse cannot be halted; the volume of the star will continue to decrease until it reaches zero. The density of the star will increase until it becomes infinite. A black hole demonstrates gravity's ultimate victory over matter.

Bizarre events occur near such a collapsing mass. As its density increases, the paths of light rays emitted from the star are bent more and more from straight lines going away from the star's surface. Eventually the density reaches such a high value that the light rays are wrapped around the star and do not leave. The photons are trapped by the intense gravitational field. The escape speed from the star is then greater than the speed of light. Any additional photons emitted after the star attains this critical density can never reach an outside observer. The star becomes a **black hole**.

In Einstein's theory of general relativity (Section 10.7), a black hole represents a severe and permanent warping of a small region of spacetime. So great is the curvature that light follows a path that returns on itself—the most acute form of the bending of light as predicted by relativity. And the theory predicts the occurrence of other bizarre events.

The Schwarzschild Radius

Let's consider the meaning of *black* in *black hole* in terms of escape speed (Section 3.4). Imagine squeezing the earth to make it smaller and denser. Its escape speed would increase. Imagine the earth compressed until its escape speed equaled the speed of light. Then nothing, *not even light*, emitted at its surface could escape into space. Nothing gets away, so to an outside observer, the earth would appear black.

How small must an object become to be dense enough to trap light? This critical size is called the **Schwarzschild radius**, after the German astrophysicist Karl Schwarzschild, who first calculated it. For the sun, the Schwarzschild radius is about 3 km—much smaller than the typical sunspot! The mass of any object gives its Schwarzschild radius directly. For 1 solar mass, it's 3 km; for 2 solar masses, 6 km; for 10 solar masses, 30 km; and so on. (The earth's Schwarzschild radius is a mere one centimeter—the size of a marble!)

A black hole is formed when any mass is collapsed to a size smaller than its Schwarzschild radius.

	Mass	greater than 3 solar masses
Black Holes	Escape speed	greater than light
	Size	depends on mass; 1 solar mass has 3-km radius
At a Glance	Density	singularity with infinite density?
	Location	most likely observed in binary systems

Note: An object of *any* size can be made into a black hole if a force compresses it enough. But an object with a mass greater than the neutron star limit of 3 solar masses *must* become a black hole after thermonuclear reactions have ceased, for there is no known source of pressure that can support it.

Einstein's theory of general relativity predicts that the matter inside a black hole will keep collapsing gravitationally until it has *no volume*. But it still has mass, so its density is infinite. This theoretical end to runaway gravitational collapse is called a **singularity**. The matter has literally squeezed itself so small that it occupies no space. Yet it's still there. General relativity points to the formation of a singularity, cloaked in the center of a black hole, as the natural end of gravitational collapse.

Journey into a Black Hole

A person falling into a black hole meets a fate an outside observer cannot even find out about—unless the outsider drops in, too. Let's take an imaginary journey into a black hole to follow the adventures of a crazy astronaut who takes the plunge, comparing this trip with what an outside observer sees of it. You and a friend start out in a spaceship orbiting a few AU from a 10-solar-mass black hole. Nothing peculiar here. The ship orbits the black hole as it would any ordinary mass. In fact, Kepler's third law and the spaceship's orbit permit you to measure the hole's mass.

Your friend volunteers to investigate. She takes with her a laser and an electronic watch. You and she synchronize watches. Once a second, according to her watch, she will flash her laser flash back at you. Down she goes! For a long time as she falls toward the black hole, nothing strange happens. But as she gets closer, stronger and stronger tidal gravitational forces stretch her out (if she falls feet first) from head to toes. (You feel such tidal forces on the earth, but the forces are too weak to bother you.) Near a black hole, however, tidal forces grow enormously. An ordinary human being would be ripped apart about 3000 km from a 10-solar-mass black hole.

Let's suppose your friend is indestructible, so she can continue her trip. Peering down, she can just make out a small black region in the sky. (Recall that a 10-solar-mass black hole has a "radius" of only 30 km.) Then she crosses the Schwarzschild radius! But

nothing new happens to her. No solid substance, no signs mark the edge of the black hole. The trip now swiftly ends for your unfortunate friend. Quickly—in about 10^{-5} second after she crosses the Schwarzschild radius—she crashes into a singularity (if it exists!). Crushed to zero volume, she is destroyed.

What of your view, back in the spaceship, of your friend's adventure? You would *never* see her final destruction; in fact, you'd not even see her fall into the black hole. As she dropped closer to the black hole, you'd notice that the light from her laser was redshifted, with the shift increasing as your friend fell closer to the black hole. General relativity predicts such a redshift. (The light must work against gravity to get to you, so it loses energy and increases in wavelength.) Also, the time between laser flashes would increase. Why? Compared with your watch, your friend's watch appeared to slow down as she entered regions of stronger gravity.

As she came closer to the Schwarzschild radius, the watches would get more and more out of sync. The times between your reception of her flashes would stretch out. In fact, a laser burst sent out just as she crossed the Schwarzschild radius would take an *infinite* time to reach you. It also would suffer an *infinite* redshift. To you, your friend's fall would seem to grow slower and slower as she got closer to the black hole, but she would never appear to fall into it. Your measurement shows time slowing down so much near a black hole that it seems to be frozen—another prediction of general relativity. In addition, the light is more and more redshifted until you can no longer detect it.

A black hole practices cosmic censorship; it prevents you from seeing an object fall into it, and you cannot know anything at all about what happens to an object inside of it.

 ## 14.8 Observing Black Holes

How to actually observe a black hole? Light emitted inside cannot get out. Light sent out close by is strongly red shifted, so it's hard to detect. In addition, a black hole is small, only a few kilometers in size. So you'll have a hard time seeing an isolated black hole.

But a black hole near any mass might be observable. Matter falling toward a black hole gains kinetic energy and heats up. It's also likely to form an accretion disk. Heated enough, the atoms in the accretion disk will be ionized. If heated to a few million kelvins or so, the material will give off x-rays. Before it is trapped in the gravitational gulf, infalling material can send x-rays into space. So a black hole close to a star can sweep up material, which can radiate before it crosses the Schwarzschild radius. Hence, x-ray sources are generally good candidates for black holes, whose accretion disks can emit x-rays. Prime candidates for black holes are binary—the x-ray source and a normal star (a potential source of infalling material) orbit a common center of mass.

Binary X-Ray Sources

Why are **binary x-ray sources** most suspect? Imagine a black hole orbiting a supergiant star. If they are very close together, their orbital period is a few days or so. The star has a huge, distended atmosphere—and material from this atmosphere can fall to the black hole (Fig. 14.23). The accretion disk around the black hole gives off x-rays. If the black hole and the star have their orbital plane in our line of sight, when the black hole goes behind the star, its x-rays will be cut off. In this case we would see an eclipsing x-ray binary system.

Such binary x-ray sources are believed to have a low-mass, main-sequence star or a high-mass, post-main-sequence star swinging around an x-ray source. Some systems exhibit x-ray eclipses; the x-ray source passes behind the normal star as we view the system. Using spectroscopic analysis of the light from the visible star, we can observe the changes in Doppler shift, so we can find out the orbital periods—they are typically a few days. These short periods indicate that the orbits are only a few times larger than those of the primary stars (Fig. 14.24). If we can determine the separation of the two objects, we can ascertain—from Kepler's third law—the sum of the masses (normal star plus x-ray source). If we can get an idea of the mass of the normal star from its lumi-

---

FIGURE 14.23

One model for a black hole as an x-ray source. In a binary system, the black hole is coupled with a supergiant star. Material from the large star flows to the black hole, where it first falls into an accretion disk. As it falls, the gas heats up to about a million kelvins and emits x-rays from the accretion disk before it ultimately enters the black hole. The arrows indicate the flow of material from the supergiant star, some of which joins the accretion disk and some of which flows out of the system. "CM" marks the center of mass of the system. The supergiant star can expand only to the outer limit shown because of its gravitational interaction with the black hole. (Adapted from a diagram by H. Gursky)

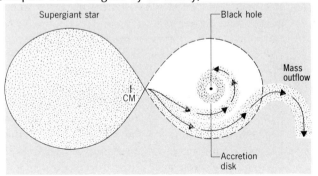

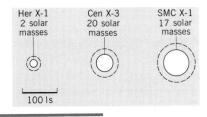

Her X-1
2 solar
masses

Cen X-3
20 solar
masses

SMC X-1
17 solar
masses

100 ls

FIGURE 14.24

Relative sizes of primary stars and the orbits (dashed circles) of their degenerate companions of some binary x-ray sources. The masses given are those of the primary star. The scale at the left indicates a span of 100 light-seconds, or about 0.2 AU.

nosity, then we can also determine the mass of the x-ray source. And if that mass turns out to be large enough (greater than 3 solar masses, the probable upper limit for a neutron star), the x-ray source must be a black hole.

Is Cygnus X-1 a Black Hole?

To prove the reality of black holes we need to observe one. A strong candidate is Cygnus X-1 in the constellation Cygnus. Cygnus X-1 emits about 4×10^{30} W in x-rays (about 10,000 times the sun's luminosity). Astronomers have identified Cygnus X-1 with an O-supergiant star, which has a surface temperature of 31,000 K. Optical observations show that the dark lines in the spectrum of the blue supergiant go through periodic Doppler shifts in 5.6 days. So the supergiant orbits with the x-ray source about a common center of mass every 5.6 days. The supergiant has a massive but optically invisible companion—Cygnus X-1.

For binaries can we find the masses of stars directly if we know the separation of the stars, their distances from the center of the mass, and the orbital tilt. In these regards, the mass of Cygnus X-1 is hard to determine. We can observe the Doppler shift in the spectrum of the visible companion, but we cannot obtain the speed of the x-ray source. And because Cygnus X-1 does not eclipse, we can't pin

down its orbital inclination. So we don't have enough information to state both individual masses with accuracy.

But we can make some reasonable estimates. Blue supergiant stars are typically 15 to 40 solar masses. Take 20 as typical. The orbital period and speed of the supergiant give us a relation between the masses of the supergiant and the x-ray source, uncertain by the amount of orbital tilt. Suppose the orbital tilt were 90°; then the mass of the companion would be 4 to 5 solar masses. But the tilt can't be that high, because the star is not eclipsing. So the true speed is higher than observed, and the mass of the companion must be higher. X-ray observations imply a tilt between 36 and 67°. These values suggest that Cygnus X-1 has a mass possibly as great as 11 solar masses or as small as 6. The most likely value is about 9 solar masses. If so, and if the limit for a neutron star is 3 solar masses, Cygnus X-1 is a black hole.

If the x-ray source in an x-ray binary star has a mass above the neutron star limit, it is a black hole.

Another strong candidate has the prosaic name of A0620-00. It is located in Monoceros, near Orion. This binary system contains an ordinary K star orbiting the center of mass every 7.75 hours at an orbital speed of 430 km/s (about 14 times the orbital speed of the earth around the sun). The distance of the K star from the center of mass has been measured: it is 0.014 AU. (The system is about 3200 ly from earth.)

What is the companion? It cannot be observed, but emission lines from the hot gas in an accretion disk surrounding it have been seen. The Doppler shift in these lines indicates that the companion has at least 3.8 solar masses—barely enough to be a black hole. Again, we have the problem of not knowing the orbital tilt of the system; the 3.8-solar-mass result comes from assuming that we observe the system edge-on. A plausible estimate for the companion is 8 solar masses. So we have a pretty firm case here for a black hole because even the minimum mass is above that for a neutron star. (If the x-ray source has a mass below the neutron star limit, then the x-ray emission could arise from an accretion disk around a neutron star.)

The Larger View

The Milky Way is populated with a incredible cast of stars, from the sun to supergiants. Yet all have finite lifetimes. Stars live, then die—the grim result of the flow of energy to space. How a star dies depends on its mass at its time of death. Our models imply that the corpse can take the form of a white dwarf, a neutron star, or a black hole—all objects that are difficult to observe.

Violence marks the death of stars, furious explosions that we can see. For stars like the sun, the death rattle involves only a small fraction of the star's mass blown into space. For massive stars, almost the entire star participates in the spectacular cataclysm of a supernova. Heavy elements are made and ejected into the interstellar medium from which new stars and their planets are born. That's the payback for the original birth.

When we look beyond the Milky Way Galaxy, we see colossal formations of stars in the sweep of space—other galaxies. Here we expect the stars in these galaxies to follow the same sequence as we have found for those in our Galaxy: births from interstellar clouds, lifetimes of fusion energy generation, then violent deaths leaving long-entombed corpses. To understand our sun and local stars gives us an awareness of distant stars and, to a large measure, much of the visible cosmos.

Key Ideas and Terms

black hole
brown dwarf
Chandrasekhar limit
neutron star
white dwarf

1. The mass of a star at the time of its *death* determines the corpse it leaves behind: brown dwarf, up to roughly 0.1 solar mass; white dwarf, up to 1.4 solar masses (the Chandrasekhar limit) or so; neutron star, 1.4 to 3 solar masses; black hole, greater than about 3 solar masses.

beta decay
degenerate gas
inverse beta decay

2. White dwarfs and neutron stars are supported against gravity by the degenerate gas pressure from electrons in a white dwarf and neutrons in a neutron star. No thermonuclear reactions are taking place in these degenerate objects.

accretion disk
binary system
conservation of
   angular
   momentum
nova
white dwarf

3. Novas occur in close binary systems when hydrogen from a companion star falls onto the surface of a white dwarf. This infalling fuel ignites explosively to produce a nova outburst. Before this material falls onto the white dwarf, it collects in an accretion disk around it.

binary system
Type I supernova
Type II supernova

4. Supernovas (Type II) are the explosions of massive stars (greater than 5–10 solar masses) that have evolved iron cores. Type I supernovas are the explosions of solar-mass stars in a binary system, perhaps involving white dwarfs close to their mass limit.

nucleosynthesis
rapid process
slow process
supernova

5. Supernova explosions make most of the elements heavier than iron and blast them into the interstellar medium (along with elements made in their lives before they become supernovas).

polarization
pulsar
supernova
supernova remnant
synchrotron
  radiation

6. The material blown off by a supernova creates a supernova remnant in the interstellar medium. The Crab Nebula is a supernova remnant, emitting light by the synchrotron process. Its emission is powered by a pulsar in the center; that pulsar was formed in the supernova explosion in the year 1054.

lighthouse model
millisecond pulsar
pulsar

7. Pulsars are rapidly rotating neutron stars; this idea is inferred most strongly from the regular timing of the fastest pulsars. Only neutron stars are small and dense enough to rotate so rapidly. Pulsar emission comes from high-speed electrons in a neutron star's intense magnetic field.

black hole
Schwarzschild
  radius
singularity
spacetime

8. A black hole forms when a mass becomes so compacted that its escape velocity is greater than the velocity of light. Black holes are very small: a 1-solar-mass black hole has a radius of 3 km; 2 solar masses, 6 km; 10 solar masses, 30 km; and so on.

binary system
black hole
Schwarzschild
  radius
singularity

9. Einstein's theory of general relativity predicts that time appears frozen near a black hole and that a singularity resides in its center. Black holes can be perceived only by their interaction with visible matter; an especially good circumstance would be a black hole–ordinary star binary system.

accretion disk
binary x-ray sources
black hole
neutron star

10. Cygnus X-1, in a binary system, is a good candidate so far for a black hole. It emits x-rays from a hot accretion disk around the suspected black hole. Most other binary x-ray sources contain neutron stars.

Review Questions

1. What kind of corpse will result from the death of a star with a mass of 2 solar masses?
2. What keeps a white dwarf from collapsing under its own gravity?
3. What is the key process that powers a nova explosion?
4. What kind of star is involved in a Type II supernova explosion?
5. What elements are made in a supernova explosion?
6. What powers the Crab Nebula?

7. What property of neutron stars makes them the prime candidates for pulsars?

8. In the lighthouse model for pulsars, how does a pulsar pulse?

9. What is the radius of a 10-solar-mass black hole?

10. What is the escape velocity from a black hole?

11. Under what conditions can a black hole be observed?

12. How does a black hole emit x-rays?

Conceptual Exercises

1. In a short paragraph, describe the primary physical characteristics of a white dwarf.

2. In a short paragraph, describe to a friend who has not studied astronomy the chief physical features of a neutron star.

3. What observational evidence do we have for the actual existence of neutron stars and white dwarfs?

4. Look around you. Of the items you see, what would not be there if supernovas didn't occur?

5. If a white dwarf has no fusion reactions, how does it produce light?

6. Make a list of the observational evidence that supports the idea of the Crab Nebula as a supernova remnant.

7. In what way does a black hole practice censorship? (*Hint:* How could information get out of a black hole?)

8. What one observational feature of pulsars links them most strongly to rapidly rotating neutron stars?

9. What is the source of the electromagnetic radiation in the pulses of pulsars?

10. How do we know that black holes exist? Evaluate the observational evidence!

11. How does the Schwarzschild radius differ from the singularity?

The View from Alpha Centauri, Nearest Star to the Sun

Adapted by Michael Zeilik from original material by Sheridan A. Simon

June 17, 2196

This is the first chance I've had to record my excitement into my computer's diary. I hope that this record will link up with others to give a clear picture of our early days here—New California Colony on Planet Armstrong, in this astounding triple-star system.

It's really strange seeing the sky from here. Alpha Centauri contains two very bright stars and one pretty dim one. We've nicknamed Alpha Centauri A as Able; Alpha Centauri B as Baker; and Proxima Centauri as Prox. Not official, and I'd really prefer names that were more romantic. Able is a G2 V star, a lot like the sun, but a bit more massive (1.1 solar masses), somewhat bigger (1.2 solar radii), and hotter and more luminous (5800 K, 1.5 solar luminosities). Our planet, Armstrong, orbits Baker, a K1 V star less massive than the sun and smaller, dimmer, and cooler (0.88 solar mass, 0.87 solar radius, 0.44 solar luminosity, and 5000 K). Prox is the puny star of this trio: a small, very dim, and cool M5 V star (0.1 solar mass, 0.03 solar radius, 0.00006 solar luminosity, and 3000 K). Actually, Prox is pretty much like most the stars in the Galaxy.

Able and Baker form a fairly close binary, with Prox orbiting both of them at a much greater distance. Prox's orbit has an average distance of 10,000 AU and a period of a million years. From Armstrong, Prox appears as a very dim, reddish star—in fact, it emits much of its light in the infrared. But it's an active star!

Frequent flares burst in its active regions, and they are hot and bright. The most energetic ones can increase the luminosity of Prox by a few times for a few hours—stellar magnetic fireworks in action!

Armstrong lies only 0.66 AU from Baker. So the surface temperature here is about the same as on the earth, though Baker is less luminous than the sun. The year is shorter: only 0.57 a terrestrial year. From Armstrong, Baker appears 1.3 times the size of the sun viewed from the earth. Because Baker is cooler than the sun, it emits more light in the red end of the spectrum and less in the blue. The colors here first looked a bit strange, but we have gotten used to that by now.

After Baker, the brightest object in our sky is Able. Able and Baker have an orbit with an average separation of 23 AU and a period of 80 years. Because the orbit is very eccentric—0.52—the stars swing as close as 11 AU and as far apart as 35 AU. Even when the stars are farthest apart, Able looks impressive. Though the disk is only one-thirtieth the size of the sun from the earth, it dazzles the eye: 500 times brighter than the earth's full moon. Able is easy to see in the day, and it keeps the night pretty bright. Real night falls only when both stars lie below the horizon.

Able is a stunning sight when the stars are closest together. Then Able glares 5000 times as bright as the earth's full moon. When alone in the sky, Able makes the "night" about as bright as a very overcast day on earth. It heats up Armstrong only a little: about 1.5 °C higher in temperature than when it was far away.

A sky with two suns is never boring. Each day both stars rise and set, though at different times. The relative position of the two stars in our sky changes over the years. For a short time every year, Able is hidden behind Baker. Now that's an eclipsing binary—to be able to see the disk of one star hide another!

T

3

Concepts of Cosmology

■

Unifying Concepts

Our sun resides in the Milky Way Galaxy; the universe contains billions of galaxies. Almost all galaxies are arranged in strings, millions of light years in size, with relative voids in between them. Within these galaxies are gas, dust, and stars. The universe originated 15 billion years ago from a cosmic explosion. The universe today is still expanding and cooling, while some material in it has formed stars, planets, and galaxies. But a large fraction of the matter in the universe is dark and invisible. Some common elements are organized in special ways to produce life on the earth and perhaps elsewhere in the cosmos.

15

The Milky Way: Home Galaxy

■

Central Concept

The structure of the Milky Way Galaxy has formed and continues to evolve from physical processes within its various parts.

■

Like a majestic cosmic pinwheel, our Milky Way Galaxy spins slowly in space. Our sun orbits it, along with more than *100 billion* other stars. The Galaxy's structure evolves at a rate so slow that we won't see any changes in our lifetimes.

In the distant past, the Galaxy looked very different, especially just after its formation 15 billion years ago from a giant cloud of gas and dust. Over time, stars were born and died, and interstellar matter gathered in a disk of gas and dust. Waves plowing through this disk periodically created a spiral shape. From our position in the Galaxy, we can't directly see its shape or read its history. But by understanding the evolution of stars and the interstellar medium, we can build a plausible model of our home Galaxy.

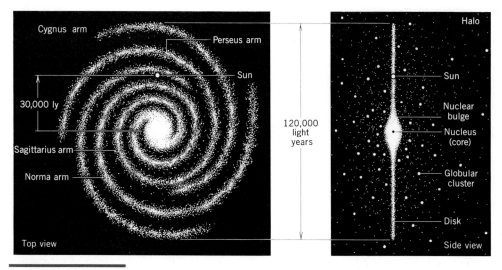

FIGURE 15.1

Schematic model of the Milky Way Galaxy, showing its main features; nucleus, halo, and disk. A top view is at the left, and a side view on the right. The sun lies in the disk, about 30,000 ly from the center. This spiral arm pattern is a reasonable guess—there may be four arms as shown; there may be two. The actual structure is probably fairly ragged. The spherical region of the globular clusters defines the halo. The nuclear bulge in the center surrounds the core, in which lies the nucleus.

15.1 Overview of the Galaxy's Structure

The Galaxy has three main parts: a central region called the nuclear bulge, a disk, and a halo (Fig. 15.1). The **disk**, the main body of the Galaxy, has a diameter of some 120,000 ly and a thickness of about 1000 ly. Population I stars and interstellar clouds of gas and dust inhabit the disk, the gas extending out farther in radius than the stars. The sun resides in the disk at a distance of approximately 30,000 ly from the Galaxy's center.

The **nuclear bulge** encases the central regions of the Galaxy, including the mysterious **nucleus**, the very heart of the Galaxy. The bulge is about 12,000 ly in diameter and 10,000 ly thick; it contains old yellowish Population I stars. The spherical **halo** encircles the nuclear bulge and the disk. Globular clusters, containing Population II stars, make up the most obvious material in the halo, which has a diameter of at least 120,000 ly.

Stars of different types inhabit each part of the Galaxy: young, metal-rich Population I stars in the disk; old, metal-rich Population I stars in the bulge, and very old, metal-poor Population II stars in the halo.

The disk contains a spiral pattern, made of spiral arms. A **spiral arm** contains many bluish OB stars.

The Galaxy At a Glance		
Diameter of disk	120,000 ly	
Diameter of the halo	300,000 ly	
Sun's distance from the center	30,000 ly	
Sun's orbital speed	220 km/s	
Mass interior to the sun	few × 10^{11} solar masses	
Total mass	few × 10^{12} solar masses	
Age	about 15 × 10^9 years	

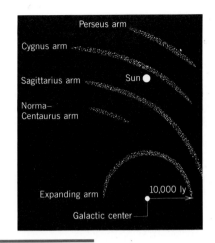

FIGURE 15.2

The layout of spiral arm segments near the sun, which lies on the inner edge of the Cygnus arm in a small branch sometimes called the Orion spur. The arms are named after the constellations in whose directions they lie as seen from the earth. The expanding arm is the closest one to the galactic center.

The overall density of material—gas, dust, and stars—inside a spiral arm is roughly 10 percent higher than in the region between arms. The disk contains most of the gas, dust, and young stars in the Galaxy. Near the sun, for example, about one-third of the matter is in gas and dust, the rest in stars. In contrast, for the Galaxy as a whole, only a few percent of all the material is in the form of gas and dust. Yet, this dust visually blocks our view of about 90 percent of the Galaxy!

The sun lies on the inner edge (Fig. 15.2) of a poorly defined structure called the *Cygnus arm*. (*Note:* Astronomers often use the word *arm* to indicate a well-defined *segment* of a larger overall spiral-looking structure.) Outward, about 10,000 ly from the sun, lies an arm segment parallel to the Cygnus arm; because the best-observed portion lies in the direction of the constellation Perseus, it is called the *Perseus arm*. At about 6000 ly interior to the sun curves the *Sagittarius arm*. Some evidence indicates that another arm segment may lie 12,000 ly from the sun toward the galactic center; it is called the *Cen-*

FIGURE 15.3

Spiral galaxy with well-defined spiral structure (NGC 2997). Note the differences between the colors in the nucleus in those in the spiral arms. (Image by D. Malin; courtesy Anglo-Australian Telescope Board)

taurus arm or the *Norma arm*. Finally, encircling the galactic center at a distance of about 10,000 ly from the center is the *expanding arm*, so called because this innermost arm appears to be expanding away from the center at roughly 50 km/s.

We surmise that our Galaxy has two major arms (interior to the sun's orbit) and perhaps four arms (exterior to it) wound around the nucleus. The arm segments are probably parts of the major arms. In particular, the arms beyond the sun seem to be extensions of the Perseus, Cygnus, and Norma arms. The spiral pattern is clearest in these outer parts.

The disk of the Galaxy contains spiral arms, which are somewhat enhanced regions of interstellar gas, dust, and young stars.

The rest of this chapter describes the techniques used to gather the information needed to build this model of the Galaxy. We note that many other galaxies have a similar spiral shape (Fig. 15.3).

15.2 Galactic Rotation

Just as the earth and planets orbit the sun, the sun and stars orbit the center of the Galaxy. Now, the sun and its nearest neighboring stars move in a variety of directions at a variety of speeds. In relation to nearby stars, the sun travels at a speed of about 20 km/s. But what motions do the sun and nearby stars share in relation to the center of the Galaxy?

We use a variety of indirect approaches to find out how fast the sun moves around the Galaxy. One good method utilizes the motions of globular clusters. The globulars seem to orbit the Galaxy in random inclinations, with a roughly spherical distribution around the nucleus (Fig. 15.4). With respect to the nucleus, the average motion of all globular clusters is roughly zero. In other words, the system of globulars has no overall rotation about the galactic center (although individual clusters move rapidly, some in one direction, some in another). So the sun's motion relative to the system of globulars is the same

as its rotational motion with respect to the Galaxy's center.

The exact value of the sun's speed around the Galaxy is very hard to determine and subject to large errors. The value usually agreed upon is 250 km/s. But enough recent observations indicate that it is probably smaller, likely 220 km/s; this book uses 220 km/s.

The sun (and other stars) orbit the center of the Galaxy; the sun's orbital speed is about 220 km/s.

The Sun's Location and Distance from the Center

How far the sun is from the Galaxy's center is also tough to find out because we cannot see the center optically. However, we can look above and below the

FIGURE   1 5 . 4

Schematic picture of the orbits of globular clusters around the galactic center. Each globular cluster moves on a highly eccentric orbit, so that most of its time it lies far from the nucleus, as expected from Kepler's second law. The far ends of these orbits define the outer limits of the halo of the Galaxy.

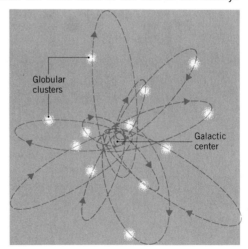

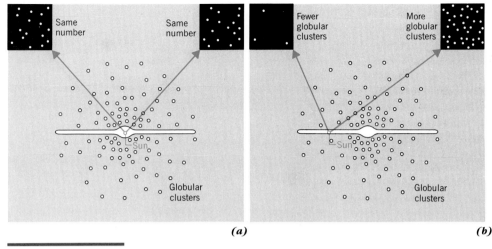

(a)   (b)

FIGURE 15.5

The position of the sun in the Galaxy inferred from the observed distribution of globular clusters in the sky. Assume that the globular clusters have a uniform distribution around the center of the Galaxy. (a) If the sun were in the center of the Galaxy, you would see roughly the same number of globulars in every direction in the sky. (b) When the sun is placed away from the center, more globulars are visible in the direction of the center than in other directions. This is the observed case.

galactic plane, where the obscuration is less, to observe objects thought to be balanced about the galactic center, such as globular clusters. Let's assume that the globular clusters have a uniform distribution in a huge sphere centered on the nucleus.

Take the simplest model: the sun lies in the center of the Galaxy (Fig. 15.5a). Trace lines of sight in a number of directions around the sky. Because we have assumed a central vantage point in a uniform distribution of objects, every line of sight you chose should intercept the same number of globular clusters. So, the expected distribution would be uniform over the sky, but that's *not* what is observed. Globular clusters in fact concentrate in the southern sky toward the constellations Sagittarius and Scorpius (Fig. 15.6). How can we explain this fact in terms of the sun's location?

Assume that the sun lies away from the center of the Galaxy (Fig. 15.5b). Now some lines of sight cut longer distances than others through the globular clusters, so more clusters are seen in these directions than along other lines of sight. The expected distribution is *not* symmetrical around the sky, but is most

concentrated in the direction of the galactic center. So with the sun away from the center, we can explain the observations and conclude that the center of the Galaxy lies in the direction of Sagittarius and Scorpius.

This technique gives us our location relative to the center. What about our distance? We can figure that out if we can find the distances to the globular clusters. The size of the sphere of globulars outlines the halo (and so the size of the Galaxy), and the distance of the sun from the center of this sphere indicates the distance of the sun from the Galaxy's center.

How to get the distances? We can use the giant variable stars in the class called RR Lyrae, which are commonly found in globular clusters. (A *variable star* is one whose luminosity changes with time.) These stars have regular periods of light variation that last half a day. RR Lyrae stars have essentially the same mean luminosity no matter what their period: 50 solar luminosities. So once an RR Lyrae star has been identified by the special shape of its light curve, we can work out its distance from a measurement of its flux and the inverse-square law for light.

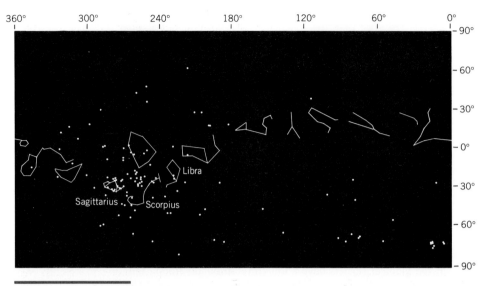

F I G U R E   1 5 . 6

The concentration of globular clusters in the region of the sky toward the center of the Milky Way. This full-sky view shows the locations of the brightest globular clusters from north to south celestial pole around the entire sky. Each dot marks a globular cluster; also indicated are the zodiacal constellations. About one-third of all known globulars are concentrated in the area just above Scorpius and Sagittarius, which amounts to only 2 percent of the entire sky. In other directions, the number of globular clusters is much less for the same area of the sky. (Diagram generated by *Voyager* software for the Macintosh, Carina Software)

By this technique, originally developed by Harlow Shapley (1885–1972) around 1915, the sun's distance from the galactic center is approximately 28,000 ly, within a range from 24,000 to 33,000 ly. I'll use 30,000 ly; but you must realize that the measured distance to the center is somewhat uncertain.

The sun orbits at about 30,000 ly (perhaps a bit less) from the center of the Galaxy.

Rotation Curve and the Galaxy's Mass

Knowing the sun's velocity and distance, we apply Newton's form of Kepler's third law (Section 3.4) to deduce the mass of the Galaxy. The result, a few hundred billion solar masses, refers *only to the mass interior to the sun's orbit*. What about the mass out-

side it? To determine this requires that we find the **galactic rotation curve**—how fast an object some distance from the galactic center revolves around it. The rotation curve is usually shown as a graph of how the orbital speed varies with distance from the Galaxy's center.

Imagine the Galaxy's mass as concentrated in the nucleus—a setup that resembles the solar system. The stars in the galactic disk revolve about the nucleus of the Galaxy much as the planets revolve around the sun. The forces are similar, so the motions should be similar. The stellar motions should follow Kepler's laws (Section 2.5), and the speeds of the stars should decrease with increasing distance from the Galaxy's center—just as, for example, the orbital speed of Mars is less than that of the earth.

In fact, the Galaxy's speeds do *not* follow Kepler's laws well, and that tells us an important fact: the major part of the mass is *not* concentrated at the center! A central concentration of mass would result

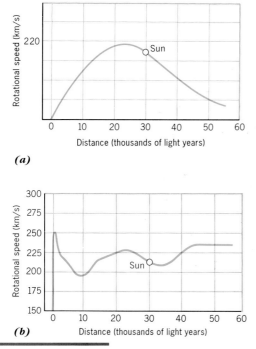

(a)

(b)

FIGURE 15.7

Rotation curves for the Galaxy. (a) Galactic rotation curve for Keplerian rotation, if most of the mass of the Galaxy were concentrated in its nucleus. Then beyond the sun's orbit, we would expect the rotational speeds to decrease with distance.
(b) Galactic rotation curve based on a combination of carbon monoxide and hydrogen observations, assuming that the sun's speed is 220 km/s and its distance from the galactic center is 30,000 ly. Note that the rotational speeds do not follow a downward trend outside the sun's orbit. (Adapted from a diagram by D. P. Clemens)

in Keplerian motion in the outer regions of the Galaxy (Fig. 15.7a). The observed rotation curve is different (Fig. 15.7b). In the outer parts of the Galaxy, the curve rises beyond the sun's orbit. It then flattens out at a distance of perhaps 50,000 ly from the galactic center. If the motions followed Kepler's laws, we would expect the rotation curve to decline, and it does *not*, even far out from the sun. The high rotation speed even at 50,000 ly implies that much of the Galaxy's material lies out beyond the sun's orbit. At least as much mass lies exterior to the sun as interior to it. The total mass of the Galaxy is roughly 10^{12} solar masses.

The rotation curve tells us the overall distribution of matter in the Galaxy, because gravity controls those orbital motions; most of the mass lies outside the orbit of the sun.

15.3 Galactic Structure

We know the mass of the Galaxy. How is it arranged? How do we infer our Galaxy spiral pattern? Bluish OB supergiants, H II regions, and Population I cepheids are some of the objects that cluster in the spiral arms. These objects are called **spiral tracers**. Added to these are the giant molecular clouds, visible by radio observations, also found in spiral arms.

We are in a bad position to observe the Galaxy's structure, for we reside in the Galaxy's dusty disk. Imagine, for example, that you are watching the halftime show at a football game. The band has set up an elaborate formation, and you, up in the stands, can easily identify it. But suppose you were down on the field, at the edge of the formation. You could eventually figure out the shape if you could find the distances to all the band players. You could make a map of their positions by plotting the distances and directions of each person. But suppose you had to do this mapping in a fog so dense that only the closest people were visible. Then you would need some other method of estimating distances—perhaps by the loudness of the sound of the instruments coming through the fog.

Optical astronomers, who try to find the distances to features that mark spiral arms, find their view blocked by interstellar dust. Radio waves get through. So radio astronomers can pick up the radio emission from clouds of gas that probably mark spiral arms. But they have more difficulty in determining distances than the optical astronomers do. The results from the two methods do not agree in many details.

Optical Observations

How to apply these spiral tracers to our Galaxy to determine the layout of its spiral arms near the sun?

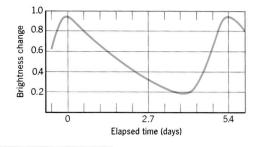

FIGURE 15.8

Light curve for a typical cepheid variable star—Delta Cephei, the prototype of the class—shows how the star's brightness varies periodically with time. Note that the rise in brightness is steep, but the decline less so. The period of brightness variation from peak to peak is about 5.4 days.

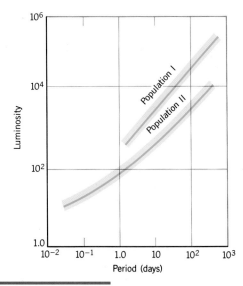

FIGURE 15.9

Simplified period–luminosity relation for cepheids, which fall into two groups—Types I and II, based on their stellar population. In both cases, the general trend is the same: the longer the period, the more luminous the star. By comparing the observed flux to the luminosity estimated from the period, you can find the distance to the cepheid with the inverse-square law for light.

We require an accurate and reliable technique for measuring the distance to each of the tracers. H II regions and supergiant OB stars trace the arms well because they lie in spiral arms, and their high luminosities make them visible over large distances. Cepheid variable stars have also been used to sketch spiral features because their distances are found by the **period–luminosity relationship**. This crucial distance-measuring technique is so important that you should know how it works (Conceptual Activity 15).

Cepheids are variable stars, whose luminosity changes periodically with time. A *light curve* is a graph of the change in a star's flux with time. A star whose light varies in a regular fashion is known as a *periodic variable*; Delta Cephei sets the standard for one such class of variables—called *cepheid variables* or *cepheids*—by the special shape of their light curves (Fig. 15.8). From Doppler shift observations, we know that cepheids actually expand and contract as they vary in luminosity.

For the cepheid variables, we find that a special relationship connects its luminosity to its period. This is the period–luminosity relationship (Fig. 15.9), which basically states that *the longer the period of light variation of a cepheid, the more luminous it is.*

We can use this relationship to determine distances to cepheids as follows: (1) find   cepheid (identifying it by its light curve); (2) measure its period of light variation, from peak to peak; (3) find the star's luminosity from the period–luminosity rela-

tionship; (4) measure the star's flux by telescopic observations; and (5) calculate its distance from the inverse-square law for light (Section 11.1).

> From the period of light variation of a cepheid, we can infer its luminosity; then if we measure its flux, we can estimate its distance.

Today we know that the stars that used to be lumped together as cepheid variables are actually *three stellar* types: Type I (classical) cepheids, Type II cepheids, and RR Lyrae stars. Type I cepheids are Population I stars, Type II are Population II. RR Lyrae stars, also Population II, are commonly found in globular clusters and in the halo. Population I and II cepheids have different period–luminosity relationships (Fig. 15.9).

The optical maps of spiral structure must be viewed with caution because the spiral arm segments did not come out the same when mapped with different objects. Investigations have found a few nearby arm segments spaced about 7000 ly apart, but a larger structure of spiral arms has not been revealed.

Radio Observations

The murkiness of interstellar dust deters the optical mapping of the Galaxy. But dust does not stop long radio waves, so radio astronomers can reach far beyond the restricted range of the optical astronomers, even to the other side of the galactic nucleus. The 21-cm line from hydrogen (Section 12.1), which comes from the concentrations of neutral hydrogen clouds in the spiral arms, is used by radio astronomers to map the Galaxy. But the best radio tracer is the carbon monoxide emission from giant molecular clouds.

To distinguish among spiral arms, we look in different directions and at different *velocities* in the same direction. Radio line emission, such as the 21-cm radiation from H I, is Doppler-shifted to different wavelengths because of the different radial velocities of the hydrogen gas clouds. These differences in velocity come mostly from the rotation of the Galaxy. So *if* we know how the Galaxy rotates (from a complete rotation curve), then we can translate 21-cm observations into a map of spiral structure. (The same technique applies to the millimeter-line emission from molecular clouds.)

We know the sun's speed around the Galaxy. When we observe line emission, we know the direction in which we are looking and can measure a radial velocity. From that radial velocity, we *infer* the rotational velocity of the cloud. We then look up this velocity on the galactic rotation curve to find the distance from the center of the Galaxy to which it corresponds. Scanning around the plane of the Milky Way, we make a step-by-step series (spaced every degree or so) of radio-line observations to get a sequence of radial velocities from Doppler shifts. We then infer distances to the H I or molecular

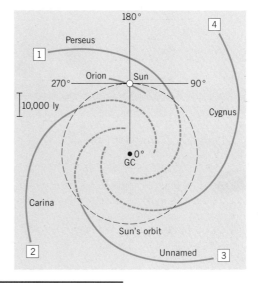

FIGURE 15.10

One model for the overall spiral structure of the Galaxy. The sun's orbit is indicated by the circle. The numbers around the outer circle indicate the galactic longitude as viewed from the sun; a longitude of 0° marks the direction to the galactic center (GC). Note the conjectured pattern featuring four arms (labeled 1, 2, 3, and 4). The scale is at the upper left. (Adapted from a figure by L. Blitz, M. Fich, and S. Kulkarni)

clouds from the rotation curve. These can be imagined as connected to trace a spiral arm and so outline the Galaxy's structure.

> The galactic rotation curve tells us that at a given position in the sky, a certain radial velocity (Doppler shift) corresponds to a specific distance from the sun.

Different investigators have drawn conflicting radio maps! The heart of the problem is that the

Spiral Tracers	Optical	OB stars, H II regions, cepheid variables
	Radio	21 cm from H I regions, millimeter from CO
At a Glance	Results	two or four main spiral arms
	Caution	limited areas mapped so far

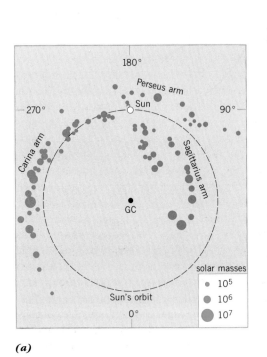

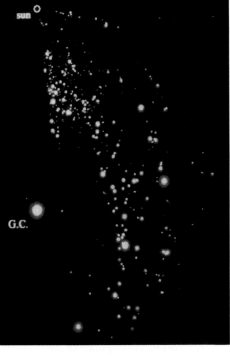

(a)   **(b)**

FIGURE 15.11

Giant molecular clouds used to outline spiral structure. The positions of the sun and the galactic center (GC) are indicated. (a) Radio observations of the positions of giant molecular clouds. The sizes of the circles indicate the masses of the cloud complex. Note how well these clouds delineate the Carina, Sagittarius, and Perseus arms. The empty areas have basically not yet been surveyed. (Observations by R. S. Cohen and colleagues.) (b) Giant molecular clouds from Sagittarius to Cygnus. The sizes of the dots represent the sizes of the clouds. The false colors show the strength of the carbon monoxide emission. (Data from the Massachusetts–Stony Brook Galactic Plane CO Survey; courtesy P. Solomon)

neutral gas clouds don't follow the simple scheme of circular rotation; in addition to their circular motion, they have their own random motions. Unfortunately, such noncircular motions lead to incorrect distances and so disrupt the unity of the spiral arm map.

Despite such problems, the radio maps *do* hint at a large-scale spiral structure. The figure of the Galaxy derived from them shows through in its broad outline. Beyond the sun, the spiral arm pattern appears more reliably, thanks to observations of giant molecular clouds at millimeter wavelengths. Very young stars are born in such clouds, so they make good tracers of spiral arms. Complementing the 21-cm neutral hydrogen line observations, the molecular data suggest four arms as the overall structure (Fig. 15.10).

This technique has clarified the spiral arm picture using giant molecular clouds with masses greater than 10^5 solar masses (Fig. 15.11). The Carina arm stands out most clearly. It stretches for more than 80,000 ly, with a giant molecular cloud roughly every 2000 ly along the segment. The Perseus and Sagittarius arms are also delineated but overall contain far less mass in molecular clouds.

Radio maps, especially from emission from giant molecular clouds, hint that the Galaxy has four major spiral arms.

15.4 The Evolution of Spiral Structure

For many years astronomers supposed that the spiral arms in our Galaxy and others were *material* arms, a coherent bunch of objects—stars, nebulas, gas, dust—somehow *physically* held together. This idea faces a serious problem: how does an arm persist for a long time? A material arm winds up if the outer parts of the arm rotate more slowly than the inner ones. So after a few rotations of the Galaxy the arms should disappear. The Galaxy has turned about 20 times since the origin of the solar system, however, and the arms are still there!

Here's an analogy. Imagine you and two friends are going to run around a track. You station yourself in the middle lane, one friend in the lane closer to the track's center, and the other in the outside lane. You start running, lined up. Now insist that the friend inside run faster than you and the one outside slower. After one or two laps, the lineup will be disrupted. The same would happen to spiral arms—if they were material arms—after a few rotations.

Astronomers have been struck by the persistence of spiral arms. We can't tell this from our Galaxy alone, for it could be that we are observing at a very special time, soon after the formation of the arms. But that would not likely be true for all galaxies. Of the brightest galaxies in the sky, more than 60 percent have a spiral form (Fig. 15.3).

How does this tell us that spiral arms are ongoing phenomena? Remember, when we look out in space, we look back in time. Consider a galaxy's spiral pattern disappearing after a few rotations by winding up. Assume that all galaxies formed at the *same* time. Then we should not see a spiral form in nearby galaxies but should see it in more distant galaxies. Why? The distant galaxies are younger (closer to the time of birth), so the spiral patterns have not yet have time to wind up.

But that's not the case! Both nearby and distant galaxies exhibit spiral patterns. Seeing so many galaxies as spirals—near and far—implies that such structures last for at least some few billions of years or that they are often renewed.

The spiral arms of the Galaxy are not material arms but are patterns that are constantly revived.

What makes spiral arms if they are not material arms? The majority view today pictures spiral arms as the result of a spiral wave of higher density moving through the Galaxy's disk. This wave produces all the signposts of a spiral arm—young stars, H II regions, lanes of dust. None of these objects last very long. As they die and the density wave moves on, new spiral arm tracers are born from the interstellar medium at the new location of the wave. So a spiral arm segment always contains objects of the same *kinds*, but not the *same* objects.

Any particular arm is a passing phenomenon. Individual objects orbit at the speed determined by gravitational forces appropriate for their distance from the center, but the *wave* pattern rotates with a constant angular speed and does not wind up. This approach is called the **density-wave model** of spiral structure. Of the models proposed to date to explain the spiral pattern, it best describes the overall scheme.

What's a *density wave*? A sound wave is a density wave. Imagine air molecules compressed together. These molecules bang into adjacent ones in the direction of their motion, which transfers the compression to the next bunch of molecules. As this compression—a density wave—travels forward, it leaves behind a trough of lower density. Two important points here: (1) a sound wave requires a source to start it, and (2) the high-density part of the wave persists even though the specific particles that make it up change at different points in the medium.

Here's an analogy. Suppose you are driving on a mountain road filled with traffic (Fig. 15.12). Everyone moves along happily at the speed limit. Ahead, an overloaded truck can go about half the maximum speed. Cars jam up just behind the truck as the car drivers wait for a clear road ahead in order to pass. When they do pass, they move along again at the speed limit, leaving the poor trucker behind. Imagine that you watched this situation from the air and concentrated on the motion of the cars. You'd see a denser region of cars just behind the truck, where they pile up for a short time; the truck moves down the road at half the average speed of the cars. You would also note that the jam-up persists, even though it does not contain the same cars. New cars get caught up in it as other cars move out.

You can think of the cars as the stars and the interstellar medium in the galactic plane, and the jam-up as the visible effect of a moving density wave (the truck). The jam-up creates in the disk a region of

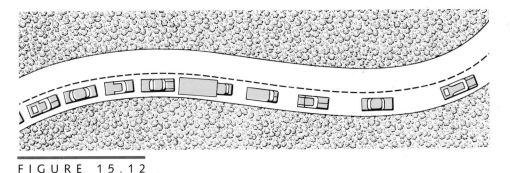

Highway bottleneck as an analogy to a density wave. The slow-moving truck has cars jammed up behind it waiting to pass. The blockage is always behind the truck, but it consists of different cars at different times. Viewed from above, the jammed-up area moves more slowly than the average speed of the cars; it is a density wave.

increased density of stars and gas—that's a spiral arm. (But remember that this increase amounts to only 10 percent more density than there is between the arms.)

This density-wave idea *assumes* that a two- or four-armed spiral density wave sweeps through the galactic plane. Although the wave's origin is not explained, once formed, it persists for a billion years or so. The gas in the disk piles up at the back of the wave. The buildup of pressure and density heats up the gas suddenly so that a shock wave forms along the front of a density wave because the density wave travels through the gas at a speed faster than that of sound. (A *shock wave* is created by moving through any medium faster than the speed of sound in that medium. For example, traveling through the air in a jet plane faster than the speed of sound creates a shock wave called a sonic boom.) The compression at the shock squeezes neutral hydrogen clouds together and initiates the collapse of the clouds to form giant molecular cloud complexes, which in turn form young stars and H II regions. The density wave forms the features associated with a spiral arm.

Spiral arms are likely maintained by density waves, which generate a spiral wave pattern by making regions of higher gravitational forces and so higher densities of matter in the Galaxy's disk.

During the short lifetimes of the newly formed OB stars, the density wave moves only a short distance. So these stars, while they last, mark the spiral arm clearly. As the density wave moves on, it provokes the formation of more stars. These take the place of the ones that have rapidly died out.

However, the model does not explain the origin of the density waves nor what sustains them. As the density waves ripple through the interstellar medium, they lose energy and should dissipate in about a billion years. But as evidenced by the abundance of spiral galaxies, the density waves—if they are the correct explanation—last longer than a billion years or are renewed several times as the old waves fade out. Some mechanism, yet unknown, keeps supplying energy to maintain the density waves or to trigger new series of them.

15.5 The Center of the Galaxy

Although dust largely obscures the center of the Galaxy, a few regions of low absorption open up optical glimpses of the central bulge (Fig. 15.13). In addition, we can surmise the nature of the stars in the nucleus from observations of the nuclei of other galaxies. These two kinds of observation imply that the nucleus contains mostly old Population I stars densely packed together. So jammed up are these

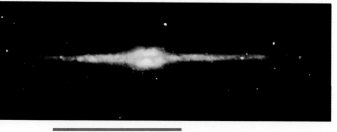

FIGURE 15.13

Wide-angle optical view of the galactic center region from earth. The area shown spans about 40°. Note the dark lanes of dust cutting through the central region of the Milky Way's plane. (Courtesy KPNO/ NOAO)

FIGURE 15.14

Wide-angle (96°) view of the Milky Way from space. This false-color image from the Cosmic Background Explorer (COBE) satellite combines views at 1.2, 2.2, and 3.4 μm and reduces the dust obscuration that blocks the optical view (Fig. 15.13). Note the well-defined central bulge, in which resides the nucleus. (Courtesy NASA)

stars that if you lived in the nucleus, the nighttime sky would be as bright as twilight on the earth.

Radio and infrared (Fig. 15.14) observations all probe the nucleus. They disclose that the heart of the Galaxy is a bizarre and active place: it contains not only many very old Population I stars, but also some very young supergiant M stars and O stars. Motions of gas here suggest a high concentration of mass at the very center—perhaps a black hole. In the very center of the Galaxy lies a radio source around 10 AU in diameter—about the distance of Saturn from the sun!

The nucleus of the Galaxy is small and contains an enormous concentration of stars and gas; it strongly emits radio and infrared radiation.

Radio and Infrared Observations

Let's look first at the continuous electromagnetic emission of the galactic center (Fig. 15.15). An intense radio source lies smack in the direction of the center. It is called *Sagittarius A* (*Sgr A* for short). The Sgr A complex combines different radio structures. Some radio emission here is from ionized gas. But some also comes from high-energy electrons traveling through a magnetic field—synchrotron emission (Section 14.4). High-resolution radio maps show that Sgr A actually consists of two separate radio sources: one, called *Sgr A East*, emits by the synchrotron process; the other, *Sgr A West*, seems more like a giant H II region. Sgr A West is associated with an agglomeration of infrared sources (to be explained shortly).

Within Sgr A West lies a compact radio source about 20 AU in size, which appears to mark the actual core of the Galaxy; it is called *Sgr A*°. Its emission is nonthermal, synchrotron emission with a power of some 10^{27} W—about 10 times the sun's luminosity. Overall, the ionized gas here, which amounts to a few million solar masses of material, rotates at about a few hundred kilometers a second.

A radio map at a wavelength of 6 cm (Fig. 15.15a) shows the thermal emission from the inner 10 ly of the Galaxy. A curious aspect of the emission is that it

The Galactic Center At a Glance		
	Size	inner 1000 ly
	Luminosity	10^9 solar luminosities
	Mass in gas	7×10^7 solar masses
	Mass in stars	6×10^9 solar masses

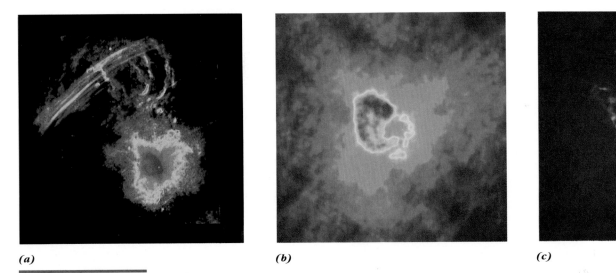

(a) *(b)* *(c)*

FIGURE 15.15

False-color radio maps (Appendix G) of the galactic center; each one shows a different area. Blue represents the regions of lowest radio emission. (*a*) This 20-cm radio map of Sgr A (lower right), shows the strongest emission (red) from the center. Note the arc of filaments extending from the upper right to the left; each filament is a few light years long. The galactic plane crosses the image from upper right to the lower left. The filaments are perpendicular to the plane and parallel to each other. The halolike emission (blue) around Sgr A is about 75 ly in diameter. The area shown here is about 250 ly by 250 ly. (Observations with the VLA by F. Yusef-Zadeh, M. Morris, and D. Chance; courtesy NRAO/AUI) (*b*) Radio map at 90 cm shows the radio emission from the Sgr A complex at center. Note that the 90-cm emission appears as a large blue halo around the entire region. The area shown here is about 90 ly by 90 ly. (Observations by A. Pedlar and colleagues; courtesy NRAO/AUI) (*c*) High-resolution flux map of the Sgr A source made from radio observations at a wavelength of 6 cm. White represents the strongest emission. Note the spiral-like structure of Sgr A. This map, which shows the thermal emission from this region, covers an area about 30 ly by 30 ly. It reveals details not visible in the Sgr A emission in (*a*) and (*b*). (Observations by K. Y. Lo and M. J. Claussen; courtesy NRAO/AUI)

has a spiral shape. Observations of a larger region here show that the ionized gas loops in a filamentary bent arc (Fig. 15.15*b*). The arc may consist of material extending from Sgr A and guided in loops by local magnetic fields. A wide-field view of the Sgr A complex at a 90 cm wavelength (Fig. 15.15*c*) reveals a halo of radio emission that arises from a mixture of thermal and nonthermal sources; the gases here have a total mass of a few thousand solar masses.

The radio emission from the nucleus appears to be mostly synchrotron, which requires high-speed electrons in an ionized gas and magnetic fields.

Sgr A also gives off radio-line emission from molecules. The observations indicate that this molecular

cloud, which may contain as much as a million solar masses of material, is in front of Sgr A East, so it is not right at the center of the Galaxy. Also, the gas and dust appear to form a ring about 15 ly in diameter. This ring tilts about 70° to our line of sight and rotates at about 100 km/s. The material on the inside of the ring is ionized, and when viewed from the earth, the emission from the ionized gas appears to have a spiral shape because of the viewing angle.

The galactic center region emits strongly at 2.2 μm (Fig. 15.16). The most intense part of this emission coincides with Sgr A. What is the source of this radiation? Simply the combined 2.2-μm emission from all the old, cool stars that inhabit the galactic nucleus. Observations of the same region near 10 μm are quite different; they show the infrared emission from dust that is heated by the radiation from stars (Fig. 15.17). Some of the heating radiation comes from the old Population I stars. But some also

FIGURE 15.16

False-color infrared (2.2 μm) image of the central 150 ly of the galactic center. The bright, white region near the center is the nucleus, which is not visible in an optical view (Fig. 15.13). The plane of the Galaxy goes from the upper right to the lower left. (Observations by R. Joyce and R. Probst at KPNO; courtesy NOAO)

FIGURE 15.17

False-color image (Appendix G) of the galactic center at the infrared wavelength of 8.3 μm. White indicates the strongest emission; red, the weakest. The size of the area covered is a little less than 2 ly by 2 ly. Note the individual sources (white and yellow regions) and the extended emission (red areas). The radio source Sgr A* lies at the top of the curved part of the bar; no strong infrared source is there. (Courtesy NASA)

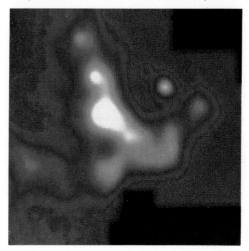

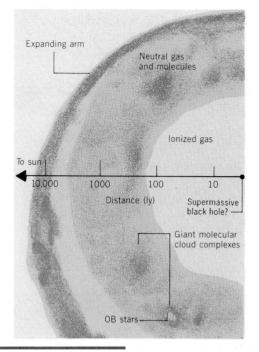

FIGURE 15.18

Schematic map of the inner regions of the Galaxy. Moving inward from the sun, we encounter the expanding spiral arm, a region of giant molecular clouds, a disk of ionized gas, and, at the core itself, perhaps a supermassive black hole.

derives from high-luminosity O stars; the condensations in these maps are probably the locations of newly formed O stars. The combined luminosity from them, in the range from 2 to 20 μm, is roughly a million times that of the sun. Over the entire infrared range, the galactic center emits about 100 million solar luminosities.

The infrared emission from the nucleus comes from a combination of old, cool stars and young, hot ones (embedded in dust).

The Inner Regions

We have a pretty good model of the overall distribution of gas and dust near the Galaxy's center (Fig. 15.18). The galactic core contains, within the Sgr A molecular cloud complex, a disk of rotating ionized gas a few hundred light years across. Within it is a smaller disk of neutral gas and dust some 30 ly across. The dusty disk lies pretty much in the plane of the Galaxy and rotates with its axis lined up to that of the general galactic rotation.

The inner part of this disk shows rapid rotational motions—and the rotational velocities *increase* closer to the core. These rotational velocities are so high (for example, ionized gas rotates at about 150 km/s only one light year from the center) that a huge concentration of mass is needed to hold the speedy gas together. To account for the rapid rotation, the mass in the core must amount to 4 million solar masses—all lumped together in a region only 0.1 ly in diameter!

This mass may be locked up in a **supermassive black hole**. If it were in the form of, say, a cluster of solar-mass stars, these stars would be located, on the average, only 1 to 2 AU from each other. Some million solar masses of material could be jammed into such a star cluster in the nucleus—and the more in the form of stars, the less mass needed in a black hole. So the idea that a supermassive black hole lurks in the heart of the Galaxy seems likely but has yet to be confirmed.

Within the nucleus of the Galaxy, a concentrated mass—perhaps a supermassive black hole—makes material orbit the center at rapid speeds.

15.6 The Halo of the Galaxy

What about the outer reaches of the Galaxy? The globular clusters outline the halo (Fig. 15.19). Population II stars not in clusters are also seen here, as well as some gas that is hot and ionized. The halo may also contain as yet undetected objects, such as very faint low-mass stars, and it may extend far beyond the edge of the disk.

Globular clusters make up the most visible part of the halo. Recall (Section 13.7) that a globular cluster has a spherical shape (some tens to hundreds of light years in diameter) and contains up to a million Popu-

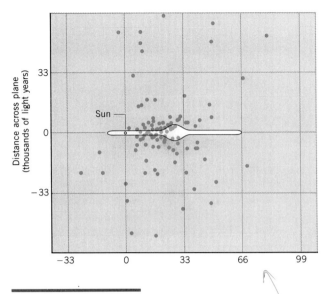

FIGURE 15.19

Edge-on view of the distribution of globular clusters (dots) about the plane of the Galaxy. The distances are in thousands of light years. The dark line shows the outer contour for the Galaxy's disk of stars. Note that the sun's position is well away from the galactic center. (Adapted from a diagram by W. E. Harris)

lation II stars, each with a little less than 1 solar mass. The globular clusters form a spherical distribution around the Galaxy's center. Their elliptical orbits bring them out to extreme distances of 40,000 ly or more from the Galaxy's nucleus. The clusters orbit at speeds some hundreds of kilometers per second, diving into and shooting out of the disk.

The outer halo of the Galaxy has no exact boundary. Observations of the placement of globular clusters indicate that the halo extends out to at least 300,000 ly, far beyond the limits of the Galaxy's disk, to the Magellanic Clouds, the two companion galaxies that are gravitationally bound to the Galaxy (Section 16.6).

Most of the visible material in the halo is in globular clusters, which outline its greatest boundary.

The rotation curve of the Milky Way indicates that the halo actually contains considerable material, perhaps much more mass than is in the rest of the Galaxy. This conclusion is reached from our knowledge that the rotation curve flattens out at distances far from the center of the Galaxy. So more mass is out there, but its form is *not* obvious. Using other spiral galaxies as a guide, the mass in the halo may be five times greater than all the mass known so far. What might it be?

Astronomers have spotted a few RR Lyrae variables above and below the galactic plane that do not belong to globular clusters. In addition to these stars, the halo may contain a large number of low-mass, faint red stars that are difficult to observe directly. Recent observations imply that a few other nearby galaxies like the Milky Way may have extensive, massive halos of faint red stars.

But the halo may also contain other objects, currently unobservable. Astronomers give the generic name of **dark matter** to this invisible material. A massive halo can help out the density-wave model for the spiral structure in the disk. The mass in the halo can stabilize the arms gravitationally and ensure that the density waves propagate for long periods of time (a billion years). A massive, invisible halo may well be the necessary ingredient to lock in a well-defined spiral structure in a galaxy.

The halo of the Galaxy contains a substantial amount of unseen material, a form of dark matter.

15.7 Evolution of the Galaxy

We have developed a fair idea of the architecture of the Milky Way Galaxy. What clues does this information provide about the birth and evolution of the Galaxy? The crucial tips lie in the chemical composition of galactic material and its dynamics. The process of galactic evolution links the chemistry with the dynamics. This linkage marks an important theme of cosmic evolution, because the evolution of the Galaxy results from the evolution of all the stuff that makes it up.

Recall (Section 13.7) that in chemical composition, Population I and Population II stars differ con-

siderably with respect to abundance of heavy elements. In general, Population II stars contain about 1 percent of the metal abundance of Population I stars. But this division turns out to be too simple. We do not find a stark division of metal abundances into just two groups. Rather, we find a continuous range of abundances, from about 3 percent to less than 0.1 percent for the ratio of metals to hydrogen. So, though the division into two populations is a useful tool, a continuous range of metal abundances exists. And when the stars are catalogued by metal abundance, we find a association with average distance from the Galaxy's disk—the larger the distance, the less the metal abundance.

Stars with a lower percentage of metals are usually found farther from the plane of the disk.

To interpret this observation, we relate metal abundance to age. How? Stars are born from clouds in the interstellar medium. Their atmospheric elemental abundance reflects that of the gas from which they formed. Also, stars inherit the orbital motions about the Galaxy of their parent gas and dust clouds. Now, massive stars evolve quickly, manufacture heavy elements, and spew back into the interstellar medium material that is enriched by the heavy elements. So as long as new stars—especially massive ones—are born, the abundance of heavy elements in the interstellar medium gradually increases as the Galaxy ages.

In general (exceptions exist!), observations show that the youngest objects (highest heavy-element abundances) hug close to the disk; the oldest objects (lowest heavy-element abundances) range far from the disk. Other objects fall in between these extremes. So the halo of the Galaxy is its oldest part, and the spiral arms its youngest.

The oldest objects in the Galaxy have the lowest metal abundances; the youngest have the highest.

We can estimate the Galaxy's age by finding the oldest stars in the halo. A comparison of theoretical models for globular cluster stars with H–R diagrams for them (Section 13.7) indicates an age of 14 to 16 billion years. That's *when* the Galaxy formed; take 15

FIGURE 15.20

The sequence of contraction and condensation of a large cloud of gas and dust to make globular clusters and the Galaxy's disk. Because of its original spin, the matter eventually makes a disk. The stars in the globular clusters form before the disk has developed. The entire process takes less than one billion years.

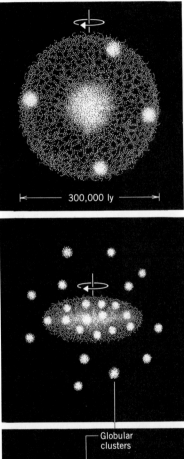

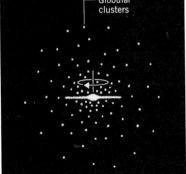

billion years as that value used in this book. Let's see *how* it might have formed. We find a process like that for the formation of the solar system and stars—gravitational contraction (Sections 8.3 and 12.3).

Imagine a tremendous, ragged low-density cloud of gas roughly as big as the Galaxy's halo today (Fig. 15.20). This cloud probably is turbulent, swirling around with random churning currents. Slowly at first, the cloud's self-gravity pulls it together, with its central regions increasing in density faster than its outer parts. Throughout the cloud, turbulent eddies of different sizes form, break up, and die away. Eventually, the eddies become dense enough to contain sufficient mass to hold themselves together. These might be hundreds of light years in size—incipient globular clusters. Each blob then splits up to form individual stars—all born about the same time with the same chemical composition.

Meanwhile, the gas contracted more and fell slowly into a disk. Why a disk? Because the original cloud had a little spin, and the conservation of angular momentum (Section 8.4) requires that it spin faster around its rotational axis as it contracts (Fig. 15.20). The temperature of the cloud slowly increases as gravitational potential energy is converted to kinetic energy. As the disk forms, its density increases, and more stars form. Each burst of starbirth leaves behind representative stars at different distances from the present disk. Finally, the remaining gas and dust settle into the narrow layer we see today. Somehow density waves appear and drive the formation of spiral arms.

During this time, massive stars were manufacturing heavy elements and flinging them back into the cloud by supernova explosions. So as stars were born in succession, each later type had more heavy elements. That enrichment continues today in the disk of the Galaxy.

Gravitational contraction condensed the Galaxy from a large cloud; the conservation of angular momentum required much of the material to form a disk.

What of the Galaxy's future? Let's speculate a bit. If we assume that no new gas is added from outside the Milky Way, then stellar evolution points to a day when most stars have become corpses. Matter that once made up the interstellar medium is locked up for good. The Galaxy will literally run out of gas; starbirth will halt. Even if density waves still endure, they have little gas to move around. When the disk of the Galaxy stops evolving, the Galaxy is essentially defunct. Globular clusters will still swing on their leisurely orbits around the core. The supermassive black hole (if there!) will survive forever. Overall, the Galaxy will be quiet and dull.

The Larger View

Looking outward, we find that local stars, gas, and dust assemble in the Milky Way Galaxy—a vast pinwheel serenely set in space. The solar system resides in the disk, about halfway out from the center. Interstellar dust blocks the view, but we have uncovered a structure with a spiral pattern in the disk. This design results from density waves rolling through the interstellar medium—waves that promote star formation. The disk is encased by a shadowy halo, containing much dark matter. Within the core of the Galaxy lies the mysterious nucleus, where a small region emits a large amount of energy—perhaps powered by a supermassive black hole. The cores of other galaxies may well contain such collapsed masses.

To construct a model of the Galaxy requires much inference from many different kinds of observations. But basically we can apply Kepler's laws and Newton's law of gravitation to work out the distribution of mass and the total mass of the Galaxy. And we apply astronomical surveying techniques and our knowledge of stars to find the layout of stars and interstellar material in three dimensions.

In basic ways, our model reflects the images of other spiral galaxies viewed with large telescopes. And this model serves as a guide to understand other disk galaxies with spiral patterns. We presume that they, too, formed long ago from the gravitational contraction of huge clouds of gas and dust. Then their material assembled into a disk, nucleus, and a halo. Stars were born within these regions, some still alive, some already corpses. When the interstellar material is used up, a galaxy is essentially dead.

Key Ideas and Terms

disk (of a galaxy)
halo (of a galaxy)
nuclear bulge
nucleus
spiral arms

1. The main parts of the Galaxy are the encircling halo, the flat disk, and the central nuclear bulge, which contains the nucleus. The sun orbits the Galaxy at a distance of some 30,000 ly from the center at a speed of 220 km/s, with some uncertainty in both values.

Population I stars
Population II stars

2. Stars of different types inhabit three principal regions: the disk contains young, metal-rich (few percent) Population I stars; the nuclear bulge, metal-poor (few tenths of a percent) Population I stars; and the halo, metal-poor Population II stars.

cepheids
H I regions

3. The disk contains at least two and possibly four spiral arms, which are composed of concentrations of young stars, gas, and dust. The sun lies on the

H II regions
molecular clouds
period–luminosity
   relationship
spiral arms
spiral tracers

inner edge of one arm, and we can see pieces of other arms toward and away from the galactic center. We delineate spiral arms by the use of tracers found in them; such as cepheids and molecular clouds. The essential problem is to determine the distance to the object observed. Cepheid variable stars show a period–luminosity relationship, which is a powerful tool to infer the distances to cepheid variables.

dark matter
galactic rotation
   curve
Kepler's laws

4. The rotation curve shows how fast objects at different distances from the galactic center orbit around it. The failure of the observed curve to follow Kepler's laws indicates that a large fraction of the Galaxy s mass lies beyond the sun. This mass has not yet been seen directly and is in the form of dark matter in the halo.

Doppler shift
H I regions
molecular clouds
21-cm line

5. Radio astronomers use the Doppler shift in the 21-cm line from clouds of atomic hydrogen and molecular radio lines (such as carbon monoxide) from molecular clouds to infer the spiral arm structure of the Galaxy. This requires the assumptions that the clouds move along near-circular orbits (they don't!) and that the rotation curve has been well observed (it hasn't!).

density-wave model
spiral arms
spiral tracers

6. Any model for the evolution of the Galaxy must explain the persistence of spiral arms. According to the density-wave model, two spiral density waves disrupt the gas of the Galaxy's disk to promote the formation of spiral arms. As the density waves plow through the disk, different material condenses into the spiral arms as the old material dissipates. The persistence of arms is really an illusion.

nucleus (of a galaxy)
black hole
supermassive black
   hole
synchrotron
   radiation

7. The nucleus of the Galaxy emits intense radio and infrared radiation. It contains supergiant M stars, young massive stars, dust, and gas rotating at high speeds. To account for the motion of the gas requires a concentration of millions of solar masses of material in the inner few light years—perhaps a supermassive black hole.

dark matter
globular clusters
halo
Population II stars

8. The Galaxy's halo extends far beyond the disk. It contains globular clusters, individual stars, some gas, and the invisible objects that make up the mass that shows up in the rotation curve.

chemical
   composition
conservation of
   angular
   momentum
conservation of
   energy
gravitational
   contraction
halo

9. The Galaxy formed from the gravitational contraction of a large, slowly spinning cloud of gas and dust. The halo formed first, then the disk. As the cloud contracted, it became hotter and denser and fragmented into smaller pieces. We can estimate the relative sequence of formation from the heavy-element content of stars: the more metals they contain, the younger they are.

Review Questions

1. What part of the Galaxy marks its central region?
2. In what part of the Galaxy does the solar system lie?
3. Are stars like the sun metal-rich or metal-poor?
4. Are Population I stars generally metal-rich or metal-poor?
5. How many major spiral arms does the Galaxy have?
6. Name two spiral arm tracers.
7. Does the Galaxy's rotation curve follow that expected from Kepler's laws and the assumption of a concentration of mass in the center?
8. To infer the Galaxy's spiral structure from H I regions and molecular cloud, what do radio astronomers measure?
9. In the density-wave model, how do spiral arms form?
10. What feature of the nucleus leads to the inference of a supermassive black hole?
11. How are globular clusters arranged in the Galaxy's halo?
12. If a star were found to have a greater percentage of heavy elements than the sun, would it be younger or older?

Conceptual Exercises

1. What limits an optical astronomer's investigation of the Galaxy's structure?
2. Why are Population I cepheids especially good spiral arm tracers?
3. Radio astronomers need the rotation curve of the Galaxy in order to use 21-cm line observations to establish its spiral structure. Why?
4. Argue that a spiral arm cannot be a physical object.
5. What kinds of celestial object are found in spiral arms?
6. What characteristics of spiral arms are accounted for by the density-wave model? In what respects is the model at present inadequate?
7. What observational evidence do we have that a large fraction of the Galaxy's mass is not in the core, nor, in fact, within the radius of the sun's orbit?
8. Relate the orbits of globular clusters and their chemical composition to the birth of the Galaxy.
9. Some of the radio emission from the nucleus of the Galaxy is nonthermal. What does that imply about the physical conditions there?
10. What observational evidence and physical argument can be used to infer that a supermassive black hole may reside in the Galaxy's core? How firm is the evidence to date?

Conceptual Activity 15 The Period–Luminosity Relation for Cepheids

Let's use a selected set of data to find a period–luminosity relation for cepheid variables stars in the Galaxy. Table F.4 gives the information; for each star, we have its period of light variation in days and its average luminosity in units of solar luminosities. Note that these cepheids are much more luminous than the sun.

Table F.4

Cepheid Variables

Star	Period (days)	Luminosity (Sun = 1)
SU Cas	2.00	960
EV Sct	3.10	1,100
CF Cas	4.90	1,800
UY Per	5.40	2,500
CV Mon	5.40	2,300
VY Per	5.50	2,800
V367 Sct	6.30	3,500
U Sgr	6.70	3,900
DL Cas	8.00	3,700
S Nor	9.80	4,500
TW Nor	10.8	2,900
VX Per	10.9	5,900
SZ Cas	13.6	8,500
VY Car	18.9	11,200
T Mon	27.0	18,600
RS Pup	41.4	22,400
SV Vul	45.0	30,200

On a piece of graph paper, lay out a horizontal axis that is the period, ranging from zero to 50 days; try 5-day intervals as the major ticks along the x-axis. For the vertical axis, use a range from zero to 35,000 solar luminosities; try major intervals of 5000. With the axes set up, now plot a point representing each star.

When all the points are plotted, you should see a clear trend. How do you describe it? Try drawing by eye a straight line through the points, close to them all, with some above and some below the line. Such a line represents roughly the cepheid period–luminosity relation for this set of stars. Note that not every star's point falls right on this line; this scatter in the data gives you some sense of the actual range within the relation falls.

16
Normal Galaxies

■

Central Concept

The structure and content of galaxies and how are they distributed throughout the universe provide clues to their origin and evolution.

■

We now reach beyond our Galaxy to other galaxies—to find a universe of galaxies. They come in three basic shapes and a wide range of sizes. Many have a swirling spiral pattern like the Milky Way, which serves as a common model. Many others are midgets and lack a distinct structure. Almost all are bound in clusters, and these clusters are grouped in tremendous clusters of clusters. The galaxies form the fundamental pieces of our modern cosmological vista and make up the skeleton of the universe.

In our journey through the universe of galaxies we really touch on the universe itself. Because the galaxies are so distant, light takes millions to billions of years to reach us—so we see them as they were, not as they are now. This look back into time at many galaxies gives us hints about their evolution and their origins.

16.1 The Discovery of Galaxies

Since the nineteenth century, astronomers have hotly debated whether the *spiral nebulas* they saw with their large telescopes were simply clouds of gas within the Milky Way or were other galaxies far beyond it. The controversy came to a head in April 1920, when Harlow Shapley and Heber D. Curtis (1872–1942) debated the point publicly. Shapley (who worked out the location of the sun in the Milky Way) believed that the spiral nebulas were distant parts of the Milky Way. Curtis opposed Shapley and claimed that these nebulas were galaxies in their own right.

Curtis argued that since the wide range of angular sizes of spirals—approximately 2° (for Messier 31, the nearest) to 10 arcminutes and less for the smallest—required a large range of distances, they could not be part of our Galaxy. Assume that all spirals have roughly the same physical diameter. Then the range in observed sizes (more or less 10:1) implies that the spirals must be enormous distances from the Galaxy. If they were the same in diameter, the range in apparent sizes means that the ones 10 times smaller must be 10 times farther off, or about 10 times the radius of the Galaxy (Fig. 16.1). Therefore, because the distances of these spirals were much greater than the size of the Milky Way, the spirals could not be members of our Galaxy.

Curtis also noted the so-called *zone of avoidance*, a region near the plane of the Galaxy where very few spirals are visible. He argued that interstellar material in the galactic plane cuts out the light from the

FIGURE 16.2

True-color image of dust in the plane of a spiral galaxy (Messier 104 in the constellation Virgo). The dark band is due to dust cutting out starlight. Note the nuclear bulge at the center, surrounded by a halo of old, yellowish stars. Our Milky Way has a similar layer of dust, creating the *zone of avoidance*—a region of the sky in and around the plane of the Milky Way where no galaxies are visible because the dust blocks out their light. (Image by J. D. Wray; courtesy McDonald Observatory)

distant spirals; we see fewer in the zone of avoidance because there we look through the dusty disk of the Galaxy. If the spirals were actually parts of the Galaxy, they would be found concentrated in the plane—rather than avoiding it—along with the stars, galactic clusters, and H II regions. As evidence, Curtis cited photographs of spirals showing dark lanes cutting through their planes (Fig. 16.2). If the Milky Way Galaxy and other spirals were similar in structure, Curtis reasoned, then our Galaxy must also have obscuring material collected in the plane.

Finally, Curtis pointed out that the spectra of spirals are *not* bright-line spectra like those of emission nebulas (such as the Orion Nebula); rather, they resemble those from a group of stars. The spectra show faint dark lines against a bright background—the same spectrum the Galaxy would show if viewed from a great distance. This proved that galaxies are basically assemblies of stars.

In 1924 the American astronomer Edwin Hubble (1889–1953) settled the dispute conclusively by the discovery of cepheid variables in the outer parts of Messier 31. He derived a distance of 490,000 ly, far beyond the farthest globular clusters that marked the outer limits of the Galaxy. (Hubble's estimate was too small; recent work on cepheids in Messier 31 establishes a distance of 2.2 million ly.)

FIGURE 16.1

Distances to galaxies estimated by their angular sizes. If galaxies have roughly the same physical size, the more distant ones will have smaller angular sizes. In the case shown here, Galaxy B is about twice as far away as the Andromeda Galaxy.

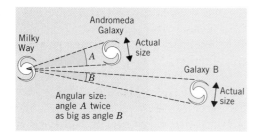

Other galaxies exist beyond the Milky Way Galaxy; some have the same kind of spiral structure; they all contain many stars.

A note about naming galaxies. Some have a number preceded by a capital *M* (such as M31 for Messier 31) and others with an *NGC* prefix; these refer to two catalogs that list galaxies. "M" is for the *Messier Catalog*, compiled by Charles Messier in the eighteenth century. It contains star clusters and nebulas in our Galaxy as well as the brighter galaxies visible from mid-northern latitudes. "NGC" stands for the *New General Catalog*, which was compiled in the nineteenth century with larger telescopes.

FIGURE 16.3

True-color image of a giant elliptical galaxy, Messier 84 in Virgo. Note the symmetry of the shape and the lack of contrasty detail or distinct structure. This giant elliptical of type E1 contains only old, yellowish stars. (Image by J. D. Wray; courtesy McDonald Observatory)

16.2 Types of Normal Galaxies

Hubble pioneered the field of extragalactic astronomy. To catalog the differences in the shapes of galaxies, Hubble in 1926 proposed a scheme for the classification of galaxies. Although this arrangement is now considered to be too simple, modern classifications still use Hubble's basic categories of *elliptical*, *spiral*, and *irregular* galaxies. Most galaxies fall into these categories—*normal galaxies* whose emission is mostly from starlight. (Chapter 17 describes the violent realm of active galaxies.)

Shapes of Galaxies

Let's look at each main type of galaxy in terms of its distinctive forms.

First, **elliptical galaxies** (Fig. 16.3)—those that exhibit an elliptical shape. Very little gas or dust appears in elliptical galaxies, and OB stars are also absent. The ellipticals generally have a reddish overall color. Hubble subdivided the ellipticals in classes from E0 to E7, according to how elliptical they appear. Imagine looking at a circular plate face-on; such is the appearance of an E0 galaxy. Now slowly tilt the plate so that it looks more elliptical and less circular. This flattening of shape presents the same views as the sequence from E0 to E7 galaxies.

Ellipticals come in a large range of sizes, from supergiants to dwarfs. The largest ellipticals, found in clusters of galaxies (Section 16.6), have diameters of a few *million* light years! In contrast, the smallest dwarf ellipticals span just thousands of light years in diameter.

In contrast to the bland shape of an elliptical galaxy, a **spiral galaxy** displays a spiral structure, usually with many segments of spiral arms (Fig. 16.4). One type of spiral has a prominent bar through the nucleus, the spiral arms winding out from the end of the bar (Fig. 16.5). Hubble termed the spirals without a bar *normal* and the others *barred*. (Our Galaxy is now known to be a barred spiral.)

Normal spiral are denoted S, and barred, SB. These are subdivided into categories a, b, and c. These types are judged by how tightly the spiral arms wind around (a, the tightest; c, the most open) and the relative size of the nucleus (a, the largest; c, the smallest). For example, the Hubble Sa is a normal spiral with a large nucleus and tightly coiled arms. A few galaxies appear to have the disk of a spiral but no arms. Hubble dubbed those S0. These are now sometimes called **lenticular galaxies** because of their lenslike shape.

We pay special attention to spiral galaxies because the Milky Way is a spiral galaxy. Recall (Section 15.4) that the density-wave model is a leading contender for an explanation of the evolution of spiral struc-

FIGURE 16.4

This large spiral galaxy with a small nucleus (NGC 253 in the constellation Sculptor) is Hubble type Sc, and one of the dustiest galaxies known. The galaxy appears to be elongated because we are viewing it only a few degrees above the plane of its disk. Two spiral arms are visible extending from the nucleus; bluish patches of star-forming regions can be seen within them. The nucleus has a yellowish cast as a result of the old stars there. (Image by D. Malin; courtesy Anglo-Australian Telescope Board)

FIGURE 16.5

Barred spiral galaxy: NGC 7479 in Pegasus. (*a*) False-color image, using red to bring out the spiral structure and bar. The blue shows the disk of the stars. (*b*) Additional processing with more colors highlights the bar, which crosses the nucleus and links the two spiral arms. (Images by R. Schild; courtesy Smithsonian Astrophysical Observatory)

(a)   *(b)*

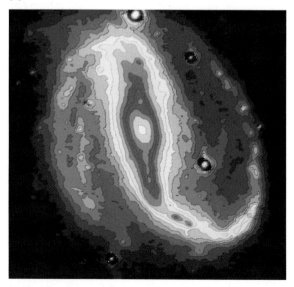

ture, the trigger for star formation from giant molecular clouds. Do observations of spiral galaxies support the density-wave concept? Generally, yes, for galaxies that show a distinct spiral structure (Fig. 16.6).

Judged by their observed shapes, galaxies come in three main types: spiral, elliptical, and irregular.

Finally, as a catch-all category, Hubble designated as **irregular galaxies,** those that were devoid of spiral structure or a symmetrical shape but were resolvable into distinct patches of stars (Fig. 16.7). These strange beasts fall into two groups. Irr I can be re-

FIGURE 16.7

Dwarf irregular galaxy: NGC 6822 in Sagittarius. Only the most luminous, bluish stars are visible; these were born fairly recently. At the right end lie a few clouds of glowing gas; these are H II regions. This galaxy is a member of the Local Group of galaxies and one of the closest to us, at a distance of 1.8 Mly. (Image by D. Malin; courtesy Anglo-Australian Telescope Board)

FIGURE 16.6

The spiral galaxy Messier 74 in the constellation Pisces. This ultraviolet view, taken in 1990 by NASA's Ultraviolet Imaging Telescope, accentuates the young, active regions of star formation, which trace out the main spiral arms. Messier 74 lies 55 Mly away; it is tilted almost face-on to our view. (Courtesy NASA/Goddard Space Flight Center)

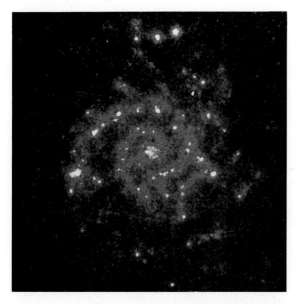

solved into OB stars and H II regions. Conspicuous dust clouds are usually absent. Irr II galaxies have the same lack of shape as Irr I. In addition, they are not resolvable into stars, have fewer visible H II regions, and usually do show prominent dust lanes.

This basic Hubble scheme does not include *all* types of galaxies. Some galaxies stand out as peculiar in shape (Fig. 16.8) and do not fit into the three general Hubble categories. Many of these peculiar galaxies turn out to have evidence of unusual activity. Some appear to be pairs of galaxies close together, interacting by tidal forces.

Luminosity Classes

The same Hubble type of galaxy, say Sb, comes in a range of luminosities (Fig. 16.9). So in analogy to stellar luminosity classes (Section 11.5), spiral and irregular galaxies also have **luminosity classes** of I, II, III, IV, and V, with I the most luminous and V the least. An Sc I galaxy, for instance, is a very luminous spiral with a small nucleus and spread-out arms. It turns out that luminosity class I galaxies are larger and more massive than class II, and so on. So class I

galaxies can be thought of as supergiant galaxies (in analogy to supergiant stars). Note that this scheme helps us to better organize the properties of galaxies.

Galaxies of the same type may differ in luminosity (and so in size and mass); generally, the more luminous are more massive and larger.

What does a census show about the main types of galaxies? It's difficult to make a complete one because faint galaxies are hard to see at great distances. One survey of a region of space out to 30 million light years (Mly) from us showed that only 34 percent of the galaxies in this volume are spirals, 12 percent ellipticals, and 54 percent irregulars. Hence, the majority of galaxies in the nearby universe are irregulars of fairly low luminosity.

To group galaxies by shape marks an initial step toward delving into the far depths of the universe. But to probe the physical properties of galaxies, their masses, sizes, and luminosities, ultimately depends on a knowledge of their distances from us—distances that are as difficult to survey as they are vast to imagine.

Other books say only 20-25% are irregular

FIGURE 16.8

The Southern Ring Galaxy, located in the constellation Volans in the southern hemisphere. The Ring lies about 270 Mly distant; its diameter is about 120,000 ly, about the same as the Milky Way Galaxy. This one galaxy represents a small class of galaxies called *ring galaxies,* which do not fit the Hubble scheme because of their peculiar shape. (Courtesy CTIO/NOAO)

16.3 Surveying Galaxies

We know the distances to most galaxies only roughly. For the most distant visible galaxies, we are lucky if we know their distances within 50 percent of their actual value. It's a hard but essential astronomical business to survey the distances in the universe of galaxies.

Galaxies **At a Glance**	Types	spiral, irregular, and elliptical (giant and dwarf)
	Relative numbers	irregulars and dwarf ellipticals most common
	Found in	clusters, numbering from a few to a few thousand galaxies
	Visible content	stars, gas, and dust
	Redshifts	increase with distance, indicating the expansion of the cosmos

FIGURE 16.9

The spiral galaxy Messier 51 and its companion. Messier 51 is classified as Hubble type Sb with a luminosity class between I and II. This photograph was processed to bring out the true colors in the different regions of the galaxy. Note how yellowish the nucleus appears compared to the spiral arms, in which the bluish regions show places of massive star formation in molecular clouds. At bottom is the companion galaxy connected to M51. (Image by J. D. Wray; courtesy McDonald Observatory)

Judging Distances

Although the distances to galaxies have continuously been revised, the essential techniques remain the same. The initial tips are that *brightness means nearness* and *smallness means farness*. Galaxies with the smallest angular size tend to be the most distant. And faint galaxies also tend to be far away. By applying these simple gauges, you can make rough estimates of the relative distances to galaxies.

For example, if you look at two galaxies, one with an angular diameter half as large as another, and if both are actually about the same size, the smaller galaxy is twice as distant as the larger one. A similar argument applies to relative brightness: if one galaxy appears to be 100 times fainter than another, it must be roughly 10 times farther away (by the inverse-square law for light, Section 11.1).

As a rough rule, the smaller and fainter a galaxy appears, the farther away it is from us.

Refining this rough first approach requires the use of the known physical properties of stars and galaxies inferred from theoretical models and careful observations. Each step in surveying the universe applies to certain objects and over a certain range of distances; astronomers work step by step to establish a distance scale using a variety of distance indicators.

Distance Indicators

To bridge the distances to other galaxies, we assume that the essential character of objects in our Galaxy (such as cepheid variable stars or supernovas) is the same as that of similar objects in other galaxies. Then, to find distances to galaxies, we must use identifiable objects (within galaxies) whose *luminosities* we know. We compare their fluxes with their luminosities to infer their distances. Even the largest telescopes have limits, and some objects are too faint to be picked up. So we want to choose the *most* luminous objects in galaxies to use as long as we know what these luminosities are.

As an analogy, imagine that you know that all street lights have the *same* luminosity, say 500 watts. Then as you look out at night at a city, you can judge

	Object	Current Limit of Use (Mly)
Distance Indicators	Red giants	3
	Cepheids	20
	Supergiant stars	30
	Novas	65
At a Glance	H II regions	80
	Supernovas (Type I)	650
	Brightest galaxy in a cluster	1500

the distances to the street lights in different locations by measuring their flux with a light meter. The inverse-square law for light (Section 11.1) then allows you to find the distance from the measured flux and the assumed luminosity—if you assume that no dust dims the light.

Starting with the closer galaxies, we apply the period–luminosity relationship (Section 15.3) to cepheids in other galaxies. That way we find the luminosities of cepheids observed in other galaxies. The cepheids, however, are useful over a limited range, for they are not especially luminous. Of all visible galaxies, only about 30 are of the right kind and close enough for us to detect their cepheids. We can use cepheids as standards out to roughly 20 Mly.

To go beyond this limit, we follow the same strategy: find fairly common bright objects, calculate their luminosities in our own or nearby galaxies, check other galaxies by methods known to be reliable, and then utilize the standard to the limits of its accuracy.

By checking luminosities of distance indicators, we can estimate the distances to galaxies, using the inverse-square law for light.

At very great distances, we can no longer see individual objects in galaxies. What next? We use the luminosities of the *galaxies* themselves! Galaxies tend to lie in clusters (Section 16.6). To ensure choosing galaxies with the same luminosity, we select one of the brightest galaxies in a cluster rather than picking one at random. The brightest galaxies in clusters have about the same luminosity, *if* you stick to the same *kind* of galaxy. Supergiant spiral galaxies serve as the most far-reaching standards—in particular, the Sc I galaxies. The luminosity of such galaxies is about 25 billion times the sun's luminosity. Visible at great distances, these galaxies are relatively easy to identify because of their distinctive shape. With current telescopes, they can be used as standards to distances of roughly 10 billion light years.

16.4 Hubble's Law and Distances

A prime discovery of twentieth-century astronomy is that the universe is *expanding*—a discovery in which Hubble played a major role. Observations with large telescopes provide two facts about galaxies: their distances and their velocities relative to us. The velocities measured are those along our line of sight only, either approaching or receding; for all distant galaxies, the velocities are those of recession. Remarkably, the distances and velocities of galaxies are tied together—firm evidence that the universe is expanding.

In honor of Hubble, that expansion is described by **Hubble's law**: *the farther a galaxy is from us, the greater its velocity of recession.* The number that relates the recessional speed and the distance is **Hubble's constant**, *H*. Finding the value of *H* relies on knowing the distances to galaxies (Section 16.3) and the Doppler shifts in their spectra (Section 9.5). Almost all galaxies show redshifts in their spectra, which indicate that they are receding from us. It turns out that general relativity (Section 10.7), when applied to models of the universe (Chapter 18), properly describes this expansion.

The universe is expanding; we know this from observations of distances and redshifts of galaxies that show a distinct trend stated by Hubble's law.

To understand Hubble's law better, try turning it around: galaxies that are moving fast are far away.

That makes pretty good sense. Because all the galaxies in the universe have been moving away from one another for the same amount of time, the ones that started out moving fastest relative to us will be the farthest away. As an analogy, consider a race with three people all starting at the same time but proceeding at different speeds. Runner Mary goes twice as fast as runner Joe, who moves twice as fast as runner Kay. After some time has elapsed, Mary will be twice as far from the start as Joe, and he will be twice as far as Kay.

) poor analogy

Redshifts and Distances

It is relatively easy nowadays to measure the redshift of a galaxy from the shift in some prominent spectral lines (such as the H and K absorption lines of calcium; Fig. 16.10). The redshift indicates the recessional velocity of a galaxy. For example, look at the galaxies pictured in Figure 16.10 and at their spectra with redshift indicated: as the redshifts get larger, the galaxies appear smaller and fainter. Now plot the distances for these galaxies versus their redshifts (Fig. 16.11). You can draw a straight line that represents the trend of these points and so find a value for Hubble's constant, which is the slope of the line (see Conceptual Activity 16). Once we have a value of Hubble's constant, we can use Hubble's law to find the distances to galaxies beyond the limits described in Section 16.3. Note that the value of Hubble's constant will have units of speed (km/s) divided by distance (a million light years)—kilometers per second per million light years, or km/s/Mly.

The value of Hubble's constant, found from measurements of galaxies' redshifts and distance, tells us the rate at which the universe is expanding.

We reverse this procedure to find distances from redshifts. Suppose you were given the redshift of a galaxy. You look up this value on the redshift axis of your plot and see the distance to which it corresponds. If the redshift is larger than measured previously, you can assume that the line you've drawn for the plotted points can be extended farther out and still be valid. Suppose a galaxy's measured redshift is

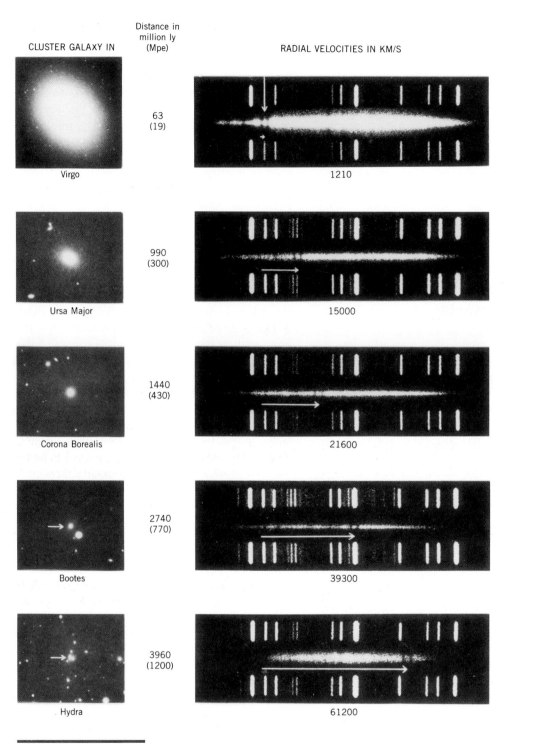

| CLUSTER GALAXY IN | Distance in million ly (Mpe) | RADIAL VELOCITIES IN KM/S |

Virgo — 63 (19) — 1210

Ursa Major — 990 (300) — 15000

Corona Borealis — 1440 (430) — 21600

Bootes — 2740 (770) — 39300

Hydra — 3960 (1200) — 61200

FIGURE 16.10

Measured redshifts and distances for selected galaxies in order of increasing redshift and distance (from top to bottom) from a galaxy in the Virgo cluster to one in the distant Hydra cluster. *Right:* spectra showing redshifts; a white arrow indicates the size of the shift in the H and K lines of calcium, the two darkest lines in the spectra. Below each spectrum is the radial velocity in kilometers per second. Above and below each spectrum are emission-line spectra that serve as wavelength markers for a source at rest. *Left:* the galaxies are in clusters of galaxies, named for the constellation in which they appear. In general, the galaxies that are farther away appear smaller than the closer ones. (Courtesy Palomar Observatory, California Institute of Technology)

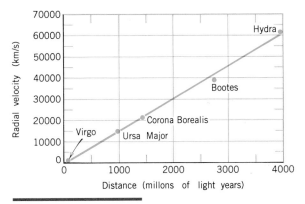

FIGURE 16.11

Hubble plot using the data from Figure 16.10. The straight line is a "best fit" through the points; its slope is 15 km/s/Mly, which is the value of the Hubble constant derived from these data.

40,000 km/s. What is its distance? If H equals 20 km/s/Mly, then we get 2×10^9 ly.

This indirect method of distance measurement rests on one crucial point: that the galaxies with known distances give an accurate value for Hubble's constant! Current estimates of Hubble's constant range from 15 to 30 km/s/Mly. The distances derived by this method can vary by a factor of 2, depending on which value of Hubble's constant is used! Beware of this range in distances calculated by this indirect method.

Many groups of astronomers have tried different strategies to find value for Hubble's constant, and still *we do not know it well.* The values tend to fall into two categories: on the smaller side, 15 km/s/Mly (with an estimated error of 10 percent); or a larger value, 27 km/s/Mly (with an alleged uncertainty of 5 percent).

Despite its importance, we do not yet have a firm value for the Hubble constant; it probably lies in the range of 15 to 30 km/s/Mly.

Note: I would like to use 20 km/s/Mly as the value of the Hubble constant in this book. However, given the range of possible values, many astronomers adopt other specific values. When I quote their work here, I will provide the value for H that was given in the original research. Don't let this variation throw you!

16.5 General Properties of Galaxies

Let's step back a bit from our cosmological vista and study the characteristics of galaxies (Table 16.1 gives typical values for the basic types), now that we have methods to find out their distances.

Sizes and Masses

Once you know the distance to a galaxy, you can find its actual size from a measurement of its angular size. The hitch here is that the definition of the edge of a galaxy is more or less arbitrary—and hard to measure well! Despite this problem, we can still make general remarks about the sizes of galaxies. Dwarf ellipticals and small irregulars tend to be the smallest galaxies—only a few thousand light years in diameter. Giant ellipticals can range up to 200,000 ly in size. To put this into perspective, imagine your height to be the size of a dwarf galaxy. Then an irregular galaxy would be about twice your size, a spiral 10 times your size, and a giant elliptical some 20 times your size!

The very largest galaxies are the *supergiant* ellipticals. These can have diameters up to 6 Mly—greater than the distance from our Galaxy to the Andromeda Galaxy (M31)! These supergiant elliptical galaxies tend to define the gravitational centers of clusters of galaxies (Section 16.6).

The light from a galaxy comes mostly from its stars, but much galactic material does not emit visible light—or does not emit at all (black holes, for example). This *dark matter* may inhabit the halos of most (all?) spiral galaxies—as we found it inhabits the halo of the Milky Way (Section 15.6).

A most widely used method of finding a galaxy's mass is measuring its rotation curve, which can be done only for spiral galaxies. The rotation curve method works as follows. A galaxy's material orbits the nucleus in a specific way, so at every distance the orbital velocity has a certain value. This rotation

Table 16.1

Typical Properties of Galaxies

Property	Spirals	Irregulars	Dwarf Ellipticals	Giant Ellipticals
Diameter (ly)	90×10^3	20×10^3	30×10^3	150×10^3
Mass (solar masses)	10^{12}	10^8	Not well known	10^{13}
Luminosity (solar luminosities)	10^{10}	10^9	10^8	10^{11}
Color	Bluish (disk), reddish (halo and nucleus)	Bluish	Reddish	Reddish
Neutral gas (fraction of mass)	5%	Greater than 15%	Less than 1%	Less than 1%
Star types	Young, metal-rich stars (spiral arms); old, metal-poor stars (halo); old, metal-rich stars (disk and bulge)	Young, metal-poor stars	Old, metal-poor stars	Old, metal-rich stars

curve comes from the orbital motions of stars, described by Newton's laws, which arise from the distribution of mass within the galaxy. (Recall the rotation curve for our Galaxy: Section 15.2).

So if we observe a galaxy's rotation curve and make up a model for that galaxy's mass distribution, we can work out the galaxy's total mass, including dark matter that we can't see directly. Spiral galaxies hit masses as high as several 10^{12} solar masses, although their rotation curves become flat without dropping off out to the extent measured so far (Fig. 16.12). That is, the value of the rotational speed remains constant even when the distance from the center increases—we have not yet found the end to the mass in the halo, even at distances of 100,000 ly. This fact has reinforced the idea that spiral galaxies have massive dark halos surrounding their starry disks. Otherwise the rotation curves should decline, as expected from Kepler's third law.

In general, giant elliptical and supergiant spiral galaxies are the most massive; dwarf elliptical and irregular galaxies the least massive.

Luminosities and Mass–Luminosity Ratios

If we know the galaxies' distances and fluxes, we can work out their luminosities from the inverse-square law for light. One trouble here is that it's not easy to measure a galaxy's total flux because a galaxy thins out gradually at its edge. In addition, corrections have to be applied for light absorption: first, for that due to dust in our Galaxy, and second, for absorption from dust in the observed galaxy itself (especially for spirals). Finally, since we view most galaxies tilted to our line of sight, we measure only a fraction of their total light output, and we must correct for this imprecision.

We can find out the luminosities of galaxies from their total fluxes and distances.

The luminosities of galaxies range from 10^5 solar luminosities for the smallest dwarf ellipticals to 10^{12} solar luminosities for the largest **supergiant (cD) elliptical galaxies**. These latter types are very rare,

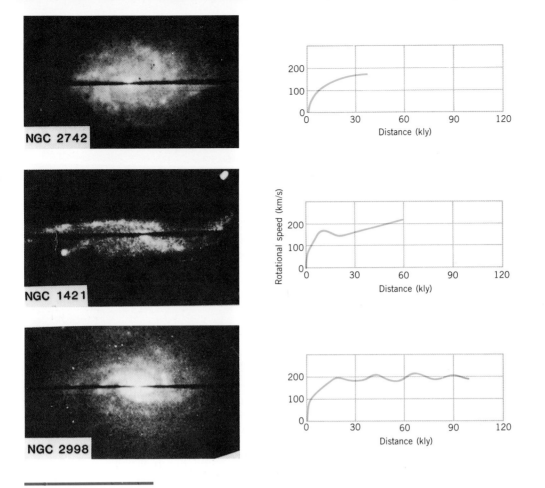

FIGURE 16.12

Actual rotation curves of the three spiral galaxies shown at left. The rotation curves are plotted with the rotational speed against distance from the center of the galaxy. (Courtesy V. Rubin)

however. The Milky Way Galaxy, if we could see it all from space, would exhibit a luminosity somewhere near 2.5×10^{10} solar luminosities.

How much matter is pumping out this power? Divide the total mass of a galaxy by its luminosity and you then have its **mass–luminosity ratio** (ab-breviated *M/L* and usually expressed in units of solar masses and solar luminosities.) Modern observations give 35 for the average mass-to-luminosity ratio for spiral galaxies, and about twice as much for giant ellipticals and lenticulars. For comparison, the *M/L* ratio for stars in the sun's neighborhood is about 1 or

	Object	Mass to Luminosity
	Sun	1.0
Mass–	Nucleus of spiral galaxy	1–3
Luminosity	Disk of spiral galaxy	2–5
Ratios	OB stars	10^{-4}
	Old red main-sequence stars	20
At a Glance	White dwarfs	100
	Open star clusters	1.0
	Globular star clusters	1.0

a bit larger; that's because these bodies are solar-type stars. A galaxy made of such stars would have the same value. If it contained only B stars, the *M/L* would be about 0.01; if all M stars, about 26. Note that dark matter adds to the mass but not to the luminosity, so it makes the ratio larger. Ellipticals have a larger *M/L* ratio than spiral galaxies because they contain a greater percentage of low-mass stars with low luminosities—main-sequence stars of class M.

The larger its mass–luminosity ratio, the more dark matter a galaxy contains.

Colors

As for stars, we can measure the colors of galaxies. The color of a galaxy depends directly on the stars it contains. For example, a galaxy with many OB stars is bluer than a galaxy with few such stars. So a galaxy's type and its color are related. Ellipticals tend to be much redder than spirals, and spirals redder than irregular galaxies. Within the spiral group, the galaxies appear redder as their nuclear bulges grow larger and their spiral arms less extensive. The progression of color from the bluer irregulars to the redder ellipticals reflects a trend in the galaxy's stellar populations. Bluer galaxies have more young, bluish stars; redder galaxies, older reddish stars. In general terms, older stars predominate in ellipticals, whereas much younger ones stand out in the irregulars.

The bluer a galaxy, the more young stars it contains.

Ellipticals contain very little gas and dust. In contrast, spirals and irregulars embrace extensive quantities of interstellar material. In spirals, the gas is most obvious as H I clouds or as H II regions cloaking OB stars. So a galaxy's color also relates to its overall content of gas and dust. The reddest galaxies, ellipticals, contain almost no gas and dust. The bluest galaxies, irregulars, contain the greatest percentage of gas and dust relative to their total mass.

Note that the sequence from ellipticals to irregulars marks a sequence from galaxies in which star-birth ceased long ago (ellipticals: smallest percentage of gas and dust) to those in which starbirth ensues very actively (irregulars: largest percentage of gas and dust).

Galaxies contain stars, gas, and dust; the mixture of gas, dust, and stars, as well as the stellar types, relates closely to a galaxy's type and structure.

16.6 Clusters and Superclusters of Galaxies

On wide-angle photos of the sky in regions where the stars thin out, you can see the tiny forms of galaxies. If you find one galaxy, you're likely to see others nearby (Fig. 16.13). You have learned that galaxies tend to come in **clusters**. In fact, it may be true that *all* galaxies belong to clusters—though many of

FIGURE 16.13

True-color image of the small group of galaxies called Stephan's Quintet. Note the various shapes. The stars visible as points of light are foreground stars in the Milky Way; fuzzy images are galaxies. Two galaxies are visible next to each other at the center; third is to the upper left; the two remaining galaxies are seen to the lower left and to the lower right. (Image by J. D. Wray; courtesy McDonald Observatory)

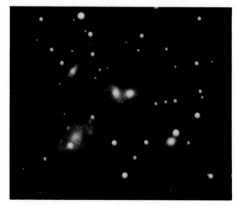

Clusters of Galaxies	Number of galaxies	about 100
	Total mass	10^{14} to 10^{15} solar masses
At a Glance	Extent	few million light years
	Distance to next cluster	few tens of millions of light years

these clusters may be a simple marriage of two galaxies. Our universe is one of clusters of galaxies!

When we measure the redshifts of galaxies, we find that the expansion of the universe really involves the recession of clusters of galaxies, *not* the individual galaxies within the cluster (and certainly *not* the stars within the galaxies). Now, if we know the value of the Hubble constant, we can estimate distances from measured redshifts. So redshifts probe the depth of the universe—they provide the crucial third dimension to the distant galaxies visible in two dimensions on the sky.

The Local Group

The cluster of galaxies to which the Milky Way Galaxy belongs is called the **Local Group of galaxies** (**Local Group** for short). This small cluster takes up a volume of space nearly 3 Mly across in its long dimension. Our Galaxy is located near one end of the Local Group, and M31 is near the other (Fig. 16.14).

As the most massive and luminous objects in the cluster, the Milky Way Galaxy and M31 dominate its motions and secure the other members gravitationally. In fact, the Galaxy and M31 orbit each other. The Local Group contains about 30 galaxies; they are mostly ellipticals. Some are quite faint; they are dwarf ellipticals. The dust in the Milky Way probably blocks our sight of other members, especially faint ones.

Within the Local Group, the *Large* and *Small Magellanic Clouds* lie closest to the Galaxy. Both are near enough that we can see individual stars. The Large Cloud lies at a distance of 160,000 ly; the Small Cloud at 200,000 ly. Both are, in fact, connected to our Galaxy by a bridge of hydrogen gas. The two clouds are physically connected by a large but thin envelope of neutral hydrogen and a thread of stars. Both are distorted by tidal interactions with our Galaxy.

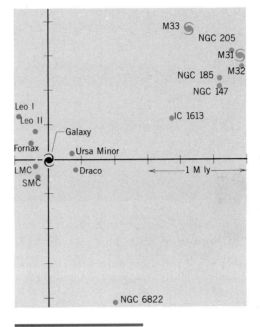

FIGURE 16.14

The most prominent galaxies in the Local Group of galaxies, as viewed looking down on the plane of the Milky Way Galaxy. The distance scale is in units of 330,000 ly. Note that most of these are small elliptical or irregular galaxies. Our Galaxy, Messier 32, and Messier 33 are the largest.

The Large Magellanic Cloud (LMC) contains stars totaling about 20 billion solar masses (Fig. 16.15). The LMC is a medium-sized galaxy in its own right, with a diameter about half that of the Milky Way and a tenth of its total mass. Its shape is irregular, with a hint of barred spiral structure (perhaps as many as six or seven ill-defined arms) but without an obvious nucleus. (Supernova 1987A exploded in the LMC: Section 14.4.) The Small Magellanic Cloud (SMC) has a total mass of about one-

FIGURE 16.15

The Large Magellanic Cloud has a hint of a horizontal bar; note the many bluish regions of young stars. (Courtesy CTIO/NOAO)

FIGURE 16.16

The Small Magellanic Cloud is an irregular galaxy with no clearly defined structure and relatively fewer young stars than the Large Magellanic Cloud. Near the right edge of the photo is a globular cluster in the Milky Way. (Courtesy CTIO/NOAO)

fourth that of the LMC and is only a little smaller—about one-third the diameter of the Galaxy. Like the LMC, the SMC has an irregular shape with a bar but no nucleus at its center (Fig. 16.16).

Like the Milky Way, M31 is a large spiral galaxy. It tilts 15° to the line of sight, and so dark lanes of dust are plainly visible along with the spiral arms marked by OB stars (Fig. 16.17). This open view provides us clear access to identify many spiral tracers in the arms (Section 15.3). A halo of globular clusters surrounds M31 like bees around a hive, in a distribution like that around our galactic system. The Andromeda Galaxy also has companions—a total of seven dwarf ellipticals that orbit it.

The Local Group, which contains about 30 galaxies spanning some 3 Mly, is dominated by the Galaxy and Messier 31.

Other Clusters of Galaxies

Other clusters range from compact ones to rather loose arrays of galaxies. The huge Coma cluster (Fig.

FIGURE 16.17

The large spiral galaxy Messier 31, showing its spiral arms. Note the dark dust lanes in the disk. One companion galaxy lies at the upper right. (Courtesy J. Riffle, Astro Works Corporation)

16.18) spreads over at least 20 Mly of space and contains thousands of galaxies. Observations of even just the brightest galaxies show how common clustering appears (Fig. 16.19a). From these observations, we find that a typical cluster contains about 100 galaxies

and is separated by some tens of millions of light years from its neighboring clusters.

The Virgo cluster stands out as one of the most stupendous in the sky (Fig. 16.19b). It covers about 7° in the sky (14 times the diameter of the moon!),

FIGURE 16.18

The central region of the Coma cluster of galaxies. This image has been processed to show the true colors of the galaxies. Note the two distinctly yellowish supergiant ellipticals, one near the center and one at the left. About 300 galaxies are visible here. The bluish star above the center lies in our own Milky Way Galaxy. (Courtesy L. Thompson, Institute for Astronomy, University of Hawaii)

which implies that its physical diameter is some 10 Mly at a distance of roughly 50 Mly. The Virgo cluster is so massive and so close that it influences the Local Group gravitationally; we are moving in the direction of the Virgo cluster.

Almost all galaxies are found in clusters, which contain from a few to thousands of galaxies spread over millions of light years.

As in the Local Group, galaxies in clusters have a range of luminosities: there are many faint galaxies and few bright ones in a cluster. Some of a cluster's mass resides in very faint galaxies. And some is dark matter of an unknown form. It's difficult to estimate the masses of these clusters because not all the material in them can be seen at visual wavelengths; so adding up all the galaxies gives a lower limit to the cluster's mass. On the other hand, if the cluster is assumed to be bound by gravity, the motions of the galaxies within it establish an upper limit on its mass. (The actual value lies between these two limits.) Masses range from 10^9 to 10^{15} solar masses.

You can use the M/L ratio to consider the amount of dark matter in clusters. Take the Coma cluster (Fig. 16.18) as an example: two supergiant galaxies lie near the center, and their huge masses collect matter around them. By studying the velocities of the galaxies in the inner 4 Mly, we estimate a mass of almost 10^{15} solar masses in this region. Yet, when we count up the luminosities of visible galaxies, we find that the M/L ratio is 300! (The typical range for clusters is 300 to 500.) Given the types of galaxies here, we expect a value of 10. So visible stars make up only a few percent of the mass in the Coma cluster.

Clusters of galaxies contain a large amount of dark matter.

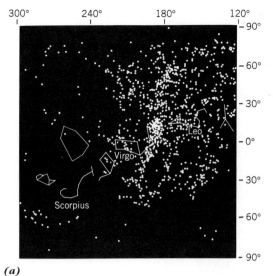

(a)

(b)

FIGURE 16.19

Clustering of galaxies and the Virgo cluster. (*a*) Map of a large angular region of the sky near Virgo. (Only a rough outline of the zodiacal constellations is shown.) Each dot represents the position of a galaxy. The dense clump in Virgo turns out to be the Virgo cluster. This map gives only two dimensions; actual clustering must be confirmed by measuring the distances to the galaxies, usually by finding their redshifts. (Map generated by *Voyager* software.) (*b*) Small part of the central region of the Virgo cluster of galaxies. The giant elliptical galaxy Messier 86 lies just to the upper right of center; to its right is another giant elliptical galaxy, Messier 84. (Courtesy KPNO/NOAO)

Superclusters and Voids

Are there clusters of clusters of galaxies—do **superclusters** exist? Yes! A **Local Supercluster** has a di-
ameter of some 100 Mly. It contains the Local Group as well as the Virgo cluster, among others, for a total mass of some 10^{15} solar masses. The Local Supercluster appears to be somewhat flattened. The

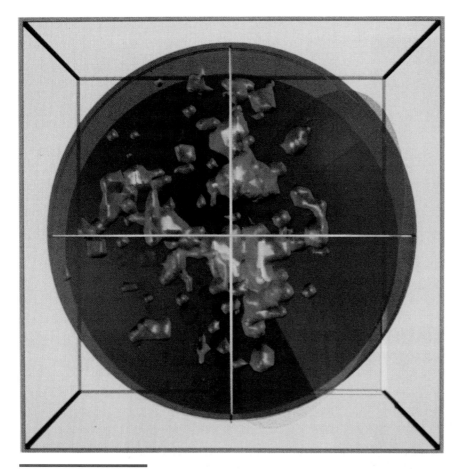

FIGURE 16.20

Distribution of rich clusters of galaxies in space. This three-dimensional, computer-generated view is edge-on to the Local Supercluster from a great distance out; the sphere has a radius of about 1000 Mly and contains 382 rich clusters of galaxies. The orange contour is the highest density of clusters per volume; the green contour, the lowest. Most of the clusters merge together with voids of space in between. (Courtesy Brent Tully, Institute for Astronomy, University of Hawaii)

center of mass lies in or near the Virgo cluster; our Galaxy and M31 lie on the outskirts. Other superclusters have sizes exceeding this—greater than 100 Mly.

Astronomers have mapped the Local Supercluster, in three dimensions, by measuring the redshifts of a few thousand galaxies (Fig. 16.20). The map shows a rich, convoluted structure that breaks into two main clouds of galaxies with streamers (thin, cigar-shaped clouds of galaxies) emerging above and below the central plane. Most of the Local Super-

cluster is empty space: 98 percent of the visible galaxies are restricted to 5 percent of the volume. These clouds of galaxies outline a disklike structure six times wider than it is thick—a true cosmic pancake of clusters of galaxies!

> Superclusters are clusters of clusters of galaxies; the Local Group is a small part of the Local Supercluster.

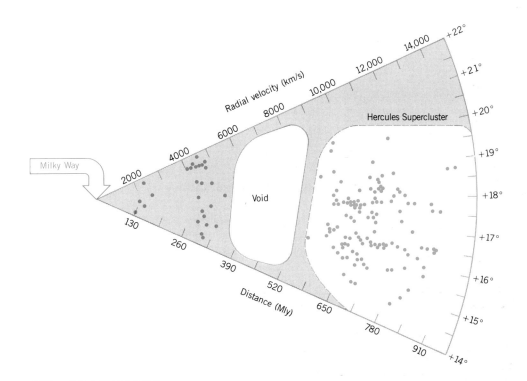

FIGURE 16.21

Slice of the Hercules supercluster, showing the radial velocities from measured redshifts (and so distance, along bottom, for *H* = 15 km/s/Mly used here) and positions on the sky for the galaxies in the supercluster, which contains a number of clusters of galaxies. The Milky Way lies at the vertex of this wedge diagram, which shows the region of space that was surveyed. Note the clumping of the galaxies and the void of space in front of the supercluster. The view spans about 8°. (Adapted from a diagram by S. A Gregory and L. A. Thompson)

Astronomers have probed a few superclusters beyond the local one. For example, the Hercules supercluster some 700 Mly distant covers more than 600 billion cubic light years of space (Fig. 16.21). The supercluster itself occupies a broad band, spreading out a distance of 400 to 600 Mly. In front of the supercluster lies a void some 300 Mly deep that separates the supercluster from foreground galaxies. The vast regions in which no obvious galaxies are found are common features of superclusters; they are called **voids**. This supercluster contains two rich clusters that seem to be linked with a curving filament of galaxies. In front of them lies a void and a chain of smaller clusters of galaxies.

The superclusters that have been well-mapped to date exhibit common features (Fig. 16.22). First,

they confirm the existence of superclusters as organized structures composed of multiple clusters of galaxies. Second, they contain large, mainly spherical voids in which no (or very few) visible galaxies exist. These voids may be an integral part of the process that forms superclusters—a process about which we have only vague ideas right now. Third, filaments or strings of galaxies appear to connect the main concentrations in superclusters—and may stretch for a *billion* light years! Fourth, the main body of superclusters has a flattened structure (mostly gently curving filaments), sometimes with a pancake shape.

Finally, keep in mind that we have examined just small slices of the cosmos in three dimensions. For instance, the Galaxy and the Local Group appear to be moving at speeds of a few hundred kilometers per

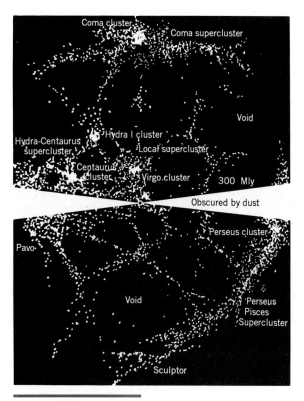

F I G U R E   1 6 . 2 2

Nearby superclusters. The Galaxy lies in the center of this schematic representation, which spans 600 Mly in space. Each dot represents a galaxy; at the cores of clusters, it is not possible to plot them individually. Our view of the regions to the left and right is obscured by dust in the plane of the Milky Way. The Local Supercluster is dominated by the Virgo cluster. Note the immense voids between us and the Coma supercluster. (Adapted from a diagram by A. P. Fairall)

second toward the Hydra–Centaurus region of the southern sky. This observation has resulted in the proposal of the "Great Attractor"—a chain of galaxies covering about a third of the southern sky. A survey of some 17,000 galaxies here reveals a grouping of galaxies at a redshift of 4000 km/s. Called the Hydra–Centaurus supercluster, its great mass may account for the motion of the Local Supercluster in that direction.

Superclusters have a flattened structure, contain filaments of galaxies and voids, and stretch over hundreds of millions of light years.

Interacting Galaxies

The spacing of galaxies in clusters is awfully close, compared to the sizes of the galaxies themselves. Consider, for instance, planets and stars in terms of sizes and spacing. In the solar system, the planets are spaced out about 100,000 times their diameters. In the Galaxy, stars are spaced out about a million times their diameters. But in a cluster of galaxies, the spacing amounts to only about 100 times a typical galaxy's diameter. (Relative to the sizes of people, the spacing is about that between the ends of a football field.) Astronomically speaking, galaxies in a cluster are jammed together. So you might wonder if they would ever pass close by one another.

Consider this additional fact: the most massive galaxies (supergiant ellipticals) are at least 10 million times more massive than the least massive ones (the dwarf ellipticals). Then it's not hard to imagine that tidal forces (Section 6.1) from the largest galaxies could disrupt the smallest ones strongly enough to destroy their structure and then pull the pieces in. Astronomers call this devouring of a smaller galaxy by a larger one **galactic cannibalism**.

Observations show that the supergiant elliptical galaxies have extensive halos, up to 3 Mly in diameter, and multiple nuclei. These properties, when linked with the motions of supergiant galaxies within clusters, suggest that these special galaxies were formed from galactic cannibalism. How? By close encounters at the centers of clusters that tidally stripped from other galaxies material that was picked up to promote the growth of a supergiant galaxy.

Galaxies do not actually have to merge to show the effects of tidal forces. A modestly close encounter is good enough. Matter would be pulled out in tongues from both sides of each galaxy; and because galaxies rotate, their material after an encounter would flow off in arc-shaped streams. So we expect that bridges of material might join two tidally interacting galaxies

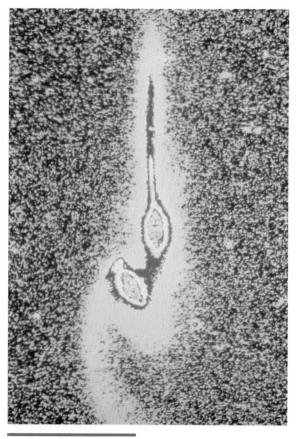

FIGURE 16.23

The galaxies NGC 4676a and 4676b, a tidally interacting pair. The false colors in this computer-enhanced photo bring out the details of the interaction, such as the bridge of material (red) between the two galaxies. Note the tails of material extending out of each. (Courtesy KPNO/NOAO)

with tails pointing away from each in opposite directions. Some galaxies with peculiar shapes that do not fall into the standard Hubble categories show indications of tidal interactions (Fig. 16.23).

Galaxies in clusters often come close enough to experience interactions from tidal forces.

Computer simulations indicate the possibility that collisions can transform spiral galaxies into elliptical ones. The interaction of stars, gas and dust, and a halo of dark matter results in the merger of the two

FIGURE 16.24

A Hubble Space Telescope (Appendix H) true-color image of the nuclear region of the galaxy NGC 1275. The white dot just above the center is the nucleus. The fainter bluish dots are young globular clusters. NGC 1275 is a strong radio source and a giant elliptical galaxy. It may well have been formed by the merger of two galaxies about 1 billion years ago, at which time these globular clusters were formed. (Image by J. Holtzman, courtesy NASA/ESA)

galaxies into a single one. The collision drives vast amounts of gas into the center of the resulting elliptical mass, which might spur the formation of a supermassive black hole there. The gas may also rapidly form into clusters of stars (Fig. 16.24).

Tidal interactions may well trigger the short, large-scale episodes of star formation seen in many irregular galaxies—sometimes called **starburst galaxies**. In particular, the gas and dust throughout a galaxy respond strongly to the tidal forces of a companion, even without an actual merger of the two, such that the gas becomes unstable and starts to clump into stars.

16.7 Intergalactic Medium and Dark Matter

Can the invisible matter hide in some kind of **intergalactic medium** in *between* the clusters of galaxies or within clusters of galaxies? If an intergalactic medium is present, it may contain both gas and dust. The gas (probably hydrogen) may be in neutral form, or it may be ionized. We can look for the intergalactic gas in the same way that we observe the interstellar gas (Section 12.1).

To get some idea of how much material might be in an intergalactic medium, imagine the following. Take the matter from all the galaxies we can see and spread it out over the entire volume of space we can observe. This spread-out material would have a density of about 4×10^{-28} kg/m^3. (That's about one hydrogen atom every 10 cubic meters—a volume roughly that of a house.) For the intergalactic medium to be significant, it would need to be about this density—equivalent to that of visible matter in the form of galaxies.

Ionized hydrogen (H II) is the most likely candidate for the intergalactic medium. Because intergalactic material does not have a high density, ionized hydrogen will basically remain ionized. If it is very hot (a few ten million kelvins), it will emit x-rays. X-ray observations of local superclusters show such sources centered on clusters. This implies that a hot gas exists in superclusters, that it is highly clumped, and that little gas exists between clusters. The gas has a temperature of some 10^8 K.

Hot plasma within clusters of galaxies is visible from its x-ray emission.

A good candidate for matter *within* clusters is (again!) ionized hydrogen gas. X-ray observations back up this idea. At least 40 clusters of galaxies are known to date to emit x-rays (Fig. 16.25). The x-ray luminosities of clusters range from 10^{36} to 10^{38} W. The sizes of the x-ray-emitting regions range from 160,000 ly to 5 Mly in diameter. They tend to lie in the central regions of the clusters.

A reasonably confirmed model for this x-ray emission is that it comes from hot, ionized gas. Such a plasma tends to settle around the center of mass of a cluster. It has typical temperatures of 10 to 100 million kelvins and densities of about one ion per cubic meter; such properties nicely explain the x-ray observations. So we have reasonable evidence of intergalactic gas in clusters—about equal to the amount

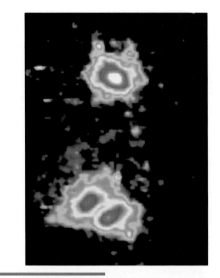

FIGURE 16.25

False-color map of the x-emission from superclusters of galaxies. Blue represents the lowest level of emission; white, the highest. The cluster at top is separated about 20 Mly from the double cluster below. The x-ray emission comes from a hot plasma within the clusters of galaxies. (Courtesy Einstein Data Bank, Harvard–Smithsonian Center for Astrophysics)

of mass in the galaxies themselves. But its density does not appear to be sufficient to bind the cluster gravitationally. Consider the Coma cluster. Its x-ray emission indicates that no more than 20 percent of the total mass in the inner 4 Mly is in the form of a hot plasma. Recall that the visible galaxies make up just a few percent; the rest is dark matter. So the x-ray-emitting material does not solve the dark matter problem for us.

Dark matter may be in the form of elusive particles such as neutrinos (especially if it turns out that neutrinos have any mass at all—even a very small one). If such dark matter commonly exists, whatever its nature, it may well control the structures of clusters and superclusters, with the luminous matter as the visible tip of the large-scale clustering of matter.

Clusters of galaxies have most of their mass in the form of dark matter.

The Larger View

This chapter made the leap from the Milky Way Galaxy to other galaxies. We find that most normal galaxies are much like our own: vast collections of stars, gas, dust—and dark matter. Galaxies come in a few main types of shapes and sizes, many with spiral patterns. Yet all are basically material orbiting a nuclear region. And their evolution is basically driven by the births and deaths of stars, as in our Galaxy, and the exchange of material with the interstellar medium.

To infer the physical properties of galaxies, we must know their distances—information that comes by indirect means because we are dealing with great distances. When distances and recessional speeds (from redshifts) are teamed up, we find that the universe is expanding—a crucial revelation of twentieth-century astronomy. We will explore the history of the expanding universe in Chapter 18—a cosmos that began in a bang and may die in a whimper.

Galaxies are not loners. They come in clusters, and sometimes pass close to each other, brushed by tidal forces or colliding and merging. The clusters are strung together to build up vast superclusters—fossils from the early times of the universe. Among the superclusters are vast voids of apparently empty space. Although we have uncovered its grand structure, the universe remains largely unknown to us. Much of the matter in clusters and superclusters is dark, invisible to current observations. Here lies a deeply buried clue about the processes by which the universe began and evolved.

Key Ideas and Terms

angular size
flux
inverse-square law for light
luminosity

1. Knowledge of the distances to galaxies is essential to finding out their properties. Distances can be estimated roughly from brightness and angular sizes. Actual measurement is difficult and relies on indirect schemes. The basic trick is to find very bright objects whose luminosities can be reasonably estimated, identify these in galaxies, compare their flux to their luminosity, and infer their distance from the inverse-square law for light.

elliptical galaxy
irregular galaxy
lenticular galaxy
luminosity classes
spiral galaxy

2. Galaxies come in three main types (based on their shapes): a, elliptical (dwarf and giant); b, spiral (normal and barred); and c, irregular. Within these classes, galaxies come in a range of luminosities and so a range of masses.

Doppler shift
Hubble's constant
Hubble's law
redshift
spectra

3. Measured redshifts (from spectra) and distances of galaxies result in Hubble's law and Hubble's constant, H. The greatest uncertainty in Hubble's constant arises from the uncertainties in distances. H has a value between 15 and 30 km/s/Mly. Hubble's law is most simply interpreted as a result of the expansion of the universe.

dark matter
Kepler's laws

4. Galaxies differ in terms of physical properties such as size, mass, luminosity, mass–luminosity ratio, and colors—all of which reflect their content of stars,

mass–luminosity ratio
rotation curve

gas, and dust. Much of the matter in many galaxies is invisible. For normal galaxies, that which is visible is mostly in the form of stars.

cluster (of galaxies)
dark matter
Local Group

5. Galaxies come in clusters, containing from two to thousands, held together (at least for a while) by the gravity between the galaxies. The Local Group contains some 30-odd galaxies, mostly of low mass and luminosity, dominated by the Milky Way and the Andromeda Galaxy. The Local Group spreads over a volume some 3 Mly in diameter. A large fraction of the material in clusters is dark matter.

Local Supercluster
superclusters
voids

6. Clusters are grouped in superclusters. The Local Group is one small piece of the Local Supercluster (which is centered on the Virgo cluster). Superclusters seem to include voids and to come in long chains of galaxies. They are the largest known entities in the cosmos.

galactic cannibalism
starburst galaxies
tidal forces

7. Galaxies within clusters can occasionally come close enough together to interact by tidal forces. Such interactions strip material from both galaxies and may promote star formation within them. The supergiant elliptical galaxies at the centers of clusters may have formed from the merger of a few galaxies.

dark matter
intergalactic medium
plasma
x-rays

8. A very thin, very hot gas exists within clusters and superclusters; within clusters, a hot, thin plasma gas exists and is observed by the x-rays it emits. This plasma does not account for the dark matter in clusters, whose nature is not yet known.

Review Questions

1. Suppose you see two spiral galaxies through a telescope, and one is larger than the other. Which one is probably closer?
2. What are two types of spiral galaxy?
3. What are two types of elliptical galaxy?
4. What astronomical objects can be used to estimate the distances to spiral galaxies?
5. Do the dark lines in the spectra of galaxies tend to show blueshifts or redshifts?
6. In general, Hubble's law states that the trend is for more distant galaxies to have what amount of a Doppler shift compared to closer ones?
7. What is the current value of Hubble's constant?
8. Which galaxies appear redder, ellipticals or spirals?
9. Which galaxies contain more gas and dust relative to their total mass, spirals or irregulars?
10. What is the Local Group of galaxies?

11. What makes up a supercluster of galaxies?
12. What observational evidence implies that a thin, hot gas lies within clusters and superclusters?

Conceptual Exercises

1. How do astronomers measure the masses of galaxies?
2. How do astronomers know that galaxies contain gas and dust?
3. At the same distance from us, would irregular galaxies appear to be larger or smaller than spiral ones?
4. How do astronomers know that other galaxies are made of stars?
5. The value of Hubble's constant ranges from 30 km/s/Mly to 15 km/s/Mly. How does this range affect the distances to galaxies inferred from redshifts and Hubble's constant?
6. Describe how supernovas can be used to estimate distances to nearby galaxies.
7. Why must intergalactic gas be both ionized and hot?
8. Describe the layout of a supercluster of galaxies in three dimensions.
9. How do astronomers see the depth of superclusters in space? What uncertainty exists in their technique?
10. What is the most common type of galaxy found in the regions of space that we can probe?

Conceptual Activity 16   Hubble's Law

Let's make a plot from Figure 16.10 that will show how astronomers visualize Hubble's law and find Hubble's constant.

Get a ruler with a millimeter scale. (Better yet, find a magnifying glass that has a millimeter scale visible through it.) Start with the top galaxy and its spectrum. Using the brightest region of the image of the galaxy, measure the galaxy's diameter in millimeters. Put this number in a table. Then examine the spectrum to the right of the galaxy. You will see a small, horizontal white arrow just below the galaxy's spectrum and above the comparison, emission-line spectrum at the bottom. Measure the length of this arrow in millimeters. The arrow shows the amount of the redshift in the H and K lines of calcium, the two darkest lines in the galaxy's spectrum. Put this number in the table next to the galaxy's diameter.

Do the same for the next three galaxies and add the information to the table. (Don't try it for the galaxy at the bottom; it's too small to give a good result.)

Now get a piece of graph paper, on which you will make two plots. On the first, plot the sizes of the galaxies, in millimeters, along the horizontal axis. The trick is to label the axis so that the largest galaxy is closest to the origin of the horizontal axis and the numbers decrease from left to right. On the vertical axis, plot the sizes of the arrows (amounts of the redshifts), starting with the smallest and going up to the largest. Once all four points have been plotted, try to draw a straight line through them. That line is a graphical form of Hubble's law, and its slope is Hubble's constant.

Of course, the units of this plot are wrong, so let's make up a second one. Here, make the horizontal axis the distance in millions of light years, from the closest to the farthest galaxies. These distances are given for each galaxy in the column between the photo of the galaxy and that of its spectrum. On the vertical axis, you want to use the radial velocity in kilometers per second; that information is given below the spectrum of each galaxy. Plot the points for the four galaxies and again draw a straight line through them. Compare it to the first graph. Do they look the same?

Now measure the slope of the line, which is the rise over the run. What value do you get? That is your value of Hubble's constant from the information in the figure. How does it compare to the range of values given in the book?

17
Active Galaxies and Quasars

■

Central Concept
Violent activity commonly occurs in the centers of some galaxies.

■

Violent events commonly occur in the nuclei of galaxies of many types. This violence can appear as jets emanating from the nuclei of active galaxies. Supermassive black holes in the nuclei of such galaxies may power these energetic, beamed outflows.

Quasars spotlight the upper range of cosmic violence. At the fringes of the observed cosmos, quasars seem to represent some of the first-formed objects in the visible universe— and the most powerful. Quasars may well produce their enormous energies by the actions of supermassive black holes in the nuclei of young, hyperactive galaxies. In fact, all active galaxies may have a common source of power that we observe in different ways. A single unified model may explain them all.

17.1 Violent Activity and Active Galaxies

As we have probed more regions of the spectrum, some galaxies—and the nuclei of many—have acquired the aspect of a compact arena of violent events. Most such galaxies have nonthermal spectra. These have been lumped into the category of **active galaxies**. But how active is "active"? Basically, emitting at least as much power as a normal galaxy from a much smaller region qualifies a galaxy as active.

An active galaxy's spectrum is mostly nonthermal and has a broad range of infrared, radio, ultraviolet, and x-ray energy outputs. (Our Galaxy does emit at these wavelengths, but not nearly as strongly as an active galaxy.) Figure 17.1 compares the spectra of some active galaxies with those of a normal galaxy. Active galaxies make up only a few percent of all known galaxies, yet they are the most powerful.

This energetic activity originates in the *nuclei* of active galaxies. So astronomers tend to lump them into one class: **active galactic nuclei**, or **AGNs** for short. The central issue for all is the same: What powers AGNs? A key clue is that the spectrum of the nucleus is nonthermal—some of the emission is synchrotron. So high-energy electrons, moving at speeds close to that of light, spiraling in intense magnetic fields, persist in the nucleus. The thermal part of their spectrum in the visual region comes from the combined light of the stars in the galaxy. Infrared thermal emission usually comes from dusty AGNs.

FIGURE 17.1

Comparison of the continuous spectra in the radio, infrared, and optical regions for a quasar (3C 273), two active galaxies (NGC 4151 and NGC 1068), and a typical normal (spiral) galaxy. Note how the quasar and the active galaxies have broad peaks in their respective emissions at far-infrared wavelengths and then flat or slightly rising spectra in the radio. (Adapted from a diagram by R. Weymann)

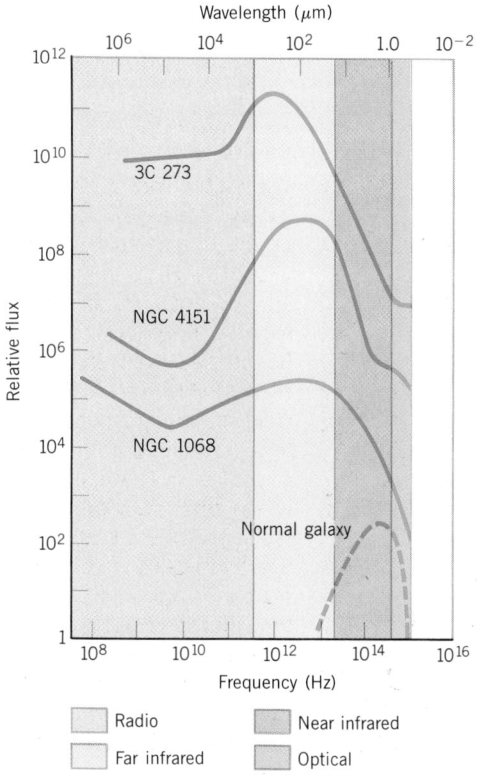

The nucleus of an active galaxy has a large energy output from a small region; its spectrum is nonthermal and comes from synchrotron emission.

Active Galaxies
_____
At a Glance

High luminosity, greater than 10^9 solar luminosities
Nonthermal emission (mostly)
Rapid variability and/or small size (a few light years at most)
Very bright nucleus with jetlike projections
Broad emission lines (sometimes); narrow emission lines (less often)

A reminder about the synchrotron process (Section 14.4), which is often polarized. Magnetic fields bend the paths of electrons, and the electrons emit electromagnetic radiation. In general, the faster the electrons travel, the more energetic (shorter wavelength, higher frequency) the radiation they emit. And the stronger the magnetic fields are, the more energetic the emission will be. As the electrons emit, they lose energy and slow down. So to keep a synchrotron source powered up requires a supply of electrons moving close to the speed of light. This process will make up an important part of our model for AGNs.

The largest class of active galaxies is composed of **radio galaxies**, which come in two principal types: **compact** and **extended**. *Compact* radio galaxies often display very small radio sources, typically no more than a few light years in size, and coincident with the optical nucleus. *Extended* radio sources, in contrast, spread far beyond the optically visible galaxy—often in two giant lobes of emission. These **radio lobes** span up to millions of light years and lie balanced on opposite sides of the nucleus. A galaxy may show both a compact source at its nucleus and the extended lobes. It then usually also has radio jets connecting the nucleus to the extended radio emission. These **nuclear jets** are assumed to be directed flows of fluid, similar to a supersonic jet engine exhaust, but with particles traveling near the speed of light.

We will deal with large redshifts in the spectra of the objects discussed in this chapter. Keep in mind that astronomers define the amount of the redshift by the *change* in wavelength of a spectral line divided by the line's wavelength *at rest* (Section 9.5).

Examples of Active Galaxies

In a universe of stunning objects, active galaxies take top honors. Let's look at a few examples to see what properties they have in common so that we can begin to construct a model of their unusual activity.

Only 65 M ly away, *Messier 87* is a fine example of an active galaxy. A supergiant elliptical galaxy, M87 dominates the Virgo cluster of galaxies. One radio source only 1.5 *light months* in diameter appears in M87's core with a group of other compact radio sources. A remarkable, optically visible jet over a length of some 6000 ly (Fig. 17.2). The wiggly jet has a luminosity of roughly 10^7 solar luminosities; its emission is polarized. It contains distinct blobs of material, each no more than a few tens of light years in size (Fig. 17.3).

M87 emits x-rays with an overall power of about 10^9 solar luminosities. The jet itself also emits x-rays, as a line of knots. Detailed radio maps of M87's jet (Fig. 17.4) have confirmed that its radio emission coincides with the optical and x-ray emissions. The radio knots line up with the optical ones. The synchrotron process within each knot produces this wide range of emission.

Nucleus

Jet

FIGURE 17.2

The central core of the elliptical galaxy Messier 87, as imaged by the Hubble Space Telescope (Appendix H) in the near-infrared. Note the steady increase in the overall brightness toward the central region, which shows that the stars are strongly concentrated in the nucleus. The bright spot at center marks the nucleus, which emits synchrotron radiation. The optical jet emerges from the nuclear region; its emission is also synchrotron in origin. The faint starlike sources scattered around the core are globular clusters orbiting within M87. (Image by T. R. Lauer, courtesy NASA/ESA)

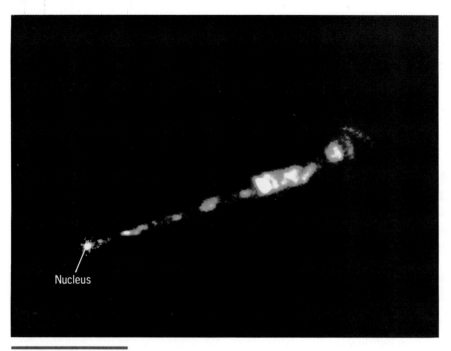

Nucleus

F I G U R E   1 7 . 3

An ultraviolet image (by the Hubble Space Telescope) of the optical jet of M87. This view shows the structure in the jet, which appears as a series of knots extending out from the nucleus of M87. The hot plasma within the jet is channeled by magnetic fields. A large, bright knot about midway along the jet shows where the jet is disrupted. The bend near the end of the jet indicates where it rams into a wall of gas. (Courtesy NASA/ESA)

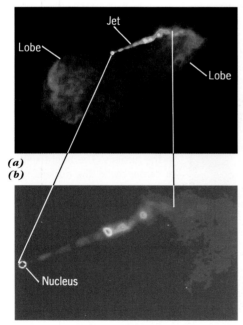

Jet

Lobe

Lobe

Nucleus

(*a*)
(*b*)

F I G U R E   1 7 . 4

The active galaxy Messier 87 (also called Virgo A) in the radio. (*a*) High-resolution, false-color radio map made at a frequency of 5 GHz. Blue indicates the weakest emission; yellow and red the strongest. Note the radio jet, which coincides with the optical one, extending toward the upper right from the nucleus and ending in the radio lobe at the right. There is no jet extending to the radio lobe on the opposite side. Some radio galaxies show this kind of asymmetry. (Observations by F. Owen; courtesy NRAO/AUI) (*b*) Radio map, in false colors, showing the details in the structure of the jet. Note the bright knots along the 6000-ly length, which ends in a weak lobe of radio emission. (Observations by F. Owen; courtesy NRAO/AUI)

FIGURE 17.5

Optical image of the galaxy Cygnus A, taken under excellent conditions. It shows the bright core of the galaxy and the galaxy's diffuse halo. (Image courtesy L. A. Thompson, Institute for Astronomy, University of Hawaii)

Cygnus A (Fig. 17.5) is one of the strongest radio sources in the sky and was one of the first to be discovered. It provides an excellent example of the typical double structure of a luminous, extended

radio galaxy. Its radio output, some 10^{11} solar luminosities, comes from two giant lobes set on opposite sides of the supergiant elliptical galaxy (Fig. 17.6). Each lobe has a diameter of 55,000 ly—about half the size of the Galaxy! They hang roughly 160,000 ly away from the central galaxy and contain a cloud of energetic electrons and magnetic fields.

The stunning radio map of Figure 17.6 reveals the link of the lobes to the nucleus of Cygnus A: a needlelike jet from the nucleus to one lobe. Note that the radio lobes have a wispy appearance, with delicate swirls in which embedded hot spots of emission. One lobe plainly marks the end of the visible jet, in which electrons travel at speeds close to that of light.

Centaurus A (NGC 5128) is the closest active galaxy to us—only some 10 million light years away. A supergiant elliptical galaxy, Centaurus A is bisected by an irregular dust lane (Fig. 17.7) in which star formation is taking place. As viewed with an optical telescope, the galaxy has a diameter of a few tens of thousands of light years. Close to the nucleus, a pair of radio lobes sit on the edges of the optical galaxy; these are some 33,000 ly in size (roughly equal to the distance of the sun from the center of the Galaxy). Centaurus A also shows a radio jet extending out to one inner radio lobe (Fig. 17.8) and a x-ray jet pouring from the nucleus. High-resolution radio observations reveal in the nucleus itself a *very* small inner jet, 4 ly long, that lines up with the larger one.

FIGURE 17.6

Radio map of Cygnus A processed to show the maximum fine detail. Note the thin jets from the nucleus and the filamentary structure in the lobes. (Observations by R. A. Perley, J. W. Dreher, and J. J. Cowan; courtesy NRAO/AUI)

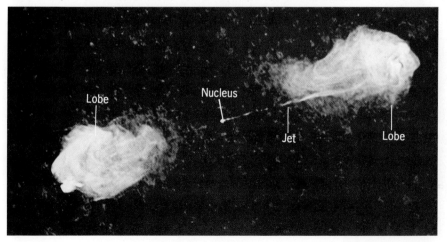

FIGURE 17.7

Centaurus A (NGC 5128), a nearby radio galaxy—a supergiant elliptical. The diffuse, circular part of this galaxy contains hundreds of billions of old, yellowish stars. Note the dust lane across the galaxy's middle, which obscures the nuclear region; at its edges are small patches of younger, bluish stars and pinkish H II regions. Infrared observations show that star formation is occurring here. (Image by D. Malin; courtesy Anglo-Australian Telescope Board)

FIGURE 17.8

False-color map of radio emission from Centaurus A: blue indicates the weakest emission; red, the strongest. Note the two inner radio lobes and the radio jet extending from the nucleus to the lobe at the upper left. The circle gives the rough size of the visible galaxy (Fig. 17.7). (Courtesy J. Burns; VLA observations by J. Burns, E. Schreier, and E. Feigelson, courtesy NRAO/AUI)

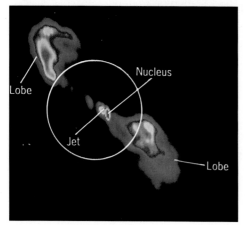

What powers the jets of these sources? Observed at a wide range of wavelengths, the spectra of the jets are nonthermal, so the synchrotron process generates their emission. High-speed electrons in magnetic fields produce the emission from the knots. The nucleus provides the high-energy electrons, which are expelled either as a fairly constant beam of particles or as a series of ionized blobs thrown out along a magnetic field. The ionized stream carries magnetic fields within it; these help to channel the flows outward.

Many radio galaxies, which appear to be ellipticals, exhibit nuclear jets, which often end in a lobe of radio emission.

Structures of Extended Radio Emission

Extended radio galaxies all appear to be double, with two lobes of emission lined up on either side of the galaxy's nucleus. These radio lobes are huge: they reach up to millions of light years in diameter. A typical lobe has a luminosity about 100 times that of the visible elliptical galaxy with which it is associated. The lobes are energy reservoirs that contain a tangle of magnetic fields and high-speed electrons, since the emission is synchrotron radiation.

How do AGNs generate these lobes? A crucial clue: the radio jets from the nucleus, aligned (more or less) with the lobes. The jets show that high-speed electrons are channeled from the nucleus into the medium around the galaxy, where they pile up to form a lobe. The nuclear energy source somehow converts gravitational energy into a jet of high-speed particles—a critical element in a model for AGNs.

Radio lobes result from the material carried by jets that plows into the gas surrounding a radio galaxy.

17.2 Seyfert Galaxies and BL Lacertae Objects

I don't want to bog you down with all the strange types in the AGN zoo. But here are two types that provide different insights in the AGN phenomenon at lower energy levels and with less drama. They offer nearby clues to build a model of AGNs, and knowledge of them predates by far what we now call active galaxies.

Seyfert Galaxies

Some nearby spiral galaxies show unusual, broad emission lines. These are now called **Seyfert galaxies** (after Carl Seyfert, who first examined them decades ago); some 600 are known to date. Seyfert galaxies are relatively close by, so they can be probed in detail. Seyferts are usually spiral or barred spiral galaxies (Fig. 17.9). Yet we have just learned that most extended radio galaxies are elliptical (Section 17.1). Overall, a few percent of all spiral galaxies (ordinary and barred) are Seyferts. They have powers that typically range from 10^9 to 10^{12} solar luminosities.

A Seyfert galaxy has a small, exceptionally bright bluish nucleus. (A normal spiral has a yellowish nucleus.) Along with this optical trademark, many Seyferts show *broad* emission lines in their spectra. How? Consider an opaque gas giving off emission lines. Some of this gas is moving toward us, so its emission is blueshifted. Other parts are moving away, and their emission is redshifted. When we observe a spectrum of the gas, the Doppler-shifted emission is added to the unshifted line at longer and shorter wavelengths. So the emission line appears wider than it would be if the gas were stationary. If we interpret the widths of the lines as from Doppler shifts produced by motions in the emitting gas, the gas in the Seyferts has velocities of *thousands* of kilometers per second.

All Seyferts show strong emission lines, which are narrow in some, broad in most (and both narrow and broad in a few). The use of "broad" and "narrow" here is relative (Fig. 17.10). In some Seyfert galaxies, the emission lines have Doppler widths as great as 10,000 km/s (broad); others have lines only 1000 km/s wide (narrow). To explain the broad emission lines from a Seyfert requires some tens to thousands of solar masses of dense, ionized gas in the nucleus moving at speeds up to 10,000 km/s. The broad lines and the narrow lines come from different regions of the gas. The region emitting the broad lines is closer to the nucleus and probably also denser than that producing the narrow ones. (The radio galaxies described in Section 17.1 occasionally display emission lines in their spectra. Often they are broad; rarely are they narrow.)

FIGURE 17.10

The emission-line spectrum of a Seyfert galaxy, showing the strongest lines. Note the especially broad lines of the hydrogen Balmer series and of iron. (Adapted from a figure by D. E. Osterbrock)

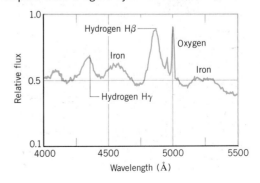

The energy source in the nucleus that generates the synchrotron emission also ionizes the gas. The nucleus provides a source of high-energy electrons and gas. Observations show that the nuclei of Seyferts are small, only a few light years in diameter. A survey of spiral galaxies shows that Seyferts tend to be in close, binary galaxy systems. Tidal forces may then induce the Seyfert activity for a short period.

Seyferts are spiral galaxies with extremely small and bright nuclei that exhibit broad emission lines; some Seyferts have compact, low-luminosity radio sources within their nuclei.

BL Lacertae Objects

The **BL Lac objects** are named after their prototype, *BL Lacertae*, and have some similarities to radio galaxies and Seyferts. As a group, the BL Lac objects have the following characteristics: (1) rapid variability (as fast as tens of minutes) at radio, infrared, and visual wavelengths, (2) extremely weak or no emission lines, (3) nonthermal continuous radiation with most of the energy emitted in the infrared, (4) strong and rapidly varying polarization, and (5) generally a starlike appearance; rarely is structure visible.

The BL Lac objects differ the most from other AGNs in that their emission varies frequently, erratically, and rapidly—in only 10 minutes in a few cases. For example, BL Lac itself fluctuates overall in luminosity by 20 times or so. Observers have noted night-to-night variations of some 20 percent in luminosity. A few BL Lac objects have changed as much as 100 times in luminosity.

These rapid variations imply that the emission region of the BL Lac objects is small, no more than a light day or so in diameter. This argument rests on the fact that light travels at a finite speed. For instance, imagine that we could suddenly turn the sun off. It would take about 3 seconds to go dark, because this is time light needs to cross the sun's diameter. Turn it back on. Then 3 seconds would elapse before it reached its normal brightness. The fastest the sun can vary in brightness is the light travel time across its diameter. (Don't confuse this time with the

8 minutes it takes light from the sun to reach the earth!) The same argument applies to all AGNs—the size of the emitting region equals the distance light can travel during the *shortest* variation of its brightness. So a one-day observed variation means a size no larger than one light day.

What also puzzles astronomers about the BL Lac objects is that most lack emission lines in their spectra! The standard model for AGNs pictures synchrotron emission that in the ultraviolet should ionize any gas near the nucleus and produce emission lines. Where are the BL Lac objects' emission lines? Are they just too faint to be seen? Some BL Lac objects, like BL Lac itself, have a faint surrounding fuzz that is a galaxy. Indications so far point to an elliptical galaxy as containing the BL Lac nucleus, although this is not yet certain.

BL Lac objects are most peculiar because of their rapid variability and lack of strong emission lines; they may be the nuclei of elliptical galaxies.

17.3 Quasars

The most distant and mysterious of the AGNs are faint, starlike objects (hence the name *quasi-stellar object*, or **quasar**) that have a spectrum of broad emission lines. One of the earliest (Fig. 17.11) was identified at the position of radio object 3C 273 (object 273 from the *Third Cambridge Catalogue* of radio sources). The most prominent emission lines of 3C 273 turned out to be those of the hydrogen Balmer series, redshifted by about 16 percent compared to their normal (at-rest) wavelengths (Fig. 17.12). If we interpret this as a Doppler shift, 3C 273 appears to be receding from us at 16 percent the speed of light!

Quasar Redshifts

More than a few thousand quasars have been identified, and redshifts have been measured for most. Most quasars have redshifts greater than 0.5. Some

FIGURE 17.11

The quasar 3C 273, the first quasar to have its spectrum deciphered. Note the optical jet sticking out of the quasar, Q. Note how fuzzy the quasar appears compared to the surrounding stars. The quasar's spectrum is shown in Figure 17.1. (Observations by S. Wyckoff, P. Wehinger, and T. Gehrels with the ESO 3.6-meter telescope; courtesy P. Wehinger)

have redshifts that exceed 4.0. If interpreted as redshifts from the expansion of the universe (Section 16.4), the light from quasars comes from such distances that it had to have originated billions of years ago. The redshifted quasars are the youngest objects we can see in the universe. Here "youngest" means objects at a point in their lives far in the past, close to the time when they first formed relative to their total lifespan. (We are also looking at the universe when it was "younger" than now—closer to the time of its birth.)

Caution: When the measured redshift approaches 1, the simple Doppler formula (Section 9.5) no longer gives the correct radial velocity. Usually, a redshift of 0.16, for example, means a radial velocity of 16 percent the speed of light. But then a redshift of 2 would suggest that a quasar was moving away at 200 percent the speed of light! And that is not possible. A modified formula must be used instead. Then, for example, a redshift of 2 means that the radial velocity equals $0.8c$. We *measure* redshift but *infer* radial velocity.

FIGURE 17.12

Spectrum of 3C 273, showing its large redshift as indicated by its emission lines. The upper spectrum is that of the quasar, the lower one a comparison spectrum that establishes the wavelength reference scale (given at bottom for a source at rest with respect to the observer). Compare the hydrogen Balmer lines in the quasar's spectrum to those in the comparison spectrum. Note how the quasar's lines are shifted to the red end of the spectrum (arrow at top). This redshift amounts to about 16 percent. (Courtesy M. Schmidt)

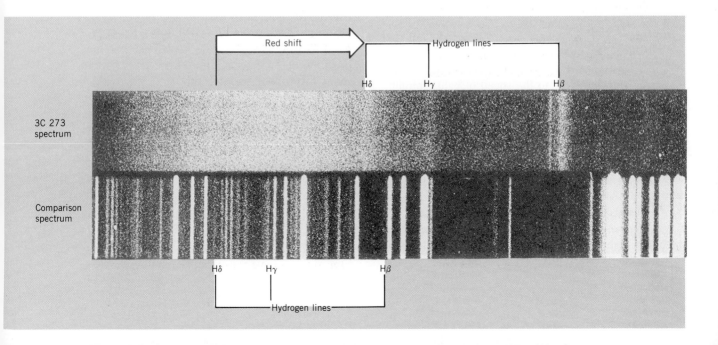

General Observed Properties

The following observed properties make quasars unique and serve as identification tags: (1) starlike appearance with a *large* redshift, sometimes associated with a radio source (only about 10 percent are known radio sources); (2) broad emission lines in the spectrum with absorption lines sometimes present (usually the redshift of the absorption lines is less than that of the emission lines); (3) often variable luminosity; (4) in those that are radio sources, aligned, double-lobed structures (like Cygnus A); and (5) nuclear jets, in the case of about 50 percent of radio-emitting quasars.

The main—and most remarkable—feature of quasars is their large redshifts. The most natural explanation of the quasars' redshifts is that they participate in the universe's expansion, as described by Hubble's law. If so, their enormous redshifts indicate that they are very far away, observed when the universe was very young. But if they are as far away as suggested by their redshifts, they must expend vast amounts of energy. For example, 3C 273's redshift of 16 percent implies a distance of some 3×10^9 ly. At this distance, 3C 273 emits about 10^{14} solar luminosities, many times that of the most luminous normal galaxies. The power of quasars typically falls in the range from 10^{12} to 10^{14} solar luminosities.

Quasars' large redshifts imply that they are very distant and so very luminous objects.

Some quasars emit radio strongly, and all emit visible light. The radio spectrum has the shape for nonthermal, synchrotron emission. So energetic electrons (moving close to the speed of light) spiraling around magnetic fields produce the radio emission. Such synchrotron emission is polarized if the magnetic fields are organized rather than chaotic. Synchrotron radiation requires a continuous supply of clouds of energetic electrons and strong magnetic fields. The continuous optical emission is in part polarized, so synchrotron emission produces some optical radiation. Given a common magnetic field, the electrons that produce the optical radiation must have higher energies than those that emit the radio radiation.

Most radio-emitting quasars also vary in radio output over periods of years. In contrast, about 20 percent of the quasars exhibit rapid variations in light and radio output with periods of days or weeks. This variability implies that the nuclear power source is very small (light days to light years), as was argued for BL Lac objects (Section 17.2) and any active galaxy with rapid changes in luminosity.

Line Spectra

All quasars have bright lines in their optical spectra—the emission lines that are used to measure a quasar's redshift (Fig. 17.13). The emission-line spectra of quasars indicate that a low-density cloud of gas is radiated with photons energetic enough to ionize some atoms. This radiation comes from the synchrotron emission from energetic electrons. So the central synchrotron source is surrounded by clouds of gas that convert ultraviolet radiation (and some x-rays) from the synchrotron source into visible emission lines.

The emission lines are typically extremely broad, which implies that the clouds from which the emission lines originate move rapidly (as was argued for Seyferts in Section 17.2). The motions of the material imply radial velocities that range up to 10,000 km/s. This region, close to the nucleus,

FIGURE 17.13

Composite spectrum of quasars showing the strongest emission lines typical for them. As for Seyfert galaxies (Fig. 17.10), note the prominent lines of the hydrogen Balmer series. (Adapted from a figure by J. S. Miller)

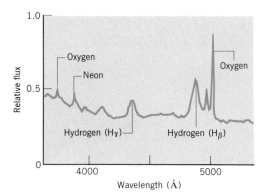

makes up the broad-line region in a model for quasar. Narrow emission lines are not as obvious as the broad ones. They have widths that correspond to Doppler shifts of some 1000 km/s. So quasars have a narrow-line emitting region farther away from the nucleus.

The spectra of quasars show a nonthermal (synchrotron) central source, broad emission lines, and narrow emission lines.

Many (but not all) quasars also have absorption lines in their spectra. Quasars with emission-line redshifts of less than 2.2 typically do not have absorption lines; those with greater redshifts have strong absorption lines. These lines are very narrow compared with the emission lines. In general, the absorption-line redshifts are *less* than the shifts for the emission lines. And, the absorption lines sometimes show more than one redshift. These absorption lines are most likely produced by cool clouds of gas closer to us along the line of sight.

17.4 Double Quasars and Gravitational Lenses

Quasars are rarely close together in the sky; yet two quasars, called 0957 + 561A and 0957 + 561B, are only 6 arcseconds apart. The emission-line redshifts of both are essentially the same: 1.41; the absorption-line redshifts in quasar 0957 + 561A are all the same and only slightly less than those of the emission lines.

This situation forced the astronomers to propose that these quasars were not separate twins but optical images of the *same* quasar. Einstein's theory of general relativity predicts that masses deflect the paths of light rays—in essence, acting like a lens (Section 10.7). If a very small, dense mass lay along our line of sight to a quasar, its image would be split into two, one above and one below the quasar's actual position. This phenomenon of image making by a mass is called a **gravitational lens effect**. Such lenses tend to make imperfect, distorted images depending on the alignment of the background and foreground masses.

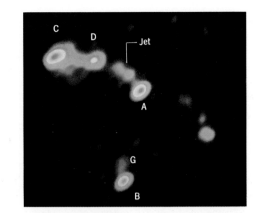

FIGURE 17.14

Map of the radio emission from the double quasar 0957+561A and B. The double quasar shows up in the radio as the two elliptical regions near the center (marked A and B). Note that there are blobs of emission (marked C and D) to the left of the upper quasar (A); these are lobes at the end of a radio jet. The label G marks the galaxy that causes the gravitational lens effect. The false colors here display red as the most intense and dark blue the least. (Observations by P. Greenfield, B. Burke, and D. Roberts; courtesy NRAO/AUI)

A radio map of the double quasar shows radio blobs (C and D in Fig. 17.14) near the northern quasar (A) not visible near the southern quasar (B). (It turns out that this emission is that of a jet flowing into two lobes!) The revelation came from exceptionally fine photographs that showed (Fig. 17.15) quasar B with a little bit of fuzz sticking out of it. This fuzz turns out to be the poorly resolved image of a faint elliptical galaxy—the gravitational lens! Since the galaxy is an extended mass, it acts like an imperfect lens and produces a complex pattern of images. The galaxy is closer to us than the quasar, for the redshift is only 0.36.

By a quirk of placement, we see only a part of the complete image. Two images are formed by a gravitational lens, an intervening, probably elliptical, galaxy about halfway between us and the quasar (Fig. 17.16). The galaxy needs a mass of some 10^{13} solar masses to bend the light enough to form the images.

This discovery has three important implications: (1) it provides another confirmation of general relativity; (2) it proves here that the quasar is more distant than the galaxy, so the quasar's redshift is cosmological; and (3) gas around the galaxy creates the

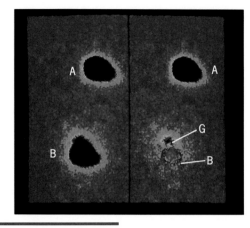

FIGURE 17.15

High-resolution optical photograph of quasars 0957+561A and B. In the pair of images on the right, the image of B has that of A subtracted from it. The little fuzz sticking out of the top of B turns out to be the poorly resolved image of the galaxy making the gravitational lens (visible as G in the radio map). The image of B is superimposed on it by the gravitational imaging. (Courtesy A. Stockton, Institute for Astronomy, University of Hawaii)

FIGURE 17.16

Schematic drawing of the optical illusion of the double quasar caused by a gravitational lens effect. Note that the bending of the quasar's light by the massive galaxy (a few billion light years away) makes two images (A and B) appear on the sky.

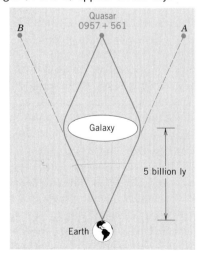

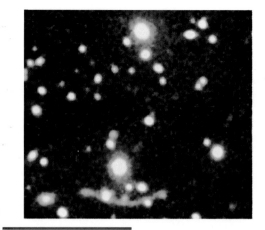

FIGURE 17.17

Giant arc in the cluster of galaxies called Abell 370. This false-color map was made at the 3.6-meter Canada–France–Hawaii telescope. The giant arc is at the bottom; near the center lies the more distant galaxy whose light is bent to make the arc. (Courtesy G. Soucail)

quasar's absorption-line spectrum; this situation may be the case for other quasars as well. Other candidates as double quasars have been discovered, strengthening these implications.

Depending on the distribution of the mass doing the lensing, the light from a distant quasar may be split into many images—or distorted into rings or incomplete rings visible as arcs. Astronomers have found other candidates along with double quasars. An image of one cluster of galaxies (Fig. 17.17) showed a huge, luminous arc whose source was quite a puzzle at first. It appears that the arc is the image of a more distant galaxy made by an imperfect gravitational lens—a massive galaxy along the line of sight. The cluster is dominated by a supergiant elliptical galaxy, and the spectrum of the arc is that of a galaxy.

The light from very distant objects, such as quasars, can be distorted by large intervening masses acting as imperfect gravitational lenses.

One of the strangest of the gravitational lens effects is the Einstein cross (Fig. 17.18). The Hubble Space Telescope (Appendix H) produced an image

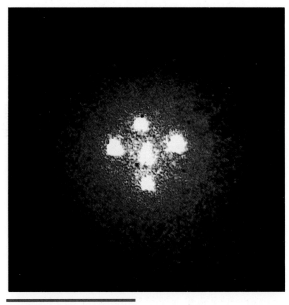

Cosmic Violence At a Glance	Object	Power Output (W)
	Nucleus of Milky Way	10^{34}
	Active galaxy	10^{39}
	Quasar	10^{42}

FIGURE 17.18

Einstein's cross (gravitational lens G2237+0305), an image taken by the Faint Object Camera on Hubble Space Telescope. The four bright outer images are from a distant quasar. (Courtesy NASA/ESA)

of a quasar that showed four distinct, bright images and a diffuse central object that is the bright core of the intervening galaxy. The quasar lies about 8×10^9 ly away. Narrow absorption lines in its spectrum originate in the halo of the galaxy, which is at a distance of only 400×10^6 ly, some 20 times closer. The mass of this galaxy causes the imperfect lensing effect.

 ## 17.5 Troubles with Quasars

If quasars are actually billions of light years away, then a typical quasar produces up to 10,000 times as much energy as an ordinary spiral galaxy. Not only do quasars blast out energy at enormous rates, but the energy comes from relatively small regions of space in the centers of quasars. Because quasars' light output varies on time scales of days to years, their energy-emitting regions cannot be larger than light days to light years.

Quasars emit enormous power from small regions of space.

High-resolution studies have revealed the radio structure of some quasars, especially those that are relatively near. These observations show that for the quasars that can be resolved, the radio structures are generally symmetrical doubles, similar to Cygnus A. And some of these show nuclear jets. In a few, the material in the jets *appears* to travel faster than light. This effect turns out to be another optical illusion, caused by jets aligned very close to our line of sight.

An example is 3C 273, which shows an optical and radio jet (Fig. 17.19a). Over a period of a few years, a blob of material appeared to separate from the jet and move across our line of sight at 10 times the speed of light (Fig. 17.19b)! This speed turns out to be another optical illusion, caused by a jet pointing close to but not exactly into our line of sight. If the material in the jet moves at close to the speed of light, it almost catches up to the radiation it emits. Time intervals are compressed, and this compression creates the illusion that the matter is traveling faster than light—*superluminal* sources.

The nuclear jets of some quasars eject material moving at close to the speed of light.

That's not all. Relativity predicts that matter moving close to the speed of light and emitting light will do so in a special way. Consider a single light bulb. It emits light equally in all directions into a sphere around it. Contrast that to the light emitted by a flashlight: here the emission is beamed, so that most of the light falls into a cone in a specific direction. Relativity says that relativistic matter will emit light

FIGURE 17.19

The radio jet of 3C 273. (a) Close-up view of the nucleus and the jet made at a frequency of 408 MHz. This false-color map has white as the strongest radio emission and blue the weakest. (Observations with the MERLIN array; courtesy Richard Davis, Jodrell Bank.) (b) Motions in the jet are visible in these very-high-resolution radio maps made at a frequency of 10.65 GHz. The contour lines show the relative amounts of the radio emission. Note how the emission moves out along the jet from 1977 to 1980; dates are given in fractions of the year. The angular scale is at the bottom in milliarcseconds (mas). (Based on observations by T. J. Pearson, S. C. Urwin, M. H. Cohen, R. P. Linfield, A. C. S. Readhead, G. A. Seielstad, R. S. Simon, and R. C. Walker at the Owens Valley Radio Observatory and NRAO)

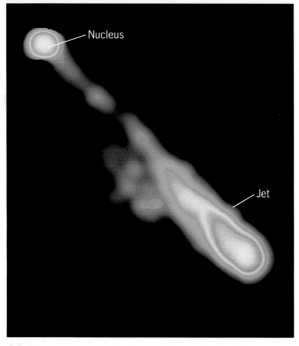

(a)

(b)

into a cone in the direction of its motion, even if, at rest, it would emit in all directions equally. This effect is called *relativistic beaming*, and it results in a source moving toward us appearing much more luminous than one moving at a large angle to our line of sight. So sources with relativistic jets pointing at us will have special observable characteristics.

Energy Sources in Quasars

What generates such vast energies in such small regions of space? The basic model implies that all or almost all of a quasar's continuous spectrum comes from synchrotron emission from high-energy electrons replenished by the central energy source. The

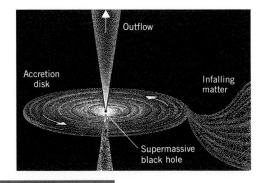

FIGURE 17.20

Schematic of one model for a supermassive black hole powering a quasar (or active galaxy). An accretion disk, made of infalling material, surrounds the black hole; it is thick close to the hole and thins out away from it. A sharp funnel forms, through which the accretion disk streams into the black hole. The disk is hot, so that gas blows off of it. The funnel directs the gas up along the rotation axis of the accretion disk; this creates a bipolar outflow from both sides of the accretion disk. (Adapted from a diagram by J. Burns and R. M. Price)

rest of the quasar acts like a transformation machine, trapping the energy of electrons and converting it into other forms, such as light.

What energy source lies in the heart of a quasar? The most developed quasar models to date invoke *supermassive black holes*—up to 100 million solar masses. This model takes after that for binary x-ray sources with black holes (Section 14.8), in which material from the normal star forms an accretion disk around the black hole before the material falls into it. A supermassive black hole in a dense nucleus is fueled by its tidal disruption of passing stars (Fig. 17.20). The stellar material forms an accretion disk and radiates as it spirals into the black hole, thus powering the quasar. Luminosities of 10^{12} solar luminosities are easily possible. To feed the black hole requires about 1 solar mass of material a year to generate such a level of power.

Supermassive black holes may produce the power of quasars by the efficient conversion of gravitational energy into other forms.

The model must also deal with the nuclear jets now known for some quasars. One way to make jets: an accretion disk can restrict the flow of ionized gas along its spin axis, either by magnetic fields or simply from the formation of a funnel of material around the black hole. This initiates narrow beams in two directions, which become visible as jets once they have gone beyond the main body of the quasar. A bipolar outflow of plasma develops, with the gas moving at close to the speed of light within the jets, which emerge at right angles to the accretion disk.

This model has many successful aspects. A supermassive black hole easily generates the level of quasar luminosity in a region of space only a few light years in size. (The Schwarzschild radius of a 10^8-solar-mass black hole is only 3×10^8 km, or about 2 AU). And it does the energy conversion (from gravitational to radiative) with high efficiency.

A Generic AGN/Quasar Model

Quasars and AGNs share some observed (and so perhaps some physical) characteristics (Table 17.1). For instance, quasars resemble Seyferts in their broad emission-line spectra. The Seyfert galaxies look like low-redshift quasars in terms of their emission spectra. So the physical conditions in the regions producing the broad-line spectra may be basically the same in Seyfert galaxies and in quasars. In addition, Seyferts visually resemble quasars: the nuclei of both are starlike, and a few nearby quasars have been shown to be surrounded by galaxylike disks.

We have observations of many nuclear jets in AGNs and quasars. The observed properties of jets in radio galaxies and in quasars are generally similar. This implies (but does not prove) that the physical conditions producing the jets are also the same.

Active galaxies and quasars share some of the same *observed* properties and so perhaps some of the same *physical* properties.

One popular idea views active galaxies as similar objects viewed from different directions, so that AGNs and quasars are driven by the same basic process.

Table 17.1

Comparison of AGNs and Quasars

Property	Radio Galaxies	Seyfert Galaxies	BL Lac Objects	Quasars
Redshifts from spectra	0.01–0.3	0.003–0.06	0.05–0.6	0.2–4+
Continuous spectrum	Nonthermal	Nonthermal	Nonthermal	Nonthermal
Emission lines	Broad and narrow (rare)	Broad and narrow	None or weak	Broad and narrow
Absorption lines	From stars in galaxy	None	None (?)	From intervening gas clouds
Shape (optical)	Elliptical	Spiral	Unclear	Starlike with fuzz
Shape (radio)	Jets and lobes	Weak emission from nucleus	Weak emission from nucleus	Jets and lobes

This generic model tries to link the nuclear engine (a supermassive black hole), an accretion disk, the regions of broad and narrow emission lines, and the jets—the key observational clues. Let's build a model with these pieces.

A generic model has a supermassive black hole as the energy source; it is surrounded by an accretion disk a few light days in diameter (Fig. 17.21). The supermassive black hole provides the power and the central synchrotron source as it swallows material from the accretion disk. Twin jets emerge from the accretion disk; far out from this central region, they plow into gas to pile up in radio lobes. Surrounding the central engine are two zones of ionized gas. The smaller (few light months in diameter) and closer one has gas moving with high random speeds (some 10,000 km/s); this region produces the broad emission lines. Another ionized region filled with clouds that have slower random motions (only a thousand kilometers a second) extends farther out, up to a few thousand light years. Here the narrow emission lines are generated. Between these two regions lies a thick, dusty ring of molecular gas. It may have a diameter of 10 to 1000 ly.

We can up the power output from this model by simply making the supermassive black hole larger and the rate of mass infall greater. For Seyfert galaxies, 10^6 to 10^7 solar masses would work; for quasars, we'd need some 10^8 to 10^9 solar masses. Other AGNs would fall within these ranges.

How we view this generic model determines what we see. Imagine a view almost edge-on (line *A* in

Fig. 17.21). The dusty ring blocks our view of the broad-line region and the central continuum source. So we see narrow emission lines, perhaps the jets, and the radio lobes, and we call the object a narrow-line Seyfert or a narrow-line radio galaxy. In contrast, if we observe the model more pole-on (line *B* in Fig. 17.21), we see the central continuum source, the broad line region, and some of the narrow-line region, perhaps the jets, and the radio lobes. We'd then call the object a quasar, broad-line radio galaxy, or a Seyfert with mostly broad and some narrow lines.

Now imagine we observed the model very close to or right along the axis of the jets. Then we would observe speeds greater than light and a very concentrated point of emission—a superluminal quasar or BL Lac object. In this orientation, we'd be looking through one radio lobe, with the other on the far side of the galaxy. Then we would have a hard time detecting both lobes. At best, we might see a faint halo of radio emission (from the lobe in front) with a strong nuclear source embedded within.

AGNs and quasars may be essentially the same kinds of object viewed from different angles.

These various orientations may then explain why sometimes we see only one nuclear jet or one much stronger than the other. The relativistic beaming ef-

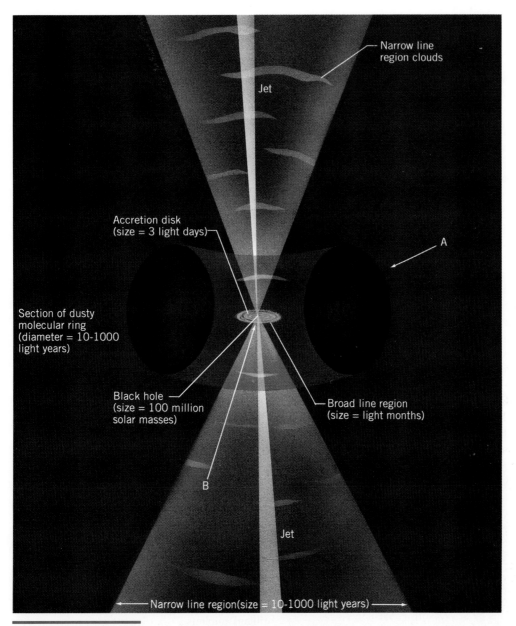

FIGURE 17.21

Generic model for an AGN or quasar. The central engine may be a
supermassive black hole surrounded by an accretion disk made from
infalling material. Close to the disk are blobs of fast-moving hot gas, which
make the broad emission lines in the spectrum. Much farther out is a dusty
ring of molecular gas. Beyond that lie blobs of slow-moving hot gas; these
make the narrow emission lines in the spectrum. If our line of sight is
along *A*, the dusty ring blocks out the broad emission line region; only the
narrow lines are visible. If our line of sight is along *B*, we see both narrow
and broad emission lines. The orientation of the AGN in space determines
our view.

fect implies that one bipolar jet close to our line of sight will appear very strong, the other much weaker (or invisible). So our generic, unified model naturally explains many aspects of AGNs and quasars.

We still have to worry about time. All these similar objects do have different redshifts, so we see them at different epochs of the universe. They probably evolve; the "how" is not clear. That is, the quasar phenomenon signals very violent activity in the nucleus of a galaxy at a very early stage in its life. Quasars have high luminosities and tend to exhibit symmetrical, two-lobed structures. The higher-luminosity active galaxies have similar radio shapes. This comparison suggests that quasars may be very distant, very active galaxies.

If we accept a model for the energy output of AGNs and quasars that imagines a black hole in the nucleus eating up huge amounts of material, the nucleus should show an intense, pointlike source of light (from a concentration of stars around the black hole). The stars orbiting the center should have high velocities, and emission lines with high Doppler shifts might be visible from the infalling matter. Observations of such effects have been reported for M87: a sharp peak of emission from M87's nucleus (such a peak is not found in the nucleus of normal elliptical galaxies), and a dramatic increase in the speeds of stars in the nucleus compared to those outside the nucleus. A model consistent with these

observations is one of a 5-billion-solar-mass black hole hiding in the inner 300 ly of the nucleus.

The Host Galaxies of Quasars

Given that a quasar may be the hyperactive nucleus of a very faraway galaxy, what about the rest of the galaxy? At cosmological distances, the disk of a quasar's parent galaxy, sometimes called the *host galaxy*, would be too small and faint to be seen easily.

We might, however, look for a quasar in a cluster of galaxies. A nice example is the quasar 3C 206, which is surrounded by some 20 faint galaxies (Fig. 17.22). The quasar's redshift and that of the galaxies are the same. This result not only shows that the quasar resides in the cluster but also proves that the quasar's redshift is cosmological, from the expansion of the universe. *what about Qso w/ different redshift*

Other observations are revealing that quasars are *from* surrounded by very faint envelopes that may well be *nearby* the hard-to-detect disks of their galaxies. For example, 3C 273 (look back at Fig. 17.11), the closest of *gal?* the high-luminosity quasars, has a fuzzy appearance in computer-processed images. The faint fuzz has emission lines in its spectrum with the same redshift as the quasar, which implies the two are the same object rather than two objects that happen to be lined up along the line of sight. In fact, almost all close quasars appear to have an extended structure associated with them—the underlying galaxy! However, we cannot tell whether the galaxies are spiral or elliptical. In general, quasars appear in host galaxies of average luminosity.

Quasars are hyperactive galaxies in which the host galaxy is usually lost in the glare of its brilliant nucleus.

Observations suggest that the host may be distorted, perhaps because of tidal interactions with a nearby galaxy. In a sample of nearby quasars, more than 30 percent appear to be interacting with another nearby galaxy. Most of these are located in a small cluster of galaxies. The quasar phenomenon may be activated by tidal interactions between galaxies in clusters.

FIGURE 17.22

The quasar 3C 206 (arrow) surrounded by galaxies, which probably all lie in the same cluster of galaxies. Because the quasar and the galaxies have the same redshift, we infer that the quasar lies in the same cluster. (Observations by S. Wyckoff, P. Wehinger, H. Spinrad, and A. Boksenberg; courtesy H. Spinrad)

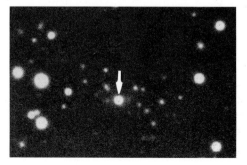

The Larger View

Most visible galaxies are normal—in energy output, in shape, and in contents of stars, gas, and dust. Yet, the nuclei of some special galaxies and all quasars are the sites of violent activity. A central engine—a supermassive black hole with a large accretion disk—spews out high-energy particles and generates electromagnetic radiation. All this action takes place in a region no more than a few light years across. Jets stretching from the nuclei show the connection of the central engine to enormous lobes of radio emission outside the galaxy, where the particles in the jets plunge into surrounding gaseous material.

By examining common aspects of AGNs and quasars, we can construct a generic model that explains many of their unusual features. This model includes a central engine of energy, hot and cool clouds of gas and dust moving at high speeds, and electrons moving in magnetic fields at close to the speed of light—particles that are channeled into nuclear jets. All this action occurs in a central core of a galaxy—one perhaps tugged by the tidal forces of another passing galaxy. Perhaps the cores of all galaxies contain supermassive black holes—some active, many dormant.

Nonthermal sources of electromagnetic radiation generated by the synchrotron process powers most of this activity. The universe has two visible aspects: a quiet one detectable by thermal emission (mostly from stars) that evolves slowly and a violent one visible from synchrotron emission that can alter swiftly.

Key Ideas and Terms

nonthermal spectrum
nucleus (of a galaxy)

1. The nucleus of the Galaxy contains a small, energetic source with a partly nonthermal spectrum. It emits radio, infrared, x-rays, and gamma rays strongly.

magnetic fields
polarization
synchrotron radiation

2. Continuous spectra can be divided into two types: thermal (blackbody) and nonthermal. Stars show thermal continuous spectra. A common form of nonthermal emission is synchrotron radiation, produced by high-speed electrons spiraling in magnetic fields.

active galactic nuclei (AGNs)
active galaxies
nonthermal spectra

3. Active galaxies have nonthermal spectra, especially from their nuclear regions; so their generic name is active galactic nuclei. Their thermal emission in the visual generally comes from their stars.

nuclear jets
radio galaxies
radio lobes
synchrotron radiation

4. Radio-emitting galaxies fall into two classes: compact (emission from the nucleus) and extended (emission usually in two lobes, widely separated from the nucleus). Often, a jet of emission extends from the galaxy's nucleus (an elliptical galaxy) to at least one of the lobes of an extended radio galaxy. The synchrotron process may be the source of the emission with high-speed electrons channeled along the jet out to the lobes.

AGNs
Doppler shift
emission lines
Seyfert galaxies

5. Seyfert galaxies, another type of AGN, are likely spiral galaxies, with bright, compact nuclei and very broad emission lines in their spectra; the broad lines arise from the Doppler shift of the emitting gas moving at high speeds.

AGNs
BL Lacertae objects
nonthermal spectra

6. BL Lacertae objects (also AGNs) have nonthermal spectra, vary rapidly in luminosity from a very small region in the nucleus, and usually have no or weak emission lines in their spectra.

Doppler effect
emission lines
quasar
redshifts
synchrotron
  radiation

7. Quasars have a starlike appearance, very large redshifts, and broad emission lines in their spectra (with narrower dark lines and emission lines). They sometimes have detectable radio emission with nuclear jets and radio lobes. A minority vary in luminosity in time intervals as short as a few days.

Doppler shift
Hubble's law
redshifts
supermassive black
  hole

8. If the redshifts in the emission lines from quasars arise from the expansion of the cosmos, quasars are billions of light years distant. Then their luminosities are huge, up to thousands of times that of normal galaxies. Yet the emitting region cannot be larger than a few tens of light years in size. We surmise that the central engine may well be a supermassive black hole.

accretion disk
AGNs
quasar
supermassive black
  hole

9. Quasars and AGNs share some of the same observational characteristics, so they might involve similar physical processes. A generic model for quasars and AGNs involves looking at the same kind of object from different directions (and at different times). The standard pieces in such a model are a supermassive black hole (surrounded by an accretion disk), eating up material in the nucleus; jets emitted perpendicular to the accretion disk; a thick, dusty ring around the accretion disk; a region of fast-moving ionized gas emitting broad lines; and a larger region of more slowly moving gas that produces narrow lines.

gravitational lens
general relativity

10. Large masses (galaxies) can gravitationally distort the light from more distant objects to produce lensing effects with various shapes, such as the double quasars and luminous arcs.

Review Questions

1. What observational property of nucleus of the Galaxy sets it apart from the rest of the Milky Way?
2. What is the common form of nonthermal emission in astronomy?
3. What is the common observational feature of AGNs?
4. What is the key observational feature of extended radio galaxies?
5. What is the key observational feature of Seyfert galaxies?
6. What sets BL Lac objects apart from other AGNs?

7. What is the key observational property of quasars?

8. How do astronomers find distances to quasars?

9. In the generic model for quasars and AGNs, what is the primary source of power?

10. In the generic model for quasars and AGNs, what is the source of the broad emission lines?

11. What can be inferred from the observation of two quasars with the same redshifts, close together in the sky?

Conceptual Exercises

1. What observational evidence do we have that the synchrotron process produces some radiation from active galaxies and quasars?

2. Contrast the evidence for violence in the Milky Way Galaxy with that for any active galaxy.

3. What evidence do we have that quasars are far away?

4. What *observational* evidence indicates that in many cases, the light of quasars passes through clouds of thin, cool gases?

5. What is thought to power a quasar?

6. What connection does extended radio emission often have with the nucleus of an active galaxy?

7. Give one way in which the observed properties of radio galaxies and quasars differ. One way in which they are similar.

8. What physical conditions are required to produce synchrotron emission in active galaxies and quasars?

9. Suppose you read an announcement in the paper tomorrow that astronomers have measured Hubble's constant to be twice the value used in this book. What would happen to the estimated distances to quasars?

10. How does the duration of variation in the light from an active galaxy indicate the size of the emitting region?

11. Emission lines from quasars and some Seyfert galaxies are very broad. What physical process can account for this observation?

18

The Cosmos: Past and Present

■

Central Concept
The universe originated at a finite time in the past and has evolved to its present state guided by fundamental physical laws.

■

Almost 20 billion years ago, our universe was born out of a cosmic fireball—the Big Bang; in the violence of this event, all that we now see was created. That, in a nutshell, is one picture of our universe's creation that is accepted by many astronomers today.

Despite its general acceptance, the Big Bang model does have shortcomings. A new theoretical scenario, called the inflationary universe, has been proposed to deal with some troublesome issues. Regardless of whether future observations will finally prove the inflationary model correct, a main conceptual basis will linger: that the universe is directly linked to elementary particles during its early history. The Big Bang signals the intimate connection of the very small with the very large.

18.1 Cosmological Assumptions and Observations

Cosmology, the subject of this chapter, is the study of the universe's nature and evolution. It relies on simple models of the universe to gain an understanding about it. Modern astronomy has arrived at models of the cosmos in which the universe *evolves*; it is dynamic, not static. How can we study the universe's evolution in detail? We need a few powerful funda-

mental starting assumptions, difficult-to-prove assertions about the nature of the cosmos.

Assumption 1 The **universality of physical laws**. This assumption covers both local (the earth and solar system) and distant regions; it means that we apply the physical laws we uncover here to all localities at all times and to the universe as a whole. Key observations support this assumption. For example, the spectra of distant galaxies contain the same atomic spectral lines as those produced by elements found on the earth. So other galaxies are made of the same elements as here, put together in the same way. Newton's law of gravitation correctly describes the motion of double stars and galaxies.

Assumption 2 The cosmos is **homogeneous** (Fig. 18.1). This means that matter and radiation are spread out uniformly, with no large gaps or bunches. You know that this assumption is not strictly true, for clumps of matter, such as galaxies and stars, do exist. But the cosmologist assumes that the size of the largest clumping—superclusters—is much smaller than the size of the universe. It's like looking at the earth from space: bumps, such as mountains, are too small to be seen, so the globe looks smooth.

Assumption 3 The universe is **isotropic** (Fig. 18.2). This idea relates to a quality of space itself rather than to the matter in it. Here's one way to think of it. Space has the same properties in all directions. So no direction or place in space can be distinguished from any other by any experiment or observation. The universe has no center in space, because there is no way to tell if you are there. No direction in space is special. For example: the expansion of the universe is the same in all directions of space. Any observer would see the same Hubble's law.

Astronomers assume that the universe is homogeneous and isotropic, and that it operates with the same physical laws throughout space and time.

What is the universe? All that is, has been, and will be. However, astronomers take a narrower view. If we consider all that can be seen with telescopes, we are considering the *observable universe*. Yet this cannot be *all* of the universe; there are objects too

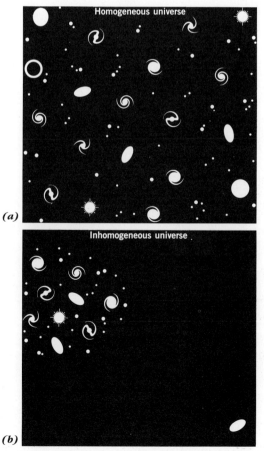

Homogeneous universe

(a)

Inhomogeneous universe

(b)

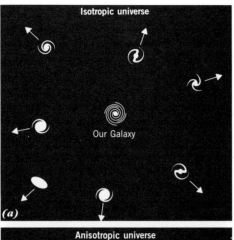

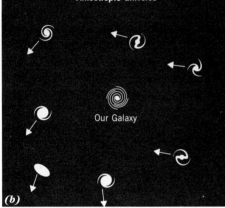

FIGURE 18.2

Isotropy and the universe. (*a*) In an isotropic universe, space has the same properties in all directions. For example, the distribution of velocities of galaxies in the expanding universe is smooth, uniform, and the same in all directions. A Hubble law is observed. (*b*) In an anisotropic universe, the expansion of the universe is different at different places in space. A Hubble law is *not* observed.

faint and too far to be seen, regions of the spectrum to which we and our instruments are so far blind, and objects detectable only by their gravitational effects. So there is more to the observable universe: a **physical universe** that includes directly observable matter as well as objects we detect by effects described by the laws of physics.

Cosmology and Relativity

What do we know about the universe? First, the universe *evolves*. Both the whole cosmos and its con-

tents change with time. Second, matter in the universe is *grouped*. Elementary particles make up protons, neutrons, and electrons, which make up atoms. Atoms make up gases (molecular and atomic) and dust particles, which form stars, planets, and us. Stars come in clusters of stars, which are found in galaxies. And galaxies are grouped in clusters of galaxies, which in turn congregate in superclusters of galaxies. And these make up the universe itself. To grasp the cosmos, we need to apply Einstein's general theory of relativity (Section 10.7) and picture

FIGURE 18.3

Comparison of geometries. (*a*) In a flat geometry, the sum of the angles of a triangle is always 180°. (*b*) In a hyperbolic (open) geometry, the sum of the angles of a triangle is always less than 180°. (*c*) In a spherical (closed) geometry, the sum of the angles of a triangle is always greater than 180° (and may be as great as 270°).

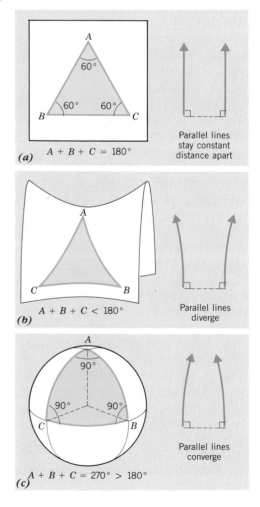

the geometry of spacetime globally rather than locally. What properties can that geometry have?

The easiest to visualize is the geometry of a flat piece of paper—Euclidean geometry. Because of Euclid's parallel-line postulate (essentially, that if two parallel lines are extended to infinity, they will remain the same distance apart and never meet), Euclidean geometry is flat. Its flatness results in the Pythagorean theorem for right triangles and the statement that the sum of the angles of any triangle *equals* 180° (Fig. 18.3*a*). This is a key property of a **flat geometry**.

Imagine a geometry in which parallel lines that *diverge* when extended. In two dimensions this geometry has properties similar to the surface of a saddle (Fig. 18.3*b*); the geometry is curved. This geometry is sometimes termed **hyperbolic geometry**. When a triangle is drawn on a hyperbolic surface, the sum of its angles is *less than* 180°. Hyperbolic geometry is infinite, because extended parallel lines never meet. Both hyperbolic and flat geometries are for this reason called **open geometries**.

Now imagine parallel lines that *converge* when extended. The surface of a sphere has this geometry (Fig. 18.3*c*), where the sum of a triangle's angles is *greater than* 180°. Such a geometry is sometimes called **spherical geometry** or **closed geometry**. Consider, as an analogy, the earth's surface with its lines of longitude and latitude. Two lines of longitude, both perpendicular to the equator (and so parallel to each other), intersect at the poles.

General relativity allows the geometry of the universe to be open or closed.

Both these non-Euclidean geometries are characterized by their curvature, which does not typify a flat, Euclidean geometry. The hyperbolic geometry has a negative curvature because it bends away from itself. It extends infinitely far. Spherical geometry, in contrast, curves in on itself. Because of its positive curvature, spherical geometry is finite but unbounded: It has a definite size, but no edge.

Consider again the earth's surface. You can travel around the earth's surface as many times as you like and in any direction you want without ever discovering a boundary. If you sent out light signals, they would eventually return to you.

We don't know for sure that our universe *is* closed—that depends on whether it contains enough matter and energy to curve it sufficiently. The density of mass (and energy) determines the curvature of spacetime. If this density has a certain critical value (or greater), then the universe curves back on itself and is closed. Einstein's general theory gives this **critical density**: it is roughly 5×10^{-27} kg/m^3. (That's about one hydrogen atom for every cubic meter of space.) We can calculate this value from the observed value of the Hubble constant.

Observations imply that the universe is now *expanding* at a rate of about 20 km/s/Mly (Section 16.4). If we assume that this rate has been constant since it began, we can calculate a time in the past when the universe had zero volume: 15 billion years ago. That time marks the origin of the universe for astronomers. What about its future? That depends on the cosmic geometry.

Let's apply the concept of escape speed (Section 3.4) to the expansion of the universe. Consider the universe and the galaxies within it. The galaxies were once all "thrown away" from one another, for we now see an expansion. Consider a galaxy at some distance from us. Matter spread within some space acts as if it were all concentrated at the center (as long as the matter is distributed uniformly). If there is enough mass within that space, the escape speed will be larger than the expansion speed. Note that "enough mass within that space" means a high enough density—the critical density.

If the density of the universe is large enough, the galaxies will not have escape speed; the expansion speeds will decrease with time and eventually reverse as gravity herds all the galaxies together. If the average density is too low, the galaxies will have more than escape speed; gravity will never bring the galaxies all together, and the expansion will continue indefinitely. Hyperbolic, flat, and spherical geometries correspond physically to the cases of greater than, equal to, and less than escape speed. Which case applies?

We can find out which case does apply by comparing the *critical density* (which we calculate from the value of Hubble's constant) to the *actual density* of matter in the universe—or at least as large a sample as we can manage. This tack runs us right back into dark matter. If we average out the luminous matter, we find that its density is at least 10 times smaller than the critical density. Taken at face value, this missing matter means that the universe is open and will expand forever. But, if some 90 percent of the matter in the universe is dark, then there is just enough to equal the critical density and close the universe.

We can determine the geometry of the universe by comparing the calculated critical density with the actual, observed density; it appears that the universe is open or contains a large fraction of dark matter.

18.2 The Big Bang Model

Ingenious theoreticians have devised a bewildering array of cosmological models. The dominant one is the *Big Bang model*, which is the standard model based on Einstein's general theory of relativity. Almost all astronomers today would agree that the Big Bang model is the standard in use. But be warned that this model is *not* final; it has problems that need to be ironed out. One attempt has been made with the inflationary universe model (Section 18.6), a revision of the standard Big Bang model.

The universe is now expanding. Imagine it running backward. The galaxies and all matter within and without them eventually smash tightly together. The extreme compression heats both matter and radiation to a very high temperature—high enough to break down all structure that was created previously, including all the elements fused in stars. The atoms break down into protons and electrons; the matter is completely ionized—a plasma. In addition, the density of the plasma is so great that photons can travel only short distances before they are absorbed. As a result, the entire universe is opaque to its own radiation.

Now, when matter is opaque to all radiation, it acts like a blackbody (Section 10.3). The radiation from a blackbody exhibits a characteristic shape in its spectrum (Fig. 18.4). In an early dense, hot state, the whole universe acts like a blackbody radiator. If you could have been there, you would have seen a bright fog all around, like sitting in the sun's interior (but with the light at even shorter wavelengths, because of the higher temperature).

The Big Bang model pictures the universe beginning in a hot, dense state, billions of years ago.

Imagine the universe expanding violently from this infernal state; hence the name *Big Bang*. As the universe expands, its overall density and temperature decrease. (This is true for the expansion of any gas. Let a gas out of a container. As it expands, the gas suddenly becomes cold.) Eventually the temperature drops so low that protons can capture electrons to form neutral hydrogen. This neutral gas is basically transparent to most kinds of light.

This event—the formation of neutral hydrogen from a plasma—marks a crucial stage in the evolution of the universe. No longer ionized, the universe becomes transparent to its own radiation; the light is freed of its close interaction with matter. The radiation freely speeds throughout space, and the expansion dilutes it. It is also redshifted, just like the light of distant galaxies, in accordance with Hubble's law: having been emitted long ago, it comes from far away.

The Big Bang model requires that a relic of low-temperature cosmic radiation pervade the universe.

The Big Bang Model	Origin	in the Big Bang, some 15–20 billion years ago	
	Evidence	expansion (redshifts), 3 K background radiation	
	Nucleosynthesis	during first few minutes	
At a Glance	Galaxy formation	during first billion years	
	Future	expansion forever (the Big Bore), or reverse and collapse (the Big Crunch)	

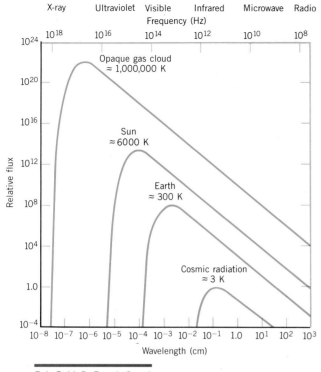

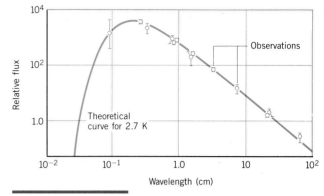

FIGURE 18.4

Comparison of the continuous spectra of blackbodies at different temperatures, from 3 K to 10^6 K. Cosmic radiation at about 3 K peaks in flux at infrared–microwave wavelengths; hotter blackbodies, such as the earth and sun, peak at shorter wavelengths. The overall shape of each spectrum is the same because these are all blackbody emitters.

FIGURE 18.5

Some observations of the spectrum of the cosmic microwave radiation. The points show measured fluxes at radio and infrared wavelengths; the solid curve corresponds to a theoretical blackbody of 2.7 K. The lines extending above and below the circles for each observation indicate the range of error in each. (Adapted from a diagram by P. J. E. Peebles)

If the universe did indeed begin in a hot Big Bang, debris (both matter and radiation) from the cosmic explosion of the Big Bang should now lie all around us. The matter's pretty obvious (except for the dark matter!). But what about the radiation produced in the Big Bang? It's cooled down to a fairly low temperature. So the radiation's peak wavelength now will be long, as we'd expect from a cool blackbody. And if the universe has expanded uniformly, the radiation's spectrum should show the telltale blackbody shape, because uniform expansion preserves the blackbody properties of the radiation.

Have we seen such cosmic radiation? Yes! Its discovery strongly supports the hot Big Bang model.

 ## 18.3 The Cosmic Background Radiation

In 1964 Arno Penzias and Robert Wilson, scientists with Bell Laboratories in New Jersey, detected an annoying excess radiation by means of a special low-noise radio antenna. Penzias and Wilson further discovered that the noise did not change in flux with the direction in the sky, the time of day, or the season. Thus it was by accident that radiation left over from the Big Bang was first observed. Many other observations since then have verified the blackbody character of these emissions, which have the blackbody temperature of 2.7 K (Fig. 18.5).

The cosmic background radiation has a spectrum with a blackbody shape and a temperature of about 2.7 K.

To clinch its cosmic origin, the radiation should be isotropic; that is, it should have the same measured flux from all directions in the sky. For if the radiation were from a hot primeval state, the isotropy of the universe at that time would set the isotropy of the radiation. All observations back up the notion that the radiation is very isotropic with a very slight alteration from our motion relative to the radiation (as you'll see in a bit). This result suggests that, in fact, the newborn universe was very nearly isotropic.

Observations show that the background radiation arrives at the earth uniformly from all directions in space.

This background radiation is usually called the **cosmic blackbody microwave radiation**: "cosmic" because it comes from all directions in space, "blackbody" because of its spectral shape, and "microwave" because its spectrum peaks at centimeter-to-millimeter wavelengths (the length of microwaves). The discovery of such radiation makes constructing models of the universe easier because it contains new information about the cosmos.

1. The radiation enables astronomers to glimpse the raw, young universe. We can conclude that the initial universe was indeed quite homogeneous and isotropic.

2. The present temperature and the isotropy of the radiation let us build a history of the universe that describes the change of temperature of matter and radiation with time.

3. The radiation's presence establishes an important marker for galaxy formation. Until the radiation and matter stopped their interaction, matter could not form any large clumps. Only after the plasma had recombined could matter form clumps that eventually became stars and galaxies (Section 18.4).

4. The slight anisotropy in the radiation provides a reference for measuring the motion of the Galaxy and the Local Group. In the direction of any such motion, the cosmic radiation is blueshifted and so appears hotter in that location in the sky. In the opposite direction of the sky, the radiation is redshifted (and so appears cooler). Observations indicate that the Local Group and Local Supercluster have a combined motion of about 600 km/s in the direction of a nearby supercluster in the constellations of Hydra and Centaurus.

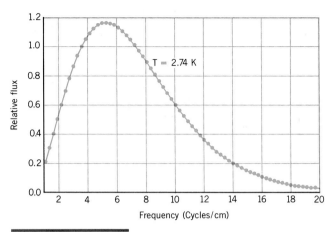

FIGURE 18.6

Observations of the cosmic microwave background radiation by the Cosmic Background Explorer (COBE) satellite. The circles are the data; the line is the expected spectrum of a perfect blackbody at a temperature of 2.74 K. Note how well the line matches the data points.

Results from the Cosmic Background Explorer (COBE) satellite confirm the pristine nature of the background radiation. They show that the spectrum is *extremely* smooth, with less than one percent deviation from that of a blackbody at a temperature of 2.74 K (Fig. 18.6)—it would be difficult to produce a better blackbody under controlled laboratory conditions! In addition, the extent over the sky is also very smooth—variations are less than a few parts in 10^5—with no obvious sign of any early clumping that might explain the clustering of matter today (Fig. 18.7). The smoothness of the radiation's distribution in the sky has profound implications for our models of galaxy formation (Section 18.5).

The smoothness of the cosmic blackbody microwave radiation indicates that the young universe was very isotropic and homogeneous.

18.4 The Standard Big Bang Model

Since the discovery of the cosmic microwave background radiation, most astronomers have come to

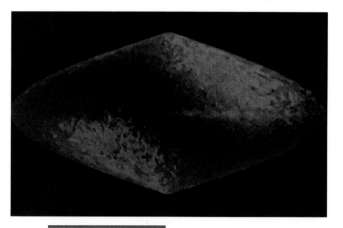

F I G U R E   1 8 . 7

Cosmic Background Explorer (COBE) satellite all-sky map of the cosmic microwave background radiation. At a frequency of 53 GHz, the radiation is very smooth; the variations in the pink and blue regions are very, very small. The colors show the earth's motion relative to the radiation, which results in a Doppler shift and so a change in temperature. The pink false color indicates a "blueshift" (higher temperature); the false color blue a "redshift" (lower temperatures). The dark blue regions are unobserved parts of the sky blocked by the earth (center) and sun (edges). (Courtesy NASA)

accept a hot Big Bang model. With the addition of knowledge on how matter behaves under hot, dense conditions, astronomers have crafted step-by-step details of what can happen in a Big Bang model.

Caution: Please don't picture the Big Bang as happening in the "center" of the universe and expanding to fill it. The Big Bang involved the *entire* universe; every place in it was *at* creation, which marked the beginning of time and space.

The Primeval Fireball

Although the present temperature of the cosmic radiation is low, the amount of energy it contributes to each cubic meter of the universe is large. If the energy in the radiation could be used to heat up all matter, the temperature would be greater than 10^{12} K. This hot beginning is often called the **primeval fireball**. From Einstein's equations of general relativity we find that a temperature of 10^{12} K corresponds to a time of about 10^{-24} second after time "zero"—the beginning of the cosmic expansion. At that time, the universe had an average density of about that of an atom's nucleus!

The primeval fireball produces such a rapid expansion that the temperature and density drop rapidly. The contents of the universe (Table 18.1)—elementary particles—change with temperature and time. Each period of time can be matched with a temperature. Roughly, the universe's thermal history divides into four eras: a **heavy-particle era**, when massive particles and antiparticles dominated; a **light-particle era**, when particles with less mass were made; a **radiation era**, when most particles had vanished and radiation was the main form of energy; and a **matter era**, in which we now live, when the energy of matter dominates that of radiation. (In a cubic meter of space, matter now averages about 1000 times the density of photons, considering their energy in the form of mass.)

Creation of Matter from Photons

Before the story unfolds, you need a little preparation about one key part: the creation of matter from photons. At some time in the primeval fireball, the energy of photons was so high that their collisions produced matter. When the energy in the colliding photons equals or exceeds the mass ($E = mc^2$) of the particles produced, energy can become matter.

The creation of matter from light involves both matter and antimatter. (**Antimatter** has the same properties as regular matter except that it has opposite electric charge. For instance, an antimatter electron has a positive charge—Table 18.1.) When matter and antimatter collide, they are annihilated and converted to photons (Fig. 18.8). In reverse, two photons (if they have enough energy) create a matter–antimatter pair when they collide. Note that this process always results in *pairs* of particles.

The photons must have at least the energy equivalent of the masses of the pair they produce. So making protons and antiprotons takes more energy than making electrons and positrons, because protons have about 1800 times more mass than electrons. This involves high temperatures: at least 10^{10} K to make a pair of electrons and at least some 10^{13} K for a pair of protons.

At high enough temperatures (and energies), photons create elementary particles in particle–antiparticle pairs.

Table 18.1

Particles in Cosmic Nucleosynthesis

Particle and Antiparticle	Symbol	Charge	Comments
Neutrino, anti-neutrino	$\nu, \bar{\nu}$	0, 0	Massless (?) particles that travel at light speed; stable (?)
Proton, anti-proton	$p, \bar{p}$	+1, −1	Nucleus of hydrogen; stable.
Electron, positron	e^-, e^+	−1, +1	Particles surrounding the nucleus of an atom; stable.
Neutron, anti-neutron	$n, \bar{n}$	0, 0	Decays to a proton and an electron in about 1000 seconds.
Photon	γ	0	Packet of radiation, electromagnetic energy.
Deuteron	^2H	+1	Nucleus of deuterium, or "heavy hydrogen"; contains 1 proton, 1 neutron; stable.
Helium-3	^3He	+2	Nucleus of an unusual type of helium; contains 2 protons, 1 neutron; stable.
Helium-4	^4He	+2	Nucleus of ordinary helium; contains 2 protons, 2 neutrons; stable.
Lithium-7	^7Li	+3	Nucleus of most abundant type of lithium; contains 3 protons, 4 neutrons; stable.
Beryllium-9	^9Be	+4	Nucleus of most abundant type of beryllium; contains 4 protons, 5 neutrons; unstable.

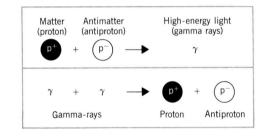

F I G U R E   1 8 . 8

Matter and antimatter annihilation to make photons; particle and antiparticle production from photons. In this case, the particles are protons; the antiparticles, antiprotons. For the conversion to work, each gamma ray must have an energy at least equal to that of the mass of a proton.

Particle production from photons played a key role in the very young universe. Let's see what the model tells about those times.

Cosmic Eras

Early on, photons have enough energy to enable the production of even the most massive elementary particles. Annihilation also takes place, and the balance between annihilation and creation marks the *heavy-particle era*. This period does not last long, because the cosmos expands rapidly and the temperature declines quickly (Table 18.2). So at the earliest times, the universe is a smooth soup of high-energy light and massive, elementary particles. The nuclei of atoms have yet to be made.

With the dropping temperature, the remaining photons lack the energy to create new heavy particles. Now only light particles—electrons—can be made from photons. The universe enters the *light-particle era*. Protons and electrons interact to make neutrons. When the temperature falls far enough, the light-particle era ends. Existing neutrons now decay into protons and electrons but no new ones are made. The ones that survive are crucial to the next step.

Only a few minutes have passed! The remaining neutrons and protons react to form nuclei of simple elements. The most important reaction involves the fusion of a neutron and a proton to form deuterium. All neutrons, except those that have decayed, end up in deuterium. Once the deuterium has been produced, further reactions create normal helium. The

Table 18.2

Sequence of Events in the Big Bang

Event	Time	Density (kg/m^3)	Temperature (K)	Comments
Origin	0	?	?	Not the province of present science; relativity fails.
Heavy-particle era	10^{-44} second	10^{97}	10^{33}	Photons make massive particles (such as protons) and antiparticles.
Light-particle era	10^{-4} second	10^{17}	10^{12}	Photons have only enough energy to make light particles and antiparticles, such as electrons and positrons. Protons and electrons combine to make neutrons.
Radiation era	10 seconds	10^7	10^{10}	Few particles left in a sea of radiation; these participate in nucleosynthesis of deuterium, helium, lithium, and beryllium.
Matter era	10^6 years	10^{-18}	3×10^3	Ionized hydrogen recombines; cosmic radiation and matter decouple.
Now	10^{10} years	10^{-31} (radiation)	3 (radiation)	Astronomers puzzle about creation.

net result is 20 to 30 percent helium by mass compared to all forms of matter. Little bits of beryllium and lithium are created. Some deuterium is left over. This sequence marks the *nucleosynthesis era*.

In its first few minutes, the universe reaches temperatures and densities suitable for the formation of deuterium and helium and a very small fraction of a few other light elements.

The cosmos now enters the *radiation era*. So far the temperature has been so high that all atoms in the universe have been ionized. After about 1 million years, the radiation's temperature has plunged to 3000 K, too low to keep matter ionized. The nuclei begin to capture electrons to form neutral atoms—a process called **recombination**, which happens in a few thousand years. As a result, the matter becomes transparent to the radiation. Suddenly light breaks through the now-transparent gas.

The radiation expands and cools down to become the 2.7 K cosmic radiation of today. The matter, however, follows a different course because of little bumps in the generally smooth distribution. Clouds of matter condense out of the primeval fireball. The

radiation era ends, and the *matter era* begins. Material clumping could not happen until the radiation and matter had decoupled. (Why not? Because the radiation exerted a pressure on the matter that kept it from gathering into clumps.) This event flags the time when galaxy formation could begin (see below).

At about 1 million years, the cosmic plasma recombined; radiation was unlocked from matter; then matter could clump, and the radiation became the cosmic background radiation.

Observations and the Big Bang

What observational evidence supports the Big Bang model? First and foremost, observations of the redshifts of galaxies indicate that the universe is expanding. Second, observations indicate that the blackbody radiation pervades the universe. Within the limits of observational errors, measurements confirm a blackbody spectrum with a temperature of 2.7 K. Measured in many directions of space, the radiation comes to the earth with a very uniform flux. The background radiation has the attributes expected of a cosmic, hot origin.

Third, the model predicts that primeval helium was formed in the first 5 minutes of the universe's history and that the helium abundance should be 20 to 30 percent by mass. The Big Bang model sets this helium abundance as the basement level; any observation of a substantially lower amount calls the model into question. Additional helium is formed in stars later on. Hence, the helium abundance we measure today must be larger than the 20 to 30 percent that was made in the Big Bang.

Unfortunately, we can assess the present cosmic helium abundance only indirectly. Good objects to search for primeval helium are the oldest stars now surviving, the Population II stars. Most Population II stars, however, are much too cool to excite helium lines in their spectra. However, indirect evidence from star models indicates that these stars have helium abundances of about 30 percent. The best measurement to date relies on the helium-to-hydrogen abundances in the H II regions in dwarf irregular galaxies. These indicate that the primordial abundance was not more than 25 percent.

> The accumulation of evidence for the hot Big Bang is strong; the model gives a simple, unified picture of the early universe.

Formation of Galaxies

Up to 1 million years after creation, gravity could not clump matter into stars and galaxies. Pressure from the radiation itself inhibited gravitational contraction. But once matter had recombined, the universe became transparent to radiation. Then radiation pressure could no longer stop gravity from doing its natural work. We refer to this period as the time of **decoupling**.

The isotropy of the cosmic background radiation implies that at the time of decoupling, the universe had a very uniform distribution of matter. But that's not the situation now. Even the most spread-out of these systems of matter—clusters of galaxies—have average densities about 100 times greater than the average density of the universe. So here's the crucial question: How did a very smooth universe become lumpy?

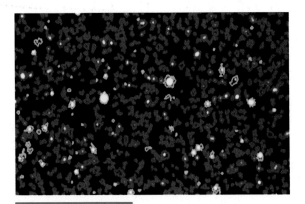

FIGURE 18.9

Long-exposure image of a small section of the southern sky (3 × 5 arcmin, or 1/200 the area of a full moon). In this processed and color-coded image, unresolved galaxies appear red, those with resolved stellar cores are blue, and stars are green. Note that many of the galaxy images overlap; more than a thousand are visible. (Courtesy J. A. Tyson, AT&T Bell Laboratories)

Before decoupling, only very large disturbances would contract. Just after decoupling, small disturbances could have condensed, along with disturbances of greater mass. So the time of decoupling marks the period during which galaxy formation could start in the young universe. Now, although clumps can be unstable, they grow slowly—so slowly that the galaxies we see could hardly have formed by now unless the disturbances were already large to start. But the smoothness of the cosmic background radiation implies that any lumps were small. That is the dilemma of galaxy formation from a smooth Big Bang.

> Galaxies formed early in the history of the universe by gravitational contraction, but we don't yet know the details of how or when.

One way to see what happens is to look very far out in space and so back in time, searching for the youngest galaxies possible. Such a search has been carried out at visual wavelengths with very long (hours!) exposures in a region of the sky above the plane of the Galaxy. Here stellar images in the Milky Way are at a minimum; what appears are distant galaxies—loads of them (Fig. 18.9)! Their distances are estimated (not measured!) to range from about 7 to 11×10^9 ly—in the same region as quasars. But these objects are extended, not pointlike, and they seem to have nuclear regions. They may be very young galaxies undergoing their first burst of star formation, just after the Big Bang.

Table 18.3

Properties of the Fundamental Forces

Force	Strength (relative to strong)	Range (m)
Strong	1	10^{-15}
Electromagnetic	1/137	Infinite
Weak	10^{-5}	10^{-17}
Gravity	6×10^{-39}	Infinite

18.5 Elementary Particles and the Cosmos

Today's standard model, the hot Big Bang model, has a number of weaknesses, such as the great dominance of matter over antimatter in the universe now, and the formation of galaxies. Both these problems and others are being tackled in imaginative theoretical ways that unite the universe of the small—elementary particles—with the universe itself. The connection occurs early in the Big Bang, before the first second has elapsed. So first some background about elementary particles and their interactions.

The Forces of Nature

Let's focus on how particles relate to each other—their interactions. We generally picture these relations as forces between particles. You are familiar with two: gravitation and electromagnetism. These forces have one property in common: they work over large distances. According to Newton's law of gravitation, the most distant galaxies exert a force on you (and you on them). The same is true of objects that have a net electrical charge. These forces differ in their relative strengths. Electromagnetic forces are *much* stronger than gravity—as you know if you've lifted a nail with a magnet. The gravitational attraction of the entire earth on the nail is weaker than the magnetic force.

Two other forces operate in the subatomic domain. One, called the **strong nuclear force**, binds the nuclei of atoms. Recall that an atom's nucleus has protons tightly packed together. The electrical force of each proton repels the others strongly, especially when so close—only 10^{-15} meter apart. The strong force, however, overwhelms this electrical repulsion and keeps the nucleus together. The other,

called the **weak nuclear force**, crops up in radioactive decay. Without it, fission would not take place. Like the strong force, the weak force operates over very short distances, 10^{-17} meter and less.

These four forces are all that are known. The strong force is the strongest of them (Table 18.3), electromagnetism is second, the weak force is next, and gravity takes the bottom as the weakest. Now, although these forces appear to operate very differently, might they have an underlying unity? That quest for a unified theory has tempted physicists for most of this century. Recently they have had some success in the struggle to find a **grand unified theory**, fondly known as *GUT*. The development of GUTs (there is more than one such theory!) has, curiously enough, modified Big Bang cosmology and helped to shore up some weaknesses.

How is that connection made? In the 1970s, theoreticians unified the weak and the electromagnetic forces—into an *electroweak* force. Buoyed by their success, they next took aim at unifying the electroweak and strong forces. They developed the concept of a new elementary particle, called a *quark* (with associated *antiquarks*). GUTs predict that the unification of strong and electroweak forces won't occur before the attainment of energies greater than some 10^5 J. Such energies cannot be made in large-particle accelerators on earth, but they do occur in the Big Bang model at a time of 10^{-35} second when the temperature was 10^{26} K and the distances between particles was 10^{-40} meter. So the conditions then may have been just right for a unification of forces. The Big Bang serves as a way to test GUTs—the whole universe acts as a high-energy particle machine!

Grand unified theories view the four fundamental forces as having an underlying unity, which operated early in the universe.

GUTs have simplified our view of elementary particles. We now believe that matter is composed of two classes of elementary particles: *quarks* and *leptons*. The *lepton* group (low-mass, "light" elementary particles) consists of particles of only six types: they include electrons and neutrinos and their antiparticles. All forces except the strong act on them. Quarks make up all the other particles, such as protons and neutrons and some hundred others. (These are high-mass, "heavy" elementary particles.) All four forces act on these particles. The aim of GUTs is to reduce all particles to one kind, interacting through one force—truly a grand unification!

GUTs and the Cosmos

We now look at the Big Bang model in a new way: the temperatures at which specific aspects of particles and their forces freeze into existence. I use the word "freeze" in the sense that when water freezes into ice (at 0° C or 273 K), its state changes abruptly. Water behaves differently in the form of ice and in the liquid or gaseous forms.

Section 18.4 brought up two freezings in the early cosmos: the formation of simple nuclei in the nucleosynthesis era (first few minutes) and of atoms during the recombination time (1 million years). GUTs predict a very special kind of freezing at 10^{-35} second, when the strong nuclear force froze out from its unification with the electroweak force. Earlier, from 10^{-43} second to 10^{-35} second, gravity had frozen out from the other three forces. Before that time, all four forces were unified.

The final freezing takes place at about one second, when the electromagnetic and weak forces split. At this time, quarks can combine to form particles such as protons, which can then react in the nucleosynthesis era. Before this freezing, the universe contained a hot gas of quarks and electrons. From this time on, the standard Big Bang model applies (Section 18.4).

A special event occurs at the separation of the electroweak force from the strong force at 10^{-35} second. Just as water releases energy in the form of heat when it freezes, the strong force freezing also released energy, but in enormous amounts. This event pumped energy into the expansion of the universe so that it grew in size by many powers of ten in just 10^{-32} second. This marks the era of inflation, which is the hallmark of the inflationary universe (Section 18.6).

These successive separations of forces solve some of the Big Bang's problems. The antimatter problem results from the fact that at 10^{-3} second, the universe contained 1 billion plus 1 protons for every billion antiprotons. Now photons create particles and antiparticles in *equal* pairs. Before the strong force freezing, that was true—equal numbers of particles and antiparticles existed. After the freezing, these particles decayed into other particles, but unequally in terms of matter and antimatter. A little more matter than antimatter was created in the decay; once this had happened, nothing could change this frozen-in asymmetry. Later, the matter annihilated all the antimatter, leaving only the excess matter in the cosmos.

GUTs and Galaxy Formation

Another aspect of this freezing relates to the problem of galaxy formation. Consider the freezing of a pond: the ice sheet does not form all at once but in patches; that is, the freezing process has defects. These defects have mass and survive for a long time—so long that at the decoupling time, when matter can clump, the defects serve as the cores (the lumps) on which gravitational instability occurs.

Neutrinos may have also helped to make galaxies. Some GUTs predict that neutrinos should have a very small mass—perhaps 0.001 percent that of an electron. Neutrinos freeze out at a time of 1 second, and those relic neutrinos should be with us today. If massless, they have little effect. If they have even a slight mass, they change the universe. (And neutrinos with mass might solve the solar neutrino problem: Section 10.5.)

Models of the clustering of dark matter, such as neutrinos, have gotten around one of the most seri-

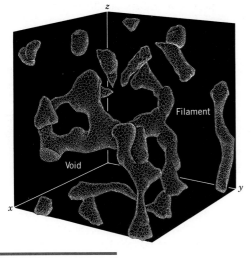

FIGURE 18.10

Large-scale structure and computer simulations of galaxy formation. Results of a theoretical calculation of the clumping of matter in an expanding universe dominated by neutrinos. The matter gathers into filaments (like superclusters of galaxies) with voids in between. This plot in three dimensions covers the equivalent of millions of light years of space. (Based on models by Joan Centralla and Adrian Melott)

ous objections to galaxy formation in the standard Big Bang model (Section 18.4). Observations rule out large disturbances in ordinary matter at the time of decoupling. But neutrino matter could be lumpy without affecting the smoothness of the background radiation. Computer simulations show that neutrinos would gather into pancakelike shapes, hundreds of millions of light years in diameter and containing some 10^{15} solar masses—about the mass of a supercluster. Where such pancakes push into each other, they would fragment into knots of ordinary material that over time could collect, cool, and condense to make galaxies. So this pancake model for galaxy formation predicts that galaxies should occur in long chains and filaments, with voids in between (Fig. 18.10). And that's what we are just beginning to see in the structure of superclusters and voids.

Other recent supercomputer models have attacked the same question, and their results show a spongelike structure of the voids and matter (Fig. 18.11). Most of the mass is assumed to be in

FIGURE 18.11

Supercomputer simulation of a sequence of possible development of large-scale clumping of the universe. This model assumes random fluctuations of density acted on only by gravity and simulates their growth from (a) to (c). The false colors represent density: blue the highest, red the lowest. (Courtesy Adrian Melott; NCSA)

(a)   *(b)*   *(c)*

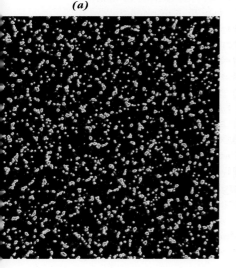

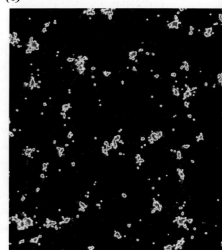

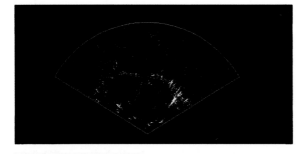

FIGURE 18.12

Results of a supercomputer simulation of the growth of small ripples in the early universe after some 13 billion years of the influence of gravity. Note the stringlike clustering of the matter (mostly in the form of cold, dark matter), with large voids in between. The results are shown as a pie diagram that can be compared to actual redshift survey maps. (Courtesy C. Park and J. R. Gott, Princeton University)

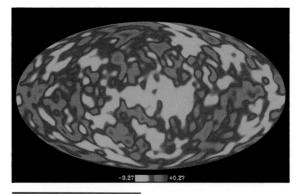

FIGURE 18.13

Cosmic Background Explorer (COBE) all-sky map made at microwave wavelengths. The Milky Way, if visible, would extend horizontally across the middle of the image. This false-color map (Appendix G) is coded so that blue indicates one-hundreth of one percent colder than the average cosmic microwave temperature; red indicates one-hundreth of one percent warmer than the average (see scale at bottom). This radiation is from the universe when it was about 300,000 years old. The red patches represent regions of higher than average density; blue, regions of lower than average density. (Courtesy NASA/Goddard Space Flight Center)

the form of cold, dark matter (even hypothetical particles!); also assumed are small, random ripples that started in the first second of the Big Bang. Small masses—less than the mass of galaxies—form first; these collide and coalesce into galaxies. The results closely imitate the patterns found in the largest redshift pie maps (Fig. 18.12). Hence, we are sensing that large-scale complexity grew from simple beginnings: random fluctuations in the early universe, amplified by the action of gravity.

> The connection of the very small to the Big Bang by GUTs enhances the Big Bang model, not only in details but also in the overall picture.

An additional COBE discovery, announced in spring 1992, provides a solution to many of these sticky issues. After processing some 70 million observations, the COBE team created a map of the sky (Fig. 18.13) that shows some of the cosmic ripples from the Big Bang—wrinkles in the cosmic tapestry that were the likely seeds for the formation of large-

scale structure. The map shows minute fluctuations in the cosmic background radiation, variations that amount to only a few parts per million at a time when the cosmos was but 300,000 years old. In regions of higher temperature, the density of matter at the time was a little higher than average; in the regions of lower temperature, the density was a bit lower. These higher-density variations were large enough to create enough gravity to attract more matter into denser clumps that eventually contracted into the first-born galaxies. The lower-density regions resulted in the voids. Because the young universe was very smooth but not perfectly so, we have a lumpy universe today.

What kind of matter created these clumps? The COBE results do not tell us that key piece of the large-scale puzzle. We will have to rely on new models that will be developed to interpret these startling observations, which reveal the most ancient and largest structures in the universe.

18.6 The Inflationary Universe

The drive to integrate particle physics with cosmology reached a new fusion recently with a variation of the Big Bang model called the **inflationary universe model**. It copes with some serious flaws in the Big Bang picture and so improves it. I'll focus on just two, the **flatness problem** and the **horizon problem**.

The Flatness Problem

Recall that the universe can have two basic geometries: open or closed. We evaluate each one by examining the ratio of the measured density (of matter and energy) to the critical density predicted from Einstein's general relativity and the value of Hubble's constant. A cosmos whose actual density is exactly the critical density is flat; the ratio is 1.

Now if at the Big Bang the ratio were 1, it would remain 1 forever. If, however, the value differed from 1 ever so slightly, the ratio would be *much* different from 1 now. Surprisingly, the ratio is believed to have a value between 0.1 and 2, very close to 1. So it must have started very close to 1; otherwise, as the universe evolved, the ratio of measured to critical density would be much different from 1. The standard Big Bang model requires a special starting condition (the geometry of the universe very, very close to flat), but does not explain how this could have happened. That is the essence of the flatness problem.

Enter the inflationary model. It deals with the early times from 10^{-45} to 10^{-30} second, when massive elementary particles dominated the universe. (After 10^{-30} second, the inflationary model melds into the standard hot Big Bang model.) During that interval, the universe undergoes a tremendous spurt of growth, by perhaps as much as 10^{50} times. That means that the distance between two particles increased by 10^{50}. To put that in perspective, the distance from the proton to the electron in a hydrogen atom is about 10^{-10} meter. If inflated by the same amount as estimated for the early universe, the distance would be 10^{40} meter—or 10^{24} ly!

The inflationary universe model interjects an early time of rapid expansion into the Big Bang model.

This inflationary period neatly and naturally solves the flatness problem. Imagine that before inflation, spacetime contained strongly curved regions. The inflation caused these to grow to flatness automatically. As an analogy, consider the curved surface of a partially inflated balloon. The surface is clearly curved, compared to the overall size of the balloon. Now rapidly blow up the balloon, keeping a close eye on its curvature. It becomes distinctly flatter. Similarly, when the universe inflates, curved regions become flat. So the ratio of the actual to critical density naturally reaches a value very close to 1, without any special assumptions.

The Horizon Problem

The uniformity of the cosmic background radiation tells us that the universe was extremely isotropic at the time of decoupling. How did this happen? The standard Big Bang model just assumes that it started that way and stayed that way. And with this assumption, the uniformity of the cosmic radiation presents a problem.

Consider a gas in a box. If you add energy to one side of the box, the temperature goes up. But that takes a little time, as the particles in the gas carry information about the addition of energy by moving around at a greater average speed and knocking into one another harder (Section 10.2). A finite time elapses before these collisions have carried the information throughout the box that energy has been added to it. Now imagine this box expanding much faster than the particles in it are, on the average, moving around. Then only a small region of the box will find out that energy has been added, and this part will have a temperature different from the rest.

The fastest that information can be communicated is the speed of light. Yet, the very early universe expanded so fast that regions of it were rapidly and widely separated. Now in a given time, a light signal can travel some maximum distance, called the *horizon distance*. For example, after one second has elapsed, light could have gone only one second of light travel time, for a horizon distance of about 300,000 km. Yet regions of the universe were separated by almost 100 times this distance from the rapid expansion. How could these regions have evolved to the same temperature when they could not communicate with each other? That is the horizon problem.

The inflationary universe model solves the horizon model by inflation. The universe evolves from a region much smaller (by 10^{50} or more) than in the standard Big Bang model. Before the inflationary era begins, the universe is much smaller than its ho-

rizon distance. All of it can reach the same temperature. Then inflation makes it much larger, preserving the uniform temperature. That way the cosmic background radiation is very uniform in the past and today.

So the basic stages of the inflationary model are as follows (Fig. 18.14). At 10^{-43} second, gravity freezes out from the other forces. At 10^{-35} second, the strong force freezes out from the electroweak force and promotes an era of inflation. From that time until 10^{-5} second, the universe was a dense soup of quarks and electrons. Then the expansion had cooled the contents enough to permit the formation of protons from quarks. From 1 second onward, the model develops as the standard Big Bang model,

---

FIGURE   18.14

The inflationary universe scenario. The horizontal axis is time in seconds; the vertical one represents *very* schematically the relative size of the universe. Just after the Big Bang, gravity on the smallest scales (supergravity) may describe the universe until a time of 10^{-45} second, when gravity freezes and GUTs describe the physical processes of particle interactions. At 10^{-35} second, inflation begins, prompted by the freezing of the strong force. Electroweak interactions control a soup of quarks and electrons until the quarks form protons, which participate in nucleosynthesis. After matter and radiation decouple, the first galaxies form in superclusters, and the standard Big Bang scenario is followed. (Adapted from a diagram by J. Burns)

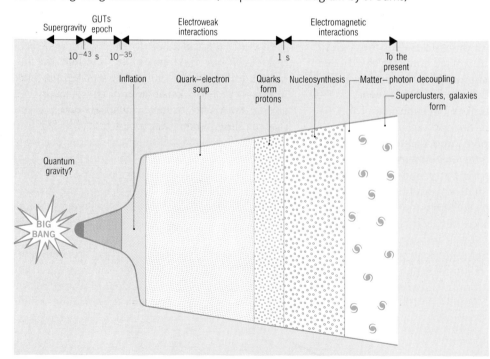

with the nucleosynthesis era, the decoupling of matter and radiation, and formation of protogalaxies and clusters.

The inflationary universe model solves some problems of the standard Big Bang model in a natural way.

Before 10^{-43} second, we have *no physical model to describe the cosmos*! A new one, which combines quantum ideas with gravity, is being worked on but is not complete. It is called *supergravity* and perhaps can describe the physical conditions then. The current goal of model physics is to unify quantum concepts with gravity—the weakest force, but the one that commands the cosmos.

The inflationary model modifies the standard Big Bang model for the story of the first second billions of years in the past. What about the far future? That depends basically on the overall geometry of the cosmos (Section 18.1). In each geometry, a special moment marks the start of the expansion. In a closed universe, the rate of expansion slows and eventually stops; then the universe contracts. In a hyperbolic universe, the rate of expansion doesn't slow as rapidly, and it never stops. Even after infinite time, the galaxies are still moving apart at a finite velocity. In the borderline case of a flat universe, the expansion slows just enough to allow it to come to a stop after infinite time of infinite expansion.

We seem to have two possible cosmic destinies. In one, the expansion persists forever, and the universe gradually thins out and cools down. That future is the Big Bore. In the other, the expansion slows down, stops, and reverses, and the universe collapses. During the time of diminishing size, the galaxies rush together into a dense conglomeration with (theoretically) zero radius—a cosmic singularity like that one that is supposed to lurk inside any black hole (Section 14.7). The cosmos then ends in a Big Crunch—and perhaps will be born again.

The Larger View

Cosmology starts with a few basic assumptions and aims to explain the physical universe in terms of cosmological models. The essential observation is that of the expansion of the universe from a finite time in the past—a deduction from Hubble's law (Chapter 16). Einstein's theory of general relativity describes the evolution of this expanding universe in the context of cosmic geometries. By applying the concept of escape speed, we realize that cosmic geometry determines the future of the universe.

The Big Bang model, confirmed by key observations of the cosmic background radiation, provides the basic modern picture for the origin and evolution of the universe. The concept for this model is basic: light photons and matter behave as particles of a gas in an expanding universe, which cools as it grows. The universe started in a hot, dense, smooth primeval fireball billions of years ago. When very hot, photons could collide to make matter and antimatter; these particles made very simple elements in the first few minutes. Somehow in this expansion, galaxies were able to form in clusters and superclusters, seeded by ripples from the Big Bang itself.

The inflationary universe model grapples modestly well with flaws in the Big Bang model. It is the best model we have to date that unites the physics of the very large with the very small. Meanwhile, we still have the problem of dark matter. If it exists in quantity—maybe 90 percent—it shapes the future of the universe: one that ends in a cosmic crunch.

Key Ideas and Terms

cosmology
homogeneous
isotropic
physical universe
universality of
   physical laws

1. Cosmological models assume the universality of physical laws and a universe that is both homogeneous and isotropic. Cosmological models must explain the evolution of the universe, the grouping of matter, and the expansion of the universe.

closed geometry
critical density
flat geometry
hyperbolic
   geometry
open geometry
spherical geometry

2. Einstein's theory of general relativity allows spacetime to have an open or closed geometry. The actual geometry determines the future of the cosmos: closed, collapse in a finite time; open, expansion forever. We can find the cosmic geometry by comparing the critical density (calculated from Hubble's constant and relativity) to the actual density. Results so far point to an open geometry, unless it is closed by dark matter.

Big Bang model
cosmic blackbody
   microwave
   radiation
primeval fireball

3. The main modern cosmological model used today is the Big Bang model (based on Einstein's theory of general relativity). The Big Bang model says that the universe began in a hot, dense state at a finite time in the past. It predicts that radiation from the past should pervade the cosmos now at low temperature.

blackbody
cosmic blackbody
   microwave
   radiation
isotropic
opaque

4. The observed properties of the cosmic blackbody radiation strongly support the Big Bang model: the radiation comes in uniformly from all directions, has a temperature of 2.7 K, and is extremely smooth in its spectrum and its spatial extent across the sky—implying that the early universe was extremely smooth.

decoupling
heavy-particle era
light-particle era
matter era
radiation era
recombination

5. The properties of the cosmic blackbody radiation, when combined with Einstein's general relativity and our present knowledge of matter, permit a detailed description of the Big Bang. Matter is made from photons and then interacts to form only light elements in the first few minutes of the universe's history.

helium abundance
nucleosynthesis
Population II stars

6. The hot Big Bang model predicts that the cosmos should have no less than 20 to 30 percent helium by mass made in the first few minutes of the universe's history. Observations tend to support this prediction, though not very strongly.

decoupling
galaxy formation
gravitational
   contraction
recombination

7. Galaxies could not form in the young universe until the temperature had dropped below a few thousand kelvins (about a million years after the expansion started) and radiation and matter had decoupled. Galaxies did form quickly and early on, but they could not have grown rapidly from small disturbances.

antimatter
grand unified
theory (GUT)
strong nuclear force
weak nuclear force

8. Grand unified theories (GUTs) of elementary particles and forces result in a new unification of elementary particles and cosmology for the Big Bang model and help to solve some of its major problems, such as the imbalance of matter and antimatter and the formation of galaxies.

flatness problem
horizon problem
inflationary universe

9. The inflationary universe model predicts that at the time of the strong force freezing out, the universe went through a stage of rapid expansion. This inflationary era solves the flatness and horizon problems of the Big Bang model.

closed universe
dark matter

10. The universe may well be dominated by dark matter in unknown form; if enough exists, it will close the universe and result in an end (the Big Crunch) billions of years in the future.

Review Questions

1. What is one basic assumption of a cosmological model?
2. What is one basic observation that a cosmological model must explain?
3. What is the basic cosmological model that is used most often today?
4. How is the cosmic background radiation a key prediction of the Big Bang model?
5. What are *two* key observed properties of the cosmic background radiation?
6. What do the COBE results regarding the cosmic background radiation imply about the nature of the early universe?
7. In the Big Bang model, how did the *first* matter form?
8. According to the Big Bang model, *when* did helium form in the early universe?
9. What prediction does the Big Bang model make about the *amount* of helium formed during cosmic nucleosynthesis?
10. When can galaxies start to form in the Big Bang model?
11. Which of the four fundamental forces dominates the universe now?
12. How do GUTs aid in the formation of galaxies?
13. What is one of the problems of the standard Big Bang model that the inflationary model seems to eliminate?
14. What is one effect of a large fraction of dark matter in the universe?

Conceptual Exercises

1. State in a few short sentences the assumptions of the Big Bang model. Which one seems the weakest to you?

2. Make a short list of the fundamental cosmological observations. Which one seems the most important to you?

3. Interpret the observations in Exercise 2 in the framework of the standard Big Bang model.

4. Give *one* observational argument for asserting that the microwave background is *cosmic* in origin.

5. How does the discovery of the cosmic microwave background radiation confirm the Big Bang model?

6. List the elements that can be made in a hot Big Bang, and give one reason for the absence of elements heavier than lithium and beryllium.

7. What observational evidence do we have that backs up the standard Big Bang model? Which seems the strongest to you?

8. How can light produce particles? In the Big Bang model, how does it happen that heavy particles are not formed out of light after a certain time?

9. In what sense is the present abundance of antimatter in the universe a problem for the Big Bang model?

10. In what sense does the Big Bang model have a flatness problem? How does the inflationary universe model cope with this problem?

19

Life in the Cosmos

■

Central Concept
Life as we know it evolved naturally over billions of years from common materials in the cosmos.

■

Life on earth seems very special. Yet, it evolved through a *natural* sequence of chemical and biological evolution. The basic material was there, but it needed to be put together in a very special way over cosmic time. The origin of life may be both natural and universal. How life developed here then gives us some insight on whether life exists elsewhere in the Galaxy.

Biological evolution provides a model (a very incomplete one!) for the origin of life on our planet. That origin followed a sequence of physical evolution, chemical evolution, and biological evolution. Each stage provides clues to whether life arose elsewhere in a broad astronomical context. There's a chance that we are alone in the Galaxy. But we might also have neighbors not far from home.

19.1 The Nature of Life on the Earth

What is life? Rather than try to define it, let's accept a useful rule of thumb: *living things are things that reproduce, mutate, and reproduce the mutations.* What does this mean?

First, that living things have an *organization* that they pass on when they reproduce (Fig. 19.1). So all living things are **organisms**. Second, reproduction may result in an offspring with a difference from its parents—one that exhibits a **mutation**. (The genes in a cell's nucleus carry the information code for how an organism is to be put together; if the genes change, the organism changes.) Mutations can provide the possibility of change, which leads to **biological evolution**, the development of organisms sometimes better adapted to their environment. Mutation and evolution define life.

Biochemical discoveries have revealed how an organism passes on its organization. Two basic types of molecule operate in all terrestrial organisms; their interaction results in life. The first are **proteins**, which make up the organism. The second are **nucleic acids**, which provide the information for the structure of the organism and the means to pass on this information in reproduction. These molecules involve carbon as a central atom in their construction.

The relative chemical composition of life as we know it reflects the general composition of the cosmos, as exemplified by stars.

Both long molecules have chief subdivisions. Proteins are built out of **amino acids** (Fig. 19.2). Nu-

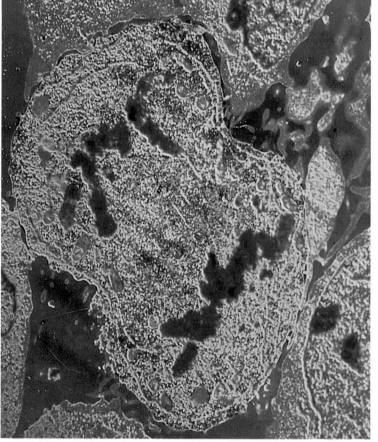

FIGURE 19.1

Human cell dividing. The main body of the cell appears yellow in this image; the genetic material appears dark pink. (Courtesy CNRL, Science Photo Library/ Photo Researchers)

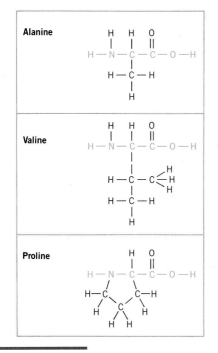

FIGURE 19.2

The chemical structure of three common amino acids found in life on the earth. Note that hydrogen, oxygen, carbon, and nitrogen—four of the most common elements in the cosmos—form the common chemical basis of these organic molecules.

cleic acids also consist of smaller subunits called *bases*. These building blocks are simple combinations of the most common chemical elements in the universe (as found in stars and the interstellar medium): hydrogen, oxygen, carbon, nitrogen, and a few others—the essential parts of **organic molecules** (Table 19.1).

Energized by sunlight, cells carry on chemical work using proteins that monitor and facilitate important chemical reactions in a cell. But how does a cell control their functions? That's the job of the cell nucleus, using the nucleic acid called **deoxyribonucleic acid**, or **DNA** for short. DNA is an enormous molecule; in your body, DNA strands contain *billions* of atoms. (Yet your whole body contains only one teaspoonful of DNA.) Chemically DNA has the bonding of sugars, phosphates (compounds of phosphorus and oxygen), and four essential materials called bases to form the DNA chain (Fig. 19.3). DNA serves as the chemical blueprint that informs the protein in cells how to function. The offspring inherit this information.

Proteins and DNA are composed of combinations of carbon, hydrogen, oxygen, nitrogen, and a few other common atoms. All terrestrial organisms have a common chemical makeup, which is closely connected to the physical evolution of the universe. The common elements of life were made in stars or created in the Big Bang.

The common chemistry of life on the earth hints that life arises naturally in the evolution of the universe.

19.2 The Genesis of Life on the Earth

Modern biology sees life's origin in nonliving materials. Life arises from the slow processes of chemical

Table 19.1

Relative Abundances of Elements (fraction of atoms)

Earth's Crust	Solar System (Sun)	Human Beings	Interstellar Medium
Oxygen	Hydrogen	Hydrogen	Hydrogen
Silicon	Helium	Oxygen	Helium
Hydrogen	Oxygen	Carbon	Carbon
Aluminum	Carbon	Nitrogen	Nitrogen
Sodium	Nitrogen	Calcium	Oxygen
Iron	Neon	Potassium	Neon

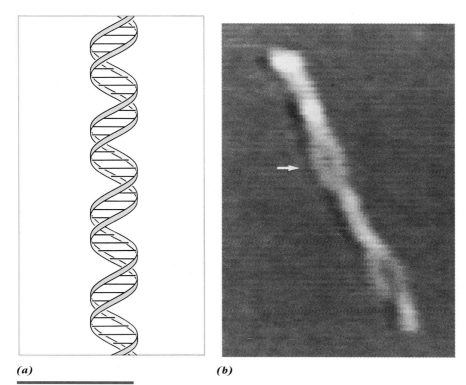

(a)   *(b)*

FIGURE 19.3

The DNA molecule. (*a*) Schematic drawing of the double-stranded structure of DNA. The two strands in the helix are cross-linked by pairs of four chemical bases: adenine, guanine, cytosine, and thymine. These materials cannot pair up at random; adenine pairs only with thymine, and guanine with cytosine. The nonrandom ordering provides the chemical basis of the DNA information coding. (*b*) Highly magnified fluorescence image of a DNA molecule. The twists visible here are not the double-helix structure; rather, they are DNA strands twisted upon themselves. The thicker sections show where the strands cross; the arrow points to a section in which both strands are visible. (Courtesy C. Bustamante, University of Oregon)

and biological evolution. What are these evolutionary processes? In *The Origin of Species* (published in 1859) Charles Darwin (1809–1882) noted that living species are adapted for survival in their environment. Some feedback mechanism selects mutations that enhance the compatibility of species with its environment. Although Darwin lacked modern knowledge of genetics, he patiently unearthed the general feedback mechanism: *natural selection*.

Put too simply, modern biologists view natural selection as a result of successful reproduction. **Natural selection** is the process by which individuals that are well adapted to a local environment not only survive but, more important, succeed in producing offspring. Their offspring tend to survive and reproduce successfully. Eventually the genes of the early well-adapted couples dominate those available to a species, and the survival and expansion of new genetic material results in evolution.

New genetic material comes from mutations. Natural selection directs the random jumble of mutation into *biological evolution*, but **chemical evolution** to make complex molecules must have come first. This progression assumes a primeval life form that started the process of biological evolution.

Mutations and the interactions of mutated organisms with their environment result in natural selection, and so biological evolution.

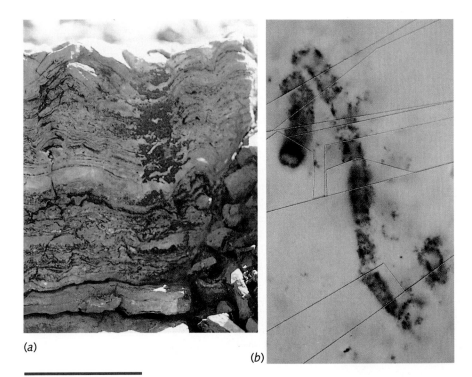

(a)

(b)

F I G U R E   1 9 . 4

Evidence of early life on earth. (a) Layered structure of sedimentary rocks about 3.5 billion years old. Once tidal mud flats, these formations might have contained colonies of photosynthetic microorganisms. (Courtesy D. I. Groves, University of Western Australia) (b) Microfossils from the Warrawoona formation near North Pole, Australia, that resemble modern bacteria. These structures are embedded in rocks that are dated at about 3.5 billion years. (Courtesy Biological Photo Research)

When did life arise on the earth? Rocks sometimes enclose fossils that show the life of the past. Radioactive dating (Section 4.3) can fix the age of the rocks and the fossils in them. Careful microscopic inspection of ancient rock samples reveals the remains of bacteria and algae from some 3.5 billion years old. These provide important clues to life's evolution on the earth. The oldest set of rocks containing microfossils lies in Australia. The rocks, which are 3.5 billion years old, show remains that resemble microorganisms and layered structures that could have been built by colonies of bacteria (Fig. 19.4a). Another group of rocks, also from Australia, contain evidence of rod-shaped structures resembling modern bacteria (Fig. 19.4b).

Ancient fossils imply that biological evolution was established on the primeval earth no more than about 1 billion years after the earth formed.

The sun reflects the average chemical composition of material in our Galaxy. But the earth has ended up with a composition quite different from that of the sun. How did this happen? Recall (Section 8.4) that the nebular model for the origin of the solar system pictures the protoearth forming from the accretion of planetesimals. The chemical compo-

FIGURE   19.5

Volcanic eruptions on Hawaii. (Courtesy C. W. Stoughton)

sition of the earth's planetesimal material depended on the temperature in the solar nebula, as given by that at 1 AU—some 600 K.

What about the earth's early atmosphere? As you saw in Section 4.6, if the earth's surface was hot at the time—perhaps even molten—the heat would vaporize volatiles within the earth, which would rise upward to the surface. (That's what happens in volcanoes today.) Volcanoes (Fig. 19.5) outgas water, carbon dioxide, hydrogen sulfide, methane, and ammonia—but no oxygen. We would expect pretty much the same materials to outgas from the primitive earth, an atmosphere that probably contained much carbon dioxide and water vapor.

The earth's early atmosphere (lacking oxygen) formed from the outgassing of volatiles from the interior.

The earth started out with an atmosphere of simple molecules. To synthesize complex molecules from simpler ones requires energy—such as ultraviolet light from the sun and lightning.

Solar ultraviolet radiation at wavelengths less than 2200 Å is absorbed by complex molecules and can cause the formation of still more complex ones. (The ozone layer now filters out most of the ultraviolet radiation. Because of the absence of free oxygen from the early atmosphere, the young earth lacked an ozone layer.) When the sun was a pre-main-sequence star, its surface temperature was less than now, but the sun was more active, emitting more ultraviolet as a fraction of its total energy. Lightning (Fig. 19.6) now accounts for almost as much energy as ultraviolet from the sun. When the earth cooled enough for rain to fall, lightning storms may have raged widely over the earth's surface. These energy sources came in spurts rather than at constant rates. They can destroy as well as synthesize molecules. The balance between creation and destruction determined the number of molecules and the kinds that could exist.

We don't know exactly how complex molecules formed. The critical aspects are energy and an atmosphere that lacks free oxygen. Laboratory experiments validate these notions. When energy (in forms such as expected on the young earth) is added to gaseous mixtures of water, carbon dioxide, methane, and ammonia, the products include amino acids, among which are some commonly found in terrestrial proteins. A key point: these experiments, whether with gaseous mixtures or solutions, naturally produce *most* of the amino acids common in protein and *none* of the amino acids *not* found in modern protein.

The organic molecules needed for life form naturally under plausible primitive earth conditions in simulations that involve adding energy to gases without free oxygen.

Chemical evolution naturally—and perhaps inevitably—leads to complex organic compounds that are the building blocks of proteins and nucleic acids. Both are needed to join in a cell. How? Quite bluntly, *we don't know*. Fossils cannot give informa-

F I G U R E   1 9 . 6

Lightning flashes in a summer storm over Tucson, Arizona. The energy released in a flash can help to make simple organic molecules. (Courtesy K. Wood, Science Source. Photo Researchers)

tion about this crucial time. And chemists have not synthesized anything as complex as a cell. What happened before this time remains a matter for speculation.

19.3 Life in the Solar System

So far we know of only one planet that harbors carbon-based life: the earth. The environments of the other planets pretty much restrict life—except, perhaps, for Mars.

The possibility of Martian life hinges on surface water. The Viking missions found the surface pressure of the Martian atmosphere to be about 0.007 earth atmosphere—much less than that on the highest mountains on the earth's surface. (In fact, you would have to go about 40 km up into the earth's atmosphere to find pressure this low.) At this low pressure, liquid water cannot exist on the surface. And Mars is a very, very dry planet, even in the polar regions.

Water flowed on Mars in the past. The planet's outflow channels attest to the flow of liquid water in the Martian past. Some astronomers imagine that at an earlier epoch, Mars had a denser atmosphere that could hold water vapor sufficient to generate rainfall. A denser atmosphere may have arisen from extensive volcanic activity, which could have resulted in the spewing spew out of large volumes of carbon dioxide and water vapor.

Recurring deluges or meltings may account for the origin of the laminated terrain found in the Martian polar regions within 10° of the poles (Fig. 19.7). There, stacks of thin plates of crustal material stand about 10 km tall and up to 200 km across. These regions appear to be the youngest, most evolved parts of the Martian surface. Because they exist only in the polar regions, where carbon dioxide and water ice form annually, the plates may be related to the influx and outgo of these molecules.

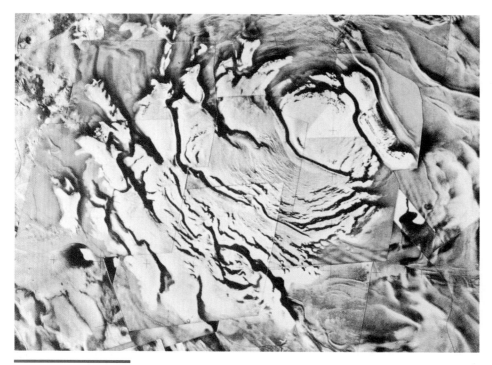

FIGURE 19.7

Laminated terrain near the south polar cap of Mars. These formations of surface soil probably contain water ice in layers. Photo shows a region about 500 km across. (Courtesy NASA)

An analysis of features on the Martian surface supports the notion that the poles have wandered extensively on the surface—a result of the shifting of the crust with respect to the spin axis. Deposits such as found near the poles now are visible near the equatorial regions, an indication that perhaps 3.5 billion years ago the poles fell in these regions. Polar wandering would bring great changes in climate to all parts of the planet.

The direct test for Martian life came from the Viking landers' biology experiments. What were these results? They proved negative: in the soil that was sampled, life does *not* exist on Mars now.

Both landers contained instruments called mass spectrometers, designed to detect and measure organic molecules that characterize life. At both landing sites, the result was the same: *no large organic molecules were found*. The instruments could detect organic compounds in a concentration of just a few parts in a billion. That's about 1 million bacteria in a sample—far below the concentration found in desert soils on the earth. The landers' three biology experiments gave a few apparently positive results.

But because of the lack of organic molecules in the soil, *chemical* reactions rather than *biological* ones might have been responsible.

Life does not now exist on Mars in the areas we have sampled.

Meteorites in the solar system provide some evidence to support the notion of the natural synthesis of organic compounds. Of the three main classes of meteorites, carbonaceous chondrites contain a fairly high percentage of carbon (a few percent). People have regularly speculated that some carbon contained in these meteorites might be organic.

In September 1969, a carbonaceous chondrite meteorite (Section 8.3) fell in Murchison, Australia. It was rushed to NASA for analysis, and very small quantities of five amino acids common to living protein were found. Was this terrestrial contamination? Probably not. Organic molecules exist in two distinct forms: right-handed and left-handed, depending on the direction of the twist of the linkage of the atoms. Almost all terrestrial organic molecules are left-handed, so earth-based contamination is expected to be left-handed. The Murchison meteorite contained just about equal quantities of right- and left-handed molecules. This evidence strongly suggests an extra-terrestrial, *nonbiological* origin of the Murchison organic molecules. Why nonbiological? When organic molecules are synthesized in a chemistry lab (rather than by an organism), they show an equal number of right-handed and left-handed forms.

The glut of complex molecules in space discovered by radio astronomers (Section 12.1) bolsters an extraterrestrial, nonbiological process for the formation of organic substances. Molecules such as form-aldehyde, hydrogen cyanide, cyanoacetylene, formic acid, methyl alcohol, and methylacetylene can play a crucial role in organic chemistry. Formaldehyde and hydrogen cyanide, for instance, can be chemically combined to make amino acids. And remember that the solar system formed from an interstellar cloud of gas and dust that probably contained such molecules.

The chemical evolution from simple compounds to complex organic substances occurs so naturally that it takes place even in the hostile environment of space.

19.4 Mass Extinctions on the Earth?

Meteorites may have an explosive impact on the evolution of life on earth. Roughly 65 million years ago, a sudden trauma swept through life here. In a short time—less than a million years, perhaps as swift as a thousand years—mass extinctions hit certain plants and animals. The fossil record shows an abrupt loss of ocean plankton, swimming molluscs and dinosaurs, and land animals with masses greater than 25 kg—most especially, the large walking dinosaurs. This end was good for us, for the mammals flourished afterward. But what happened then to promote these **mass extinctions** of life forms?

Many ideas have been proposed. One seems to be gaining the weight of reasonable evidence: that of the impact of an asteroid-sized body that caused environmental stress, resulting in selective, world-wide extinctions. An object some 10 km in diameter (about the size of a small city) with a mass of some 10^{14} kg could easily penetrate the earth's atmosphere and strike the surface at 11 km/s. Its impact would release some 10^{23} J, equivalent to 10^{14} tons of TNT. (Similar impacts shaped the large basins on the moon.) Temperatures at the impact point would hit 20,000 K. The vaporized object and ground would shoot hot gas into the air. A blast wave would rocket out at 35,000 km/h, leveling everything for a few hundred kilometers around. Some material ejected by the impact would plume into space, condense, and shower back onto the earth.

Astronomical evidence suggests that an object this large collides with the earth once every 100 million years or so. About a hundred such objects are known; they tend to have sizes of a few tens of kilometers. Systematic searches by a few astronomers are hunting down more. In January 1991, one asteroid plunged by within 170,000 km of the earth—only half the distance between the earth and moon! Collisions among these near-earth asteroids may well supply most of the meteorites that hit our planet. (For such collisions to occur, the debris must cross the orbit of the earth.)

How might such an impact have influenced the earth's environment? The blast could have deposited small dust particles (less than 1 μm in size) high in the earth's atmosphere, where they would have remained for months. It may have caused a darkness like night for well over a month. Winds could have circulated the dust globally. These particles would have cut out a significant fraction of the sunlight reaching the earth's surface, sharply reducing photosynthesis (especially by plants in the oceans) and the general temperature. Animals especially sensitive to temperature changes would not have been able to adapt and so would have disappeared.

This idea sounds plausible from an astronomical view. Does it have any solid evidence? The main clue comes from the composition of a clay layer deposited about this time. Below it (before it in time), we find the usual range of fossils from the age of the dinosaurs. Above it (afterward), certain fossils no longer appear. The layer itself has a composition that is enriched (relative to the earth's crust) in noble metals such as iridium and gold. The overabundance of iridium in this layer appears to be a worldwide phenomenon. One source of this enrichment could be material from a large asteroid. Its impact would have mixed asteroidal material with terrestrial, to result in the abundances found in the clay layers.

Asteroidal-size bodies colliding with the earth may have had dramatic repercussions on the evolution of life.

One way to confirm this notion is to track down the fossils of such impacts—craters on the earth. But much of the earth's surface is covered with water. And erosion wipes out impact craters fairly fast. Still, workers have uncovered more than a hundred old craters, a few with diameters of about 100 km—just about the size expected from the impact of a 10-km asteroid.

19.5 Life in the Milky Way Galaxy

Might life exist elsewhere in the Galaxy? The huge number of stars in the Galaxy implies planets elsewhere, if the nebular model of planetary formation is correct (Section 8.4). Even if the probability of life is slim, the number of possible habitats may still be very large. Civilizations of living creatures must evolve; that's part of cosmic evolution. So the number of technological civilizations in the Galaxy changes with time. At any one time, the number of civilizations depends on the rate at which these civilizations are born and how long they last. We'll focus on **technologically advanced civilizations**. (*Technologically advanced* applies to creatures who can manipulate their environment at least to the extent that we can, so they have electricity, radios, telescopes, and so on. They may even read books on astronomy.)

Here's an analogy. Suppose you are locked in a dark room filled with candles; a friend gropes about and lights one candle every 15 minutes—four per hour. Suppose each candle burns for one hour. How many candles are lit at any given time? During the first hour, the number increases from one to two to three to four. But just as the fifth one is lit, the first one goes out. As the sixth is lit, the second goes out. One goes out as each new one is lit, leaving four candles burning at any time. If you think about it, you see that the number of observed candles is equal to the rate of candle lighting times the *lifetime* of one candle.

So, if you know the average lifetime of a single candle and the rate at which the candles begin their life, you can anticipate the number lit at any time. The same reasoning applies to the number of civilizations in the Galaxy at any time: the rate of formation of technological civilizations times their lifetime results in the number of technological civilizations now.

Well, neither of these numbers is easy to estimate! And any estimate has a large range of uncertainty. Let's tackle pieces of the first, the rate of formation of technological civilizations. To begin, we need the rate of star formation averaged over the age of the Galaxy. Then the probability that once a star has formed, it will possess planets. Next, the probability that the star will exist long enough for life to form. Then the number of planets in the region around the star with a suitable range of temperatures and the probability that a planet will develop life. Finally, the probability that biological evolution will ultimately lead to intelligent life. If we estimate these, we end up with about 1 civilization born per year—if we assume optimistically that life will evolve on a planet with the right conditions and that intelligence must evolve.

Now for the most crucial factor: How long can an advanced, technological civilization survive? By our own example to date, the lifetime of a technological civilization may be only a few thousand years. But if it is possible that every advanced civilization steers clear of its problems, it should survive as long as the parent star. The lifetimes of civilizations encircling a G-type star may be about 10^{10} years. But their lifetimes may also be much, *much* shorter.

The result depends critically on the lifetime—how long our candle remains lit. If we assume that we are at the brink of destruction, then the lifetime of a civilization is a mere 1000 years and intelligent civilizations are rare: their current number about 1000. The end of the Cold War supports a more optimistic view. If we survive as long as the sun shines, then lifetime is roughly 10^{10} years, and the current number some 10^{10}. Then, many a star in the Galaxy has fostered a technological civilization!

The number of technological civilizations in the Galaxy now depends on the rate of their birth and their lifetimes; the estimate can range from 1 (us!) to a billion (a multitude!).

19.6 Neighboring Solar Systems?

We assume that life exists on planets. But how to find planets? Because a planet shines faintly by reflected light from its parent star, a planet's gleam would be lost in the stellar glare. So we cannot *directly* observe other planets outside the solar system with earth-based telescopes. Instead of searching for the images from very large planets, we can hunt for the motion around the *center of mass* of the planet–star system. Because of this seesaw effect, the visible star will wobble from side to side about the center of mass if a massive planet orbits it (Fig. 19.8). From the observed stellar wobble and an estimate of the stellar mass, we can estimate the mass of the invisible planetary companion. We use the same method that we applied to measure binary star masses (Section 11.6).

The observations required to detect planets around nearby stars are *extremely* difficult to make. The wiggles we seek are only about 0.001 arcsecond, or about one one-hundredth the size of a star's

image viewed from the earth. Such minuscule changes are dramatically affected by changes in the telescopes themselves over time. One analysis of the errors in such observations concluded that *no* good evidence supports the existence of Jovian-mass planets. In fact, the lowest measured mass of any *stars* from this technique is found in a binary system called Wolf 424. One star has a mass of 0.06 solar mass, the other 0.05 solar mass. But these are far more massive than any planet.

Note that this discussion refers to *Jovian-sized* planets. We have little hope of detecting terrestrial-sized planets from the earth. Not only are the gravitational effects smaller, because of their smaller mass, but the effects also would have to be disentangled from the effects of larger bodies in the same planetary system. Although detection of Jovian-sized planets is conceivable, but difficult (even with the Hubble Space Telescope), the direct detection of terrestrial planets around other stars is even harder.

Doppler techniques seem to offer a more promising approach to detection of planetary companions to stars. Techniques developed in recent years give high-precision radial velocity measurements of stars, and so the ability to detect small variations in speed from low-mass companions. A computer finds the Doppler shift that best matches a standard spectrum to the observed star—in effect measuring the average shift of thousands of lines, some too weak to see with the eye. One study measured radial velocities for more than 1500 stars during an 8-year period. One solar-type star, some 100 ly distant, seems to have a variation in speed of 500 m/s with a period of 84 days (Fig. 19.9). The analysis suggests a companion with a mass 10 times the mass of Jupiter or a little less—perhaps a brown dwarf.

Another technique measures Doppler shifts to a precision of about 10 m/s—the speed of a good sprinter! Over several years, a few selected stars show velocity variations indicating a possible planetary companion between 1 and 10 Jupiter masses. But only one, Gamma Cephei, has been followed so far for an entire orbital period. Its companion may have a mass of about 2 Jupiter masses.

We have limited observational evidence for the existence of Jovian-sized planets and none yet for terrestrial planets around other stars.

FIGURE 19.8

The path of a star in space with orbital motion around the center of mass of the star–planet system. The planet is not visible to direct telescopic view, but the wobble of the star's motion is detectable.

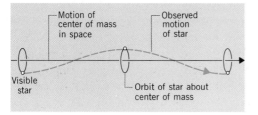

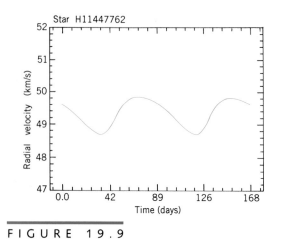

FIGURE 19.9

Cycle of the radial velocity from the Doppler shift observed for the star named HD 114762. Note from the horizontal axis that one complete period takes 84 days. This cyclical pattern of the Doppler shift is typical of that for a binary system. (Adapted from a diagram by D. W. Latham, T. Mezah, R. P. Stefanik, M. Mayor, and G. Burki)

 ## 19.7 Where Are They?

Whether we should search for others depends on how many technologically advanced civilizations exist in the Galaxy *now*. If the number is large, then on the average such civilizations must be closer together than if the number is small. The key element in estimating this number is the *lifetime* of technological civilizations. For example, if the lifetime is about 100 years (which is about as long as a technologically advanced human culture has existed on earth), then the average distance between galactic civilizations is roughly 10,000 ly. That makes communication practically impossible. Why? If we tried to signal by radio—for example, by sending out a message just as soon as our technology permitted—our civilization would have died while our words were in transit. Under known physical laws, communication is possible only if the number of civilizations is large and their lifetimes are long.

If we are it, we don't have any neighbors within 10^5 ly—the size of the Galaxy. If the number of civilizations is very large, say 10^9 to 10^{10}, then our neighbors are only a few tens of light years away. If the number is 10^6—the compromise guess—then our neighbors live a few hundred light years away—just within radio reach. Note that each of these choices implies a value for lifetime. If the current number is very small, then the lifetime is at most a few hundred years. We are then probably on the verge of extinction. If the number is very large, then the lifetime is very long, some 10^9 to 10^{10} years. Civilizations then last as long as their suns.

We have no direct evidence for other advanced civilizations in the Galaxy. But from cosmic evolution, we expect them. So where are they? Are we alone?

You could take a pessimistic stance and argue that there are other advanced civilizations in the Galaxy—the "we are alone" view, which is based partly on computer calculations of the evolution of the earth's atmosphere. These models show that a most delicate balance must be maintained to keep temperatures in a moderate range. If the earth had an orbit of 0.95 AU, the greenhouse effect would run away and turn us into a Venus. On the other hand, if the earth orbited at 1.01 AU from the sun, glaciation would have iced up the earth 1.7 billion years ago. Our planet would never have gotten warm enough to foster the evolution of life. The key point is this: early conditions on the earth may have been so special, balanced between freezing and steaming, that the chances of a similar balance occurring elsewhere in the Galaxy are very small.

Another version of the pessimistic view has two points. One, that biological evolution from one-celled creatures to those that think is so improbable that we are the only intelligent species to exist. Two, that low-speed interstellar space travel is easy and

"Where Are They?" At a Glance		
	Fact 1	We are here, but we don't know for how long.
	Fact 2	No life has been found on Mars.
	Fact 3	No positive results from radio searches.
	Fact 4	No extraterrestrial aliens are here now.
	Conclude	We *are* alone.

cheap, especially for robot probes. A civilization just somewhat more advanced than ours could probe the Galaxy. Yet, we have no evidence so far that such visits have taken place in the solar system. So such civilizations do not exist. (This argument is stronger, but it has the "absence of evidence is not evidence of absence" problem.)

The evidence so far suggests that we may very well be alone.

If we are alone, there's not much point in speculating about alien civilizations elsewhere. We can imagine, though, a future for humankind. Using pretty much current technology, we could build large, self-contained spacecraft—essentially space colonies (Fig. 19.10). These vehicles could be constructed out of raw materials found on the moon or in asteroids.

Space colonies could convey people on journeys between stars at speeds of 0.01 the speed of light. Imagine that we send colonies out across the Galaxy. When each arrived at a suitable star, it could build a new colony ship from local resources and send it out to the next star. A few centuries might elapse from arrival to the next departure, and another few centuries to glide to the next star. How long does it take to travel across the Galaxy? Take 100,000 ly for the diameter of the Galaxy. At 0.01 the speed of light, the time in years is 100 times the distance in light years, or 10 million years. This star-hopping process could carry people across the Galaxy and allow them to colonize it in roughly 10 million years. Then we would find the answer to "Where are they?" They may well be us.

FIGURE 19.10

Artist's view of a space colony in the twenty-first century. Mirrors and windows control the sunlight to simulate night and day. The rotation of the spacecraft would simulate gravity. (Model as envisioned by G. K. O'Neill; courtesy NASA)

The Larger View

The models of the physical universe in Chapter 18 did not deal with life in the universe. Because we are here with complex brains, we have the power to build mental pictures of the cosmos—models that disclose our deepest desire to fathom the farthest reaches of space and time. That cosmos contains life—at least on this planet earth! To understand the universe in its fullness, we must apprehend the place of life within it.

Astronomical ideas underpin our understanding of physical evolution that set the stage for the evolution of life. For life as we know it, we need a planet (the earth), a star (the sun), and the proper elements (hydrogen, carbon, nitrogen, oxygen, and some others). Where did these come from? The earth, from the dust of the interstellar medium; the sun, from the gases, when an

interstellar cloud contracted. Some dust and gas are the material lost by earlier stars, mostly in their violent ends. These explosions and normal fusion reactions in stars manufactured the chemical elements, except hydrogen and some helium, which were made in the first few minutes of the Big Bang. The long line of cosmic evolution produced us. But that does not mean that evolution ends—it will continue.

If life is typical, it may be common. If so, our Galaxy and the entire universe teem with life. The evidence, though, does not indicate any neighbors yet. The search continues, though not with great intensity. Of all the concepts of cosmology, the most startling and significant may be that we are very much alone—right now—still coming of age in the Milky Way.

Key Ideas and Terms

deoxyribonucleic acid (DNA)
mutation
natural selection
organic molecules
organism
proteins
nucleic acids

1. Life as we know it consists of organisms that reproduce, may mutate (undergo changes in their genetic structure), and reproduce those mutations. These organisms contain proteins (the material of their construction) and nucleic acids (the blueprint of their construction). The chemical composition of organic molecules reflects the chemistry of the cosmos. Natural selection shapes mutations and so drives biological evolution; mutations are changes in the structure of DNA.

biological evolution
chemical evolution
organisms

2. Fossils show that simple organisms existed on the earth at least 3.5 billion years ago and that significant evolution took billions of years more. Life arose from nonlife. Physical and chemical evolution preceded biological evolution.

chemical evolution
energy
organic molecules
ultraviolet light

3. Laboratory experiments adding energy (sparks, ultraviolet light) to simple compounds in gaseous form (carbon dioxide, water, methane, ammonia, and so on—but no oxygen) result in the natural synthesis of amino acids and other organic molecules.

energy
outgassing
volatiles

4. The early atmosphere of the earth resulted from the outgassing of volatiles from the interior. The composition was mainly carbon dioxide and water vapor, but no free oxygen. The addition of energy (from lightning, ultraviolet light) could promote the synthesis of simple organic compounds.

amino acids
carbonaceous chondrites

5. No life has been found on Mars. Some meteorites (those rich in carbon and water) contain amino acids made by nonbiological, chemical processes in the early solar system. Mass extinctions may have shaped the evolution of life on

mass extinctions
organic molecules

earth. It is doubtful that life exists anywhere else in the solar system.

technologically
  advanced
  civilizations

6. We can roughly estimate the number of technologically advanced civilizations in the Galaxy now; that number could range from 1 (us) to 100 billion; the greatest uncertainty arises from our lack of knowledge of the lifetimes of such civilizations. The number of technologically advanced civilizations now sets the average distance between them; if that number is 1 million, the average distance is a few hundred light years.

binary systems
center of mass
Doppler shift

7. Some nearby stars may have dark companions; these may be planetary systems or very low mass stars (brown dwarfs), but the evidence is not strong. We have *not* yet observed earthlike planets around nearby stars.

Review Questions

1. What is the most abundant element in the human body?
2. What molecule carries information about the structure of a cell?
3. What process adapts mutations to the environment?
4. How old are the oldest fossils of simple organisms?
5. What might have been the composition of the earth's early atmosphere?
6. What role might ultraviolet light from the sun have played in the formation of organic molecules?
7. In what other material in the solar system has organic material been found?
8. What were the overall results from the Viking landers' examination of the Martian surface for life?
9. What two factors control the estimate of the number of civilizations in the Galaxy now? Which is the more important?
10. What two astronomical techniques can be used to detect planets around other stars?
11. If the number of technologically advanced civilizations in the Galaxy is large, what is their average distance apart?

Conceptual Exercises

1. Life on earth centers on carbon. Where did the carbon come from, and how did it get to the earth?
2. How did supernovas play a crucial role in the origin of life?
3. What are the chances for life elsewhere in the solar system now?
4. Criticize the book's estimate of the number of technological civilizations in the Galaxy and come up with your own.
5. In what sense is it astronomically correct to say that we are "children of the stars"?
6. Eventually, when the interstellar gas is depleted, star formation in the Galaxy will stop. What effect will that have on an estimate of the number of civilizations in the Galaxy?
7. How have the results from the Viking landers reinforced the pessimistic view that we may be alone in the Galaxy?

The View from Your Imagination

Adapted by Michael Zeilik from original material by Sheridan A. Simon

■

Imagine a universe that differs from ours in only one way: the speed of light. Slow it down! Now it's 300,000 km/s. Make it a mere 4 mm/y. (The capital letters in this book are about 4 mm high.)

Imagine you are located in Washington, D.C., looking out the window in a room in the Capitol building. You have a powerful optical telescope that allows you to see the whole surface of the earth. What would you see with light at a crawl? Remember, you have only the information that light conveys to you. Let's say that it's 2 meters to the nearest wall. At a speed of 4 mm/y, it would take photons 500 years to get to your eyes. However, the Capitol building did not exist 500 years ago (neither did the United States of America!), so you'd see no wall—or ceiling or floor.

Outside the window, no city of Washington. Light from the grounds 200 meters away takes 50,000 years to reach you. The landscape would be vacant of people. At distances of tens of kilometers, you'd see—with an optical telescope—extinct species of mammals, including the ancestors of modern elephants, camels, and lions. A few hundred kilometers away (in Virginia and Pennsylvania), a telescope would reveal dinosaurs stalking the land. At a distance of about 260 km, the dinosaurs disappear suddenly—wiped out in the mass extinctions 65 million years ago.

You can see a great variety of extinct species stretching out to the region where the Mississippi River ought to be, over 1000 km away. But no river! The photons reaching you from this region left 250 million years ago—before the Mississippi River had developed. Farther out, no more land animals are to be seen, though the plant life is thick for several hundred kilometers more. By the time you begin to scan the Pacific Coast, you are seeing regions 5000 km away. The coastal territory appears as it did more than a billion years ago. Out in the Pacific Ocean, the only life is single-celled creatures.

Look farther away. The region 10,000 km distant looks as it did 2.5 billion years ago. Oxygen had not yet been manufactured by photosynthetic organisms, and the atmosphere contained very little of it. At 14,000 km, signs of life vanish. At 18,000 km, your view stops. Here you have reached the time 4.5 billion years ago when the earth formed. The scene is hot and molten, the earth's surface is still in the process of formation, battered by infalling meteorites. The whole pageant of the earth's history stretches out in your telescopic view, ending in a very small region of blazing light that marks the birth of our planet.

Light really does travel fast, but the universe is huge. And so a telescope acts as a time machine, revealing a universe that unwinds in time—always to the past. And the final glare of glory, beyond which an optical telescope cannot see, marks the birth of the cosmos in the Big Bang.

APPENDIX
A
Units

Powers of Ten

Astronomers deal with quantities that range from the microcosmic to the macrocosmic. To avoid having to read and write numbers such as 149,597,890,000 meters (the astronomical unit, which is the average earth–sun distance), we use powers-of-ten notation. That simply means that we write a number as a fraction between one and ten times the appropriate power of 10 rather than with a string of leading or trailing zeros.

A positive exponent tells how many times to multiply by ten. For example:

$$10^1 = 10$$
$$10^2 = 100$$
$$10^3 = 1000$$

and so on. A negative exponent, in contrast, tells how many times to *divide* by 10. For instance:

$$10^{-1} = 0.1$$
$$10^{-2} = 0.01$$
$$10^{-3} = 0.001$$

in analogy to the positive powers. The only trick here is to remember that 10 to the zero power is 1:

$$10^0 = 1$$

In powers-of-ten notation the earth–sun distance is 1.4959789×10^{11} meters.

Significant Figures

Significant figures are the meaningful digits in a number. In experimental or observational results, the accuracy of the procedure or technique limits the number of significant figures. No more than that limit can be stated as the true value of the result (which should always have a range of error attached to it). In arithmetical calculations, we are sometimes interested in an approximate result to some specified degree of accuracy. Approximate calculations using just one or two significant figures prove useful in science, especially in astronomy, because they are sufficient to illustrate a point or make a general case, or because the quantities are simply not well known.

Here are some examples:

0.0045676	five significant figures
4.5676×10^{-1}	five significant figures
4.5×10^{-4}	two significant figures
2.17	three significant figures

What about the number 2,170,000? How many significant figures? It is not clear whether there are three or seven significant figures. But when we use powers-of-ten notation, the ambiguity disappears: 2.17×10^6 has three significant figures; 2.170000×10^6 has seven significant figures.

The general rule in doing calculations is that the result cannot have more significant figures than the *smallest* number of significant figures in the quantities that are used in the computation. That is, the results cannot be any better than the one with the fewest significant figures—no better than the weakest link. It may have even fewer—for example, if the calculation involves subtraction of nearly equal numbers.

Rounding

To round off numbers, look at the last digit. If it is greater than 5, round up; round down if it is less than 5. If it is 5, round up if the next-to-the-last digit is even, and down if it is odd. Most of the numbers in the main text of this book have been rounded off to two significant figures.

The English and Metric Systems

You are probably familiar with the fundamental units of length, mass, and time in the English system: the yard, the pound, and the second. Other common units in this system are often strange multiples of these fundamental units. For example: the ton is 2000 pounds, the mile is 1760 yards, and the inch is $\frac{1}{36}$ of a yard. Almost all the world uses the much more rational metric system, which uses powers-of-ten relationships between units, making them easy to multiply and divide.

The fundamental metric system units are as follows:

Length

1 meter (m)

Mass

1 kilogram (kg)

Time

1 second (s)

This is often called the *meter-kilogram-second*, or *mks* system. An older system, often used by astronomers, is the *centimeter-gram-second*, or *cgs* system, with the following fundamental units:

Length

1 centimeter (cm)

Mass

1 gram (g)

Time

1 second (s)

The most recent international standard for physical units is called the *Système International* (in French) or *SI* units. This system is based on the mks system with many modifications. Many astronomers do not yet use SI units, though they should, because they are now the international standard. I have tried to stick with SI units in this book but have occasionally used units of convenience, such as atmospheres for pressure. Table A.1 gives most of the fundamental SI units, and Table A.2 some conversions.

The SI system uses powers of 10 expressed as prefixes to describe the order of magnitude of the quantity. Some multiples of the metric system and their associated prefixes are as follows:

10^{-15}	femto-
10^{-12}	pico-
10^{-9}	nano-
10^{-6}	micro-
10^{-3}	milli-
10^{-2}	centi-
10^{-1}	deci-
10	deka-
10^2	hecto-
10^3	kilo-
10^6	mega-
10^9	giga-
10^{12}	tera-
10^{15}	peta-

Table A.1

Selected SI Basic Units

Property	Units
Length	meter (m)
Time	second (s)
Mass	kilogram (kg)
Current	ampere (A)
Temperature	kelvin (K)
Force	newton (N)
Work and energy	joule (J)
Power	watt (W)
Frequency	hertz (Hz)
Charge	coulomb (C)
Magnetic field	tesla (T)
Pressure	pascal (Pa)

Table A.2

Conversion to SI Units

Length
    1 ft = 12 in. = 30.48 cm
    1 yd = 3 ft = 91.44 cm
    1 light year = 9.461×10^{15} m
    1 Å = 0.1 nm = 10^{-10} m

Area
    1 m^2 = 10^4 cm^2
    1 km^2 = 0.3861 mi^2
    1 $in.^2$ = 6.4516 cm^2
    1 ft^2 = 9.29×10^{-2} m^2
    1 m^2 = 10.75 ft^2

Volume
    1 m^3 = 10^6 cm^3
    1 liter (L) = 1000 cm^3 = 10^{-3} m^3
    1 gal = 3.786 L = 231 $in.^3$

Time
    1 h = 60 min = 3.6 ks
    1 day = 24 h = 1440 min = 86.4 ks
    1 year = 365.24 days = 31.56 Ms

Speed
    1 km/h = 0.2778 m/s = 0.6125 mi/h
    1 mi/h = 0.4470 m/s = 1.609 km/h

Density
    1 g/cm^3 = 1000 kg/m^3 = 1 kg/L

Force
    1 N = 0.2248 lb = 10^5 dynes
    1 lb = 4.4482 N

Pressure
    1 Pa = 1 N/m
    1 atm = 101.325 kPa = 14.7 $lb/in.^2$ = 760 mm Hg

Energy
    1 kW · h = 36 MJ
    1 eV = 1.602×10^{-19} J
    1 erg = 10^{-7} J

Power
    1 hp = 550 ft · lb/s = 745.7 W
    1 W = 1.341×10^{-3} hp

Magnetic field
    1 G = 10^{-4} T
    1 T = 10^4 G

Here are some relationships between the metric and English system.

Length

1 kilometer (km) = 1000 meters = 0.6214 mile

1 meter (m) = 1.094 yards = 39.37 inches

1 centimeter (cm) = 0.01 meter = 0.3937 inch

1 millimeter (mm) = 0.001 meter = 0.03937 inch

1 mile = 1.6093 km

1 inch = 2.5400 cm

Mass

1 metric ton = 10^6 g = 1000 kg = 2.2046×10^3 pounds

1 kilogram (kg) = 10^3 g = 2.2046 pounds

1 gram (g) = 0.0353 oz = 0.0022046 pound

1 milligram (mg) = 0.0001 g = 2.2046×10^{-6} pound

1 pound = 453.6 g

1 ounce = 28.3495 g

Temperature Scales

The freezing and boiling points of water (at sea level) set the fundamental points on common temperature scales. In the United States, the Fahrenheit (F) system dominates. Water freezes at 32° F and boils at 212° F. Most of the world uses the Celsius (C; formerly centigrade) system, in which water freezes at 0° C and boils at 100° C.

Astronomers mostly employ the Kelvin (K) system, which uses the same size degrees as the Celsius system, but starts from absolute zero, the temperature at which random molecular motion reaches a minimum (0 K = −273.15 °C). At absolute zero, no more thermal energy can be extracted from a body. Water freezes at about 273 K and boils at about 373 K. Note that the degree mark is not used with temperatures on the Kelvin scale and that temperatures are often referred to as "kelvins," not "degrees kelvin."

To convert between systems, recognize that 0 K = −273 °C = −460 °F and that Celsius and Kelvin degrees are larger than Fahrenheit degrees by 180/100 = 9/5. The relationships between systems are expressed as follows:

$$K = °C + 273$$

$$°C = \frac{5}{9} (°F − 32)$$

$$°F = \left(\frac{9}{5} °C\right) + 32$$

Note that kelvins are never negative, because absolute zero is the lowest point on any temperature scale.

Astronomical Distances

Although astronomers do use the metric system, they encounter distances so large that other measures are often used as well. In the solar system, the natural scale is the average distance from the earth to the sun, the astronomical unit (AU), which is about 1.5×10^8 km.

Beyond the solar system, the AU is too small compared to stellar distances. Here astronomers use the parsec and the light year. One parsec (pc) equals 206,265 AU or about 3.1×10^{13} km. A light year (ly) is the distance that light travels in one year, about 9.5×10^{12} km. Note that 1 pc is about 3.3 ly; you can remember it roughly as 3 ly.

For even larger distances, astronomers most often talk in large multiples of parsecs: a kiloparsec (kpc) is 10^3 pc, and a megaparsec (Mpc) is 10^6 pc. Equivalent units in light years are abbreviated kly and Mly.

Other Physical Units

The SI unit of force is the *newton* (N). It is the force needed to accelerate an object having a mass of one kilogram by one meter per second squared. Since weight is a force, the newton is also a unit of weight. One newton is about the weight of a (quarter-pound) hamburger. However, in the metric system it is more common to express weight as mass (kilograms) rather than in newtons.

The SI unit of energy is the *joule* (J). It takes about one joule to lift an apple off the floor to over your head. Table A.3 compares the energy of some familiar actions and some astronomical ones.

Power, the rate of energy transfer, has units of joules per second. Astronomers often use the term *luminosity* in place of power. A convenient unit for luminosity or power is the *watt* (W), defined as one joule per second. The electrical power used by light

Table A.3

Comparative Energy Outputs

Energy Source	Total Energy (J)
Big Bang	10^{68}
Radio galaxy	10^{55}
Supernova	10^{46}
Sunlight (1 year)	10^{34}
Volcanic explosion	10^{19}
H-bomb	10^{17}
Thunderstorm	10^{15}
Lightning flash	10^{10}
Baseball pitch	10^2
Typing (per key)	10^{-2}
Flea hop	10^{-7}

bulbs is stated in watts, such as a 100-W light bulb. A person has a luminosity of about 100 W!

When talking about stars, astronomers often measure quantities relative to the sun—for example, in units of a solar mass, a solar luminosity, or a solar radius. For example, the mass of Sirius is 2.3 solar masses, its radius is 1.8 solar radii, and its luminosity is 23 solar luminosities.

Astronomers use both the *gauss* (G) and *tesla* (T) as the unit of magnetic field strength (1 T = 10^4 G); the tesla is the SI unit. The earth's magnetic field is about 5×10^{-3} T, or 0.5 G. It serves as the most familiar basis of comparison.

The SI unit of pressure is the *pascal* (Pa), which is one newton per meter squared. You can get a feel for a pascal next time you check the air in your car's tires: a typical pressure of 28 pounds per square inch (psi) equals 196 kilopascals (kPa). Astronomers also use the *atmosphere* (atm) as a pressure unit; one atmosphere is about 100 kPa.

APPENDIX
B
Planetary Data

Table B.1

Planetary Rotation Rates and Inclinations

Planet	Sidereal Rotation Period (equatorial)	Inclination of Equator to Orbital Plane
Mercury	58.65 days	0°
Venus	243.01 days (retrograde)	117° 18′
Earth	23 h 56 min 4.1 s	23° 27′
Mars	24 h 37 min 22.6 s	25° 12′
Jupiter	9 h 50.5 min	3° 07′
Saturn	10 h 14 min	26° 44′
Uranus	17 h 14 min (retrograde)	97° 52′
Neptune	16 h 3 min	29° 34′
Pluto	6.39 days (retrograde)	98°

Table B.2

Distances and Periods of the Planets

Planet	Semimajor Axis (AU)	Semimajor Axis (× 10^6 km)	Sidereal Period (tropical years)	Sidereal Period (days)	Synodic Period (days)	Orbital Eccentricity
Mercury	0.387	57.9	0.24	87.97	115.9	0.206
Venus	0.723	108.2	0.615	224.7	583.9	0.007
Earth	1.000	149.6	1.000	365.26		0.017
Mars	1.524	227.9	1.881	686.98	779.9	0.094
Jupiter	5.203	778.4	11.86	4332	398.9	0.048
Saturn	9.522	1424	29.46	10,761	378.1	0.054
Uranus	19.20	2872	84.01	30,685	369.7	0.048
Neptune	30.07	4499	164.8	60,195	367.5	0.007
Pluto	39.72	5943	248.6	90,471	366.7	0.253

Note: A tropical year is the year of seasons. For definitions of sidereal and synodic period, see the Glossary.

Table B.3

Physical Data of the Planets

Planet	Average Radius (km)	Radius (earth radii)	Albedo	Temperature (K)	Escape Speed (km/s)
Mercury	2,439	0.38	0.06	100–700	4.3
Venus	6,052	0.95	0.76	700	10.4
Earth	6,378	1.00	0.40	250–300	11.2
Moon	1,738	0.27	0.07	120–390	2.4
Mars	3,397	0.53	0.16	210–300	5.0
Jupiter	71,492	11.19	0.51	110–150	59.6
Saturn	60,268	9.45	0.50	95	35.6
Uranus	25,559	4.01	0.66	58	21.3
Neptune	25,269	3.96	0.62	56	23.8
Pluto	1,140	0.18	0.50	40	1.2

Table B.4

Masses and Densities of the Planets

Planet	Mass (earth masses)	Mass (kg)	Density (kg/m^3)	Surface Gravity (earth = 1)
Mercury	0.0562	3.30×10^{23}	5430	0.38
Venus	0.815	4.87×10^{24}	5240	0.91
Earth	1.000	5.974×10^{24}	5520	1.00
Moon	0.012	7.35×10^{22}	3340	0.16
Mars	0.1074	6.42×10^{23}	3940	0.39
Jupiter	317.9	1.899×10^{27}	1330	2.54
Saturn	95.1	5.68×10^{26}	690	1.07
Uranus	14.56	8.66×10^{25}	1270	0.90
Neptune	17.24	1.03×10^{26}	1640	1.14
Pluto	0.0018	1.1×10^{22}	2100 (?)	0.06

Table B.5

Atmospheric Gases of the Major Planets

Planet	Gases (in order of relative abundance)
Mercury	Sodium, potassium, helium, hydrogen
Venus	Carbon dioxide, carbon monoxide, hydrogen chloride, hydrogen fluoride, water, argon, nitrogen, oxygen, hydrogen sulfide, sulfur dioxide, helium
Moon	Helium, argon
Earth	Nitrogen, oxygen, water, argon, carbon dioxide, neon, helium, methane, krypton, nitrous oxide, ozone, xenon, hydrogen, radon
Mars	Carbon dioxide, carbon monoxide, water, oxygen, ozone, argon, nitrogen
Jupiter	Hydrogen, helium, methane, ammonia, water, carbon monoxide, acetylene, ethane, phosphine, germane
Saturn	Hydrogen, helium, methane, ammonia, acetylene, ethane, phosphine, propane
Titan	Nitrogen, methane, ethane, acetylene, ethylene, hydrogen cyanide
Uranus	Hydrogen, helium, methane
Neptune	Hydrogen, helium, methane, ethane
Pluto	Methane

Table B.6

Satellites of the Terrestrial Planets

Planet	Satellite	Distance ($\times 10^3$ km)	Orbital Period (days)	Radius (km)	Mass (planet = 1)	Bulk density (kg/m^3)
Earth	Moon	384	27.32	1738	0.012	3340
Mars	Phobos	9.38	0.3189	$14 \times 11 \times 9$	1.5×10^{-8}	2200
Mars	Deimos	23.46	1.262	$8 \times 6 \times 6$	3.1×10^{-9}	1700

Table B.7

Major Satellites of Jupiter

Name	Distance ($\times 10^3$ km)	Distance (Jupiter radii)	Orbital period (days)	Radius (km)	Mass (planet = 1)	Bulk density (kg/m³)
Metis	128	1.79	0.294	20	0.5×10^{-10}	
Andrastea	129	1.80	0.298	$12 \times 10 \times 8$	0.11×10^{-10}	
Amalthea	181	2.55	0.498	$135 \times 83 \times 75$	38×10^{-10}	3000
Thebe	222	3.11	0.674	55×45	4×10^{-10}	
Io	422	5.95	1.77	1815	4.7×10^{-5}	3550
Europa	671	9.47	3.55	1569	2.5×10^{-5}	3040
Ganymede	1,070	15.10	7.15	2631	7.8×10^{-5}	1930
Callisto	1,883	26.60	16.7	2400	5.7×10^{-5}	1830
Leda	11,094	156	239	8	0.03×10^{-10}	
Himalia	11,480	161	251	93	50×10^{-10}	~1000
Lysithea	11,720	164	259	18	0.4×10^{-10}	
Elara	11,737	165	260	38	4×10^{-10}	
Ananke	21,200	291	631R	15	0.2×10^{-10}	
Carme	22,600	314	692R	20	0.5×10^{-10}	
Pasiphae	23,500	327	735R	25	1×10^{-10}	
Sinope	23,700	333	758R	18	0.4×10^{-10}	

Note: Distances are from the center of Jupiter; "R" indicates retrograde rotation.

Table B.8

Major Satellites of Saturn

Satellite	Distance ($\times 10^3$ km)	Distance (Saturn radii)	Orbital Period (days)	Radius (km)	Mass (planet = 1)	Bulk density (kg/m^3)
Atlas	137.67	2.28	0.602	20×10		
Prometheus	139.35	2.31	0.613	$70 \times 50 \times 40$		
Pandora	141.70	2.35	0.628	$55 \times 45 \times 35$		
Epimetheus	151.42	2.51	0.694	$70 \times 60 \times 50$		
Janus	151.42	2.51	0.694	$110 \times 100 \times 80$		
Mimas	185.54	3.08	0.942	196	8.0×10^{-8}	1170
Enceladus	238.02	3.95	1.370	250	1.3×10^{-7}	1240
Tethys	294.66	4.88	1.887	530	1.3×10^{-6}	1260
Telesto	294.66	4.88	1.887	$17 \times 14 \times 13$		
Calypso	294.66	4.88	1.887	$17 \times 11 \times 11$		
Dione	377.40	6.26	2.736	560	1.85×10^{-6}	1440
Helene	377.40	6.26	2.736	$18 \times 16 \times 15$		
Rhea	527.04	8.74	4.517	765	4.4×10^{-6}	1330
Titan	1221.8	20.25	15.945	2575	2.36×10^{-4}	1880
Hyperion	1481.1	24.55	21.276	$205 \times 130 \times 110$	3×10^{-8}	
Iapetus	3561.3	59.02	79.330	730	3.3×10^{-6}	1210
Phoebe	12.952	214.7	550.4R	110	7×10^{-10}	

Note: Distances are from the center of Saturn; "R" indicates retrograde rotation.

Table B.9

Major Satellites of Uranus, Neptune, and Pluto

Planet	Satellite	Distance ($\times 10^3$ km)	Orbital Period (days)	Radius (km)	Mass (planet = 1)	Bulk Density (kg/m^3)
Uranus	Ariel	191.0	2.520	579	1.8×10^{-5}	1650
	Umbriel	266.3	4.144	586	1.2×10^{-5}	1440
	Titania	435.9	8.706	790	6.8×10^{-5}	1590
	Oberon	583.5	13.463	762	6.9×10^{-5}	1500
	Miranda	129.41	1.413	240	0.2×10^{-5}	1260
Neptune	Triton	354.8	5.877R	1350	2.1×10^{-4}	
	Nereid	5513	360	170	2.0×10^{-7}	
Pluto	Charon	19.1	6.387	600	0.2	2080

APPENDIX C

Physical Constants, Astronomical Data

Physical Constants

Gravitational constant

$$G = 6.673 \times 10^{-11} \ \text{N} \cdot \text{m}^2/\text{kg}^2$$

Speed of light in a vacuum

$$c = 2.9979 \times 10^8 \ \text{m/s}$$

Planck's constant

$$h = 6.62608 \times 10^{-34} \ \text{J} \cdot \text{s}$$

Wien's constant

$$\sigma_{\text{w}} = 0.02898 \ \text{m} \cdot \text{K}$$

Boltzmann's constant

$$k = 1.3806 \times 10^{-23} \ \text{J/K}$$

Stefan–Boltzmann constant

$$\sigma = 5.6697 \times 10^{-8} \ \text{W/m}^2 \cdot \text{K}^4$$

Electron mass

$$m_{\text{e}} = 9.10939 \times 10^{-31} \ \text{kg}$$

Proton mass

$$m_{\text{p}} = 1.6726 \times 10^{-27} \ \text{kg} = 1836.1 \ m_{\text{e}}$$

Neutron mass

$$m_{\text{n}} = 1.6749 \times 10^{-27} \ \text{kg}$$

Mass of hydrogen atom

$$m_{\text{H}} = 1.6735 \times 10^{-27} \ \text{kg}$$

Astronomical Data

Astronomical unit

$$\text{AU} = 1.4959789 \times 10^{11} \ \text{meters}$$

Parsec

$$\text{pc} = 206{,}264.806 \ \text{AU}$$
$$= 3.2616 \ \text{ly}$$
$$= 3.0856 \times 10^{16} \ \text{meters}$$

Light year

$$\text{ly} = 9.46053 \times 10^{15} \ \text{meters}$$
$$= 6.324 \times 10^4 \ \text{AU}$$

Sidereal year

$$y = 3.155815 \times 10^7 \ \text{seconds}$$

Mass of sun

$$M_{\text{sun}} = 1.989 \times 10^{30} \ \text{kg}$$

Luminosity of sun

$$L_{\text{sun}} = 3.90 \times 10^{26} \ \text{W}$$

Solar constant

$$S = 1370 \ \text{W/m}^2$$

Radius of sun

$$R_{\text{sun}} = 6.96 \times 10^5 \ \text{km}$$

Mass of earth

$$M_{\text{earth}} = 5.9742 \times 10^{24} \ \text{kg}$$

Equatorial radius of earth

$$R_{\text{earth}} = 6.37814 \times 10^3 \ \text{km}$$

Mass of moon

$$M_{\text{moon}} = 7.348 \times 10^{22} \ \text{kg}$$

Radius of moon

$$R_{\text{moon}} = 1.738 \times 10^3 \ \text{km}$$

Hubble constant

$$H \sim 20 \ \text{km/s/Mly}$$

APPENDIX D

Nearby Stars

Name	Parallax (arcsec)	Distance [pc (ly)]	Spectral Type	Proper Motion (arcsec/y)	Apparent Visual Magnitude	Absolute Visual Magnitude
Sun			G2 V		−26.7	4.85
α Cen C (Proxima)	0.772	1.3 (4.2)	M5.5e V	3.85	11.0	15.5
α Cen A	0.750	1.3 (4.3)	G2 V	3.68	−0.01	4.37
α Cen B		1.3 (4.3)	K1 V		+1.33	5.71
Barnard's star	0.546	1.8 (6.0)	M3.8 V	10.3	+9.5	13.2
Wolf 359	0.419	2.4 (7.7)	M5.8e V	4.70	+13.5	16.7
BD+36° 2147 (Lalande 21185)	0.397	2.5 (8.2)	M2.1e V	4.78	+7.5	10.5
Luyten 726-8A	0.387	2.7 (8.8)	M5.6e V	3.36	+12.5	15.5
Luyten 726-8B (UV Ceti)		2.7 (8.8)	M5.6e V		+13.0	16.0
Sirius A	0.377	2.6 (8.6)	A1 V	1.33	−1.46	1.42
Sirius B			wd		+8.3	11.2
Ross 154	0.345	2.9 (9.4)	M3.6e V	0.72	+10.5	13.1
Ross 248	0.314	3.2 (10.4)	M4.9e V	1.60	+12.3	14.8
ε Eri	0.303	3.3 (10.8)	K2e V	0.98	+3.73	6.14
Ross 128	0.298	3.3 (10.9)	M4.1 V	1.38	+11.1	13.5
61 Cyg A	0.294	3.4 (11.1)	K3.5e V	5.22	+5.22	7.56

Note: An *e* after the spectral type indicates emission lines in the spectrum; *wd* indicates a white dwarf. The apparent and absolute visual magnitude are over the visual range of the spectrum. These terms, and proper motion, are defined in the Glossary.

(CONTINUED)

Name	Parallax (arcsec)	Distance [pc (ly)]	Spectral Type	Proper Motion (arcsec/y)	Apparent Visual Magnitude	Absolute Visual Magnitude
61 Cyg B			K4.7e V		+6.03	8.37
ε Ind	0.291	3.4 (11.2)	K3e V	4.70	+4.68	7.00
Luyten 789-6A	0.290	3.4 (11.2)	M5e V	3.26	12.2	15.0
Luyten 789-6B			?		?	15.6
Procyon A	0.285	3.5 (11.4)	F5 IV	1.25	+0.37	2.64
Procyon B			wd		+10.7	13.0
BD+59° 1915A (S 2398 A)	0.282	3.6 (11.6)	M3 V	2.29	+8.9	11.2
BD+59° 1915B (S 2398 B)			M3.5 V		+9.7	11.9
CD−36° 15693	0.279	3.6 (11.7)	M1.3e V	6.90	+7.35	9.58
G 51-15	0.278	3.6 (11.7)	M6.6 V	1.27	+14.8	17.0
τ Ceti	0.273	3.6 (11.8)	G8 V	1.92	+3.5	5.72
BD+5° 1668	0.266	3.8 (12.3)	M3.7 V	3.77	+9.82	11.9
Luyten 725-32 (YZ Ceti)	0.261	3.8 (12.5)	M4.5e V	1.32	+12.0	14.1
CD−39° 14192 (Lacaille 8760)	0.260	3.8 (12.5)	K5.5e V	3.46	+6.66	8.74
Kapteyn's star	0.256	3.9 (12.7)	M0 V	8.72	+8.84	10.9
Kruger 60 A	0.253	4.0 (12.9)	M4	0.86	+9.85	11.9
Kruger 60 B			M5e		+11.3	13.3

Note: An *e* after the spectral type indicates emission lines in the spectrum; *wd* indicates a white dwarf. The apparent and absolute visual magnitude are over the visual range of the spectrum. These terms, and proper motion, are defined in the Glossary.

APPENDIX
E
The Brightest Stars

Star	Name	Apparent Visual Magnitude	Spectral Type	Absolute Visual Magnitude	Distance [pc (ly)]	Proper Motion (arcsec/y)
α CMa A	Sirius	−1.46	A1 V	+1.4	2.7 (8.7)	1.32
α Car	Canopus	−0.72	A9 II	−2.5	23 (74)	0.034
α Boo	Arcturus	−0.04	K1.5 III	0.2	10 (34)	2.28
α Cen A	Rigil Kentaurus	−0.01	G2 V	+4.4	1.3 (4.3)	3.68
α Lyr	Vega	+0.03	A0 V	+0.6	7.7 (25)	0.35
α Aur AB	Capella	+0.08	G6 III G2 III	−0.4	13 (41)	0.43
β Ori A	Rigel	+0.12	B8 Ia	−8.1	430 (1400)	0.004
α CMi A	Procyon	+0.38	F5 IV−V	+2.7	3.4 (11)	1.25
α Ori	Betelgeuse	+0.5	M2 Iab	−7.2	430 (1400)	0.028
α Eri	Achernar	+0.46	B3 V	−1.3	21 (69)	0.108

(CONTINUED)

Star	Name	Apparent Visual Magnitude	Spectral Type	Absolute Visual Magnitude	Distance [pc (ly)]	Proper Motion (arcsec/y)
β Cen AB	Hadar	+0.63	B1 III	−4.4	98 (320)	0.030
α Aql	Altair	+0.77	A7 V	+2.3	5.1 (16)	0.662
α Tau A	Aldebaran	+0.86	K5 III	−0.3	18 (60)	0.200
α Sco A	Antares	+0.92	M1.5 Ia	−5.2	160 (520)	0.024
α Vir	Spica	+1.0	B1 V	−3.2	67 (220)	0.054
β Gem	Pollux	+1.14	K0 III	+0.7	10.7 (35)	0.629
α PsA	Fomalhaut	+1.16	A3 V	+2.0	6.9 (22.6)	0.373
α Cyg	Deneb	+1.25	A2 Ia	−7.2	460 (1500)	0.005
β Cru	Becrux	+1.28	B0.5 III	−4.7	140 (460)	0.042
α Leo A	Regulus	+1.35	B7 V	−0.3	21 (69)	0.248

APPENDIX
F

Periodic Table
of the Elements

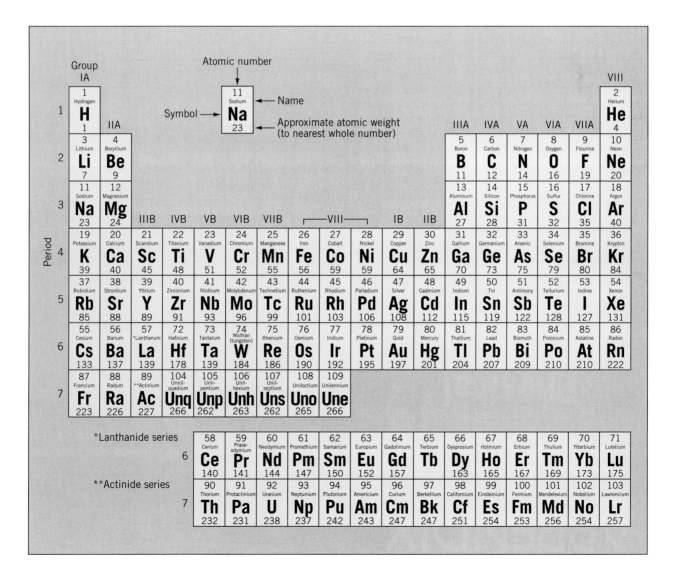

APPENDIX
G
Understanding Flux Maps

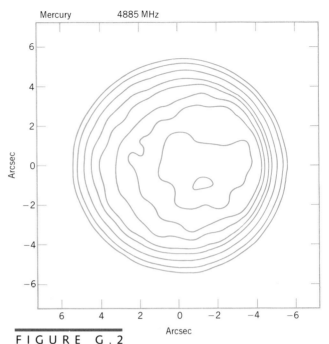

FIGURE G.2

Contour map of the radio flux from the planet Mercury at a wavelength of 6 cm (4885 MHz). Note the peak near the center. The two axes give the angular scale in arcseconds.

Many of the figures in this book are **flux maps**, pictures of how the flux of some kind of radiation (radio, visible, infrared) varies over a given region of the sky. Flux is the amount of energy arriving every second per unit area at a telescope from a source. Flux maps may show a lot of wavy, connected lines labeled by numbers; these are called *contour maps*. Or such information may be displayed as a smooth range of grays or colors.

FIGURE G.1

Map showing one day's atmospheric pressure over the continental United States. The contour lines connect places with the same pressure readings, which are given in millibars. A "High" marks a local region of a peak value in the surface pressure; a "Low" a region of minimum value.

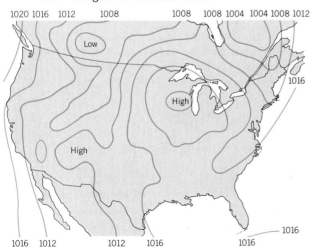

Here's an analogy that may be familiar: a weather map of atmospheric pressure across the United States, with high and low pressure systems indicated (Fig. G.1). What do the contours tell you? First, consider how a pressure contour is drawn. The weather stations around the United States report their local pressures. Each is put on the map. Then a contour line is drawn that connects all stations giving the same reading, provided the line does *not* cross a station of higher or lower reading. Then another contour (of higher or lower pressure) is drawn in, and so on. Notice that contour lines *cannot* cross. Why? Because if they did, it would mean that the *same* place has two *different* pressures, and that's impossible!

Second, note that there are places where the pressures hit a maximum (high) or a minimum (low). At the center of each such area is a last contour surrounding the region of highest or lowest pressure. For example, you can imagine the center of a high as the peak of a pressure mountain. The contours around the peak tell you how the pressure falls from the peak. If the contour lines are close together, the pressure drops quickly over a short distance (the fall-

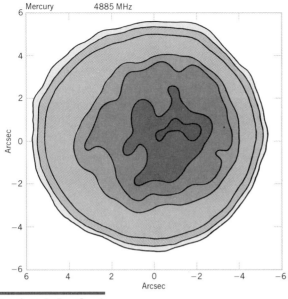

FIGURE   G.3

Combination contour map and gray-scale map of the radio emission from Mercury at 6 cm (4885 MHz). The darker the region, the greater the flux. The contours here are somewhat different from those in Figure G.2 because of a somewhat different processing procedure. Again, note the central peak of emission.

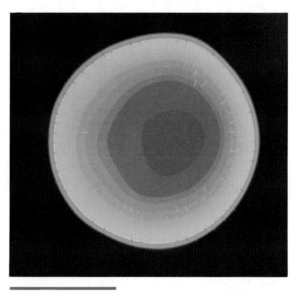

FIGURE   G.4

False-color map of the radio emission from Mercury at 6 cm (4885 MHz). The color coding is dark red for the strongest flux and blue for the weakest. Note the peak near the center, which is also a peak of the temperature about a meter below the surface. The computer processing has smoothed out much of the wiggling in the various flux levels. The short radial lines show the direction and size of the polarization of the radio emission. (Photo by M. Zeilik)

off is steep). If the contour lines are spread out, the pressure drops slowly; you have a ridge.

By using image processing, astronomers can display astronomical data in a variety of formats, depending on what aspects they want to emphasize. The simplest one is a contour map like the weather map, which shows high and low regions clearly. Figure G.2 gives the distribution of radio emission at the frequency of 4885 MHz from the disk of the planet Mercury. Note the high region of emission offset a bit from the center. The scale is the angle on the sky, given in arcseconds. A variation of this is a *gray-scale map*, in which the density of the filled areas gives an indication of the flux—the darker the shading, the higher the flux. Figure G.3 shows a somewhat different processing of the radio emission from Mercury with both contour lines and a gray scale.

Finally, the most elaborate is a **false-color map**, in which the computer generates different colors for different levels of flux (Fig. G.4). Again, we are displaying the radio emission from Mercury, with blue showing the weakest emission, and dark red the strongest. (This map has a somewhat different look because the image processing aimed at smoothing out the small variations in the observed emission.) These materials are colorful, but you must recognize the colors as processing artifacts rather than as visual properties of the object observed. (For example, a TV weather map may show a range of temperatures and give them a color code that runs from blue for the lowest and red for the highest.) By contrast, the colors in a *true-color map* are those you would perceive by eye.

Many of the four-color images in this book are false-color maps; keep that in mind as you try to understand the information they display. Beware that the color coding may differ from map to map! The choice of the coding is aesthetic as well as scientific.

APPENDIX
H
Telescopes

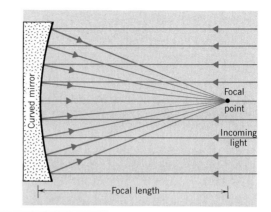

FIGURE H.2

Smoothly curved mirror reflects incoming light (from a source at a distance) to a focus. Astronomical mirrors have their reflecting material on their front surfaces and follow a parabolic curve to make a sharp focus.

You are probably most familiar with optical telescopes—those that manipulate light detectable by the eye. The **optics** of a telescope determines how the direction of light is controlled. When traveling through empty space or a uniform medium, light moves in a straight line. Optics views light as particles moving along straight-line paths, called *light rays*. Using lenses and mirrors, we can change the direction of light rays.

When light crosses the boundary from one transparent material to another (from air to glass, for example), its direction generally changes. This bending of light rays is termed **refraction**. A mirror with a smooth surface returns light by bouncing back rays—a process called **reflection**. The goal of optics is to make images by refraction and reflection. An **image** occurs when light rays are gathered together in the same relative alignment as when they left an object.

A smoothly curved piece of glass is a **lens** (Fig. H.1), which uses refraction to bring rays from a very distant point source to a point image at its focus. The lens makes an image of an object of finite size by focusing parallel rays from each point of the source onto a separate point in the image. For objects at large distances, the distance from the lens to the image is the same for all objects. This distance is termed the **focal length**. A smoothly curved mirror—whose surface follows the curve of a parabola—brings all the light to a focus, using reflection to make an image (Fig. H.2).

Basically, a telescope gathers light and makes an image at a focus. A telescope needs a lens or mirror, called the **objective**, to bring light to a focus. An-

FIGURE H.1

Lens forming a point image from a point source. Because its surface is smoothly curved, a lens forms a sharp focus at the *focal point*. The distance from the lens to the focal point, for a distant source, is the *focal length*.

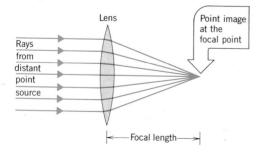

FIGURE H.3

The optical layout of a simple refracting telescope. An objective lens gathers the incoming light and brings it to a focus. An eyepiece allows viewing by eye of the magnified image.

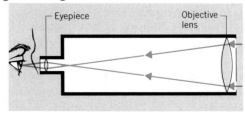

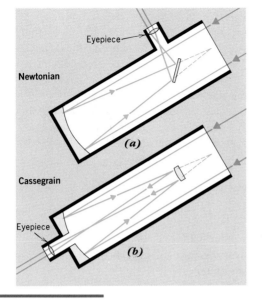

FIGURE   H . 4

Designs of reflecting telescopes. The main problem is viewing the image made by the objective mirror. Newton placed a small, flat mirror at an angle to reflect the light out to one side, where an eyepiece is positioned (*a*). This is called a *Newtonian reflector*. A common modern design uses a small convex mirror to reflect the light back down through a hole in the objective mirror for viewing (*b*). This is called a *Cassegrain reflector*.

other lens, called an **eyepiece**, placed just beyond the focus allows visual examination of the image; or, a camera can be placed at the focus to photograph the image. There are basically two types of telescope, distinguished by their objectives: **refracting telescopes** (or *refractors*) use a lens (Fig. H.3), and **reflecting telescopes** (or *reflectors*) use a mirror (Fig. H.4). All large telescopes built now are reflectors; advances in technology allow them to be smaller, more rigid, and cheaper to construct.

The Keck telescope is the largest ground-based optical telescope. Located on Mauna Kea, Hawaii, at an elevation of 4145 meters (Fig. H.5), the Keck has innovation designs—especially its 10-meter mirror, which consists of 36 hexagonal segments (Fig. H.6) with a total weight of 14.4 tons. Each segment has a diameter of 1.8 meters and a thickness of 75 mm. The segments are individually controlled by an active system to maintain the mirror's shape under a variety of stresses. The dome and the telescope's structure minimize local *seeing* to take advantage of the site's natural potential—occasionally as good as 0.3 arcsecond. A twin Keck will be built next to the current one.

FIGURE   H . 5

The dome of the Keck telescope on Mauna Kea. (Courtesy California Association for Research in Astronomy)

FIGURE H.6

Partial assembly of the Keck telescope's mirror: 9 of the 36 segments of the mirror are in place. (Courtesy California Association for Research in Astronomy)

Functions of a Telescope

Gathering Light

Whether a reflector or a refractor, a telescope has one primary function: to gather light. A telescope is basically a light funnel, collecting photons and concentrating them at a focus. How much light a telescope can collect depends on the *area* of its objective, which is proportional to the square of the diameter. So a mirror with twice the diameter of another can gather four times as much light. That's why astronomers want large telescopes—big mirrors have plenty of light-gathering power.

To Resolve Fine Detail

The next most important function of a telescope is to separate, or *resolve*, objects that are close together in the sky. This ability is called **resolving power**, usu-

ally expressed in terms of the minimum angle between two points that can be clearly separated. Resolving power depends directly on the diameter of the objective and is inversely proportional to the wavelength of the light (because of the wave properties of light). For the same wavelength, the resolving power depends directly on the objective's diameter. So a mirror *twice* the size of another has *twice* the resolving power; that is, it can resolve objects half as far apart in angular distance.

For example, a 10-cm telescope has a resolving power of 1.4 arcseconds at visual wavelengths. This means that if the telescope is aimed at two stars that are more than 1.4 arcseconds apart, you will see two separate star images. If the angular separation of the stars is less than 1.4 arcseconds, you will see a single, elongated image.

The theoretical resolving power of a 500-cm telescope is 0.02 arcsecond. But ground-based telescopes never attain such performance. The resolving power of a big telescope is limited by the earth's atmosphere. Stars twinkle from turbulence in the air. The motion of blobs of air distorts and blurs images seen through a telescope. It is a rare night when star images are smaller than 1 arcsecond. Astronomers call this quality of the atmosphere **seeing**.

To Magnify the Image

The *least* important of a telescope's functions is its **magnifying power**, the apparent increase in the size of an object compared with visual observation. Magnifying power depends on the focal length of the objective and the focal length of the eyepiece. Changing the eyepiece on a telescope changes its magnifying power as follows: the *shorter* the focal length of the eyepiece, the *greater* the magnifying power. For example, if you put in an eyepiece with *half* the focal length of a previous one, you *double* the magnifying power. But there is no point in using a magnifying power any greater than is necessary to see clearly the smallest detail in the image, which is determined by the resolving power. Extremely high magnification merely makes the fuzziness worse.

Invisible Astronomy

Astronomers rarely observe with their eyes these days. Some light-sensing device, called a **detector**, is usually placed at the focus. The detector may be a photographic plate (light-sensitive materials on glass rather than film). Or it may be an electronic detector similar to a television camera. The detector may even be sensitive to electromagnetic radiation outside the visual range, say in the ultraviolet or infrared. Your eye senses only a tiny sliver of the total electromagnetic spectrum. Electromagnetic radiation outside the visible region of the spectrum is often produced in astronomical objects by processes different from those that generate visible light. Invisible astronomy detects these photons.

A common type of ground-based invisible astronomy is radio astronomy. A radio dish (Fig. H.7) functions like an optical reflecting telescope. Essentially, it's a radio-wave bucket with a detector (a radio receiver) at the focus of the dish, which reflects and concentrates radio waves in much the same way a mirror does in a reflecting telescope. Radio receivers detect the incoming radio waves and an amplifier translates the signal into a voltage that can be recorded. A computer then generates a map of the radio flux over a region of the sky (Appendix G).

Radio telescopes have one major drawback: poor resolving power. Resolving power depends on both the diameter of the objective and the wavelength of the gathered light. Radio waves are much longer than visible light, typically 100,000 times as long. So if an optical telescope and a radio telescope had the same diameter, the radio telescope would have 100,000 times *less* resolving power and much less ability to show details in radio sources. Clearly, radio telescopes need to be large to have decent resolving power, but single dishes cannot be made on earth as large as is required.

Astronomers have devised a clever method for making a small radio telescope function as if it were a large one. Imagine two radio telescopes placed, say, 10 km apart. By synchronizing the signals received by both, these instruments can be made to act like a single dish with a diameter of 10 km, but only for a thin strip across the sky. (That's because they act like two small pieces at the opposite ends of a strip of a larger radio dish.) To get good resolving power in a small, more or less circular region of the sky requires an array of coordinated radio telescopes. The most advanced such telescope, called the *Very Large Array* (VLA), is in New Mexico. The VLA (which can have antennas spread out over a few tens of kilometers) has a resolving power at centimeter wavelengths equivalent to that of a moderate-sized optical telescope. Such devices are called **radio interferometers**.

How good can the resolution get? In the technique known as very-long-baseline interferometry (VLBI), the signals received by very distant antennas (even located on different continents) are recorded on magnetic tape and combined later in a computer. The maximum baseline extends to the diameter of the earth, so a resolving power of 2×10^{-4} arc-second is possible. In the future, radio telescopes in space, separated by larger baselines, will give even better resolution; some proposals envision one element of an interferometer on the moon—a separation some 30 times greater than possible on the earth. Resolutions could then approach 10^{-6} arc-second.

The United States is constructing the *Very-Large-Baseline Array* (VLBA) to be centered in New Mexico. Resembling the VLA in function, the VLBA will incorporate VLBI techniques in a coordinated array that will span the United States from the Virgin Islands to Hawaii. Present plans call for ten 25-meter antennas (similar to those of the VLA) with separations as great as a few thousand kilometers. The receivers will be synchronized by atomic clocks, and the data recorded on magnetic tapes and shipped to a central computer in New Mexico for processing.

FIGURE H.7

Section of a radio interferometer: looking down one arm of the Very Large Array (VLA) in New Mexico. Each antenna has a diameter of 25 meters. The antennas move along tracks to different positions along the array. (Photo by M. Zeilik)

Space Astronomy

Our atmosphere effectively absorbs large blocks of the electromagnetic spectrum, especially ultraviolet, x-rays, gamma rays, some infrared, and short-wavelength (millimeter) radio waves. One way to get around atmospheric absorption is to go above it—space astronomy. Most space astronomy now is done by earth-orbiting telescopes.

Low-energy infrared astronomy was the goal of the *Infrared Astronomical Satellite* or *IRAS*. A collaboration among the United States, the United Kingdom, and the Netherlands, IRAS mapped the sky at far-infrared wavelengths, from 12 to 100 μm, with a 57-cm telescope. The satellite functioned for 9 months until it ran out of the liquid helium that kept the infrared detectors cooled to 2.5 K. IRAS scanned almost all the sky close to three times. It detected more than 200,000 infrared sources, records of which are kept in a computer archive.

All ultraviolet astronomy must be done from space, for the absorbing layer of the atmosphere is higher than balloons or aircraft can reach. Probably the most successful ultraviolet telescope is the *International Ultraviolet Explorer* (*IUE*), launched in 1978 and still operational—far beyond its expected life. IUE has a 0.45-meter telescope and detectors that work in the wavelength range from 1150 to 3200 Å. The observations are accessible to all astronomers in a computer data bank.

A productive, early x-ray telescope was the *Einstein Observatory*, launched in 1978 and used up to 1981. This 58-cm telescope produced high-resolution x-ray images in the wavelength range from 3 to 50 Å. Although the telescope is now defunct, its observations are accessible in a computer archive of magnetic tapes. The current functional x-ray telescope is called *ROSAT* (the *Roentgen Satellite*), launched in 1990; it is a joint project of the United States, Germany, and the United Kingdom. It can provide high-resolution, x-ray images covering areas about 5 arcminutes square. ROSAT's first priority is a complete survey of x-ray sources in the sky.

In April 1991, the shuttle *Atlantis* deployed the Gamma Ray Observatory (GRO) of NASA in earth orbit at an altitude of 450 km. GRO is designed to detect, with great sensitivity, gamma rays over a wide range of energies (and even some of the most energetic x-rays). It has four separate instruments, each designed for a specific type of gamma-ray observation. GRO is the second in a series of space telescopes known as NASA's Great Observatories program. The GRO has been named the *Compton Gamma-Ray Observatory*, after Arthur Holly Compton, an American physicist who received a Nobel prize in physics.

The first Great Observatory was launched on April 24, 1990, when the shuttle *Discovery* transported the *Hubble Space Telescope* (*HST*) into space. The astronauts deployed it the next day. Its primary goal was to make observations of faint objects with high resolution, especially at ultraviolet wavelengths. But astronomers were shocked to find out that the main mirror is flawed. Simply put, the overall shape of the 2.4-meter primary mirror is too shallow, from edge to center, by about 2 μm. So the images of stars cannot be brought into good focus. According to the original specifications, most of the starlight should have fallen into an image with a radius of 0.1 arcsecond. The actual performance is a radius of 0.7 arcsecond—about that of a ground-based telescope at an exceptional site.

In the short term, two HST instruments have suffered the most: the Wide Field and Planetary Camera (WF/PC) and the Faint Object Camera (FOC). Both instruments were to have turned out high-resolution optical images. That won't come about in some instances, though the FOC can still do imaging in the ultraviolet that is not possible from the ground. However, the WF/PC can turn out good images (resolution of about 0.2 arcsec) if just the central part of the image is used. And a mission is planned to bring corrective optics to the HST to fix these problems. Meanwhile, the instruments designed for optical and ultraviolet spectroscopy will largely be able to achieve their goals.

APPENDIX

Seasonal Sky Charts

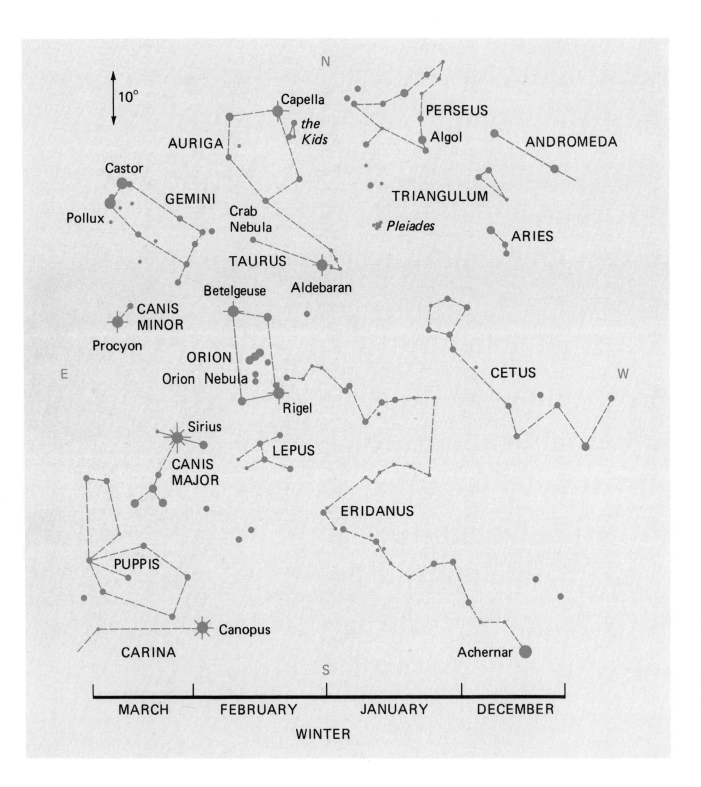

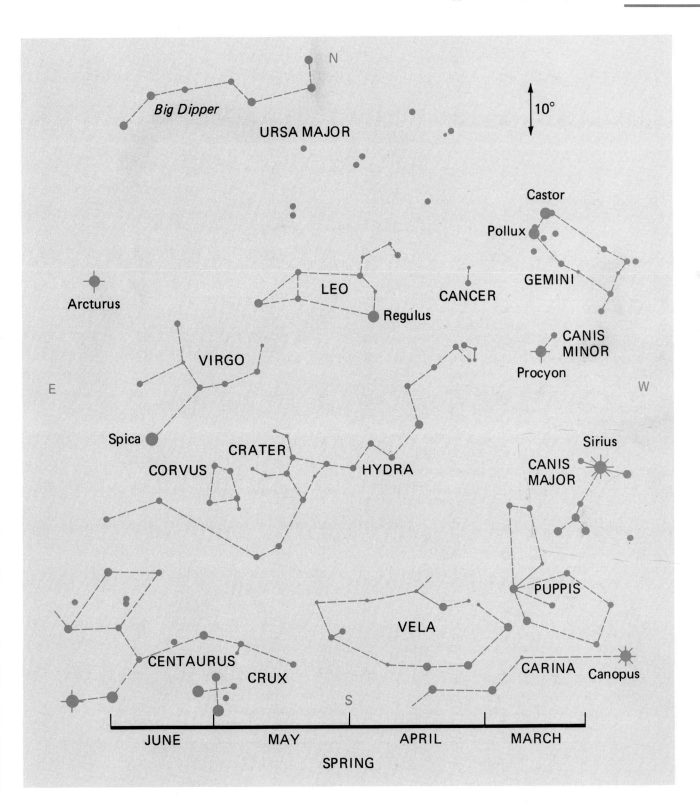

SPRING

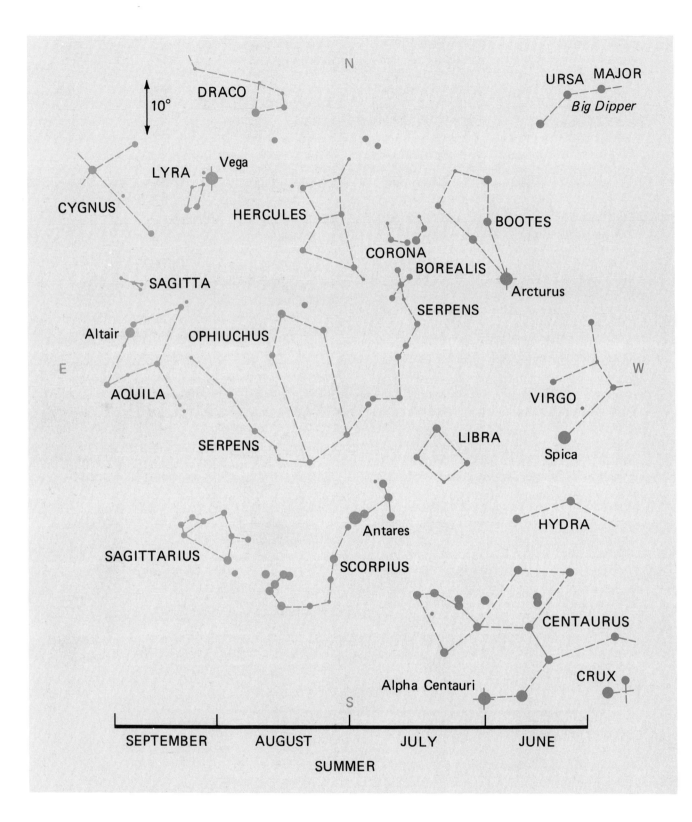

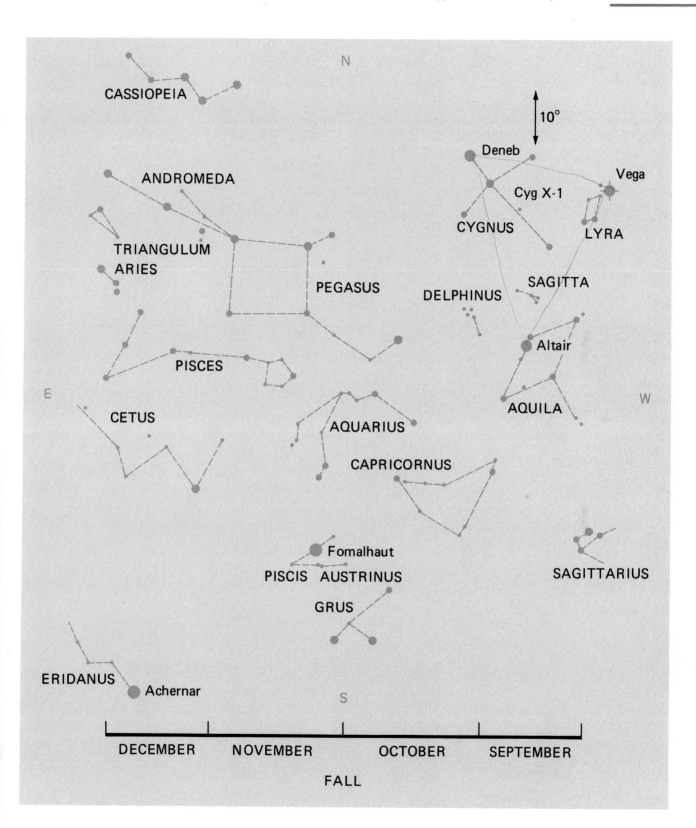

Glossary

absolute magnitude An astronomical measure of the brightness a star would have if it were placed at a standard distance of 10 parsecs (32.6 light years) from the sun.

absorption (dark) lines Discrete colors missing in a continuous spectrum because of the absorption of those colors by atoms.

absorption-line spectrum Dark lines superimposed on a continuous spectrum.

acceleration The rate of change of velocity with time.

accretion The colliding and sticking together of small particles to make larger masses.

accretion disk A disk made by material accumulating around a central mass (such as a star); the conservation of angular momentum results in the disk shape.

active galactic nuclei (AGNs) The nuclei of galaxies that have a nonthermal continuous spectrum over a wide range of wavelengths and signs of unusual, energetic activity such as radio jets. Examples: Seyfert galaxies, BL Lac objects, quasars.

active galaxies Galaxies characterized by a nonthermal spectrum and a large energy output compared to a normal galaxy. See **AGNs**.

active regions On the sun (and other stars), areas in which magnetic fields are concentrated; these generate sunspots and flares.

albedo The ratio of reflected light to incoming light for a solid surface, where complete reflection gives an albedo of 1.0 or 100 percent.

alpha particle A helium nucleus emitted in the radioactive decay of heavy elements.

altitude In the horizon system, the angle along a vertical circle from the horizon to a point on the celestial sphere; the angular height above the horizon of any celestial object.

amino acids The building blocks of proteins, consisting mostly of carbon, nitrogen, oxygen, and hydrogen atoms.

angstrom (Å) A unit of length, equal to 10^{-10} meter, often used to measure the wavelength of visible light.

angular diameter The apparent diameter of an object in angular measure; the angular separation of two points on opposite sides of the object.

angular distance The apparent angular spacing between two objects in the sky.

angular momentum The tendency for bodies, because of their inertia, to keep spinning or orbiting.

angular separation The observed angular distance between two celestial objects, measured in degrees, minutes, and seconds of angular measure.

angular speed The rate of change of angular position of a celestial object viewed in the sky.

anorthosite A basaltic mineral composed of calcium and sodium with aluminum silicate; the predominant mineral of the lunar highlands.

antimatter Elementary particles with opposite electric charge compared to that of ordinary matter; atoms made of positive electrons and negative protons.

aphelion The point on the orbit of a body orbiting the sun that is farthest from the sun.

apogee The point in its orbit at which an earth satellite (artificial or the moon) is farthest from the earth.

apparent magnitude The brightness of a star (or any other celestial object) as seen from the earth; an astronomical measure of the object's flux.

arroyo A dry channel carved in the ground by sporadic water flows.

asteroid Minor planet; one of several thousand very small members of the solar system that revolve around the sun, generally between the orbits of Mars and Jupiter; most are made of rocks and metals and have irregular shapes.

asteroid belt The region lying between the orbits of Mars and Jupiter, containing the majority of asteroids.

astronomical unit (AU) The average distance between the earth and the sun: 149.6×10^6 km or 8.3 light minutes.

astrophysics Physical concepts applied to astronomical objects.

astrophysical jets Collimated beams of material (usually ions and electrons) expelled from astrophysical objects, such as the nuclei of active galaxies. See nuclear jets.

asymptotic giant branch (AGB) The path on the H–R diagram that a star traverses during the phase of evolution when it has both a hydrogen-burning and a helium-burning shell, after its main-sequence phase.

atmosphere A gaseous envelope surrounding a planet, or the visible layers of a star; also a unit of pressure (abbreviated *atm*) equal to the pressure of air at sea level on the earth's surface.

atmospheric composition The abundances of the constituents of an atmosphere, usually measured at the surface of a planet.

atmospheric escape The process by which particles with greater than escape speed and unhindered by collisions leave a planet for space.

atmospheric extinction The decrease in light caused by passage through the atmosphere.

atmospheric pressure The force per unit area from all layers of an atmosphere above a certain level; usually measured at a planet's surface.

atmospheric reddening Preferential scattering of blue light over red by air particles, which results in an object (such as the setting sun) appearing to be redder than it actually is.

atom The smallest particle of an element that exhibits the chemical properties of the element.

AU Abbreviation for astronomical unit.

aurora Light emission from atmospheric atoms and molecules excited by collisions with energetic, charged particles from the magnetosphere.

autumnal equinox The fall equinox; see equinox.

axis One of two or more reference lines in a coordinate system; also, the straight line through the poles about which a body rotates.

azimuth Angular position along the horizon, measured clockwise from north.

Balmer series The set of transitions of electrons in a hydrogen atom between the second energy level and higher levels; also the set of absorption or emission lines corresponding to these transitions that lies in the visible part of the spectrum, the first of which is the hydrogen-alpha (H_α) line.

barred spirals A subclass of spiral galaxies that have a bar across the nuclear region.

basalt An igneous rock, composed of olivine and feldspar, that makes up much of the earth's lower crust.

basins Large, shallow lowland areas in the crusts of terrestrial planets created by asteroidal impact or plate tectonics.

belts Regions of convective downflow and low pressure in the atmosphere of a Jovian planet.

beta decay A process of radioactive decay in which a neutron disintegrates into a proton, an electron, and a neutrino.

beta particle An electron or positron emitted by a nucleus in radioactive decay.

Big Bang model A picture of the evolution of the universe that postulates its origin, in an event called the Big Bang, from a hot, dense state that rapidly expanded to cooler, less dense states.

binary accretion model A model for the origin of the moon in which the moon and the earth form by accretion of material from the same cloud of gas and dust.

binary galaxies Two galaxies bound by gravity and orbiting a common center of mass.

binary stars Two stars bound together by gravity that revolve around a common center of mass.

binary x-ray source A binary system containing a star and an x-ray emitter, which is usually a neutron star or black hole surrounded by a hot accretion disk giving off x-rays.

biological evolution The natural development from simple to often complex organisms with the passage of time, generated by mutations that change the offspring's gene structure and directed by natural selection of individuals well-adapted to the environment.

bipolar outflows High-speed outflows of gas in opposite directions from a young stellar object (YSO); probably the result of a magnetized accretion disk around the YSO, collimating its stellar wind.

black dwarf The cold remains of a white dwarf after all its thermal energy has been exhausted.

black hole A mass that has collapsed to such a small radius that the escape speed from its surface is greater than the speed of light; thus light is trapped by the intense gravitational field.

blackbody or **blackbody radiator** A (hypothetical) perfect radiator of light that absorbs and reemits all radiation incident upon it; its light output depends only on its temperature.

blackbody spectrum The continuous spectrum emitted by a blackbody; the flux at each wavelength is given by the formula known as Planck's law.

BL Lacertae (BL Lac) objects A type of active galaxy (AGN) whose nonthermal emission from the nucleus varies rapidly (one day or so) and is highly polarized.

blueshift A decrease in the wavelength of the radiation emitted by an approaching celestial body as a consequence of the Doppler effect; a shift toward the short-wavelength (blue) end of the spectrum.

Bohr model of the atom A simple picture of atomic structure in which electrons have well-defined orbits (energy levels) about the nucleus of the atom.

bolometric magnitude The magnitude of an object measured over all wavelengths of the electromagnetic spectrum.

bolometric luminosity The total energy output per second at all wavelengths emitted by an astronomical object.

Boltzmann's constant The number that relates pressure and temperature, or kinetic energy and temperature in a gas; the gas constant per molecule.

bow shock wave The shock wave created by the interaction of the solar wind with a planetary magnetosphere.

breccias Rock and mineral fragments cemented together; a common part of the lunar surface and some meteorites.

bright nebula See **diffuse nebula.**

bright-line spectrum See **emission-line spectrum.**

brightness An ambiguous term, usually meaning the energy per unit area received from or emitted by an object—its flux—but sometimes used to refer to an object's luminosity.

brown dwarf A very-low-mass object (roughly 0.08 solar mass or less) of low temperature and luminosity that never becomes hot enough in its core to ignite thermonuclear reactions.

C-type asteroids Dark asteroids with low albedos (around 0.04) that probably contain a large percentage of carbon materials; probably undifferentiated in structure and akin to comets.

canali Italian term for "channels," used by Giovanni Schiaparelli to describe dark linear features seen on the surface of Mars.

capture model A model of the moon's origin proposed in the mid-1950s that pictures the moon as having been captured by the earth's gravity, after which it spiraled in toward the earth, reversed orbital direction, and spiraled outward.

carbon–nitrogen–oxygen (CNO) cycle A series of thermonuclear reactions taking place in a star's core, in which carbon, nitrogen, and oxygen aid the fusion of hydrogen into helium; it is a secondary energy production process in the sun, but the major process in high-mass, main-sequence stars.

carbonaceous chondrites A class of dark, crumbly meteorites that contain chondrules embedded in a matrix material with a large percentage of carbon (about 4 percent).

Cassegrain reflector A design for a reflecting telescope in which the secondary mirror directs the beam to a focus through a hole in the center of the primary mirror.

Cassini's division A gap about 2000 km wide in Saturn's rings, discovered in 1675 by Giovanni Cassini; now known to contain many small ringlets.

cD galaxies See **supergiant elliptical galaxies.**

celestial coordinate system A system for specifying positions on the celestial sphere, similar to the system of latitude and longitude used to specify positions on the earth.

celestial equator An imaginary projection of the earth's equator onto the celestial sphere; declination is zero along the celestial equator.

celestial meridian See **meridian.**

celestial pole An imaginary projection of the earth's pole (north or south) onto the celestial sphere; a point about which the apparent daily rotation of the stars takes place.

celestial sphere An imaginary sphere of very large radius, centered on the earth, on which the celestial bodies appear to be fastened and against which their motions are charted.

cell Most common form of living organism on earth; component of larger organisms.

center of mass The balance point of a set of interacting or connected bodies.

central force A force directed along a line connecting the centers of two objects.

centripetal acceleration The acceleration of a body toward the center of a circular path.

centripetal force A force required to divert a body from a straight path into a curved one directed toward the center of the curve.

cepheid variables (cepheids) Giant stars that vary in luminosity as a result of a regular variation in size and temperature; a class of variable stars for which the star Delta Cephei is the prototype.

Chandrasekhar limit The maximum mass for a white dwarf star, about 1.4 solar masses; this amount leads to the highest density and smallest radius for a star made of a degenerate electron gas; if more than this mass is present, the star cannot be supported by the electron gas pressure and so collapses gravitationally.

charge-coupled device (CCD) A small chip of semiconductor material that emits electrons when it absorbs light; the electrons are trapped in small regions called pixels; the pattern of charges is read out in a way that preserves the image striking the chip.

chemical condensation sequence The sequence of chemical reactions and condensation of solids that occurs in a low-density gas as it cools at specific densities and temperatures; an important process in the solar nebula.

chemical evolution The natural development from simple to complex molecules, such as proteins and nucleic acids, which are the building blocks of life.

chondrite An undifferentiated stony meteorite characterized by the presence of small, round silicate granules (chondrules).

chondrules Round silicate granules lacking volatile elements; found in chondritic meteorites, or chondrites, they are believed to be primitive solar system materials.

chromosphere The part of the sun's atmosphere just above the photosphere. Hotter and less dense than the photosphere, it creates the flash spectrum seen during eclipses.

circular velocity The speed at which an object must travel to maintain uniform circular motion around a gravitating body.

circumpolar stars For an observer north of the equator, the stars that are continually above the northern horizon and never set; for a southern observer, the stars that never set below the southern horizon.

closed geometry See **spherical geometry.**

cluster of galaxies A grouping of galaxies containing a few to thousands of galaxies.

CNO cycle See **carbon–nitrogen–oxygen cycle.**

collisional de-excitation Loss of energy by an electron of an atom in a collision so that the electron drops to a lower energy level.

collisional excitation Forcing an electron of an atom to a higher energy level by a collision.

coma The bright, visible head of a comet, consisting of gas and dust expelled from the nucleus.

comets Bodies of small mass that revolve around the sun, usually in highly elliptical orbits. In the dirty snowball model, comets consist of small, solid particles (probably of rocky material) embedded in frozen gases.

compact (radio) galaxies Active galaxies that have a small, strong radio source in their nuclei.

comparative planetology The study of the planets that compares their physical properties to uncover common processes in their formation and evolution; earth serves as the comparative model for the terrestrial planets.

composition The chemical elements and/or compounds that make up a body, usually expressed in percentages.

compound A substance composed of the atoms of two or more elements bound together by chemical forces.

condensation The growth of small particles by the sticking together of atoms and molecules.

condensation sequence See **chemical condensation sequence.**

conduction Transfer of thermal energy by particles colliding into one another.

conjunction The time at which two celestial objects appear closest together in the sky.

conservation of angular momentum The principle stating that if no torques are applied, the total angular momentum of an isolated system is constant.

conservation of energy A fundamental principle in physics that states that the total energy of an isolated system remains constant regardless of internal changes that may occur.

conservation of momentum The physical principle stating that in the absence of outside net forces, the total momentum of a system is constant.

constellation An apparent arrangement of stars on the celestial sphere, usually named after an ancient god, hero, animal, or mythological being; now an agreed-upon region of the sky containing a group of stars.

continental drift The theory that the present continents were at one time a unified landmass that broke up, such that the fragments drifted apart, driven by plate tectonics.

continuous spectrum A spectrum showing emission at all wavelengths, unbroken by either absorption lines or emission lines.

contour map A diagram showing how the intensity of some kind of radiation varies over a region of the sky; lines on such a map connect points of equal flux. Closely spaced lines mean that the flux changes rapidly over a small distance, widely spaced lines mean it changes more slowly.

convection The transfer of energy by the moving currents of a fluid.

coplanar Lying in the same plane.

core (of the earth) The central region of the earth; it has a high density, is probably liquid, and is thought to be composed of iron and iron alloys.

core (of the Galaxy) The inner few light years of the nucleus; contains a small, nonthermal radio source and fast-moving clouds of ionized gas.

core (of the sun) The inner 25 percent of the sun's radius, where the temperature is great enough for thermonuclear reactions to take place.

core–mantle grains Interstellar dust particles with cores of dense materials (such as silicates) surrounded by a mantle of icy materials (such as water).

corona The outermost region of the sun's atmosphere, consisting of thin, ionized gases at a temperature of about 10^6 K; large magnetic loops give the corona its structure.

coronal holes Regions in the sun's corona that lack a concentration of high-temperature plasma; here magnetic field lines extend out into interplanetary space and mark the source of the solar wind.

coronal interstellar gas High-temperature interstellar plasma made visible by its x-ray emission.

cosmic blackbody microwave radiation Radiation with a blackbody spectrum at a temperature of about 2.7 K permeating the universe. These remains of the primeval fireball in which the universe was created constitute observational confirmation of the Big Bang model.

cosmic rays Charged atomic particles moving in space with very high energies (the particles travel close to the speed of light); most originate beyond the solar system, but some low-energy cosmic rays are produced in solar flares.

cosmology The study of the nature and evolution of the physical universe.

cosmos The universe considered as an orderly and harmonious system.

crater A circular depression of any size, usually caused by the impact of a solid body or by a surface eruption.

cratered terrain Landscape with an abundance of craters, which implies that it is old and unevolved.

crescent The phase during which a moon or planet has less than half its visible surface illuminated.

critical density In cosmology, the density that marks the transition from an open to a closed universe; the density that provides enough gravity to just bring the expansion to a stop after infinite time.

crust The thin, outermost surface layer of a planet; on the earth, it is composed of basaltic and granitic rocks.

curvature of spacetime See **spacetime curvature.**

cyclone Spiral flows in a planet's atmosphere produced by convection and the planet's rotation.

D lines (of sodium) A pair of dark lines in the yellow region of the spectrum, produced by sodium.

dark cloud An interstellar cloud of gas and dust that contains enough dust to blot out the light of stars behind it (as seen from the earth).

dark matter Matter that cannot be detected by current telescopes; found in halos of galaxies and in clusters of galaxies. The form of the dark matter is unknown, but it may amount to most of the mass in the universe.

dark-line spectrum See **absorption-line spectrum.**

declination In the equatorial coordinate system, the angle measured north or south along an hour circle from the celestial equator to a point on the celestial sphere.

decoupling The time in the universe's history when the density and opacity became low enough that matter and light stopped interacting.

deferent An ancient geometric device used to account for the apparent eastward motion of the planets; a large circle, usually centered on the earth, that carries around a planet's epicycle.

degenerate electron gas An ionized gas in which nuclei and electrons are packed together as much as possible, filling all possible low-energy states, so that the law relating pressure, temperature, and density in ordinary gases no longer applies.

degenerate gas pressure A force exerted by degenerate matter that depends mostly on how dense the matter is and very little on its temperature.

degenerate neutron gas Matter made up of neutrons packed together as tightly as possible.

density The amount of mass per volume in an object or region of space.

density-wave model A model for the generation of spiral structure in galaxies; spiral density waves (similar to sound waves) are pictured as plowing through the interstellar matter and sparking star formation.

deoxyribonucleic acid (DNA) The basic genetic material of life as we know it; the DNA molecule is very large, consisting of subunits called nucleotides.

detector Any device sensitive to electromagnetic radiation placed at the focus of a telescope.

differential rotation The tendency of a fluid, spherical body to rotate faster at the equator than at the poles.

differentiated Describing the layered interior of a planet or a moon, generally with the less dense materials atop more dense ones.

diffraction The spreading out of light waves as they pass the edge of an opaque body.

diffuse (bright) nebula A cloud of hot ionized gas, mostly hydrogen, with an emission-line spectrum.

dipole field A magnetic field configuration like that of a bar magnet with opposed north and south poles.

dirty snowball comet model A model for comets that pictures the nucleus as a compact solid body of ices (mostly water) mixed with sooty rocky matter; the ices turn into gases as a comet nears the sun, creating the head and tail.

disk (of a galaxy) The flattened wheel of stars, gas, and dust outside the nucleus of a spiral galaxy.

DNA See **deoxyribonucleic acid.**

Doppler shift A change in the wavelength of waves from a source reaching an observer when the source and the observer are moving with respect to each other along the line of sight; the wavelength increases (redshift) or decreases (blueshift) according to whether the motion is away from or toward the observer; the frequency also changes.

double quasar An optical illusion, produced by a gravitational lens effect, of two adjacent images on the sky of a single quasar.

dust tail The part of a comet's tail containing dust particles, pushed out by radiation pressure from sunlight.

dwarf A star of relatively low light output and relatively small size; a main-sequence star of luminosity class V.

dynamo model A model for the generation of a planet's (or star's) magnetic field by the circulation of conducting fluids in its core (or convective zone).

eccentric An ancient geometric device used to account for nonuniform planetary motion; a point offset from the center of circular motion.

eccentricity The ratio of the distance of a focus from the center of an ellipse to its semimajor axis.

eclipse The phenomenon of one body passing in front of another, cutting off its light.

eclipsing binary system Two stars that revolve around a common center of mass, the orbits lying edge-on to the line of sight, so that each star periodically passes in front of the other to cause a decrease in the amount of light.

ecliptic From the earth, the apparent yearly path on the celestial sphere of the sun with respect to the stars; also, the plane of the earth's orbit.

effective temperature The temperature a body would have if it were a blackbody of the same size radiating the same luminosity.

Einstein ring An optical illusion created by highly symmetrical gravitational imaging of a radio source.

electromagnetic radiation A self-propagating electric and magnetic wave, such as light, radio, ultraviolet, or infrared radiation; all types travel at the same speed and differ in wavelength or frequency, which relates to the energy.

electromagnetic spectrum The range of all wavelengths of electromagnetic radiation.

electron A lightweight, negatively charged subatomic particle.

electron volt (eV) A convenient unit of energy for atomic physics; the energy gain of an electron when it is accelerated by one volt: $1 \text{ eV} = 1.60 \times 10^{-19}$ J.

element A substance that is made of atoms with the same chemical properties and cannot be decomposed chemically into simpler substances.

ellipse A plane curve drawn so that the sum of the distances from a point on the curve to two fixed points is constant.

elliptical galaxy A gravitationally bound system of stars that has rotational symmetry but no spiral structure; contains mainly old stars, with little gas or dust.

elongation The angular separation of an object from the sun as seen in the sky.

emission (bright) lines Light of specific wavelengths or colors emitted by atoms; sharp energy peaks in a spectrum caused by downward electron transitions from a discrete quantum state to another discrete state.

emission nebula See **diffuse nebula;** a hot cloud of mostly hydrogen gas whose visible spectrum is dominated by emission lines.

emission-line spectrum A spectrum containing only emission lines.

energy The ability to do work.

energy level One of the possible quantum states of an atom, with a specific value of energy.

epicycle A small circle whose center lies on a larger one (the deferent) used by ancient astronomers, such as Ptolemy, to account for the westward retrograde motion and other irregular motions of the planets.

equant An ancient geometrical device invented by Ptolemy to account for variations in planetary motion; essentially an eccentric in which the center of the circle is not the center of uniform motion.

equatorial bulge The excess diameter, about 43 km, of the earth through its equator compared with the diameter through its poles; any planet will have a large equatorial bulge if it has a fluid interior and rotates rapidly.

equatorial system A celestial coordinate system based on the celestial equator and the celestial poles, in which two angles, right ascension and declination, are used to specify the position of a point on the celestial sphere.

equilibrium A state of a physical system in which there is no overall change.

equilibrium temperature The temperature that has been achieved when a body emits the same energy per second that it absorbs.

equinox Time of year of equal length of day and night; the two times of the year when the sun crosses the celestial equator. Spring (vernal) equinox occurs about March 21, and fall (autumnal) equinox about September 22.

escape speed The minimum speed a body must achieve to break away from the gravity of another body and never return to it.

escape temperature The temperature that particles in an atmosphere require to enter space with the escape speed.

Euclidean (flat) geometry The geometry in which only one parallel line can be drawn through a point near another line; the sum of the angles in a triangle drawn on a flat surface is always 180°.

event A point in four-dimensional spacetime.

evolution Changes in the physical properties of a body with time.

evolutionary track On a temperature–luminosity (H–R) diagram, the path made by the points that describe how the temperature and luminosity of a star change with time.

excitation The process of raising an atom to a higher energy level.

exosphere The topmost region of a planet's atmosphere, from which particles in the atmosphere can escape into space.

expansion of the universe The expansion of spacetime, as indicated by markers such as the distances between clusters of galaxies; thought to be a result of the Big Bang.

exponential decay A process, such as radioactive decay, for which the rate of change is directly proportional to the quantity present, so that the quantity remaining is given by an exponential function of time.

extended radio galaxies Active galaxies that show extended radio emission, usually in the form of two lobes on either side of the nucleus.

extinction The dimming of light when it passes through some medium, such as the earth's atmosphere or interstellar material.

extragalactic Outside the Milky Way Galaxy.

eyepiece A magnifying lens used to view the image produced by the main light-gathering lens or mirror of a telescope.

f-ratio The ratio of the focal length of a lens or mirror to its diameter.

false-color map A computer-processed image in which a color code represents various levels of a physical property, such as flux or temperature.

first dredge-up Convection acting to bring to a star's surface material processed by hydrogen burning during the star's first episode as a red giant.

first quarter The phase of the moon a quarter of the way around its orbit (eastward) from new moon; it looks half-illuminated when viewed from the earth.

fission See **nuclear fission.**

fission model The earliest of the major models for the origin of the moon, suggesting that a young, rapidly spinning, molten earth lost a piece that spiraled out into orbit and cooled down to form the moon.

flat geometry See **Euclidean geometry.**

flatness problem The failure of the unmodified Big Bang model to account for the facts that the geometry of the universe is flat (or very close to it) and remains flat as the universe evolves.

fluorescence The process by which a high-energy photon is absorbed by an atom and reemitted as two or more photons of lower energy.

flux The amount of energy flowing through a given area in a given time.

flux map A two-dimensional image of the flux emitted over the angular extent of an astronomical object.

focal length The distance from a lens (or mirror) to the point at which it brings light to a focus for a distant object.

focus (pl., foci) The point to which light is gathered by a telescope; one of two points that define an ellipse.

forbidden line An emission line from an atom produced by a transition with a low probability of occurrence in laboratory conditions, more common in interstellar space.

force A push or pull; a measure of the strength of a physical interaction.

forced motion Any motion under the action of a net force.

Fraunhofer lines The name given to strongest absorption lines in the spectrum of a star, especially the sun.

free-fall Gravitational collapse under the condition of no resisting internal pressure caused by collisions among the particles.

free-fall collapse time The time for a cloud to collapse gravitationally under free-fall conditions; uniquely a function of the initial density of the cloud.

frequency The number of waves that pass a particular point in some time interval (usually a second), usually given in unit of hertz; one hertz (Hz) is one cycle per second.

full moon The phase during which the moon is opposite the sun in the sky and looks fully illuminated as seen from the earth.

fusion See **nuclear fusion.**

G-band Absorption band in stellar spectra caused by metals.

galactic cannibalism A model for galactic interaction in which more massive galaxies strip material, by tidal forces, from less massive galaxies.

galactic center (core) The innermost part of the Galaxy's nuclear bulge; contains a cluster of stars, ionized gas clouds, and a rotating ring of material; may include a supermassive black hole.

galactic (open) cluster A small group (about ten to a few thousand) of gravitationally bound stars of Population I, found in or near the plane of the Galaxy.

galactic equator The great circle along the line of the Milky Way, marking the central plane of the Galaxy.

galactic latitude The angular distance north or south of the galactic equator.

galactic longitude The angular distance along the galactic equator from a zero point in the direction of the galactic center.

galactic rotation curve A description of how fast an object some distance from the center of a galaxy revolves around it; the orbital speed as a function of distance.

galaxy (normal) A huge assembly of stars (between a million and hundreds of billions), plus gas and dust, that is held together by gravity; the Milky Way Galaxy, our own galaxy, containing the sun.

Galilean moons The four largest satellites of Jupiter (Io, Europa, Ganymede, Callisto), discovered by Galileo with his telescope.

gamma ray A very-high-energy photon with a wavelength (10^{-12} m) shorter than that of x-rays.

gamma-ray bursters Astronomical sources of short bursts of gamma rays, thought to originate from highly magnetic neutron stars.

gas (plasma) tail The part of a comet's tail that consists of ions and molecules; it is shaped by its interaction with the solar wind.

gauss (G) A physical unit measuring magnetic field strength.

general theory of relativity The idea developed by Albert Einstein that mass and energy determine the geometry of spacetime and that any curvature of spacetime shows itself by what we commonly call gravitational forces; Einstein's theory of gravity.

geocentric Centered on the earth.

geomagnetic axis The axis that connects the earth's magnetic poles; it is inclined about 12° from the geographic spin axis and does not pass through the earth's center.

giant impact model A scenario for the moon's origin in which a Mars-sized object strikes the young earth a glancing blow; materials from the colliding object and the earth form a disk around the earth, out of which the moon accretes.

giant molecular clouds Large interstellar clouds, with sizes up to tens of light years and containing 100,000 solar masses of material. Found in the spiral arms of the Galaxy, giant molecular clouds are the sites of massive star formation.

giant star A star whose luminosity places it above the main sequence on the Hertzsprung–Russell diagram; such stars have luminosities between 10 and 1000 times that of the sun and radii typically 10 to 100 times the sun's radius.

gibbous The phase of the moon occurring between first quarter and and last quarter, when the moon is more than half illuminated as seen from the earth.

globular cluster A gravitationally bound group of about 10^5 to 10^6 Population II stars (of roughly solar mass), symmetrically shaped, found in the halo of the Galaxy and orbiting the galactic center.

gnomon A sun dial device for measuring solar time; most simply, a stick stuck vertically into the ground whose shadow is used to indicate the sun's position with respect to the horizon.

grand unified theories (GUTs) Physical theories that attempt to unite the elementary particles and the four forces in nature as the actions of one particle and one force.

granite The type of rock making up the continental regions of the earth's crust and the highlands of other terrestrial planets.

gravitation In Newtonian terms, a force between masses that is characterized by their acceleration toward each other; the size of the force depends directly on the product of the masses and inversely on the square of the distance between them; in Einstein's terms, the curvature of spacetime.

gravitational bending of light The effect of gravity on the usually straight path of a photon.

gravitational contraction The unhindered contraction of any mass from its own gravity.

gravitational field The property of space having the potential for producing gravitational force on objects within it; characterized by the acceleration of free masses.

gravitational instability The tendency for a disturbed region in a gas to undergo gravitational collapse.

gravitational lens effect The bending effect of a large mass on light rays so that they form an imperfect image of the source of light.

gravitational mass The mass of an object as determined by the gravitational force it exerts on another object.

gravitational potential energy Potential energy related to a body's position in a gravitational field.

gravitational redshift The change to longer wavelengths that marks the loss of energy by a photon moving from a stronger to a weaker gravitational field.

great circle Shortest distance between two points on a sphere.

greenhouse effect The production of an increased equilibrium temperature at the surface of a planet that occurs when the opacity of its atmosphere in the infrared results in the trapping of outgoing heat radiation; increasing the opacity will likely increase the equilibrium temperature.

greenhouse gases Any gas in an atmosphere that promotes a greenhouse effect by virtue of being transparent to visible light but opaque to infrared.

ground state The lowest energy level of an atom.

GUTs See **grand unified theories.**

H I region A region of neutral hydrogen in interstellar space; emits photons at a wavelength of 21 cm.

H II region A zone of hot, ionized hydrogen in interstellar space; it usually forms a bright nebula around a hot, young star or cluster of hot stars.

H-alpha line The first line (at 6563 Å) of the Balmer series, the set of transitions in a hydrogen atom between the second energy level and levels with higher energy; it lies in the red part of the visible spectrum.

half-life The time required for half the radioactive atoms in a sample to disintegrate by fission.

halo (of a galaxy) The spherical region around a galaxy, not including the disk or the nucleus, containing globular clusters, some gas, a few stray stars, and an invisible form of matter.

head–tail (radio) galaxy An active galaxy whose radio lobes have been swept back to form a tail because of interaction with the surrounding medium.

heat The flow of thermal energy from a hotter body to a cooler one.

heavy-particle era In the hot Big Bang model, the time up to 0.001 second when gamma rays collided to make high-mass particles, such as protons.

heliocentric Centered on the sun.

heliocentric parallax An apparent shift in the positions of nearby stars (relative to more distant ones) from the changing position of the earth in its orbit around the sun; the size of the shift can be used to measure the distances to close stars. See **trigonometric parallax.**

helium burning Fusion of helium into carbon by the triple-alpha process; requires a temperature of at least 10^8 K.

helium flash The rapid burst of energy generation with which a star initiates helium burning by the triple-alpha process in the degenerate core of a low-mass red giant star.

hertz (Hz) A physical unit of frequency equal to 1 cycle per second.

Hertzsprung–Russell (H–R) diagram A graphic representation of the classification of stars according to their spectral class (or color or surface temperature) and luminosity (or absolute magnitude). The physical properties of a star are correlated with its position on the diagram, so a star's evolution can be described by its change of position on the diagram with time (see **evolutionary track**).

highlands Regions of higher than average elevation on a planet or satellite; usually refers to the older, cratered region of the moon's surface.

homogeneous Having a consistent and even distribution of matter, the same in all parts.

horizon The intersection with the sky of a plane tangent to the earth at the location of the observer.

horizon distance In cosmology, the maximum distance that light can travel in some epoch of the universe.

horizon problem In the Big Bang model, the problem that arises because the early universe expanded too rapidly to permit distant regions to communicate with each other.

horizon system The celestial coordinate system based on the horizon plane and the zenith, in which altitude and azimuth are the two angles specifying a point on the celestial sphere.

horizontal branch A portion of the Hertzsprung–Russell diagram reached by Population II stars of low mass after the red giant stage and typically found in a globular cluster; it ranges from yellowish to reddish stars all having about the same luminosity (about 100 times that of the sun).

hour circle A great circle on the celestial sphere passing through the celestial poles.

H–R diagram A Hertzsprung–Russell diagram.

Hubble's constant The proportionality constant relating velocity and distance in the Hubble law; the value—now around 75 km/s/Mpc (20 km/s/Mly)—changes with time as the universe expands.

Hubble's law The description of the expansion of the universe according to which the more distant a galaxy lies from us, the faster it is moving away; the relation $v = Hd$, between the expansion velocity (v) and distance (d) of a galaxy, where H is the Hubble constant.

hydrogen burning Any fusion reaction that converts hydrogen (protons) to heavier elements, such as helium.

hydrostatic equilibrium An equilibrium characterized by the absence of mass motions, when pressure balances gravity.

hyperbola A curve produced by the intersection of a plane with a cone; the shape of the orbit of a body with more than escape speed.

hyperbolic geometry An alternative to Euclidean geometry, constructed by N. I. Lobachevski on the premise that more than one parallel line can be drawn through a point near a straight line; the sum of the angles of a triangle drawn on a hyperbolic surface is always less than 180 degrees.

ideal gas law The pressure of a gas is directly proportional to its temperature and density, the property of an ordinary gas.

igneous rock A rock formed by the cooling of molten lava.

image Light rays gathered at the focus of a lens or mirror in the same relative alignment as the real object.

image processing The computer manipulation of digitized images to enhance aspects of interest.

impact cratering The formation of craters by the impact of solid objects onto a surface.

inertia The resistance of an object to a force acting on it because of its mass.

inertial mass Mass determined by subjecting an object to a known force (not gravity) and measuring the acceleration that results.

inferior conjunction For a planet orbiting interior to another, the alignment with the sun when the interior

planet lies on the same side of the sun as the outer planet.

inflationary universe model A modification of the Big Bang model in which the universe undergoes an early, brief interval of rapid expansion.

instability strip The region on the H–R diagram of cepheid and certain other variable stars; these stars are burning helium in their cores and are expanding and contracting.

intense bombardment An early time in the solar systems history when many solid masses orbited between the planets and collided with them, making abundant impact craters.

intergalactic medium The gas and dust found between the galaxies; appears to be mostly in the form of a high-temperature plasma.

interferometer See **radio interferometer.**

interstellar dust Small, solid particles in the interstellar medium.

interstellar gas Atoms, molecules, and ions in the interstellar medium.

interstellar medium All the gas and dust found between stars.

inverse beta decay The process in which electrons and protons are forced together to form neutrons and neutrinos; the reverse process of neutron decay.

inverse-square law The relationship that describes how a physical quantity (such gravitational force) decreases with the inverse square of the distance.

inverse-square law for light The decrease of the flux of light with the inverse square of the distance from the source.

ion An atom that has become electrically charged by the gain or loss of one or more electrons.

ionization The process by which an atom loses or gains electrons; collisions and absorption of energetic photons are the two most common processes in astrophysics.

ionization energy The minimum energy required to ionize an atom.

ionized gas A gas that has been ionized so that it contains equal numbers of free electrons and charged ions; a plasma.

ionosphere A layer of the earth's atmosphere ranging from about 100 to 700 km above the surface in which oxygen and nitrogen are ionized by sunlight, producing free electrons.

iron meteorites One of the three main types of meteorite, typically made of about 90 percent iron and 9 percent nickel, with a trace of other elements.

irregular galaxy A galaxy without spiral structure or rotational symmetry, containing mostly Population I stars and abundant gas and dust.

isotopes Atoms with the same number of protons but different numbers of neutrons.

isotropic Having no preferred direction in space.

jet streams Latitudinal, coherent wind flows at high speed in the upper atmosphere of the earth or another planet.

joule (J) A physical unit of work and energy.

Jovian planets Planets with physical characteristics similar to Jupiter: large mass and radius, low density, mostly liquid interior.

kelvin The unit of temperature in the SI system.

Kepler's laws of planetary motion The three laws of planetary motion, propounded by Johannes Kepler, which describe the properties of elliptical orbits with an inverse-square force law.

Keplerian motion Orbital motion that follows Kepler's laws.

kiloparsec (kpc) One thousand parsecs.

kinetic energy The ability to do work because of motion.

Kirchhoff's rules Empirical descriptions of the physical conditions under which the main types of spectrum originate.

KREEP A lunar material composed of potassium (K), rare-earth elements (REE), and phosphorus (P).

last quarter The phase during which the moon, three-quarters of the way around its orbit from new moon, looks half-illuminated when viewed from the earth.

latitude Angular distance north or south of the equator.

law of gravitation See **gravitation.**

laws of motion Newton's three laws that describe the nature of forced and natural motion.

lens A curved piece of glass designed to bring light rays to a focus.

lenticular galaxies Galaxies of Hubble type S0, having a disk like a spiral galaxy but no spiral arms, and no gas or dust.

lepton An elementary particle that has low mass, such as an electron.

light curve A graph of a star's changing brightness with time.

light rays Imaginary lines in the direction of propagation of a light wave.

light year (ly) The distance light travels in a year, about 3.1×10^{13} km.

lighthouse model For a pulsar, a rapidly rotating neutron star with a strong magnetic field; the rotation provides the pulse period, and the magnetic field generates the electromagnetic radiation.

light-gathering power The ability of a telescope to collect light as measured by the area of its objective.

light-particle era In the hot Big Bang model, the interval from 0.0001 to 4 seconds when gamma rays could collide to make low-mass particles, such as electrons.

line profile The variation of a spectral line's flux as a function of wavelength; greatest at the center of the line.

local celestial meridian An imaginary line through the north and south points on the horizon and the zenith overhead.

Local Group (of galaxies) The gravitationally bound group of about 30 galaxies to which our Milky Way Galaxy belongs.

local standard of rest A frame of reference that participates in the average motion of the sun and nearby stars around the center of the Milky Way Galaxy.

Local Supercluster The supercluster of galaxies in which the Local Group is located; spread over 100 million light years, it contains the Virgo cluster.

long-period comets Comets with orbital periods greater than 200 years.

longitude Angular distance, east or west, along the equator; on the earth, the reference longitude is that of Greenwich, England; a similar reference is used on other planets.

lowlands Regions of below-average elevation on a planet or satellite; often refers to the younger, impact basins of the moon; usually basaltic.

low-velocity stars Stars with close to circular orbits in the plane of the Galaxy; they travel at less than 60 km/s with respect to the sun.

luminosity The total rate at which radiative energy is given off by a celestial body, over all wavelengths; the sun's luminosity is about 4×10^{26} W.

luminosity class The categorization of stars that have the same surface temperatures but different sizes, resulting in different luminosities; based on the widths of dark lines in a star's spectrum, giant stars having narrower lines than dwarf stars of the same surface temperature.

lunar eclipse The cutoff of sunlight from the moon, when the moon lies on the line between the earth and sun so that it passes through the earth's shadow; a lunar eclipse can occur only at full moon, and when the full moon lies very close to the ecliptic.

lunar occultation The passage of the moon in front of a star or planet.

lunar soil The fine particles created by the bombardment of the lunar surface by meteorites that, with larger rock fragments, compose the lunar soil.

Lyman series Transitions in a hydrogen atom to and from the lowest energy level; they involve large energy changes, corresponding to wavelengths in the ultraviolet part of the spectrum. Also, the set of absorption or emission lines corresponding to these transitions.

M-type asteroids Differentiated asteroids with albedos of about 10 percent and resembling metals in their reflective properties.

magnetic field The property of space having the potential to exert magnetic forces on bodies within it.

magnetic field lines A graphic representation of a magnetic field showing its direction and, by the degree of packing of the lines, its intensity.

magnetic flux The number of magnetic field lines passing through an area.

magnetic reconnection The sudden connection of magnetic field lines of opposite polarity; a process that releases energy stored in the magnetic field.

magnetometer A device to measure the strength of a magnetic field.

magnetosphere The region around a planet in which particles from the solar wind are trapped by the planet's magnetic field.

magnifying power The ability of a telescope to increase the apparent angular size of a celestial object.

magnitude An astronomical measurement of an object's brightness; larger magnitudes represent fainter objects.

main sequence The principal series of stars in the Hertzsprung–Russell diagram. Such stars are converting hydrogen to helium in their cores by the proton–proton process or by the carbon–nitrogen–oxygen cycle; this is the longest stage of a star's active life.

main-sequence lifetime The duration of time that a star spends fusing hydrogen to helium in its core; the longest phase of its evolution.

major axis The larger of the two axes of an ellipse.

mantle The major portion of the earth's interior below the crust, made of a plastic rock probably composed of olivine.

mare (pl., maria; Latin for "sea") A lowland area on the moon that appears darker and smoother than the highland regions, probably formed by lava that

solidified into basaltic rock about 3 to 3.5 billion years ago.

mare basalts Basaltic rocks found on the surface of the lunar maria; tend to be the youngest (most recently formed) rocks on the lunar surface, with ages around 3.2 billion years.

mascons High-density concentrations of mass beneath the lunar maria; they have been detected by their effect on the orbits of moon-orbiting satellites.

mass A measure of an object's resistance to change in its motion (inertial mass); a measure of the strength of gravitational force an object can produce (gravitational mass).

mass extinction The sudden disappearance from the fossil record of a large number of species; thought to have resulted from catastrophic changes in the earth's environment.

mass loss The rate per year at which a star loses mass, usually by a stellar wind.

mass–luminosity ratio For galaxies, the ratio of the total mass to the luminosity, usually given in solar units; a rough measure of the kind of stars in the galaxy. A large ratio implies a large fraction of dark matter.

mass–luminosity relation An empirical relation, for main-sequence stars, between a star's mass and its luminosity, roughly proportional to the third power of the mass; stars of other types have a different numerical value of the power.

matter era In the Big Bang model, the time interval from about one million years after the Big Bang to now, when matter dominates the universe.

maximum elongation The greatest angular distance (east or west) of an object from the sun.

mechanics A branch of physics that deals with forces and their effects on bodies.

megaparsec (Mpc) 1 million parsecs, or about 3.3 million light years.

megaton An explosive force equal to that of 1 million tons of TNT (about 4×10^{15} J).

meridian (celestial) An imaginary line drawn through the north and south points on the horizon and through the zenith.

metal-poor stars Stars with metal abundances much less than the sun's.

metal-rich stars Stars with metal abundances like that of the sun (about 1 to 2 percent of the total mass).

metallic hydrogen A state of hydrogen, reached at high pressures, in which it is able to conduct electricity.

meteor The bright streak of light that occurs when a solid particle (a meteoroid) from space enters the earth's atmosphere and is heated by friction with atmospheric particles; sometimes called a falling star.

meteor shower A rapid influx of meteors that appear to come out of a small region of the sky, called the radiant.

meteor stream A uniform distribution of meteoroids along an orbit around the sun.

meteor swarm Meteoroids grouped in a localized region of an orbit around the sun; the source of meteor showers.

meteor trail The visible path of a meteor through the atmosphere created by ionization of the air and vaporization of the meteoroid.

meteorite A solid body from space that survives a passage through the earth's atmosphere and falls to the ground.

meteorite fall A meteorite seen in the sky and recovered on the ground.

meteorite find A recovered meteorite that was not seen to fall.

meteoroid A very small solid body moving through space in orbit around the sun.

micrometeorites Very small meteorites (up to 1 μm in diameter) that cool off and solidify before they hit the ground.

midoceanic ridge The almost continuous submarine mountain chain that extends some 64,000 km through the earth's ocean basins.

Milky Way The band of light that encircles the sky, caused by the blending of light from the many stars lying near the plane of the Galaxy; also sometimes used to refer to the Galaxy to which the sun belongs.

millisecond pulsar Any pulsar with a pulse period of a few milliseconds.

minute of arc One-sixtieth of a degree $\left(\dfrac{1°}{60}\right)$.

molecular clouds Large, dense, massive clouds in the disk of a spiral galaxy; they contain dust and a large fraction of gas in molecular form.

molecular maser Microwave amplification by stimulated emission of radiation from a molecule.

molecule A combination of two or more atoms bound together electrically; the smallest part of a compound that has the chemical properties of that substance.

momentum The product of an object's mass and velocity.

month See **sidereal month; synodic month.**

mutation A basic change in gene structure.

nadir The point on the celestial sphere directly below the observer, opposite the zenith.

nanometer (nm) A billionth of a meter (10^{-9} m); common unit of wavelength measurement for light.

narrow-tailed radio galaxies Radio galaxies that show U-shaped tails behind them; they are fast-moving galaxies in clusters of galaxies.

natural motion Motion without forces.

natural selection The process by which individuals with genes that produce characteristics best adapted to their environment have greater genetic representation in future generations.

nebula (English pl., nebulas; Latin for "cloud") A cloud of interstellar gas and dust.

nebular model A model for the origin of the solar system in which an interstellar cloud of gas and dust collapsed gravitationally to form a flattened disk out of which the planets formed by accretion.

neutrino An elementary particle with no (or very little) mass and no electric charge that travels at the speed of light and carries energy away during certain types of nuclear reaction.

neutron A subatomic particle with about the mass of a proton and no electric charge; one of the main constituents of an atomic nucleus; the union of a proton and an electron.

neutron star A star of extremely high density and small size (10 km) that is composed mainly of very tightly packed neutrons; cannot have a mass greater than about 3 solar masses.

new moon The phase during which the moon is in about the same direction as the sun in the sky, appearing almost completely unilluminated as seen from the earth.

newton (N) The SI unit of force.

Newtonian reflector A reflecting telescope designed so that a small mirror set at a 45° angle in the top center of the tube brings the focus out the side of the tube.

Newton's laws of motion The three laws describing motion from Newton's viewpoint; they are the inertial, force, and reaction laws.

nonthermal radiation Emitted energy that is not characterized by a blackbody spectrum; in astronomy, commonly refers to synchrotron radiation.

noon Midday; the time halfway between sunrise and sunset when the sun reaches its highest point in the sky with respect to the horizon.

north magnetic pole One of the two points on a star or planet from which magnetic lines of force emanate and to which the north pole of a compass points.

north pole star Any fairly bright star that lies close to the north celestial pole in the sky; the current star is Polaris.

nova (English pl., novas; Latin, "new") A star that has a sudden outburst of energy, temporarily increasing its luminosity by hundreds to thousands of times; now regarded as the thermonuclear outburst on a degenerate star in a binary system. Used in the past to refer to some stellar outbursts that modern astronomers call supernovas.

nuclear bulge The central region of a spiral galaxy, containing mostly old Population I stars.

nuclear fission A process that releases energy from matter: a heavy nucleus hit by a high-energy particle splits into two or more lighter nuclei whose combined mass is less than the original, the missing mass being converted into energy.

nuclear fusion A process that releases energy from matter by the joining of nuclei of lighter elements to make heavier ones; the combined mass is less than that of the constituents, the difference appearing as energy.

nuclear jet Directed streams of plasma extended from the nuclei of active galaxies; the material in the jet moves at speeds close to that of light.

nucleic acid A huge helix-shaped molecule, commonly found in the nucleus of cells, that is the chemical foundation of genetic material.

nucleosynthesis The chain of thermonuclear fusion processes by which hydrogen is converted to helium, helium to carbon, and so on, through all the elements of the periodic table.

nucleus (of an atom) The massive central part of an atom, containing neutrons and protons, about which the electrons orbit.

nucleus (of a comet) Small starlike point visible in the head of a comet; a solid, compact (diameter a few tens of kilometers) mass of frozen gases embedded in black rocky material.

nucleus (of a galaxy) The central portion of a galaxy, composed of old Population I stars, some gas and dust, and, for many galaxies, a concentrated source of nonthermal radiation.

number density The number of particles per unit volume.

OB association Loose groupings of O and B stars in small subgroups; they are not bound by gravity and so dissipate in a few tens of millions of years.

OB subgroup A small collection of about 10 stars of types O and B, a few tens of light years across, within an OB association.

objective The main light-gathering lens or mirror of a telescope.

observable universe The parts of the universe that can be detected by the light they emit.

occultation The eclipse of a star or planet by the moon or another planet.

Oort cloud A cloud of comet nuclei in orbit around the solar system, formed at the time the solar system formed; the reservoir for new comets.

opacity The property of a substance that hinders (by absorption or scattering) light passing through it; opposite of transparency.

open cluster Same as **galactic cluster.**

open geometry See **hyperbolic geometry.**

opposition The time at which a celestial body lies almost opposite the sun in the sky as seen from the earth; the time at which it has an elongation of 180.

optics The manipulation of light by reflection or refraction.

orbital angular momentum The angular momentum of a revolving body; the product of a body's mass, orbital speed, and the distance from the system's center of mass.

orbital inclination The angle between the orbital plane of a body and some reference plane; in the case of a planet in the solar system, the reference plane is that of the earth's orbit; in the case of a satellite, the reference is usually the equatorial plane of the planet; for a double star, it is the plane perpendicular to the line of sight.

orbital period The amount of time an astronomical body takes to complete one revolution around another body (or the center of mass).

organic Relating to the branch of chemistry concerned with the carbon compounds of living creatures.

organic molecules Molecules containing carbon found in living creatures on earth.

organism An ordered living creature.

outgassing The release of gases from nongaseous materials; extrusion of gases from the body of a planet after its formation.

ozone layer (ozonosphere) A layer of the earth's atmosphere about 40 to 60 km above the surface, characterized by a high content of ozone, O_3.

parabola A geometric figure that describes the shape of an escape speed orbit.

parallax The change in an object's apparent position when viewed from two different locations; specifically, half the angular shift of a star's apparent position as seen from opposite ends of the earth's orbit.

parent meteorite bodies Small solid bodies, a few hundreds or thousands of kilometers in size, thought to be the source of nickel–iron meteorites; formed early in the history of the solar system and broke up through collisions.

parsec (pc) The distance an object would have to be from the earth to achieve a heliocentric parallax of 1 second of arc; equal to 3.26 light years. A kiloparsec is 10^3 parsecs, and a megaparsec is 10^6 parsecs.

pascal (Pa) Unit of pressure in the SI system.

perigee The point in its orbit at which an earth satellite is closest to the earth.

perihelion The point at which a body orbiting the sun is nearest to it. the sun

period The time interval for some regular event to take place; for example, the time required for one complete revolution of a body around another.

period–luminosity relationship For cepheid variables, a relation between the average luminosity and the time period over which the luminosity varies; the greater the luminosity, the longer the pulsational period.

periodic comets Comets that have relatively small elliptical orbits around the sun, with periods of less than 200 years.

periodic (regular) variables Stars whose light varies with time in a regular fashion.

perturbations Small changes in the motions of a mass because of the gravitational effects of another mass.

phases of the moon The monthly cycle of the changes in the moon's appearance as seen from the earth; at new, the moon is in line with the sun and so not visible; at full it is in opposition to the sun and we see a completely illuminated surface.

photodissociation The breakup of a molecule that has absorbed light with enough energy to split the molecular bonds.

photometer A light-sensitive detector placed at the focus of a telescope; used to make accurate measurements of small photon fluxes.

photometry Measurement of the intensity of light.

photon A discrete amount of light energy; the energy of a photon is related to the frequency f of the light by the relation $E = hf$, where h is Planck's constant.

photon excitation Raising an electron of an atom to a higher energy level by the absorption of a photon.

photosphere The visible surface of the sun; the region of the solar atmosphere from which visible light escapes into space.

physical universe The parts of the universe that can be seen directly plus those that can be inferred from the laws of physics.

pixel The smallest picture element in a two-dimensional detector.

Planck curve The continuous spectrum of a blackbody radiator.

Planck's constant The number that relates the energy and frequency of light; it has a value of 6.63×10^{-34} J · s.

planet From the Greek word for "wanderer"; any of the nine (so far known) large bodies that revolve around the sun; traditionally, any heavenly object that moved with respect to the stars (in this sense, the sun and the moon were also considered planets by ancient astronomers).

planetary nebula A thick shell of gas ejected from and moving out from an extremely hot star; thought to be the outer layers of a red giant star thrown out into space, the core of which eventually becomes a white dwarf.

planetesimals Asteroid-sized bodies that, in the formation of the solar system, combined with one another to form the protoplanets.

plasma A gas consisting of ionized atoms and electrons.

plate tectonics A model for the evolution of the earth's surface that pictures the interaction of crustal plates driven by convection currents in the mantle.

polar caps Icy regions at the north and south poles of a planet.

Polaris The present north pole star; the outermost star in the handle of the Little Dipper.

polarization A lining up of the planes of vibration of light waves.

polarized light Light waves whose planes of oscillation are the same.

Population I stars Stars found in the disk of a spiral galaxy, especially in the spiral arms, including the most luminous, hot, and young stars, with a heavy-element abundance similar to that of the sun (about 2 percent of the total); an old Population I is found in the nucleus of spiral galaxies and in elliptical galaxies.

Population II stars Stars found in globular clusters and the halo of a galaxy; somewhat older than any Population I stars and containing a smaller abundance of heavy elements.

positron An antimatter electron; essentially an electron with a positive charge.

potential energy The ability to do work because of position; it is storable and can later be converted into other forms of energy.

power Total energy emitted or produced per second.

PP chain See **proton–proton chain.**

precession of the equinoxes The slow westward motion of the equinox points on the sky relative to the stars of the zodiac because of the wobbling of the earth's spin axis.

pre-main-sequence star The evolutionary phase of a star just before it reaches the main sequence and starts hydrogen core burning.

pressure Force per unit area.

primary The more luminous of the two stars in a binary system.

prime focus The direct focus of an objective without diversion by a lens or mirror.

primeval fireball The hot, dense beginning of the universe in the Big Bang model, when most of the energy was in the form of high-energy light.

principle of equivalence The fundamental idea in Einstein's general theory of relativity; the statement than one cannot distinguish between gravitational accelerations and accelerations of other kinds, or, equivalently, a statement about the equality of inertial mass and gravitational mass. A consequence is that gravitational forces can be made to vanish in a small region of spacetime by choosing an appropriate accelerated frame of reference.

prominences Cooler clouds of hydrogen gas above the sun's photosphere in the corona; they are shaped by the local magnetic fields of active regions.

proper motion The angular displacement of a star on the sky from its motion through space.

protein A long chain of amino acids linked by hydrogen bonds.

protogalaxies Clouds with so much mass that they are destined to collapse gravitationally into galaxies.

proton A massive, positively charged elementary particle; one of the main constituents of the nucleus of an atom.

proton–proton (PP) chain A series of thermonuclear reactions that occur in the interiors of stars, by which four hydrogen nuclei are fused into helium; this process is believed to be the primary mode of energy production in the sun.

protoplanet A large mass formed by the accretion of planetesimals; the final stage of formation of the planets from the solar nebula.

protoplasm The fluid within a living cell.

protostar A collapsing mass of gas and dust out of which a star will be born (when thermonuclear reactions turn on); its energy comes from gravitational contraction.

pulsar A radio source that emits signals in very short, regular bursts; thought to be a highly magnetic, rotating neutron star.

quantum (pl., quanta) A discrete packet of energy.

quantum state The quantum description of the arrangement of electrons in an atom; allowed quantum states are filled starting with those of lowest energy first.

quark An elementary particle that makes up others, such as protons.

quasar or quasi-stellar object An intense, pointlike source of light and radio waves that is characterized by large redshifts of the emission lines in its visible spectrum.

radar mapping The surveying of the geographic features of a planet's surface by the reflection of radio waves from the surface.

radial velocity The component of relative velocity that lies along the line of sight.

radiant The point in the sky from which a meteor shower appears to come.

radiation Usually refers to electromagnetic waves, such as light, radio, infrared, x-rays, ultraviolet; also sometimes used to refer to atomic particles of high energy, such as electrons (beta radiation) and helium nuclei (alpha radiation).

radiation belts In a planet's magnetosphere, regions with a high density of trapped solar wind particles.

radiation era In the Big Bang model, the time in the universe's history during which the energy in the universe was dominated by radiation.

radiative energy The capacity to do work that is carried by electromagnetic waves.

radio galaxies Galaxies that emit large amounts of radio energy by the synchrotron process, generally characterized by two giant lobes of emission situated on opposite ends of a line drawn through the nucleus and connected to it by jets. They are two types of radio galaxy: compact and extended.

radio interferometer A radio telescope that achieves high angular resolution by combining signals from at least two widely separated antennas.

radio jets Astrophysical jets that are visible to radio telescopes; generally refers to the jets extending from the nuclei of active galaxies.

radio lobes Large regions of radio emission seen outside radio galaxies and containing magnetic fields and high-speed electrons; often connected to the nucleus by a jet.

radio recombination line emission Sharp energy peaks at radio wavelengths caused by low-energy transitions in atoms from one very high energy level to another nearby level following recombination of an electron with an ion.

radio telescope A telescope designed to collect and detect radio emissions from celestial objects.

radioactive dating A process that determines the age of an object by the rate of decay of radioactive elements within the object; for rocks, gives the date of the most recent solidification.

radioactive decay The process by which an element fissions into lighter elements.

rapid process (r-process) The route to the formation of very heavy elements by the rapid addition of neutrons to a nucleus followed by beta decay.

ray On the moon or other satellite, a bright streak of material ejected from an impact crater.

recombination The joining of an electron to an ion; the reverse of ionization.

recombination line Emission line from an electron following the process of recombination.

red giant A large, cool, old star with a high luminosity and a low surface temperature (about 2500 K), which is largely convective and has fusion reactions going on in shells; lies on the red giant branch or asymptotic giant branch in an H–R diagram.

red variables A class of cool stars variable in light output.

reddening The preferential scattering or absorption of blue light by small particles, allowing a greater proportion of red light to pass directly through.

redshift An increase in the wavelength of the radiation received from a receding celestial body as a consequence of the Doppler effect; a shift toward the long-wavelength (red) end of the spectrum.

reflecting telescope A telescope that has a uniformly curved mirror as a primary light gatherer.

reflection The return of a light wave at the interface between two media.

reflection nebula A bright cloud of gas and dust that is visible because of the reflection of starlight by the dust.

refracting telescope A telescope that uses glass lenses to gather light.

refraction Bending of the direction of a light wave at the interface between two media, such as air and glass.

regolith The loose, pulverized surface soil of the moon or similar material on a planetary surface.

relativistic Doppler shift Wavelength shift from the radial velocity of a source as calculated in special relativity, so that very large redshifts do not imply that the source moves faster than light.

relativistic jet A beam of particles moving at speeds close to that of light; such jets contained charged particles channeled by magnetic fields.

resolving power The ability of a telescope to separate close stars or to pick out fine details of celestial objects.

retrograde motion The apparent anomalous westward motion of a planet with respect to the stars, which occurs near the time of opposition (for an outer planet) or inferior conjunction (for an inner planet).

retrograde rotation Rotation from east to west.

revolution The motion of a body in orbit around another body or a common center of mass.

rift valley A depression in the surface of a planet created by the separation of crustal plates.

ring system The complete set of rings around a planet; these consist of smaller ringlets.

ringlets The subsections of a planetary ring system, containing many small particles orbiting together and obeying Kepler's laws.

Roche lobe In a binary star system, the region in the space around the pair in which the stars' gravitational fields provide a path from one to the other.

rotation The turning of a body, such as a planet, on its axis.

rotation curve The graphic expression of the relation between the rotational speed of objects in a galaxy and their distance from its center.

S-type asteroids Asteroids whose albedos and spectra indicate a surface made of silicates.

scarp A long, vertical wall running across a planet's surface; common on Mercury.

scattering (of light) The change in the paths of photons without absorption or change in wavelength.

Schwarzschild radius The critical size that a mass must reach to be dense enough to trap light by its gravity, that is, to become a black hole.

scientific model A mental image of how the natural world works, based on physical, mathematical, and aesthetic ideas.

seafloor spreading The lateral motion of the ocean basins away from midoceanic ridges.

second dredge-up The theoretical process by which convection brings the products of helium burning to the surface of a massive star during the second time it becomes a red giant.

second of arc 1/3600 of a degree (0.000278°), or 1/60 of a minute of arc (0.0167 arcmin).

secondary The less luminous of the two stars in a binary system.

seeing The unsteadiness of the earth's atmosphere that blurs telescopic images.

seismic waves Sound waves traveling through and across the earth that are produced by earthquakes.

seismometer An instrument used to detect earthquakes and moonquakes.

semimajor axis Half the major axis of an ellipse; distance from the center of an ellipse to its farthest point.

sequential star formation A model for the formation of massive stars from a giant molecular cloud in which the hot, ionized gas around a small cluster of OB stars creates a shock wave that initiates the collapse of another part of the molecular cloud to give birth to another small cluster of OB stars until the cloud's material is used up.

Seyfert galaxy A galaxy having a bright, bluish nucleus showing broad emission lines in its spectrum; Seyferts probably are spiral galaxies; one type of AGN.

shepherd satellites Small moons that confine a planet's ring in a narrow band; one is located on the inside edge of the ring, one on the outside.

sexagesimal system A counting system based on the number 60, such as 60 minutes in an hour, or 60 minutes of arc in one degree.

shield volcano A large volcano with gentle slopes formed by the slow outflow of magma.

shock wave A discontinuity created in a medium when an object travels through it at a speed greater than the local sound speed.

short-period comets Comets with orbital periods less than 200 years, probably captured from longer period orbits by an encounter with a major planet.

sidereal month The period of the moon's revolution around the earth with respect to a fixed direction in space or a fixed star; about 27.3 days.

sidereal period The time interval needed by a celestial body to complete one revolution around another with respect to the background stars.

signs of the zodiac The 12 traditional angular divisions of 30° each into which the ecliptic is divided; each corresponds to a zodiacal constellation.

silicate A compound of silicon and oxygen with other elements, very common in rocks at the earth's surface.

singularity A theoretical point of zero volume and infinite density to which any mass that becomes a black hole must collapse, according to the general theory of relativity.

slow process (s-process) The route to the formation of very heavy elements by the slow addition of neutrons to the nucleus followed by beta decay.

solar core Region of the sun's interior in which temperatures and densities are high enough for fusion reactions to take place.

solar cosmic rays Low-energy cosmic rays generated in solar flares.

solar day The interval of time from noon to noon.

solar eclipse An eclipse of the sun by the moon, caused by the passage of the new moon in front of the sun.

solar flare Sudden burst of electromagnetic energy and particles from a magnetic loop in an active region; probably triggered by magnetic reconnection.

solar mass The amount of mass in the sun, about 2×10^{30} kg.

solar nebula The thin disk of gas and dust, around the young sun, out of which the planets formed.

solar wind A stream of charged particles, mostly protons and electrons, that escapes into the sun's outer atmosphere (from coronal holes) at high speeds and streams out into the solar system.

solstice The time at which the day or the night is the longest; in the Northern Hemisphere, the summer solstice (around June 21) is the time of the longest day and the winter solstice (around December 21) the time of the shortest day; the dates are reversed in the Southern Hemisphere.

south magnetic pole A point on a star or planet from which the magnetic lines of force emanate and to which the south pole of a compass points.

space A three-dimensional region in which objects move and events occur and have relative direction and position.

space astronomy Astronomy done by instruments in orbit above the earth's atmosphere.

spacetime Space and time unified; a continuous system of one time coordinate and three space coordinates by which events can be located and described.

spacetime curvature The bending of a region of spacetime because of the presence of mass and energy.

special theory of relativity Einstein's theory describing the relations between measurements of physical phenomena as viewed by observers who are in relative motion at constant velocities.

spectral line A particular wavelength of light corresponding to an energy transition in an atom.

spectral sequence A classification scheme for stars based on the strength of various lines in their spectra; the sequence runs O-B-A-F-G-K-M, from hottest to coolest.

spectral type (or class) The designation of the type of a star based on the relative strengths of various spectral lines.

spectrograph or spectrometer An instrument used in conjunction with a telescope to obtain and record a spectrum of an astronomical object; a spectrograph has a photographic plate, a spectrometer another type of spectral detector, such as a CCD.

spectroscope An instrument for examining spectra; also a spectrometer or spectrograph if the spectrum is recorded and measured.

spectroscopic binary Two stars revolving around a common center of mass that can be identified by periodic variations in the Doppler shift of the lines of their spectra.

spectroscopic distance A technique for estimating stellar distances by comparing the fluxes of stars with their actual luminosities, as determined by their spectra.

spectroscopy The analysis of light by separating it by wavelengths (colors).

spectrum (pl., spectra) The array of colors or wavelengths obtained when light is dispersed, as by a prism; the amount of energy given off by an object at every different wavelength.

speed The rate of change of position with time.

spherical (closed) geometry An alternative to Euclidean geometry, constructed by G. F. B. Riemann on the premise that no parallel lines can be drawn through a point near a straight line; the sum of the angles of a triangle drawn on a spherical surface is always greater than 180°.

spicules Spears of hot gas that reach up from the sun's photosphere into the chromosphere.

spin angular momentum The angular momentum of a rotating body; the product of a body's mass distribution, rotational speed, and radius.

spiral arm Part of a spiral pattern in a galaxy; an enhanced region of gas, dust, and young stars.

spiral galaxy A disk galaxy with spiral arms; the presumed shape of our Milky Way Galaxy, which may contain a central bar.

spiral nebulas An older term applied to spiral galaxies as they appeared visually through a telescope.

spiral tracers Objects that are commonly found in spiral arms and so are used to trace spiral patterns: for example, Population I cepheids, H II regions, molecular clouds, and OB stars.

spontaneous emission The emission of a photon by an excited atom in which an electron falls to a lower energy level.

sporadic meteor A meteor that apparently occurs at random and so is not associated with a known shower.

stadium (pl., stadia) An ancient Greek unit of length, probably about 0.2 km.

standard candle An astronomical object of known luminosity used to estimate distances to galaxies.

star model See **stellar interior model.**

starburst galaxies Active galaxies that show evidence of large-scale formation of massive stars.

starspots Dark, magnetic active regions in the photospheres of stars, analogous to sunspots.

steady-state model A theory of the universe based on the perfect cosmological principle, in which the universe looks basically the same to all observers at all times; largely discredited by current observations of the cosmic microwave background radiation.

Stefan–Boltzmann law The relation, for a blackbody radiator, between temperature and energy emitted per unit area of surface; the flux varies as the fourth power of the temperature.

stellar color The color of a star as determined by measuring its flux at two different wavelengths.

stellar corona The hot (1 million K), thin outer atmosphere of a star in analogy to the sun's corona; visible by the emission of x-rays.

stellar flares Flares associated with active regions on stars in analogy to solar flares.

stellar interior model A table of values of the physical characteristics (such as temperature, density, and pressure) as a function of position within a star for a specified mass, chemical composition, and age, calculated from theoretical ideas of the basic physics of stars.

stellar lifetime The total duration of a star's lifetime from birth to death; the sun's lifetime is expected to be about 10 billion years.

stellar parallax See **heliocentric parallax.**

stellar spectral sequence See **spectral sequence.**

stellar surface temperature The temperature of the photosphere of a star.

stellar wind The outflow, both steady and sporadic, of gas at hundreds of kilometers per second from the corona of a star.

stony meteorite A type of meteorite made of light silicate materials.

stony-iron meteorite A type of differentiated meteorite that is a blend of nickel–iron and silicate materials from the core–mantle boundary of a now-fragmented parent meteor body.

straight line The shortest distance between two points in any geometry.

strong nuclear force One of the four forces of nature; the strong force acts over short distances to keep the nuclei of atoms together.

summer solstice See **solstice.**

sunspot A temporarily cool region in the sun's photosphere, associated with an active region, having a magnetic field of some 0.1 T (a few thousand gauss).

sunspot cycle The 11-year number cycle and a 22-year magnetic polarity cycle of the formation of sunspots.

superclusters Clusters of clusters of galaxies; superclusters appear as long, filamentary chains with apparent voids in between.

supergiant A massive star of large size and high luminosity; red supergiants are thought to be the progenitors of Type II supernovas.

supergiant elliptical (cD) galaxies The largest and most massive elliptical galaxies, sometimes with more than one nucleus. Found only at the cores of rich clusters of galaxies, they may have been formed by mergers of galaxies.

supergravity A model that combines quantum ideas with gravity to describe the unification of all forces of nature during the first 10^{-43} second of the universe's history.

superior conjunction A planetary configuration in which an inner planet lies in the same direction as the sun but on the opposite side of the sun as viewed from the outer planet.

superluminal motion Motion apparently faster than the speed of light.

supermassive black hole A black hole with a mass of a million solar masses or more; probably powers active galaxies and quasars.

supernova (English pl., supernovas) A stupendous explosion of an old star, which increases its brightness hundreds of millions of times in a few days. Type II results from the core implosion of a massive star at the end of its life; Type I may originate from a carbon-rich white dwarf in a binary system.

supernova remnant An expanding gas cloud from the outer layers of a star blown off in a supernova explosion; detectable at radio wavelengths; moves through the interstellar medium at high speeds.

surface temperature See **stellar surface temperature.**

superwind A very strong stellar wind.

synchronous rotation The behavior of a solid body that rotates at the same rate that it revolves; for a satellite, this means that it keeps the same face to its parent planet.

synchrotron radiation Radiation from an accelerating charged particle (usually an electron) in a magnetic field; the wavelength of the emitted radiation depends on the strength of the magnetic field and the energy of the charged particles.

synodic month The time interval between similar configurations of the moon and the sun; for example, between full moon and the next full moon; about 29.5 days.

synodic period The interval between successive similar lineups of a celestial body with the sun, for example, between oppositions.

tail (of a comet) The extended stream of plasma and dust expelled from the nucleus of a comet.

technologically advanced civilizations Civilizations that are at or above the level of technology we now have.

temperature A measure of the average random speeds of the microscopic particles in a substance.

terrestrial planets Planets similar in composition and size to the earth: Mercury, Venus, Mars, and the moon.

tesla (T) In the SI system, a unit of measure of magnetic flux.

theoretical resolving power A telescope's resolving power based on its optics alone; ground-based optical telescopes have a resolving power limited by the seeing of the atmosphere.

thermal energy The average kinetic energy of the particles that make up a body.

thermal equilibrium Steady-state situation characterized by an absence of large-scale temperature changes.

thermal pulses Bursts of energy generation from the triple-alpha process in the shell of a red giant star.

thermal radiation Electromagnetic radiation from a body that is hot; often characterized by a blackbody spectrum.

thermal speed The average speed of the random motion of particles in a gas.

threshold temperature The temperature at which photons have enough energy to create a given type of particle–antiparticle pair.

tidal force The difference in gravitational force between two points in a body, caused by a second body, which may result in the deformation of the second body.

tidal friction Friction caused by the tidal motion of the water in the ocean basins.

time A measure of the flow of events.

ton (metric) 1000 kg.

torque A twisting force applied through a lever arm.

total energy The sum of all forms of energy attributed to an isolated body or to a system of bodies; usually just the sum of the kinetic and potential energies.

transition (in an atom) A change in the electron arrangements in an atom, which involves a change in energy.

transition region In the sun's atmosphere, the region of rapid temperature rise lying between the chromosphere and the corona.

transverse fault A crack in a solid surface where the ground has moved sideways.

transverse velocity The component of an object's velocity that is perpendicular to the line of sight.

trigonometric parallax A method of determining distances by measuring the angular position of an object as seen from the ends of a baseline having a known length. See **heliocentric parallax.**

triple-alpha reaction A thermonuclear process in which three helium atoms (alpha particles) are fused into one carbon nucleus.

Tully–Fisher relation The relation between the luminosity of a galaxy and the width of its 21-cm emission line; the greater the luminosity, the wider the line.

turbulence Irregular and sometimes violent convective motion.

turnoff point The point on the H–R diagram of a cluster at which the main sequence appears to terminate at the high-luminosity end.

T-Tauri stars Newly formed stars of about 1 solar mass; usually associated with dark clouds; some show evidence of flares and starspots, others have circumstellar disks; one type of young stellar object (YSO).

21-cm line The emission line, at a wavelength of 21.11 cm, from neutral hydrogen gas; it is produced by atoms in which the directions of spin of the proton and electron change from parallel to opposed.

Type I, Type II supernovas Classification of supernovas by their light curves and spectral characteristics: Type I show a sharp maximum and slow decline with no hydrogen lines; Type II have a broader peak and a very sharp decline after 100 days with strong hydrogen lines in the spectrum.

ultraviolet telescope A telescope optimized for use in the ultraviolet region; must be used above the earth's atmosphere.

uniformity of nature The assumption that astronomical objects of the same type are the same throughout the universe.

universal law of gravitation Newton's law of gravitation. See **gravitation.**

universe The totality of all space and time; all that is, has been, and will be.

universality of physical laws The assumption, borne out by some evidence, that the physical laws understood locally apply throughout the universe and perhaps to the universe as a whole.

upland plateaus Large highland masses on the surface of Venus; granitic, analogous to lunar highlands and terrestrial continents.

Van Allen radiation belts Belts of charged particles (from the sun) concentrated and trapped in the earth's lower magnetosphere.

variable star Any star whose luminosity changes over a short period of time.

velocity The rate and direction in which distance is covered over some interval of time.

vernal equinox The spring equinox. See **equinox.**

vertical circle On the celestial sphere, any great circle through the zenith.

virial theorem The statement that the gravitational potential energy is twice the negative of the kinetic energy of a system of particles in equilibrium.

visual binary Two stars that revolve around a common center of mass, both of which can be seen through a telescope so that their orbits can be plotted.

visual flux The flux from a celestial object measured across the visual part of the electromagnetic spectrum.

visual luminosity The luminosity from a celestial object measured across the visual part of the electromagnetic spectrum.

voids Regions between superclusters that contain no visible concentrations of luminous matter; they have a somewhat spherical shape and extend over millions of light years.

volatiles Materials, such as helium and methane, that vaporize at low temperatures.

volcanic model The formation of craters as cones left over from lava eruptions.

volcanism The geological process in which molten material and associated gases (produced by internal heating) rise through a planets crust to its surface.

watt (W) A unit of power; one joule expended per second.

wavelength The distance between two successive peaks or troughs of a wave.

weak nuclear force The short-range force that operates in radioactive decay.

weight The total force on some mass produced by gravity.

weightlessness The condition of apparent zero weight, produced when a body is allowed to fall freely in a gravitational field; in general relativity, weightlessness signifies motion on a straight line in spacetime.

white dwarf A small, dense star that has exhausted its nuclear fuel and shines from residual heat; such stars have an upper mass limit of 1.4 solar masses, and their interior is a degenerate electron gas.

Widmanstätten figures Large crystal patterns seen in iron meteorites when they are polished and etched; formed by slow cooling of the material.

Wien's law The relation between the wavelength of maximum emission in a blackbody's spectrum and its temperature; the higher the temperature, the shorter the wavelength at which the peak occurs.

winter solstice See **solstice**.

x-rays High-energy electromagnetic radiation with a wavelength of about 10^{-10} meter.

x-ray burster An x-ray source that emits brief, powerful bursts of x-rays; probably occurs from accretion onto a neutron star in a binary system.

young stellar object (YSO) Any young star past the protostar stage; usually hidden in dust and often associated with molecular clouds and bipolar outflows.

ZAMS Acronym for zero-age main sequence.

Zeeman effect The splitting of spectral lines because of strong magnetic fields.

Zeilik An American form of a Ukrainian family name.

zenith The point on the celestial sphere that is located directly above the observer at 90° angular distance from the horizon.

zero-age main sequence (ZAMS) The position on the H–R diagram reached by a star once it has come to derive most of its energy from thermonuclear fusion reactions rather than from gravitational contraction.

zodiac The traditional 12 constellations through which the sun travels in its yearly motion, as seen from the earth; a thirteenth constellation, Ophiuchus, is actually part of the zodiac.

zodiacal light Sunlight reflected from dust in the plane of the ecliptic.

zone A region of high pressure convective uplift in the atmosphere of a Jovian planet.

zone of avoidance A region near the plane of the Galaxy in which very few other galaxies are visible because of obscuration by dust.

INDEX

apparent Magnitude?